SPECTROSCOPY AT RADIO
AND MICROWAVE FREQUENCIES

Spectroscopy at Radio and Microwave Frequencies

SECOND EDITION

D. J. E. INGRAM

M.A., D.Phil., D.Sc. (Oxon), Hon. D.Sc. (Clermont–Ferrand)

Professor of Physics and Head of Physics Department
University of Keele, Staffordshire

Springer Science+Business Media, LLC

1967

Published in the U.S.A. by
PLENUM PRESS
a division of
PLENUM PUBLISHING CORPORATION
227 West 17th Street, New York, N.Y. 10011

First published by
Butterworth & Co. (Publishers) Ltd.

First Impression 1955
Reprinted 1957
Second Reprint 1959
Second Edition 1967

ISBN 978-1-4899-6180-8   ISBN 978-1-4899-6361-1 (eBook)
DOI 10.1007/978-1-4899-6361-1
© Springer Science+Business Media New York 1967
Originally published by Butterworth & Co. (Publishers) Ltd. 1967.
Softcover reprint of the hardcover 2nd edition 1967

Suggested U.D.C. number: 535·33 : 539·143·43
Suggested additional number: 543·422·8
Library of Congress Catalog Card Number 67–26701

Set in Monotype Baskerville type
Made and printed in Great Britain by
Robert MacLehose and Co. Ltd, The University Press, Glasgow

PREFACE TO SECOND EDITION

A VERY large amount of research on spectroscopy at radio and microwave frequencies has been carried out in the ten years following the publication of the first edition of this book. A considerable proportion of this research work has developed along the lines considered and discussed in the first edition, but there have also been some major new advances in experimental techniques, theoretical treatment and practical applications. It was therefore felt that an attempt should be made, at this fourth printing, to include some of these more recent advances in the subject.

During the next two or three years, however, further extremely important new developments are likely to occur in various branches of radio and microwave spectroscopy (such as the design of rapid recording spectrometers to follow continuous reactions). For this reason it seemed unwise to revise the whole of the book at this particular time, but rather to wait a few more years until the full implications of such new technical advances have been evaluated. The first nine chapters of this edition therefore remain substantially the same as before; but they are followed by three new chapters which have been added as a form of appendix to the main body of the book, and are meant to be read in conjunction with it.

Three factors have been mainly responsible for the progress that has taken place during the last ten years: (i) an improvement and refinement of experimental techniques, such as the production of extremely homogeneous magnetic fields; (ii) the development of new forms of theoretical treatment and analysis, such as the application of molecular orbital treatment and ligand field theory to electron resonance problems; (iii) new ideas on methods of observation, such as the advent of the 'double resonance' techniques, and on other fields of practical application, such as the production of 'masers' and the investigation of biochemical and biophysical problems by electron resonance techniques. The research and results that have followed from these advances can probably be best brought together under the three separate headings: 'High Resolution Nuclear Resonance'; 'Recent Advances in Electron Resonance' which includes the very wide application of this technique to free radical studies, as well as the more refined treatment of transition group complexes by molecular orbital methods; and Double Resonance—Masers and Lasers'. These headings are in fact used for the three additional chapters at the end of the book. Much of the work is of course interrelated, and the presentation of the recent advances in spectroscopy at radio and microwave frequencies under these different headings is mainly a matter of convenience, and is not meant to imply any rigorously defined or separated lines of development.

KEELE

PREFACE TO FIRST EDITION

In view of the growing interest in spectroscopy at radio and micro-wave frequencies, and the increasing number of its applications to both physics and chemistry, it was thought that a general outline of the subject for non-specialists might be of some value. Research in this field is still expanding, but is now sufficiently developed for a critical review to be made both of its main applications and of the techniques that are used in this wavelength region.

A broad approach has been taken, and the similarity and inter-relation of the different branches have been stressed, as well as their general setting in spectroscopy as a whole. In this way it is hoped that the book will be of interest to many research workers and students who, although not directly concerned with the subject, would like to obtain a general picture of its methods and applica-tions. At the same time considerable space has been given to the design of experimental apparatus and equipment, so that those wishing to set up such spectroscopes should be able to find much useful and detailed information.

The range of frequencies covered by these measurements varies from under 1 Mc/s to the verge of the infra-red (500,000 Mc/s) and the three branches of gaseous microwave spectroscopy, electron paramagnetic resonance, and nuclear radiofrequency resonance are all considered at some length. Detailed mathe-matical treatment is avoided so that the interpretation of the results should be clear to those without specialized knowledge of quantum mechanics, although sufficient theory is given to correlate the measurements with fundamental physical or chemical properties.

The various applications of spectroscopy at these frequencies are considered in the last chapter, in relation both to fundamental research and to determinations of a more practical nature. An attempt has been made to keep a balance between these two sides throughout the book, as may be illustrated by the results obtained from paramagnetic resonance. On the one hand these have intro-duced a very direct method for the determination of such parameters as nuclear spins and moments; and on the other, they have opened up wide fields of application in the study of free-radical chemis-try, irradiation damage, and the impurities and carriers of semi-conductors. For this reason it is felt that the book should be of considerable interest to those engaged in pure or applied research.

I would like to acknowledge the help I have received in the many talks with Dr. B. Bleaney, Dr. K. W. H. Stevens and others at the Clarendon Laboratory, and also the assistance of Mr. J. E. Bennett in reading through the manuscript and proofs. Especial thanks are also due to Professor E. E. Zepler for his continuous advice and interest.

SOUTHAMPTON

CONTENTS

CONTENTS

6. RESULTS AND THEORY OF PARAMAGNETIC RESONANCE

7. FERROMAGNETIC RESONANCE. FREE RADICALS AND F-CENTRES

xi

10. HIGH RESOLUTION NUCLEAR RESONANCE

11. RECENT ADVANCES IN ELECTRON RESONANCE

ACKNOWLEDGEMENTS

The illustrations listed below were reproduced from the sources stated. Acknowledgement is made to authors, publishers and manufacturers for their permission. Figure numbers given refer to this work.

Figure
3. BLEANEY, B. and PENROSE, R. P. *Proc. Roy. Soc.* A. 189 (1947) 358
8. Raytheon Manufacturing Co. Ltd.
9. POUND, R. V. *Rev. sci. Instrum.* 17 (1946) 490
11. VAN VOORHIS *Microwave Receivers.* M.I.T. Radiation Laboratory Series No. 23. McGraw-Hill (1948)
12. JOHNSON, C. M., SLAGER, D. M. and KING, D. D. *Rev. sci. Instrum.* 25 (1954) 213 (Also *Table* 2.2)
13. Sylvainia Ltd.
14. Sylvainia Ltd.
15. TORREY, H. C. and WHITMER, C. A. *Crystal Rectifiers.* M.I.T. Radiation Laboratory Series. McGraw-Hill (1948)
17. STRUM, P. D. *Proc. Inst. Radio Engrs, N.Y.* 41 (1953) 875
20. Sperry Gyroscope Co.
21. HEDRICK, L. C. *Rev. sci. Instrum.* 24 (1953) 566
33. GORDY, W. and KING, W. C. *Phys. Rev.* 93 (1954) 407
36. BLEANEY, B., LOUBSER, J. H. N. and PENROSE, R. P. *Proc. phys. Soc.* 59 (1947) 185
42. STRANDBERG, M. W. P., WENTINK, T. and KYHL, R. L. *Phys. Rev.* 75 (1949) 270
43. GILLIAM, O. R., JOHNSON, C. M. and GORDY, W. *Phys. Rev.* 78 (1952) 140
44. GESCHWIND, S. *Ann. N.Y. Acad. Sci.* 55 (1952) 751
45. JEN, C. K. *Phys. Rev.* 74 (1948) 1396
47. BAGGULEY, D. M. S. and GRIFFITHS, J. H. E. *Proc. phys. Soc.* A. 65 (1952) 594
48. BERINGER, R. and CASTLE, J. G. *Phys. Rev.* 78 (1950) 581
51. KING, W. C. and GORDY, W. *Phys. Rev.* 93 (1954) 407
54. DAKIN, T. W., GOOD, W. E. and COLES, D. K. *Phys. Rev.* 70 (1946) 560
55. TOWNES, C. H., HOLDEN, A. N. and MERRITT, F. R. *Phys. Rev.* 74 (1948) 1113
56 and 57. GORDEN, J. P., ZEIGER, H. J. and TOWNES, C. H. *Phys. Rev.* 95 (1954) 282
61. BLEANEY, B. and INGRAM, D. J. E. *Proc. Roy. Soc.* A. 205 (1951) 336; and *Proc. Roy. Soc.* A. 208 (1951) 143
63. BLEANEY, B. and SCOVIL, H. E. D. *Proc. phys. Soc.* A. 63 (1950) 1369
64. GRIFFITHS, J. H. E. *Nature* 158 (1946) 670
65. TROUNSON, E. P., BLEIL, D. F. and WANGSNESS, R. K. *Phys. Rev.* 79 (1950) 542
67. GRIFFITHS, J. H. E., OWEN, J. and WARD, I. M. *Nature* 173 (1954) 439
68. KUSCH, P. *Physica* 17 (1951) 339
70 (a). BLOEMBERGEN, N., PURCELL, E. M. and POUND, R. V. *Phys. Rev.* 73 (1948) 679
70 (b). POUND, R. V. and KNIGHT, W. D. *Rev. sci. Instrum.* 21 (1950) 219
71. BLOCH, F., HANSEN, W. W. and PACKARD, M. *Phys. Rev.* 70 (1946) 474
72. GUTOWSKY, H. S., McCALL, D. W. and SLICHTER, C. P. *J. chem. Phys.* 21 (1953) 279
73. PURCELL, E. M. *Physica* 17 (1951) 282

ACKNOWLEDGEMENTS

74 and *75*. ANDREW, E. R., BRADBURY, A. and EADES, R. G. *Nature* 185 (1959) 1802 and 1803

78. BADEN, A. R., GUTOWSKY, H. S., WILLIAMS, G. A. and YANTWICH, P. E. *J. Amer. chem. Soc.* 78 (1956) 2385 and 2386

79. COREY, E. J., BURKE, H. J. and REMERS, W. A. *J. Amer. chem. Soc.* 77 (1955) 4941

80 and *81*. ARNOLD, J. T. *Phys. Rev.* 102 (1956) 144 and 148

82. WEINBERG, I. and ZIMMERMAN, J. R. *J. chem. Phys.* 23 (1955) 748

83. PHILLIPS, W. D. *J. chem. phys.* 23 (1955) 1363

84. SHOOLERY, J. N. *Disc. Faraday Soc.* 19 (1955) 220

85. ANDERSON, W. A. *Phys. Rev.* 102 (1956) 163

86 and *87*. WATERS, G. S. and FRANCIS, P. D. *J. sci. Instrum.* 35 (1958) 90 and 91

88. ELLIOT, D. J. and SCHUMACHEN, R. T. *J. chem. Phys.* 26 (1957) 1350

92. FAULKNER, E. A. *J. sci. Instrum.* 39 (1962) 135

94 and *95*. KLEIN, M. P. and BARTON, G. W. *Rev. sci. Instrum.* 34 (1963) 756 and 758

97. YAMAZAKI, I., MASON, H. S. and PIETTE, L. *J. biol. Chem.* 235 (1960) 2445

98. BRAY, R. C. *Biochem. J.* 81 (1961) 191

99. VENKATARAMAN, B. and FRAENKEL, G. K. *J. Amer. chem. Soc.* 77 (1955) 2707

101. WERTZ, J. E. and VIVO, J. L. *J. chem. Phys.* 23 (1955) 2441

103. DEGUCHI, Y. *J. chem. Phys.* 32 (1960) 1585

104. REITZ, D. C., DRAUNIEKS, G. and WERTZ, J. E. *J. chem. Phys.* 33 (1960) 1880

105 and *106*. HOLMBERG, R. W. and LIVINGSTON, R. *J. chem. Phys.* 33 (1960) 542

109. FESSENDEN, R. W. and SCHULER, R. J. *J. chem. Phys.* 39 (1963) 2518.

110. JOHNSTON, C. W., VISCO, R. E., GUTOWSKY, H. S. and HARTLEY, A. M. *J. chem. Phys.* 36 (1962) 1581

119, *120* and *121*. HUTCHINSON, C. A. and MAGNUM, B. J. *J. chem. Phys.* 34 (1961) 909–912

122. BRAY, R. C., PALMER, G. and BEINERT, H. *J. biol. Chem.* 239 (1964) 2668

133. PORTIS, A. M. *Phys. Rev.* 91 (1953) 1071

137. FEHER, G. *Phys Rev.* 105 (1957) 1122

149. NATHAN, M. I., DUMKE, W. P., Burns, G. Dill, F. H. and Lasher, G. *Appl. Phys. Lett.* 1 (1962) 63

150. CROWE, J. W. and CRAIG, R. M. *Appl. Phys. Lett.* 4 (1964) 57

152. BROSSEL, J. and BITTER, F. *Phys. Rev.* 86 (1952) 312

1

INTRODUCTION

1.1 THE NATURE OF SPECTROSCOPY

SPECTROSCOPY as a general title can be applied to measurements throughout the whole electromagnetic spectrum, 'spectroscopy at radio and microwave frequencies' being one of its specialized branches. It is therefore best to consider first the underlying principles and general nature of the subject as a whole.

Spectroscopy is essentially a determination of energy levels of molecules, atoms and nuclei. The experimental methods employed to do this consist, fundamentally, of measurements of frequency, because the energy difference between the two levels involved in a transition determines the frequency of the radiation which is emitted or absorbed by the particular molecule, atom or nucleus being investigated. The fundamental equation of spectroscopy is thus the quantum condition

$$h\nu = E_1 - E_2$$

where h is Planck's constant

ν is the frequency of the radiation

E_1 and E_2 are the initial and final energy states or levels of the system emitting or absorbing the radiation.

The problem of the experimental spectroscopist is thus to measure the frequency of the emitted or absorbed radiation as accurately as possible, this entailing consideration of such questions as line widths, intensities and standards of frequency. Once the experimental data has been obtained for as many transitions as possible in a given system, it is then the task of the theoretical physicist to produce a consistent model, or theory. This must predict a comprehensive picture of the energy level system, such that transitions between different energy levels will produce the experimentally measured energy differences and frequencies. The classical example is Bohr's theory of the hydrogen atom. In this case the experimental data consisted of the frequencies of the different spectral lines forming the Lyman, Balmer and Paschen series; and, by applying Planck's quantum condition to the motion of the electrons, Bohr was able to produce a model which predicted a complete system of energy levels, such that the differences between them accounted for the observed frequencies of the different spectral series.

When this theoretical explanation of the energy level system has been obtained it can be used, in turn, to give information concerning the forces and interactions which exist inside the particular molecule, atom or nucleus being studied. In this way the experimental data derived from the frequencies of spectral lines is used to check, modify and expand theories of atomic and molecular forces.

1.2 The Concept of Energy Levels

The different forces and interactions, which produce and modify the energy level system, will vary enormously, according to the frequency range being investigated, but the idea of a 'ground state' and 'excited levels' remains the same whether protons and neutrons inside a nucleus are being considered, or the interactions between different atoms in a polyatomic molecule. In any system containing electrons or nucleons there will be a certain configuration which has a minimum energy (e.g. in the particular case of the inert gases, the electrons line up their spins and orbital momentum to give a zero resultant which has the minimum amount of energy. A similar effect occurs in the 'magic number' nuclei of nuclear theory.) If the configuration of the molecule, atom or nucleus now changes from that of the ground state (e.g. an electron or neutron changes its spin or orbital momentum) then the new configuration will have an energy higher than that of the ground state. In passing from the ground state to this excited state the system will absorb radiation, the frequency of which is given by the quantum condition

$$\nu = \text{Energy difference}/h$$

There will, of course, be many different excited states, corresponding to different possible configurations of the electrons or nucleons, and the spectral lines, arising from transitions between them, may lie in quite different parts of the electromagnetic spectrum (e.g. in a $CuSO_4 . 5H_2O$ crystal the blue colour is due to transitions between excited states of different orbital momentum and the ground state, which cause absorption in the optical region; whereas absorption lines also occur in the microwave region, these being due to transitions between energy levels for which the orbital motion is effectively quenched, and only the electron spin is changing). Measurements in different parts of the spectrum can thus be compared and collated, and the complete energy level system deduced from a combination of different experimental results.

In the above considerations it has been assumed that all the energy levels have definite and precise values, and that the transitions, and spectral lines to which they give rise, have zero frequency spread. In practice every form of spectral line has a finite width, and the problem of reducing the widths of absorption or emission lines is one that the spectroscopist is always having to face. This width arises from an indefiniteness in the actual energy value of the different energy levels, and can be caused by various factors such as thermal vibrations, collisions with other molecules, or interactions with the magnetic or electric fields of surrounding atoms or nuclei. The reduction of line width to as small a value as possible is always desirable, as the main transition being investigated is often split by weaker interactions, to give rise to a fine or hyperfine structure which can only be resolved if the line is sufficiently narrow (i.e. the indefiniteness of the energy levels must not be greater than the splitting produced by the fine or hyperfine structure interaction). The fine structure obtainable in a spectral line can give very valuable additional information. This is especially so in the case of spectroscopy at microwave frequencies, where the different energy levels concerned are caused by molecular or solid state interactions, but perturbation of these by interactions with the nucleus, give rise to a hyperfine structure on the spectrum, which can be highly resolved and give considerable nuclear data. This is additional to the direct information on the molecular or solid state interactions obtained from the frequency of the main transitions.

1.3 The Electromagnetic Spectrum

The great range of magnitude in the forces and interactions which spectroscopy can study may be seen in the range of frequencies covered by the electromagnetic spectrum. This extends from γ-rays of frequencies around 10^{21} c/s to radio waves of frequencies of the order of 10^6 c/s, a difference in magnitude of 10^{15}. In the same way, the strength of the forces and interactions studied by spectroscopy varies over a range of 10^{15}, from nuclear forces binding neutrons and protons together, to the molecular and solid state forces which hold atoms in their respective positions. These forces and interactions, and the energy levels to which they give rise, are measured by various different units. The electron-volt is one of the most common, this being the energy acquired by one electron as it falls through a potential of one volt. The potential is measured in practical units of voltage, and thus one electron-volt is equivalent to

$$4 \cdot 8025 \cdot 10^{-10} \times \frac{10^8}{c} = 1 \cdot 602 \cdot 10^{-12} \text{ erg.}$$

In these units the forces investigated at either end of the electromagnetic spectrum vary from ten million electron volts or more,

3

associated with hard γ-ray spectra from nuclear disintegrations, down to a millionth of an electron-volt or so, corresponding to the molecular and solid state forces studied by microwave and radio-frequency absorption.

The cm⁻¹ and the degree are other units commonly used in the measurement of energy level spacings, though neither are strictly units of energy themselves. They are related to the electron-volt by the following equations and numerical factors.

$$eV = \frac{hc}{\lambda} = kT$$

or:

$$1 \text{ electron-volt} = 8{,}065 \text{ cm}^{-1}$$
$$1 \text{ cm}^{-1} \qquad = 1{\cdot}436° \text{ K}$$

See Appendix I for values of the physical constants.

The wave-number, or cm⁻¹, is a unit which is most commonly used in optical spectroscopy, and it is also usual to employ this when giving energy level splittings in the microwave region. The fact that the cm⁻¹ is of the same order as the degree is very useful in enabling quick calculations to be made to determine which energy levels will be populated at any given temperature. Thus, at room temperature there will be few levels populated which are more than 250 cm⁻¹ higher than the ground state. For configurations where the excited

The electromagnetic spectrum

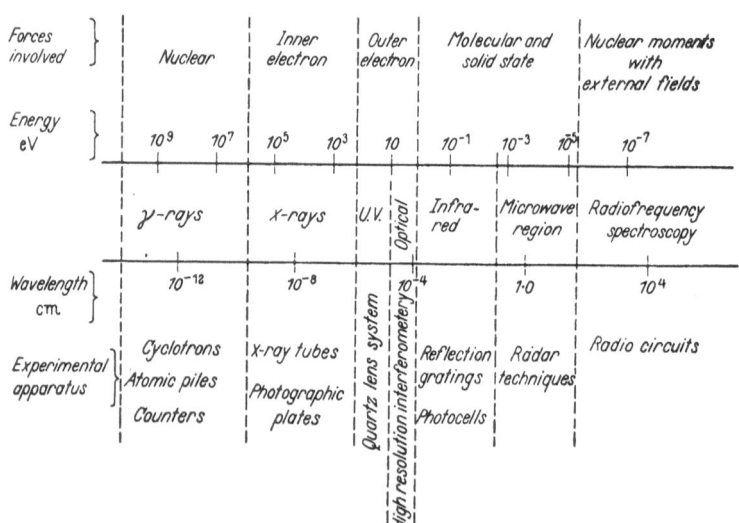

Figure 1

4

levels are relatively close to the ground state (e.g. Co^{2+} in a crystal-line lattice), the use of low temperatures enables the effects of the excited levels to be considerably reduced.

The relation between the energy change involved, and the frequency of the spectral lines obtained, is summarized in *Figure 1*, where the energy change is given in electron volts, and the wavelength of the radiation obtained is given in centimetres. The types of forces and interactions, which give rise to the different orders of energy change are shown at the top of the figure, though there is often an overlap between them. Thus, starting at the high energy end, there are γ-rays of ten million electron volts or more, with wavelengths of the order of 10^{-11} cm, these being produced during changes inside the nucleus, and their study enables nuclear energy level systems to be built up and information obtained concerning nuclear forces and interactions.

Then come X-rays with energies in the hundred thousand electron-volt range, and wavelengths of the order of 10^{-8} cm (or 1 Å). These are produced by changes in the states of the inner electrons close to the nucleus, and a study of their spectra enables the forces binding these electrons to be investigated (a classical example of this being the work of Moseley on the K line X-ray spectra). Then, overlapping the X-ray region comes the ultra-violet region, produced by transitions occurring in the outer electrons, and this in turn merges into the optical region. It has been the study of these spectra that has given such detailed information on the forces binding the electrons to the nucleus, and which has provided such a fruitful field for testing the older and modern quantum theories. The energies associated with transitions in this region have fallen to the order of 10 electron volts, with corresponding wavelengths of 10^{-5} cm. Then comes the infra-red region stretching from 10^{-4} to 10^{-2} cm, with energies corresponding to the interatomic binding of molecules, being only of the order of a fraction of an electron-volt. This region has been of immense interest to the chemist, since resolution of the spectra at these wavelengths gives detailed information concerning the chemical binding of molecules, from their vibrational and rotational states.

This region merges into the microwave region in that the same kinds of forces and interactions are involved, although the experimental techniques are widely different. Gaseous microwave spectroscopy is almost entirely concerned with the rotational spectra of molecules, or some other similar molecular motion, such as inversion. The theory and interpretation of these spectra is thus the same as in the infra-red region, the masses of the atoms and their bond lengths being such that absorption occurs at wavelengths of the order of 1 cm. Although, basically, the results obtained from the two regions are very similar, the microwave region has the great advantage that very much higher accuracy and resolution is possible, and hence a vast amount of information on nuclear properties is also obtained from the hyperfine structure of the spectrum. Microwave absorption in solids is more concerned with the forces and interactions inside the crystalline lattice, these being rather different from the interatomic binding forces of gaseous molecules. Investigation of the spectra in this case yields information on the fields present in the crystal, together with a great deal of data on the magnetic properties of ions in the solid state. The very high resolution available also enables nuclear properties to be measured accurately via the hyperfine splittings, and it is somewhat complementary to the gaseous work in this respect, as the elements that can be investigated by paramagnetic resonance are those which are difficult to investigate by gaseous spectroscopy.

Finally, the region of the electromagnetic spectrum occupied by normal radio waves can be utilized to obtain absorption spectra, usually by the interaction of magnetic fields with nuclear spins, thus separating their energy levels by amounts of the order of a ten-millionth of an electron-volt. These experiments on nuclear resonance and induction have also given much information on solid state theory.

It can be seen from this brief summary that much data can be obtained, by spectroscopic means, at all frequencies throughout the electromagnetic spectrum, although the experimental methods and techniques vary greatly as the wavelength changes.

1.4 Methods of Spectroscopy

The methods employed to obtain and record spectroscopic data vary considerably from one region of the spectrum to another, but the basic principles and technique remain the same. The essential requirements for the observation of absorption spectra at any wavelength are:

1. A source of radiation
2. Some form of absorption cell or specimen holder
3. A detector with which to measure the intensity of the radiation after it has passed through the absorption cell.

When emission spectra are being studied the source of radiation is replaced by an energy source, this produces emission of radiation from the target, and the emitted wavelengths can then be detected in the same way as for absorption spectra. If the source of radiation or energy is not monochromatic then some means of selecting different wavelengths is also needed, the prism being a classical example.

Thus in the optical region the source of radiation is usually some form of gas discharge, and after passing through the absorption cell, the different wavelengths are separated by means of a prism or diffraction grating, and can either be studied by eye or recorded on a photographic plate. This technique extends, with modifications, into both the ultra-violet and infra-red regions. In the former it becomes necessary to evacuate the system and use quartz prisms and windows to avoid continuous absorption, while in the infra-red the problems of detection become increasingly difficult.

At the high energy end of the frequency scale, sources of energy of known and precise value are usually employed to produce emission spectra. Thus electrons can be accelerated through known potentials to strike a target, and the energies which are necessary, before different series of X-ray lines are emitted, can be measured for any particular element. Similarly, in producing γ-rays, electrons or heavier particles can be accelerated to very high energies by cyclotrons, synchrotrons, etc., and the frequencies of the γ-ray spectra produced, when they hit a particular target, can then be deduced from energy relations. In this wavelength region various types of counters are used to detect and record the radiation, instead of photographic plates, which are normally employed from X-rays down to the near infra-red.

At the other end of the spectrum (i.e. microwave region) absorption spectra are always studied. This is because of the very small value of the quantum at these wavelengths, since it becomes very difficult indeed to detect any emitted quanta above the general background of radiation. The coefficients of absorption are also very much smaller than in the optical region, but the techniques are correspondingly more sensitive and highly monochromatic sources of radiation are available. There is thus no need for a dispersive element in the system, corresponding to a prism or diffraction grating; the source, absorption cell, and detector being the only essential parts of the apparatus. Emission spectra can, however, now be studied by masers (see Chapter 12).

It is of interest to note that the advance of spectroscopy has been dependent, to a large measure, on the discovery of sources and detectors of radiation at new wavelengths. The ultra-violet and infra-red regions were gradually opened up as new detectors were produced, and the recent extension of spectroscopy to very high and very low frequencies has been due to the new sources of energy and radiation that have become available. On the one hand very high energy accelerators have been built, and on the other, all the war-time research on radar has led to very stable and reliable microwave oscillators. Microwave spectroscopy is also fortunate in that a large amount of the rest of the apparatus can be taken over, more or less directly, from the developments of war-time research, an enormous amount of work having been carried out on microwave techniques.

A brief capitulation of the different sources of radiation and detectors employed for the different regions of the spectrum is included in *Figure 1*.

1.5 THE MICROWAVE REGION

Following this brief introduction to the electromagnetic spectrum as a whole, the microwave region may now be considered by itself in greater detail. The term 'microwave radiation' is taken to refer to radiation with a wavelength between 1 mm and 30 cm. This is an arbitrary choice of course, but experimentally this region is one in which the techniques of velocity modulation and waveguide propagation have taken over from normal radiofrequency methods; and its high frequency end is bordering on the far infra-red region. These wavelengths correspond to energy level changes of the order of a ten-thousandth of an electron-volt, or 1 cm^{-1}, and are very small compared with the normal atomic transitions giving rise to optical spectra. They are, in fact, produced by the weaker forces of a molecular or solid state character, which bind the different atoms together to form molecules, or a crystalline lattice.

A study of the spectra obtained in this region will therefore give detailed information concerning interatomic binding forces, and on the magnitude and symmetry of the fields of solid state interactions. Owing to the extremely high accuracy available (frequencies can be measured to 1 part in 10^7, and the line width reduced to 50 kc/s, i.e. $\Delta f / f$ is 1 in 10^5 or better) much more detailed and exact information can be obtained than from infra-red spectroscopy, and chemical bond lengths and angles can be determined to a very high degree of precision. It is probably true to say, however, that the greater successes of microwave spectroscopy have been in the determination of nuclear data, for which it is particularly well suited, owing to the very great resolution obtainable at these wavelengths. Such things as nuclear spins, magnetic moments, and electric quadrupole moments can all be determined and accurately measured from the hyperfine splitting that they produce in the spectrum. The mechanism responsible for the hyperfine structure is the same as that in the infra-red region, but the resolution obtained is so very much greater that much more accurate measurements are possible. Instead of carefully plotting out the intensity distribution of spectral lines from a photographic plate, as is usually necessary in optical hyperfine structure, the individual position of the different lines (which are often separated by twenty times their width, or more) can be very accurately measured by comparison with standard frequency markers, or proton resonance pips, if magnetic fields are used.

The apparatus and techniques employed for experiments in the microwave region will be considered in detail in the next two chapters, which are followed by descriptions of actual microwave spectroscopes. The results obtained will also be discussed with the corresponding theoretical analysis, under the different headings of gaseous spectroscopy and paramagnetic resonance measurements, in Chapters 5, 6 and 7. Here a brief summary of the research so far carried out in the microwave region is given. The work can be divided into three main groups; first, gaseous microwave spectroscopy, which is concerned with molecules in a gaseous state, and usually transitions occurring in their rotational energy. Secondly, paramagnetic resonance absorption, which is concerned with ions in a crystalline lattice, the energy levels of which can be affected by an external magnetic field. Thirdly, grouped together, are experiments on ferromagnetic resonance (really a special case of paramagnetic resonance), measurements on free radicals and on F-centres, and electron resonance in metals and semi-conductors, these all being somewhat specialized problems that have been studied with considerable success by microwave absorption techniques.

A very considerable amount of the experimental apparatus and technique is the same in all cases, and is considered in detail in the following chapters, the main differences being in the actual 'absorption cell', and the frequency control of the klystron.

1.6 Gaseous Spectroscopy

The first experiments on gaseous spectroscopy in the microwave region were carried out in 1933 by CLEETON and WILLIAMS[1], who showed much skill in obtaining any results with the techniques then available. They investigated the absorption of ammonia vapour around a wavelength of 1·5 cm, since theoretical predictions had shown that the inversion absorption frequency for the two positions of the nitrogen atom should be in this region. The apparatus was very direct and simple, the ammonia gas being contained in a rubberized cloth bag at atmospheric pressure, and placed in a beam formed by focusing the radiation with parabolic brass mirrors. To produce radiation of this wavelength they had to construct very small split-anode magnetrons, with anode radii of less than ½ mm. The results that they obtained showed a broad absorption band centred on a wavelength of 1·25 cm, the relatively high pressure of the gas preventing any further resolution.

Apart from this one experiment, spectroscopy in the microwave region lay dormant until after the war, when the enormous advance in microwave techniques opened up an entirely new field of research. The invention and development of the klystron and magnetron had

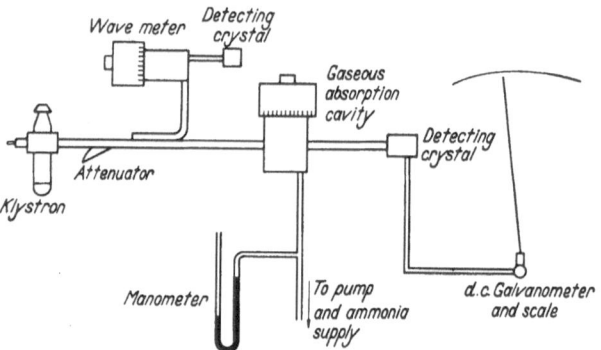

Figure 2. Block diagram of original NH₃ microwave spectroscope

produced highly robust oscillators which could generate radiation of much greater power level than was necessary for spectroscopical purposes. The frequency of these could also be accurately controlled and measured to nearly the same limits as for radiofrequency determinations (i.e. 1 part in 10^7), and all the technique of microwave propagation in waveguides, and detection of the radiation, had been studied in great detail for the needs of radar. There was thus a new tool of research available to the physicist, and one that had been developed to a high degree of precision. The invaluable assistance of all this war-time research on radar is one of the best examples of the way in which results of applied technology have directly benefited the research worker in pure science. Since the flow of ideas and techniques is usually in the opposite direction, it is gratifying to see this outstanding case of a return!

It was thus not surprising that work on spectroscopy in the microwave region started in several different laboratories as soon as the war was over, and ammonia gas was again the first object of attack. BLEANEY and PENROSE[2] were the first to publish their results, and a block diagram of their apparatus is shown in *Figure 2*.

The radiation is produced by the klystron valve, and passed down a waveguide run to the cavity resonator, which acts as the absorption cell. A directional coupler and wavemeter also lead off from the main run so that the output power and frequency of the klystron can be monitored. A silicon-tungsten crystal was employed as a detector at the output of the cavity, this being connected directly to a sensitive galvanometer, as no a.c. method of frequency sweep was used at first. In this way the output power from the cavity resonator was measured directly by the galvanometer, a square-law relation existing between detected current and output power. The output was

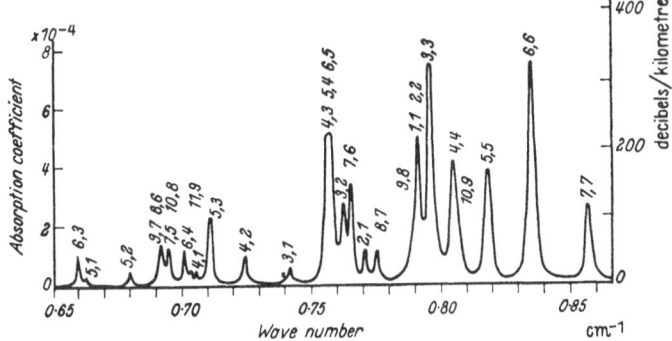

Figure 3. *The spectrum of ammonia near* 0·8 cm⁻¹

noted for each different frequency setting of the klystron, the cavity resonator always being kept tuned to the klystron frequency, and hence a plot of absorption against frequency was obtained, their results being shown in *Figure 3*. The use of a low pressure is necessary to overcome collision broadening, and it can be seen that the transitions due to the effect of different rotational states have been resolved out. These will be discussed in more detail in the chapter on gaseous spectroscopy, here it may just be noted that the different rotational states of the ammonia molecule slightly alter the distance between the nitrogen atom and the plane of the hydrogen atoms, and hence slightly alter the inversion frequency, giving a fine structure to the spectrum. The different lines are labelled in the diagram by their respective rotational quantum numbers. Because they were using a cavity resonator, and saturation broadening therefore became important at pressures below 0·2 mm Hg, they were not able to take the next step and resolve out the hyperfine structure of these lines, which is due to interaction with the nuclear electric quadrupole moment.

Similar work was being carried out at the Westinghouse Research Laboratories in America by GOOD[3], using a waveguide type of absorption cell, two-and-a-half metres long. With this he was able to reduce the pressure below 10^{-2} mm Hg, with no trouble from saturation broadening, and a few months after Bleaney and Penrose's original paper he was able to announce the discovery of hyperfine structure in the wings of the absorption lines. From that time forward an increasingly larger number of workers have been engaged on research in gaseous microwave absorption, and the list of compounds now studied runs into many hundreds. The work can be divided into two main parts, first that concerned with chemical problems, and therefore mainly used to measure bond lengths and angles; and secondly, that aimed at obtaining new nuclear data, such as spins and quadrupole moments, from the hyperfine structure. These two aspects will be considered in detail in Chapter 5.

1.7 Solid State Spectroscopy

A considerable amount of research has now been carried out on microwave absorption in solid state compounds, and the subject to which most attention has been given is paramagnetic resonance absorption. In this, the energy levels of a paramagnetic ion are investigated as it is sitting in the crystalline lattice instead of the energy levels of a molecule in a free gaseous state.

Paramagnetic ions are chosen because it is possible to vary their energy levels by the application of an external magnetic field, and hence their splitting can be adjusted to cause absorption of the particular microwave frequency being employed. The results obtained from an analysis of these spectra give considerable information on the forces and interactions existing in the solid state, and the very well resolved hyperfine structure also permits the determination of much nuclear information. A brief outline of the theory underlying paramagnetic resonance may be given as follows. The very strong electric fields existing inside a crystalline lattice act on the energy levels of the free paramagnetic ion to remove most of their orbital degeneracy. In many cases a single level two-fold degenerate in electron spin is left as the ground state, with all the other levels about 10^4 cm^{-1} higher. As the splitting to these other levels is so large only the ground state is populated at ordinary temperatures, and this may then be resolved into its two components by applying an external magnetic field. The separation between the two levels is then given by $g\beta H$, where β is the Bohr magneton, H the value of the magnetic field, and g is known as the 'spectroscopic splitting factor', and would be equal to 2·0 if the electron was completely free. If this splitting is now adjusted, by variation of the magnetic field strength, to be equal to the energy of a quantum of the microwave radiation, then the radiation will be absorbed and more spins excited to the higher state. Hence the condition for observing resonance becomes

$$h\nu = g\beta H$$

In practice the g value differs from 2·0, because the electron is not entirely free and is still affected by the binding forces. It is, in fact, from measurements of the g value that considerable data concerning the electronic configuration and higher excited levels can be obtained.

The apparatus employed is very similar to that used in the gaseous absorption work, but with the addition of an external magnetic field. The 'absorption cell' now takes the form of a cavity resonator in all cases, and this sits in between the pole pieces of a powerful electromagnet. The resonator is also often cooled down to low temperatures; this narrows the absorption line by increasing the spin-lattice relaxation time, and also increases the signal strength by reducing the number of ions in the excited levels, according to the Boltzmann factor. When using a.c. methods of detection the sweep frequency is applied to the magnetic field rather than to the klystron frequency, but apart from these points, the techniques of waveguide matching, frequency determination, signal detection and display are very similar to those of the gaseous spectroscopy.

The first experiments conducted on paramagnetic absorption were by ZAVOISKY[4] in the U.S.S.R., and CUMMEROW and HALLIDAY[5] in America. A large scale systematic survey of paramagnetic salts of the iron transition group was then undertaken by the Oxford workers, and the Clarendon Laboratory has continued to be the main centre for research on these lines, although an increasingly larger number of other laboratories are now engaged on this type of work. The study of the relatively broad absorption lines in the pure paramagnetic salts gave a great deal of interesting information, since separate energy levels could now be studied in detail, and very much more refined magnetic and solid state theories were possible. The widths and shapes of the lines were also useful in giving a measure of the various interactions occurring in the solid state, and on the relaxation processes between the ions and the surrounding lattice.

The biggest step forward came with the discovery of hyperfine structure in the spectrum. This was first obtained by PENROSE[6], who diluted some copper ammonium sulphate with the isomorphous magnesium salt and grew crystals with the copper ions very much farther apart than usual. By thus reducing the spin-spin interaction, he was able to reduce the line width to such an extent that the four lines, due to the four possible orientations of the copper nucleus in the field of the electrons, were separated out. From then on a large number of diluted crystals have been studied, and the hyperfine structure for many different nuclei obtained and analysed. The paramagnetic resonance work has the advantage that the resolution of the hyperfine structure is very much greater than by other methods, and its interpretation is much more simple. It arises from the direct interaction of the nuclear magnetic moment with the magnetic field of the electrons, and the single electronic level is split into $(2I + 1)$ levels, which give rise to $(2I + 1)$ hyperfine lines, I being the nuclear spin. These levels are perturbed to a varying extent by effects of the nuclear electric quadrupole interaction, but the determination of the nuclear spin is very much simpler than in the corresponding gaseous absorption case where the nuclear interaction is via the quadrupole moment. The theory of both these effects will be considered in detail in later chapters.

Much of the work on paramagnetic absorption has been correlated with other low temperature measurements on the same salts, such as specific heat and susceptibility determinations. In particular, it has been used to determine the best compounds suitable for aligning nuclei at very low temperatures and it is a subject which has linked together magnetic, low temperature and nuclear physics in a remarkable way.

1.8 Ferromagnetic Resonance

The phenomenon of ferromagnetic resonance absorption was first discovered by GRIFFITHS[7] in 1946, and may be regarded as a special case of paramagnetic resonance. Very wide absorption lines are usually obtained, and the reason for this is still unknown, as the strong exchange interaction should have a very marked narrowing effect. The apparatus employed is more or less identical with that used in paramagnetic resonance absorption, the ferromagnetic material takes the form of a thin foil and becomes one end of the resonator. The measurements are complicated by the demagnetizing factors which have to be allowed for, although recent work on colloidal suspensions has obviated this in some cases. The 'g values' obtained are usually very close to 2·0, but their divergence from the free-spin value, and the width and shape of the absorption lines has enabled various conclusions to be drawn concerning the interactions present in the ferromagnetic state. The theory is somewhat complicated, and as yet the study of ferromagnetic absorption has not yielded such strikingly new data as in the case of the gaseous work, or that on paramagnetic absorption.

1.9 FREE RADICALS AND F-CENTRES

Paramagnetic resonance is probably the most direct and sensitive method available for the detection and investigation of free radicals, as the presence of their unpaired electrons is just the condition required for such resonance absorption. A considerable amount of work is now in progress on the detection of free radicals by this means, and usually a narrow absorption line is obtained, with a g value very close to that of a free-electron spin. The techniques required are more or less identical with those of normal paramagnetic resonance.

Paramagnetic resonance may also be used to study F-centres, and other cases of 'trapped electrons' formed by irradiation. This subject, together with the recent work on electron resonance in metals and semi-conductors, promises to open up quite a new field of research, and all three are discussed in some detail in Chapters 7 and 11.

1.10 MEASUREMENTS AT RADIOFREQUENCIES

Spectroscopy at radiofrequencies was first used in the case of molecular and atomic beams, and only applied to the solid state some years later, the work being initiated by BLOCH[8] and PURCELL[9]. This type of resonance absorption can be regarded as the nuclear counterpart of electron paramagnetic absorption, the nuclear, instead of the electron spin changing its orientation during a transition. The difference in energy associated with this change of nuclear magnetic moment in the applied field is very much smaller than for the electron case, however, and the quantum required for resonance therefore falls in the radiofrequency, and not the microwave region.

The interaction of the nuclei with the rest of the substance is usually very small, and hence extremely narrow absorption lines are obtained. This not only affords very high precision, in the determination of nuclear gyromagnetic ratios and magnetic moments, but also means that nuclear resonance can be used as a very accurate method of measuring magnetic field strengths (see Section 13 of Chapter 4). Measurements on nuclear resonance have also been used to obtain detailed information on the interaction between the nucleus and the crystalline lattice, and surprisingly long relaxation times have been observed showing that the nuclear spin system acts very much like a well-insulated body sitting inside the lattice framework. The experimental techniques are similar in principle to the microwave case, the cavity resonator being replaced by an inductance and condenser, with the sample to be investigated sitting in the radiofrequency magnetic field of the inductance. Absorption of the radiation can then be observed by a drop in the effective Q of the circuit.

Another method, named nuclear induction, employs a second coil to pick up the induced fields at resonance; the details of these and other methods with the general theory, and results of measurements at radiofrequencies, will be found in Chapter 8. In recent years, considerable advance in nuclear resonance has taken place in the study of highly resolved spectra from the liquid state; the details and applications of this 'High Resolution Nuclear Resonance' are discussed more fully in Chapter 10.

REFERENCES

[1] CLEETON, C. E. and WILLIAMS, N. H. *Phys. Rev.* 45 (1934) 234

[2] BLEANEY, B. and PENROSE, R. P. *Nature* 157 (1946) 339

[3] GOOD, W. E. *Phys. Rev.* 69 (1946) 539

[4] ZAVOISKY, E. *J. Phys. U.S.S.R.* 9 (1945) 211

[5] CUMMEROW, R. L. and HALLIDAY, D. *Phys. Rev.* 70 (1946) 433

[6] PENROSE, R. P. *Nature* 163 (1949) 992

[7] GRIFFITHS, J. H. E. *ibid.* 158 (1946) 670

[8] BLOCH, F., HANSEN, W. W. and PACKARD, M. *Phys. Rev.* 69 (1946) 127

[9] PURCELL, E. M., TORREY, H. C. and POUND, R. V. *ibid.* 69 (1946) 37

2

THE PRODUCTION AND DETECTION
OF MICROWAVES

2.1 Introduction

Before considering the various kinds of microwave spectroscope that have been used in practice, this chapter and the next will be devoted to a general introduction of microwave techniques, with special emphasis on those points which are of particular importance in designing and constructing a spectroscope.

In all forms of absorption spectroscopy the source of the radiation plays a most prominent part, and this is especially so in the microwave region where monochromatic radiation is required, no dispersive element being incorporated in the apparatus after the absorption cell to separate out the different wavelengths. Sources of monochromatic radiofrequency radiation have been available for many years, and very high frequency stability is made possible by the use of quartz-crystal oscillators; but the successful production of microwave radiation has necessitated the use of an entirely new type of electron interaction.

The fundamental drawback in attempting to use conventional valve circuits for the production of ultra-high-frequencies is that the transit time of the electrons becomes equal to an appreciable part of a high frequency cycle and the necessary phase relations can no longer be obtained in the valve. This can be partially overcome by constructing valves with very small electrode spacings, but their manufacture requires extremely high precision work, and the small spacings of necessity mean that only low powers can be handled, otherwise thermal heating will distort the internal structure too much. Microwave triodes have been built to operate as high as 3,000 Mc/s, but, generally speaking, at all frequencies above 1,000 Mc/s it becomes necessary to employ oscillators based on the principle of 'velocity modulation', instead of 'space charge control' which is the underlying mechanism in conventional valves.

2.2 Velocity Modulation

The method of 'velocity modulation' makes use of the finite velocities of the electrons and the differences between them to produce a system of electron bunches, which will react with the external circuits and feed radiofrequency power into them. Since nearly all microwave spectroscopes employ klystrons as their radiation sources, a brief summary of the theory of velocity modulation, as it occurs in the klystron valve, is given below.

The essential features of a klystron are illustrated in *Figure 4*, an electron beam being accelerated from the electron gun to the first

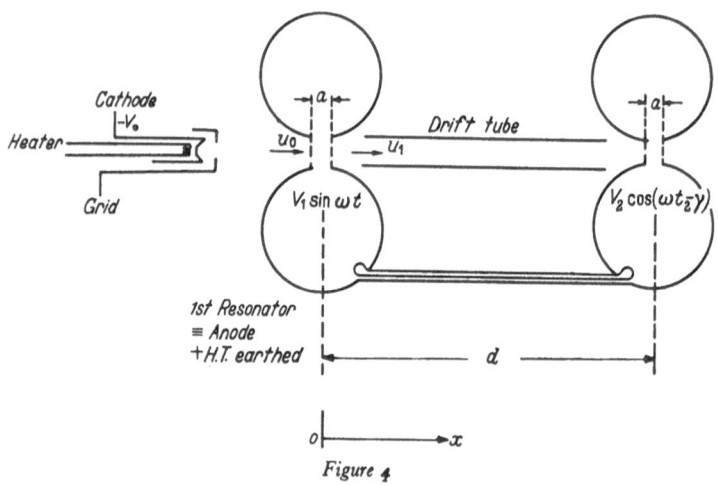

Figure 4

resonator, and then, after crossing this, it passes down a field-free region known as the 'drift tube' to cross the grids of a second resonator. A beam of electrons of uniform velocity u_0 thus enters the first gap, where

$$\tfrac{1}{2}mu_0^2 = eV_0 \qquad \qquad \dots (2.1)$$

and it can be shown[1] that, if a radiofrequency field, $V_1 \sin \omega t$, already exists across the gap then the emergent velocity of an electron will be:

$$u_1 = u_0 (1 + \tfrac{1}{2}\beta . \alpha_1 . \sin \omega t_0) \qquad \qquad \dots (2.2)$$

where β is the 'gap modulation coefficient', α_1 the depth of modulation, and t_0 the time that the electron crosses the centre of the gap. By considering the motion of the electron down the drift tube its time of arrival t_2 at the second resonator is obtained as:

$$\omega t_2 - \theta = \omega t_0 - \tfrac{1}{2}\theta . \beta . \alpha_1 . \sin \omega t_0 \qquad \qquad \dots (2.3)$$

where the transit angle $\theta = \omega d/u_0$. The principle of continuity of charge can then be applied to the electron motion as a whole, and an expression for the variation in bunched current at the second gap derived[1].

I.e.
$$i_2 = i_0 \cdot \frac{dt_0}{dt_2} = i_0/[1 - \tfrac{1}{2}\theta \cdot \beta \cdot \alpha_1 \cdot \cos \omega t_0]$$
$$= i_0\left[1 + 2\sum_{n=1}^{\infty} \mathcal{J}_n\left(n \cdot \frac{\theta\alpha_1\beta}{2}\right) \cdot \cos n\left(\omega t_2 - \theta\right)\right] \quad \ldots (2.4)$$

The current at the 'catcher resonator' thus consists of a d.c. component plus a.c. components of frequencies equal to ω and all its harmonics, the amplitude of these varying as Bessel functions of the

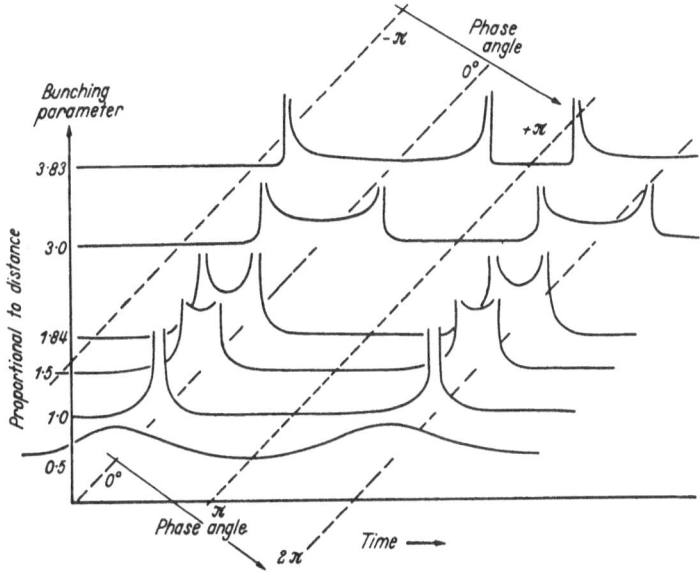

Figure 5. Electron bunching

first kind. In general, the catcher resonator is tuned to the fundamental frequency, and since its geometry is somewhat complicated, its higher resonant frequencies will not be in a geometrical progression, and only the term in $\mathcal{J}_1(\tfrac{1}{2}\theta \cdot \alpha_1 \cdot \beta)$ is effective in producing oscillations in the resonator.

The optimum conditions for bunching are best seen graphically, and in *Figure 5* the 'bunching parameter' $(\tfrac{1}{2}\theta \cdot \alpha_1 \cdot \beta)$ is plotted against the 'phase factor' $(\omega t_2 - \theta)$, the axes thus being proportional to distance and time. To calculate the actual optimum values, the power taken from the beam by the resonator must be derived, and this is given by:

$$\text{Power taken} = i_0\beta V_2 \cdot \cos (\theta - \gamma) \cdot \mathcal{J}_1(\tfrac{1}{2}\theta \cdot \alpha_1 \cdot \beta) \quad \ldots (2.5)$$

and the maximum efficiency of the klystron is then

$$\eta = \beta \cdot \alpha_2 \cdot \mathcal{J}_1(\tfrac{1}{2} \cdot \theta \cdot \alpha_1 \cdot \beta) \qquad \ldots (2.6)$$

The function $\mathcal{J}_1(x)$ is plotted in *Figure 6*, and is seen to have a maximum value of $0\cdot584$, when the value of its argument is $1\cdot84$.

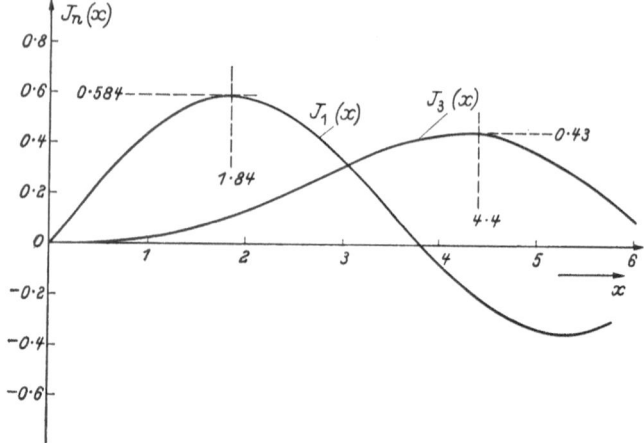

Figure 6. Plot of Bessel functions $\mathcal{J}_1(x)$ and $\mathcal{J}_3(x)$

The maximum efficiency of the klystron is thus $58\cdot4$ per cent, and is obtained when the bunching parameter is made equal to $1\cdot84$. In a similar way, it can be seen from the graph of $\mathcal{J}_3(x)$, also plotted in *Figure 6*, that the maximum efficiency of a klystron tripler would be 43 per cent, and the bunching parameter must then be equal to $4\cdot4$.

The klystron can be turned from an amplifier into an oscillator by just feeding back a proportion of the power in the catcher, and adjusting the transit angle so that:

$$\theta = \frac{\omega d}{u_0} = (4n - 1) \cdot \frac{\pi}{2}. \qquad \ldots (2.7)$$

In practice the functions of both resonators are usually performed by one, and a reflex klystron is employed in which the electron stream is repelled back across the same gap by a more negative reflector electrode. It can be shown that provided the repelling field is linear with distance, the previous analysis applies throughout. The conditions for maximum output power are then obtained by combining equations (2.6) and (2.7), to give

$$1\cdot84 = \frac{\theta \cdot \alpha_1 \cdot \beta}{2} = (4n - 1) \cdot \frac{V_1}{V_0} \cdot \frac{\pi\beta}{4} \qquad \ldots (2.8)$$

20

The maximum value of bunched current will therefore occur at higher values of V_1 for modes with the smaller mode number (n). It therefore follows that maximum power will be obtained when the reflex klystron is operating on the mode with greatest negative reflector voltage. The frequency of the klystron can be changed by about 10 per cent by mechanically distorting the cavity size, but a small shift can also be produced by altering the reflector voltage. This changes the time of transit slightly, increasing the reactive component in the admittance of the beam to the resonator, and hence the frequency of resonance shifts to cancel this. In practice the useful frequency range that can be covered by electronic tuning is of the order of ± 25 Mc/s.

2.3 PRACTICAL MICROWAVE SOURCES

The essential requirement in microwave spectroscopy is a source which can be controlled to give a known frequency, and which can be tuned to cover a reasonable frequency range. The power output need only be of the order of milliwatts, as higher power levels often cause saturation effects in the absorbing compounds. It is just these features which make the klystron an ideal source for work in microwave spectroscopy, in comparison with the magnetron, which can give so much greater power, but is difficult to tune and hold at a precise frequency.

Work in gaseous spectroscopy has been concentrated mainly in the K-band region, which had been developed for radar purposes, and at lower wavelengths, where klystrons soon became commercially available in the United States. Experiments on paramagnetic resonance have been normally carried out at the 'spot frequencies' offered by the war-time X, K and Q bands (3 cm, 1·25 cm and 8 mm wavelengths); since magnetic field variation replaces that of frequency, to a large extent, no complete wavelength coverage is essential. A list of the different klystrons, that have been used for micro-wave spectroscopy, is given in *Table 2.1* and others covering the range between 0·4 cm and 3·0 cm are produced by Varian Associates Ltd.

Table 2.1. Klystrons used in Microwave Spectroscopy

Klystron	Manufacturer	Wavelength Range	Power	Operating Voltage
723 A/B	B.T.L. / Raytheon	3·1–3·5 cm	28 mW	500 V
K.302	E.E.V.	3·16–3·22	15	430
CV.323	U.K. Services	3·0 –3·4	50	2,000
2K.50	B.T.L.	1·20–1·36	20	450
2K.33	Raytheon	1·20–1·36	40	1,800
QK.277	Raytheon	1·00–1·11	18	2,250
QK.227	Raytheon	0·60–0·71	5	3,000

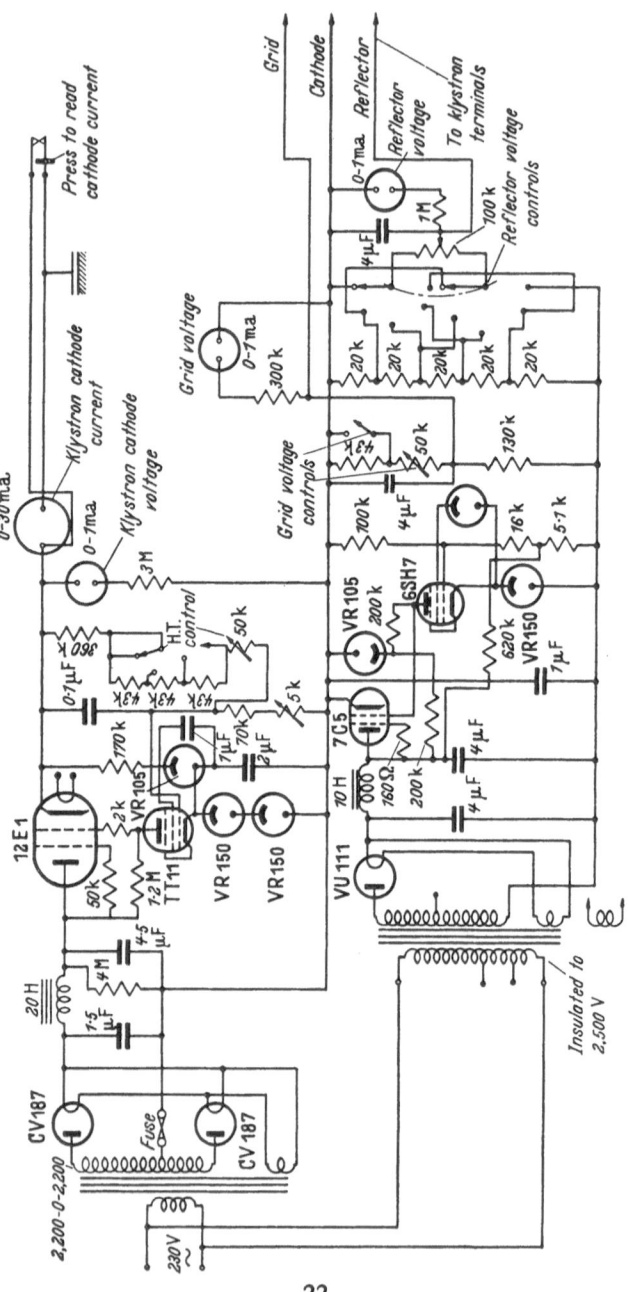

Figure 7. Klystron power pack

22

It can be seen that most of these require several kilovolts H.T., and this must be very well stabilized to prevent changes in frequency. Circuit diagrams of power packs devised to supply high voltage klystrons for radar purposes are given in Chapter 2 of the Radiation Laboratory Series *Techniques of Microwave Measurements*. In these the main H.T. supply is electronically stabilized, and the reflector and grid voltages are taken off from resistor chains across it. *Figure 7*

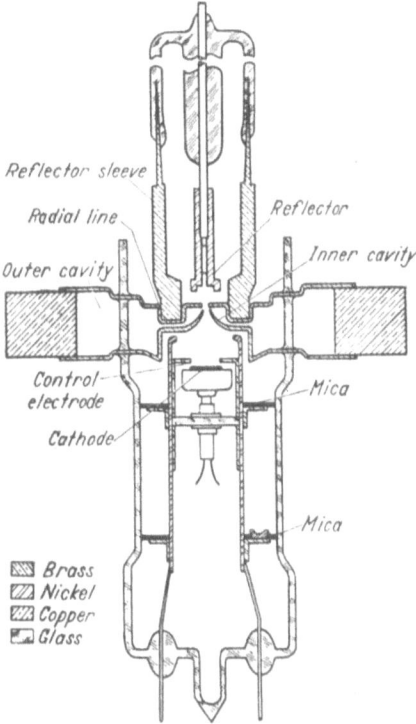

Figure 8. Cross-sectional diagram of 2K33 klystron

shows a circuit in which the supply to the reflector and grid are separately stabilized, and for some type of work it is necessary to supply all the voltages from batteries. The heater current is always taken from batteries, to avoid 50 c/s pick-up.

Figure 8 shows the cross-sectional diagram of a 2K33 klystron, and it can be seen how the beam is focused past the narrow gap formed by the edges of the 'inner cavity'. This is connected to the 'outer cavity', containing the metal-to-glass seal, and has a direct waveguide output taken from it.

2.4 Automatic Tuning

There are two cases in particular where automatic tuning of the klystron is a necessity. The first is when the absolute frequency must be held constant for a considerable time, such as when a lock-in amplifier is being employed and the absorption line is being swept through very slowly in order to have as narrow a bandwidth as possible. Any slight change in the frequency during this period will be produced as a change in amplitude on the recorded signal, when a

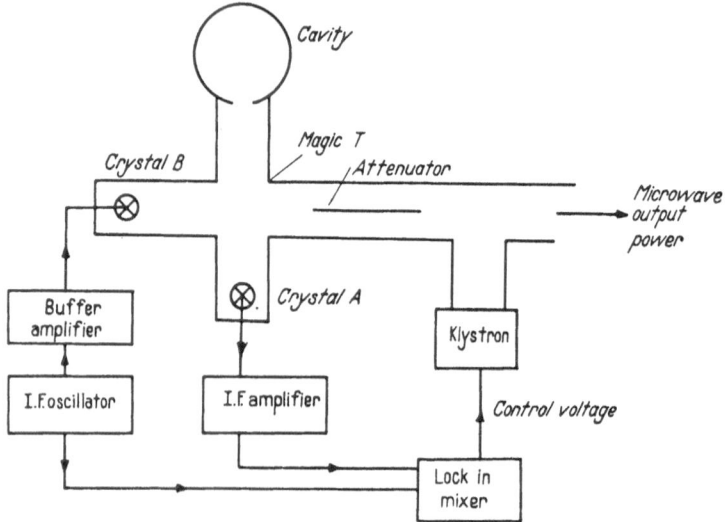

Figure 9. *Pound stabilizer circuit*

tuned cavity is incorporated in the waveguide circuit. The other condition, where automatic frequency control is required, is when a superheterodyne method of detection is being employed, and it is necessary to keep the frequency difference between two klystrons at the intermediate frequency value, although the absolute value of the frequency is not so important.

In order to keep the klystron locked to an absolute value of frequency a Pound stabilizer circuit is usually employed. This locks the klystron to a high Q cavity resonator, and derives an error signal when the frequency of the klystron is different from that of the cavity, this error signal being used to alter the potential of the klystron reflector and bring its frequency back to that of the cavity. A block diagram of this type of circuit is shown in *Figure 9*, which is taken from Pound's original paper[2]. In this, an intermediate-frequency amplifier feeds crystal B with a 30 Mc/s signal, and if the cavity is not resonant at the klystron frequency some of the microwave radiation which has been reflected back from it, on to crystal B, will be returned with amplitude-modulated sidebands spaced at 30 Mc/s difference above and below the signal frequency. These will then be mixed with some of the direct microwave power in crystal A to produce an intermediate-frequency signal. This is fed to a control circuit which produces a d.c. potential proportional to its amplitude, this, in turn, is applied to the klystron reflector, the correct sign being obtained by mixing in some of the signal from the 30 Mc/s oscillator in a phase-sensitive detector. The advantage of this method is that the d.c. output voltage can be obtained directly from the phase-sensitive detector, provided that the insulation of the phase mixer circuit is sufficient.

In the second type of automatic frequency control circuit, for use with superheterodyne systems, the error signal is obtained from differences with the intermediate frequency instead of differences with the frequency of a cavity resonator. A typical circuit that can be employed to do this is shown in *Figure 10*. It is designed so that the

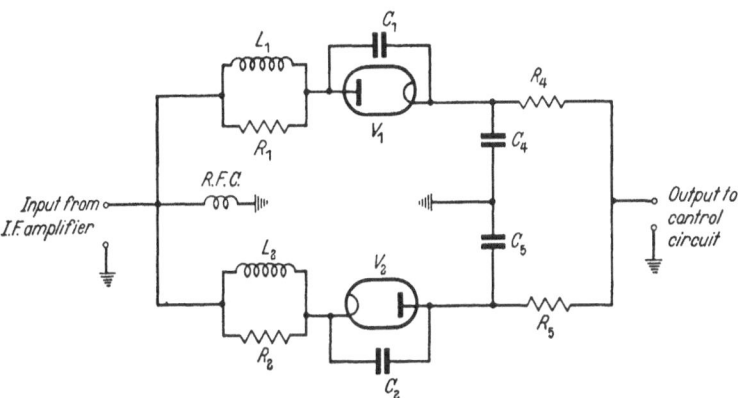

Figure 10. A.F.C. discriminator circuit

combination of L_1 and C_1 with the capacity of the diode V_1 will resonate at a higher frequency than the correct intermediate frequency, whereas the combination of L_2, C_2 and V_2 resonates at a lower frequency than the correct intermediate frequency value. Direct current potentials are thus built up across C_4 and C_5 of magnitude and sign corresponding to the deviation of the input frequency from the correct value. This d.c. potential is then used to retune the klystron by altering the reflector potential, a typical control circuit for a low-voltage klystron being shown in *Figure 11*, which is effectively a d.c. amplifier; the correct operating range is obtained by adjusting the potentiometer R_2.

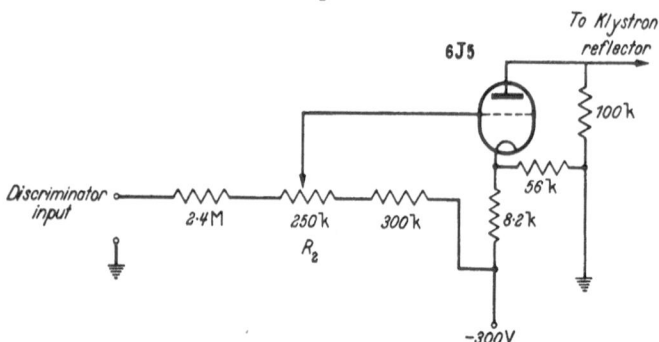

Figure 11. Klystron control circuit (300 volt H.T.)

The particular cases in which either Pound stabilization, or automatic control of the intermediate frequency, are employed will be seen when considering individual spectroscopes in detail.

2.5 OTHER SOURCES OF RADIATION

The advantages of the reflex klystron for producing stable, easily controlled, low-power microwave radiation are so great that very little attempt has been made to use any other source of radiation. The initial experiment of CLEETON and WILLIAMS[3] was performed using a magnetron of their own construction, and the use of a pulsed magnetron for relaxation-time measurements has also been suggested. In some recent work[4] with millimetre waves, harmonics obtained from pulsed magnetrons have also been used to provide radiation of sufficient power at these frequencies, but in general the higher power output of the magnetron is not required, and the greater difficulty in tuning is a great disadvantage.

The lowest wavelength available from klystrons, at the moment, is

Table 2.2. Mm-wavelength Power from Klystron Harmonics*

Driver Klystron	Power Output	Fundamental mm	Harmonic Number	Harmonic Wavelength mm	Conversion Loss dB
2K33	35 mW	12·5	2	6·25	18
			3	4·17	35
			4	3·12	45
			5	2·50	60
QK290	35 mW	9·5	2	4·75	20
			3	3·17	40
			4	2·38	55
			5	1·90	70

* Figures are taken from; Johnson, C. M., Slager, D. M. and King, D.D., *Rev. Sci. Inst.* 25 (1954) 213.

about 5 mm, and for work below this some form of crystal harmonic generation must be employed[5]. Such crystals are mounted across the waveguide and feed into a smaller waveguide which will pass the harmonic wavelength, but not the fundamental, as is illustrated in *Figure 12*. The amplitude of the harmonics produced by this means increases markedly with the input power to the crystal, and, for this reason, a 2K33 klystron operating at 1·25 cm wavelength, is normally used.

A recent experimental analysis by Johnson, Slager and King shows that, down to 3 mm wavelength, the harmonics of a 2K33 give more power than those from any lower-wavelength klystron. Below 3 mm, more power is available from the harmonics of the QK290, but it is not so steady as that from the 2K33. The results are summarized in *Table 2.2*, which is taken from their paper.

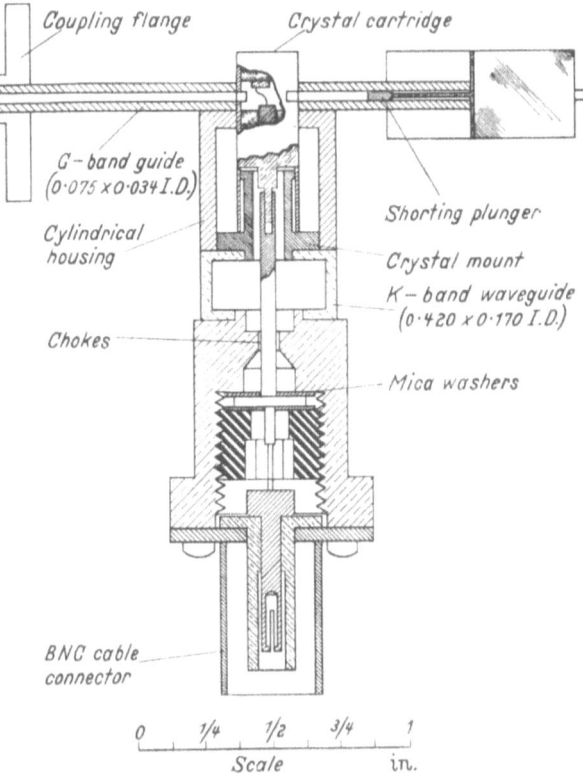

Figure 12. Details of harmonic generator which utilizes cut-out 1N26 or 1N31 crystals

The lowest wavelength yet used to observe spectral lines by these methods appears to be 0·77 mm, this being obtained by harmonic generation from a K-band klyston [6] (considerably reduced by 1966).

2.6 DETECTION OF THE RADIATION—SILICON CRYSTALS

In nearly all the work on microwave spectroscopy a silicon-tungsten crystal of one form or another is used to detect the radiation. BERINGER[7] obtained high sensitivity using a bolometer in a balanced-bridge circuit, and experiments have been conducted with a travelling-wave detector[8] to compare the signal-to-noise ratio with that from a crystal, but the great majority of workers have employed standard silicon-tungsten crystals.

These crystals, on which intense research work was carried out during the war, consist of a flat disk of silicon, with a carefully

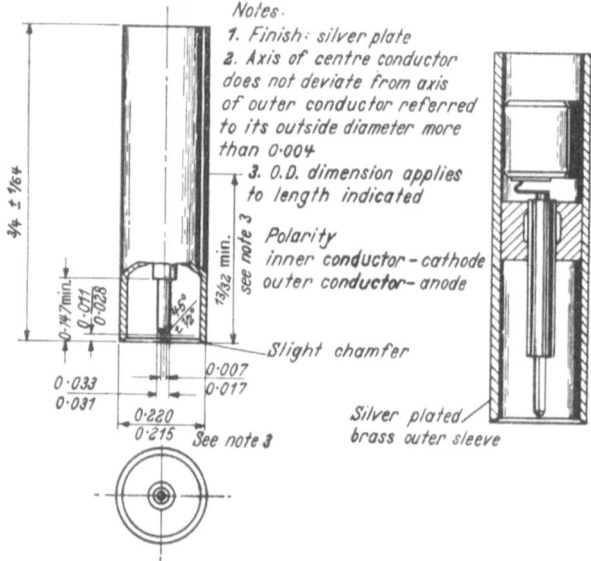

Figure 13. Outline specification type 'C' coaxial type Sylvania silicon diode. Drawing on right shows cutaway view of the type 'C' diode

prepared surface, mounted rigidly inside a cartridge. A spring whisker of tungsten wire is mounted on another metal plug, this being advanced up the case by a micrometer screw until a sensitive spot contact has been found. The whole is then sealed into position to make permanent contact. The method of construction is shown in *Figure 13*, which illustrates one of the shielded coaxial type cases, the cartridge being designed to provide a matched termination for a 65 ohm line. The crystal is mounted in a holder such that the outer shield makes direct contact with the waveguide, while the inner conductor fits into a probe which passes across the waveguide and ends as the centre conductor of a coaxial plug, the rectified d.c. or intermediate frequency being taken off from here on a coaxial line. A typical K-band crystal holder is illustrated in *Figure 14*, and further details of crystals and crystal holders are given in Section 3.6.

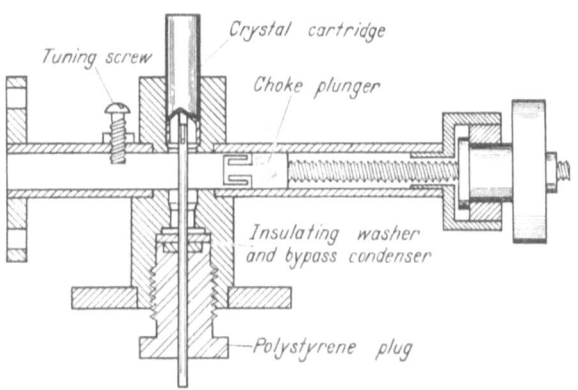

Figure 14. K-band crystal holder

So far as the actual detecting action of the crystal is concerned there are three important parameters which describe its behaviour, these are 'conversion loss', 'noise figure' and 'impedance'. The conversion loss is, in effect, a measure of the efficiency of detection, and measures the fact that not all the available microwave input power is converted into power at the intermediate frequency or d.c. This loss is high for low input power and rectified current, but falls rapidly to a more or less constant value of about 6 dB for crystal currents in excess of 0·5 mA.

This is not the only parameter which determines the over-all sensitivity of the detector, however, the noise generated in the crystal being a very important factor. This varies with both the magnitude of the rectified current and the intermediate frequency of the output. For a given intermediate frequency the noise temperature of the crystal varies more or less linearly with the rectified current, so that the combination of conversion loss and noise of the crystal produce an over-all noise figure which has a broad minimum, as illustrated in *Figure 15*. This is drawn for the particular case of a 1N23B crystal (3 cm band) at an intermediate frequency of 30 Mc/s, and the optimum power level is seen to be about 0·5 mA of rectified crystal current, which corresponds to an input power level of about 1 milliwatt.

The matching of the impedance of the crystal is important if maximum sensitivity is to be attained, and both its radiofrequency and intermediate frequency impedance vary somewhat with the strength

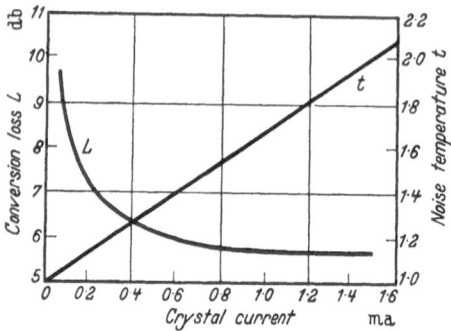

Figure 15. Conversion loss and noise temperature for a crystal

of the input signal. For rectified currents above 0·5 mA, the intermediate frequency impedance remains reasonably constant, and does not vary much with frequency, having the same kind of value down to d.c., this usually being of the order of 400 ohms.

2.7 DIFFERENT DETECTING SYSTEMS

The different methods of detecting microwave spectra can be classified into three main groups, (1) d.c. detection, (2) Video detection, (3) Superheterodyne detection.

The d.c. method of detection was the first used in both gaseous spectroscopy and paramagnetic resonance experiments, being very simple and direct. The output from the detecting crystal is fed straight to a sensitive galvanometer via suitable matching resistors, and the magnitude of the signal is thus given by the galvanometer deflection, the crystal usually working on the square-law part of its characteristic. The variation of output signal is then plotted as either the frequency of the klystron, or the magnetic field strength, is changed, an absorption line being shown as a reduction in the galvanometer deflection. This method is very convenient for wide lines of strong intensity, but cannot be used for weak signals or those of narrow line width—which are usually the ones of interest.

The sensitivity of detection can be greatly increased by employing a.c. methods, this not only increases the signal-to-noise ratio, but also enables the absorption lines to be displayed on a cathode-ray tube. There are various means by which a modulation can be applied to the microwave radiation, these falling into two groups; (i) either the frequency of the klystron can be modulated, or (ii) the conditions affecting the absorbing substance can be changed, by varying electric or magnetic fields. The simplest of these methods is the 'crystal-video method of detection' used for gaseous spectroscopy; in this the frequency of the klystron is swept by applying a modulating voltage to the reflector electrode, and if the frequency of the klystron is near a spectral line, the modulation will cause the frequency to sweep to and fro across the line. In this way the detected microwave signal will consist of a function representing the shape of the absorption line, recurring at twice the modulating frequency. The ratio of the amplitude of the modulation to the width of the spectral line will determine whether the shape of the line itself forms the repeated 'pulse', or whether only part of this is produced. The bandwidth of the following amplifier stages must be large enough to pass most of the frequencies which are needed when a Fourier analysis of the line shape is made, this depending, of course, both on the frequency and the amplitude of the modulating voltage, and on the actual width of the absorption line.

As an example of this one may quote the values used in a normal paramagnetic resonance spectroscope. Here the modulation is usually applied to the magnetic field, the frequency of the klystron being held constant. To produce a large enough depth of field modulation to sweep through two or more hyperfine structure lines at once, it is necessary to use low modulating frequencies as the inductance due to the magnet poles is very high. A frequency of 50 c/s is thus usually chosen, which means that each spectral line will appear as a signal on the detected microwave power at a frequency of 100 c/s (the line being traced out as the field is swept both forwards and backwards). With a total sweep of about 200 gauss and a representative line width of 5 gauss, an amplifier with a bandwidth of 10 kc/s is found to be sufficient to pass the lines with little distortion of shape. (The coefficients of frequency components above 10 kc/s have values less than 2 per cent of the fundamental.) The absorption spectrum may be displayed directly on a cathode-ray tube by applying the output from the video-amplifier to the Y plates, and feeding the time base from the source of modulating frequency, a phase-shifting circuit being incorporated so that the spectra obtained on the forward and backward sweep can be made to coincide. A block diagram of this direct crystal-video method of detection is shown in *Figure 16*.

The third method of detection is one using a microwave local oscillator to produce a superheterodyne system. Superheterodyne methods of detection have been used very widely in all fields of microwave spectroscopy, but the gain in sensitivity over the direct video method is not so great as is normally obtained by the use of superheterodyning. The method also requires very efficient stabilization of the local oscillator frequency and the circuit details become

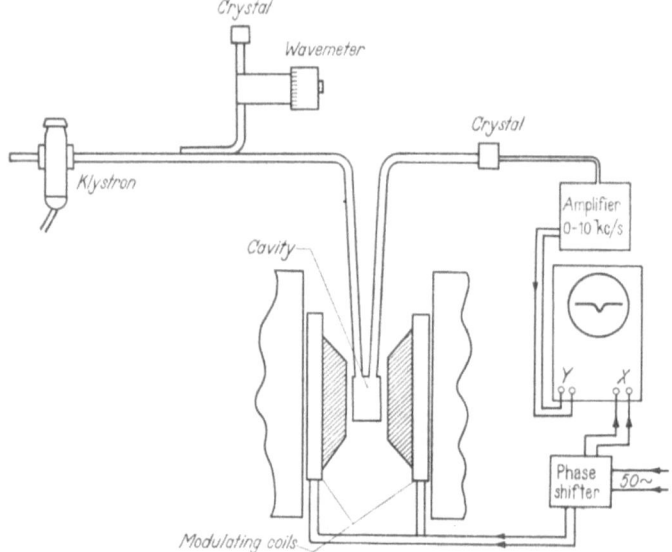

Figure 16. Crystal-video method of detection

rather complicated. The power from the local oscillator is usually fed into the waveguide detecting arm via a directional coupler, or magic-T, and mixes with the signal frequency in a silicon-tungsten crystal, from which the intermediate frequency is taken to the amplifier chain. A frequency of 30 Mc/s or 45 Mc/s is usually chosen as the intermediate frequency, the choice of this value being due to a combination of several factors.

The excess noise generated by the crystal, for a given rectified current, varies inversely with frequency over a range at least as wide as 1 c/s to 24.10^9 c/s, which might suggest that the best results would be obtained by using as high an intermediate frequency as possible. This is counterbalanced by the fact that the noise of a given type of intermediate frequency amplifier increases with frequency, and also that the conversion loss of the crystal is dependent on the rectified current, with a result that there is a broad optimum value for intermediate frequency centred at about 30 Mc/s (i.e. for a crystal current of about 0·5 mA the excess noise produced by the crystal will be smaller than that of the amplifier for frequencies above 30 Mc/s). The actual optimum value will vary according to the particular crystal and amplifier employed; a full discussion of this is given in a paper by STRUM[9], the results of which are summarized in

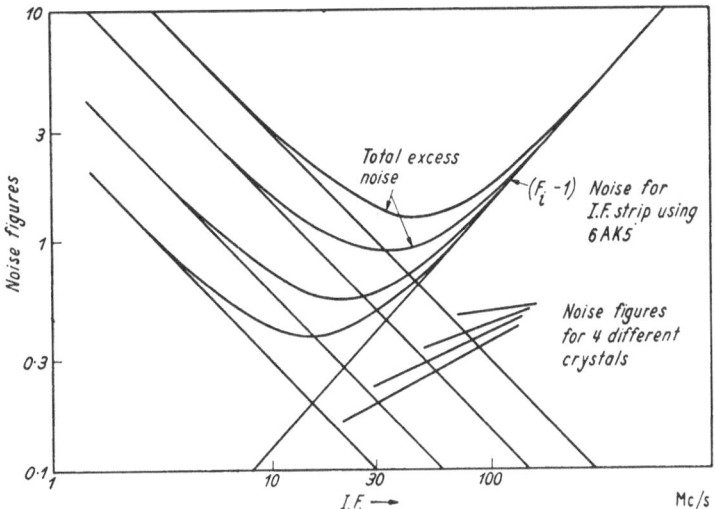

Figure 17. Total mixer noise due to I.F. and crystal noise

Figure 17; where the curves of noise figure against intermediate frequency for four different crystals are added to that of the low-noise intermediate frequency amplifier, to produce curves representing the total noise figure of the detecting system. The main problem associated with this technique is to keep the difference in frequency between the signal and the local oscillator source exactly equal to the intermediate frequency; and some type of automatic frequency control and discriminator circuit, similar to those already referred to, must be employed.

Spectroscopes employing superheterodyne detection usually incorporate a balanced bridge system, so that the detected intermediate frequency is only obtained when absorption takes place. The bridge element is normally a 'magic-T' or 'hybrid-ring' (discussed in detail in the next chapter), a spectroscope incorporating such is shown in *Figure 18*. When correctly matched and balanced, no signal will be obtained in the fourth detecting arm; but any absorption which now takes place will unbalance the bridge, and cause power to be fed into the fourth arm which will mix with that from the local oscillator to give an intermediate frequency signal. One of the disadvantages of this system is that the magnitude of the power reflected into the fourth arm (δP) is not equal to the actual power absorbed

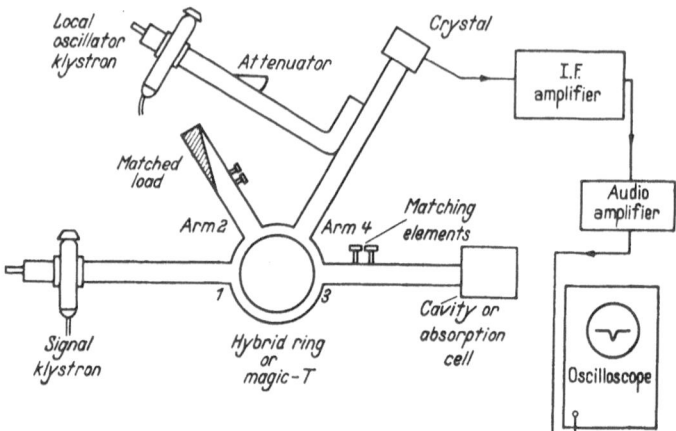

Figure 18. Balanced bridge for superheterodyne detection

(ΔP), but is only a fraction of it, and is given by the expression derived by GORDY[10]:

$$\frac{\delta P}{\Delta P} \approx \frac{\Delta P}{P}$$

where P is the total power in the absorption cell or cavity.

Several other forms of balancing are possible instead of the most direct one described above, and these are illustrated in the spectroscopes designed by GESCHWIND[11] and BAGGULEY[12] which are discussed in detail in Chapter 4. The necessity for a second klystron as a local oscillator has also been avoided in one method of superheterodyne detection by feeding some of the original radiation into an arm containing a crystal which is fed from an oscillator at the intermediate frequency. This modulates the original microwave power which can then be used to beat with the signal from the far end of the absorption cell or cavity, the power in its sidebands acting as the local oscillator.

Details of actual spectroscopes, with their different detecting systems, including Stark and other forms of modulation, are given in Chapter 4. More recent advances are given in Chapter 11.

2.8 DISPLAY SYSTEMS

There are various methods whereby the signal can be displayed or recorded after it has been detected and amplified. If all of an absorption line is covered by the modulating sweep (the modulation being applied either on the klystron reflector, or to the magnetic or electric field strength) then the line can be very easily displayed directly on a cathode-ray screen. In this case a video amplifier, of bandwidth sufficient to pass the line shape without distortion, is used to amplify the detected signal, being preceded by an intermediate frequency amplifier when superheterodyne detection is employed. The output of the video amplifier is then fed directly across the Y plates of the oscilloscope, the time base being fed at the modulating frequency via a phase shifter, so that the positions where the absorption line is traced out by the forward and backward sweeps of the modulation can be made to coincide on the oscilloscope screen. This method is nearly always used when initially lining up any apparatus, or when searching for lines of medium amplitude, as it requires no careful matching or balancing and gives a direct visual trace of the spectrum being investigated. In most cases, however, this method of presentation requires a relatively large bandwidth, as the lowest modulation frequency is usually determined by the persistence of the oscilloscope screen, and in order to pass a line shape undistorted this means a bandwidth of 2 kc/s or more is required for the video amplifier.

One method of reducing the bandwidth, and thus increasing the signal-to-noise ratio, is to employ some form of 'phase sensitive detection'. The great advantage of this method is that the signal-to-noise ratio is only dependent on the bandwidth of the last recording stage, the noise components of previous stages being eliminated in the mixer. The essence of this form of detection is that the absorption line, or more generally, its differential, is obtained as a slow change in amplitude on the modulating frequency (the rate of increase of the average value of the magnetic field strength, or the klystron frequency through the absorption line being very slow). This signal is then mixed with one that is obtained from the same modulating source via a variable phase-shifting network. The resultant output is the product of a signal of constant amplitude and frequency, but adjustable phase, with one which has an identical frequency as the first, but of very slowly varying amplitude, which is the differential of the absorption line.

A mathematical analysis shows that the output from the mixer is then a d.c. component, plus the low frequency signal corresponding

35

to the absorption line, the amplitude of this being proportional to cos (phase angle between the two input signals). It follows that the bandwidth of the recording stage is the only one of importance, as any noise voltages in preceding stages not identical with the carrier frequency will produce beat voltages which are rejected by the low pass filter. The same result can be obtained by viewing the reference voltage as a gating pulse[13]. In most cases the signal is finally traced out by a pen-recorder, and a switched R–C network of filters preceding this allows different bandwidths to be selected, values lower than 1 c/s often being employed.

Another advantage of this method is that it is very easy to separate out the two different effects which occur at resonance—i.e. fall in amplitude of microwave power ('absorption') and change in reactive component ('dispersion'). It can be shown[14] that the two effects are comparable in magnitude, and hence the resultant unbalance signal from the microwave-bridge is generally a mixture of both phase and amplitude unbalance. For small signals it can also be easily shown that these two are 90° out of phase, and hence, by adjusting the phase of the reference signal, applied to the phase-sensitive detector, either one or the other can be selected and passed on to the recording instrument.

Several different types of circuit have been used in phase-sensitive detectors, and a paper by SCHUSTER[15] gives details of a high stability detector operating at 30 c/s. The mixing can be done in one valve using two electrodes, or in two balanced valves, or in a rectifier bridge system. One possible type of phase-sensitive detector, as used for work in paramagnetic resonance, is illustrated in *Figure 19*, and a d.c. amplifier stage is usually incorporated between the output of the mixer and the actual recorder.

The differential of the absorption line is usually recorded in such cases, as a small amplitude of modulation is employed and the mean frequency, or field value, then swept very slowly through the resonant value. If the bridge is adjusted for phase unbalance the differential of the dispersion curve will be produced (i.e. two peaks with a larger inverted one in between). Normally the amplitude unbalance is recorded as the apparatus is usually more stable against changes in amplitude than changes in phase.

2.9 MEASUREMENT OF FREQUENCY

There are essentially two different methods by which the frequency of a given microwave signal can be determined—either by measuring its wavelength with a micrometer, or by comparing its frequency directly with the harmonic of a known standard.

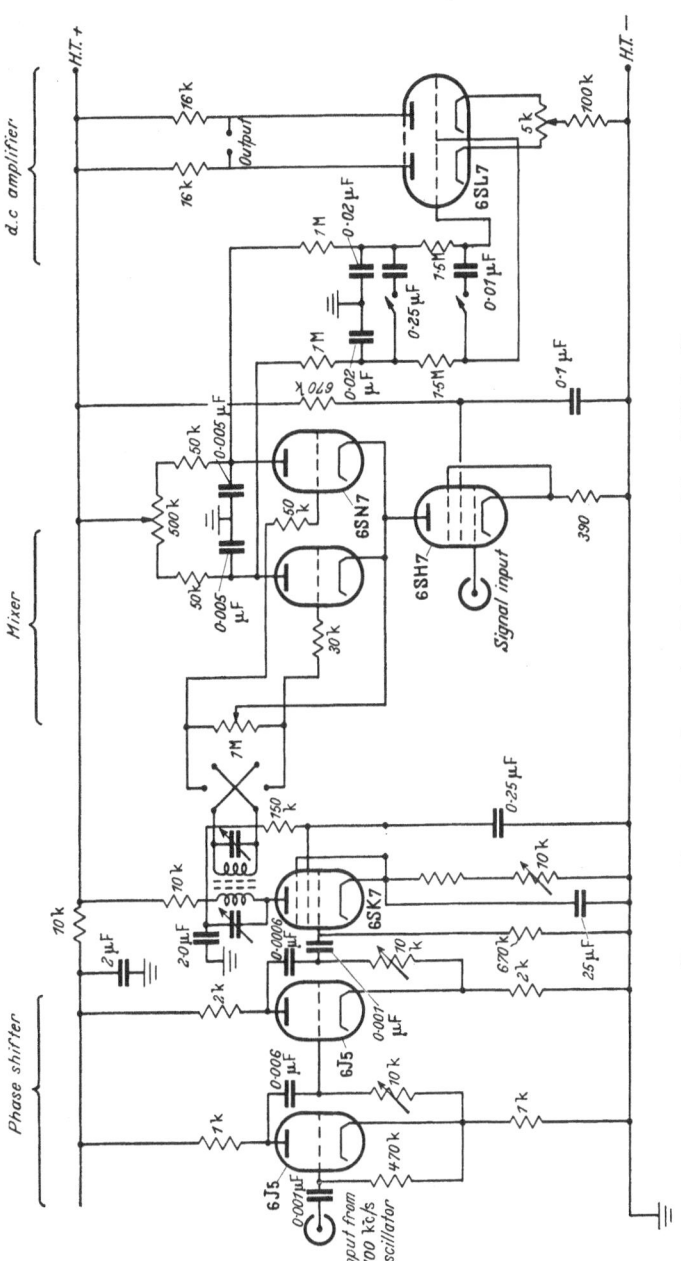

Figure 19. Phase-sensitive detector (100 kc/s, based on Schuster's [15] circuit)

37

In order to measure the actual wavelength a cavity resonator wavemeter is used, the end of which can be moved to and fro to obtain two or more positions of resonance. The free-space wavelength of the radiation is then given by

$$\frac{1}{\lambda_{f.s}^2} = \frac{1}{\lambda_g^2} + \frac{1}{\lambda_c^2}$$

where λ_g is the wavelength in the guide, or resonator, i.e. the length actually measured, and λ_c depends entirely on the geometry of the cavity. Cavity wavemeters are discussed in detail in the next chapter, here it may be noted that their Q values are of the order of 20,000, setting a limit on the accuracy of determining λ to about 1 in 10^4 or,

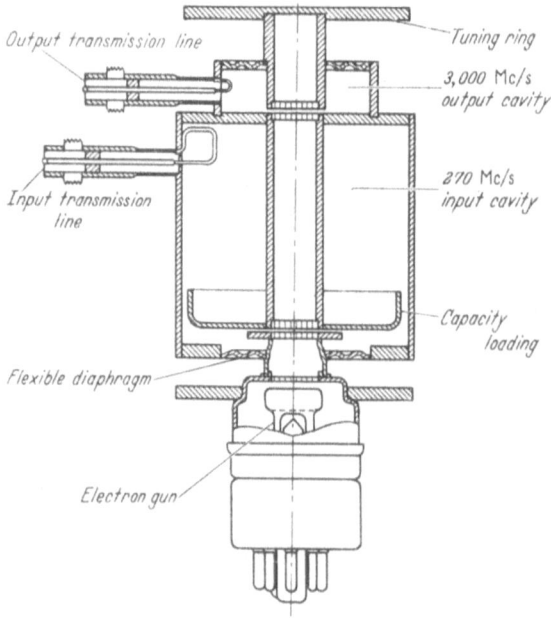

Output transmission line

Tuning ring

3,000 Mc/s output cavity

270 Mc/s input cavity

Input transmission line

Capacity loading

Flexible diaphragm

Electron gun

Figure 20. Schematic diagram of typical klystron multiplier

in the best cases, 1 in 10^5. Very great care must also be taken to keep the dimensions of the cavity within fine limits, or variations in λ_c will add to the error. The free-space wavelength measured by these wavemeters can be corrected for the dielectric constant of air, and when divided into the velocity of light will then give the microwave frequency. It can be seen that frequencies in the 1·25 cm wavelength band (24,000 Mc/s) can be measured to an accuracy of about ±1 Mc/s by this means.

For a lot of work in microwave spectroscopy, especially when taking measurements on hyperfine structure, it is often necessary to determine the frequency within closer limits than these. To do this, use is made of the standard broadcast frequencies which are kept accurate to 1 part in 10^8 or 10^9 (e.g. Rugby at 200 kc/s M.S.F. at 5 Mc/s and 10 Mc/s). A quartz crystal in a thermostatically controlled oven is kept at the same frequency as one of these standard signals, by tuning out the beat notes between the two, and this frequency is then multiplied up by various stages to a value of about 500 Mc/s. This can then be fed directly to a silicon crystal placed across the waveguide, to produce harmonics which beat with the microwave power; or an intermediate stage in the form of a klystron multiplier may be incorporated, to multiply the frequency further by the process of velocity modulation, as already described. For microwave frequencies up to 20,000 Mc/s it is usually sufficient to feed the 500 Mc/s signal straight to the crystal in the guide, provided the output power of the last stage is reasonably high. For frequencies above 20,000 Mc/s some form of klystron multiplier is usually required, however, as the power falls off very rapidly in the higher crystal harmonics. As an illustration of this type of tube a Sperry klystron multiplier is shown in *Figure 20*, and the detailed circuit diagram of another possible multiplier chain is given in *Figure 21*

One of the harmonics of the final input frequency to the crystal multiplier then beats with the microwave radiation, to produce a beat frequency which can be picked up by a normal radio receiver. Further standard reference signals can be obtained by feeding some 25 Mc/s power as modulation on the input frequency to the crystal, and then a radio receiver with range up to 12·5 Mc/s will always be able to detect a beat note with one of the crystal harmonics. The particular harmonic can be ascertained by measuring the microwave frequency by a cavity wavemeter, to give a value within ±5 Mc/s.

The output to the radio receiver can either be taken from a second separate crystal mixer, or can, in some cases, be taken from the same crystal as is producing the harmonics, the one crystal thus acting as both multiplier and mixer. In this case neither side of the crystal is earthed directly to the waveguide, but one is fed from the standard frequency multiplier chain, while the other leads to the input of the radio receiver. If the receiver has been accurately calibrated and is checked against a standard broadcast frequency, it can be used to interpolate the frequency reading between the 25 Mc/s points directly. Alternatively the modulation on the reference signal can be obtained from a calibrated oscillator, and the frequency of this can be varied until zero beat is obtained between the microwave power and

39

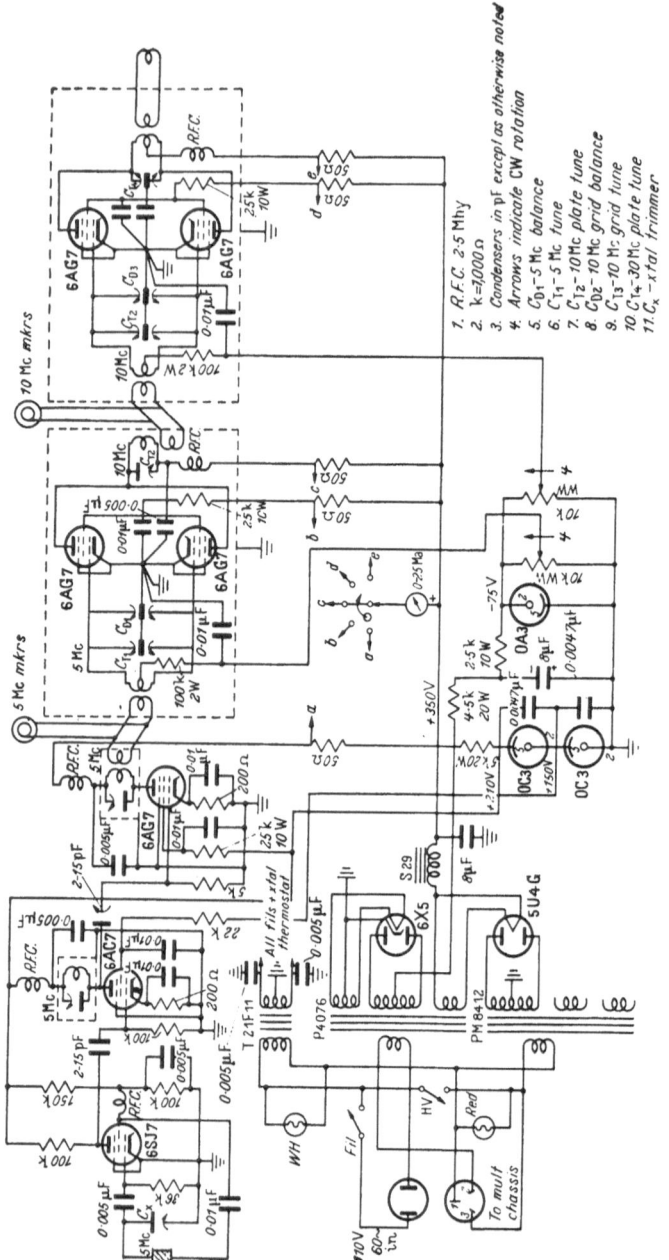

Notes:

1. R.F.C. 2·5 Mhy
2. k=1,000 Ω
3. Condensers in pF except as otherwise noted
4. Arrows indicate CW rotation
5. C_{D1}-5 Mc balance
6. C_{T1}-5 Mc tune
7. C_{T2}-10 Mc plate tune
8. C_{D2}-10 Mc grid balance
9. C_{T3}-10 Mc grid tune
10. C_{T4}-30 Mc plate tune
11. C_x-xtal trimmer

Figure 21 (Part 1)

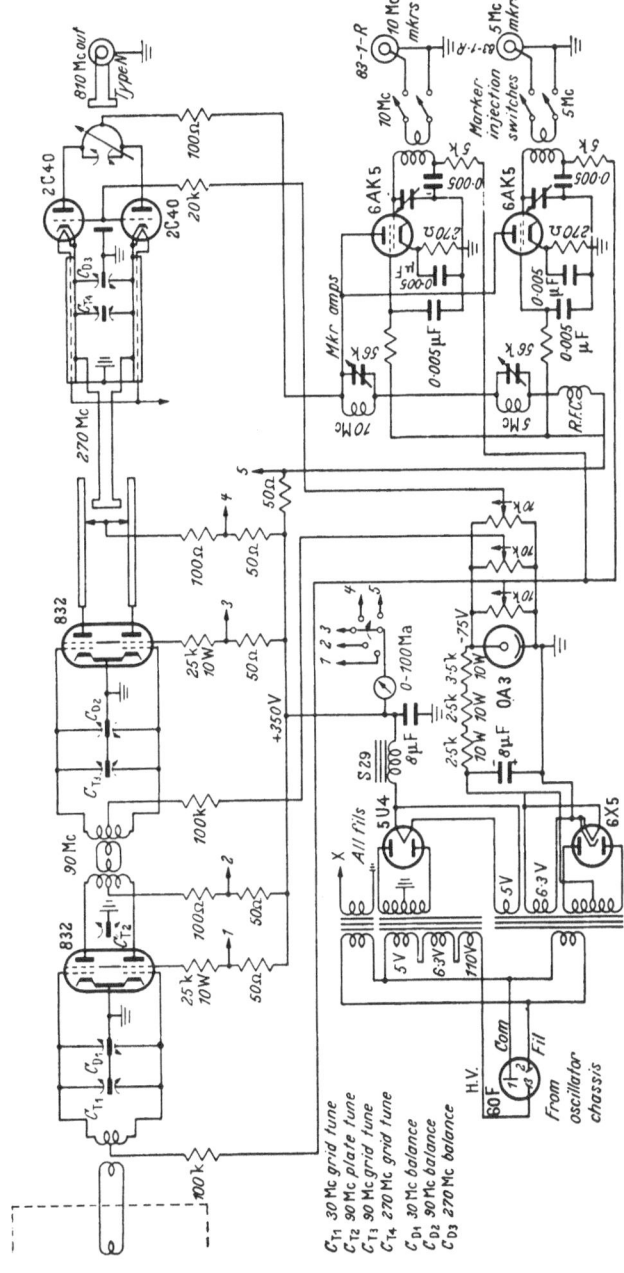

Figure 21 (Part 2)

Circuit diagram of frequency standard

C_{T1} 30 Mc grid tune
C_{T2} 90 Mc plate tune
C_{T3} 90 Mc grid tune
C_{T4} 270 Mc grid tune

C_{D1} 30 Mc balance
C_{D2} 90 Mc balance
C_{D3} 270 Mc balance

the crystal harmonic. The frequency measurement can be greatly facilitated if the reflector of the klystron is swept. The radio receiver can then be set at a given frequency f_1, and the output from its detector stage applied across the Y plates of an oscilloscope, the time base being fed with the same frequency as the sweep on the klystron reflector. Two peaks will then be obtained on the oscilloscope screen, corresponding to the points when the klystron frequency passes

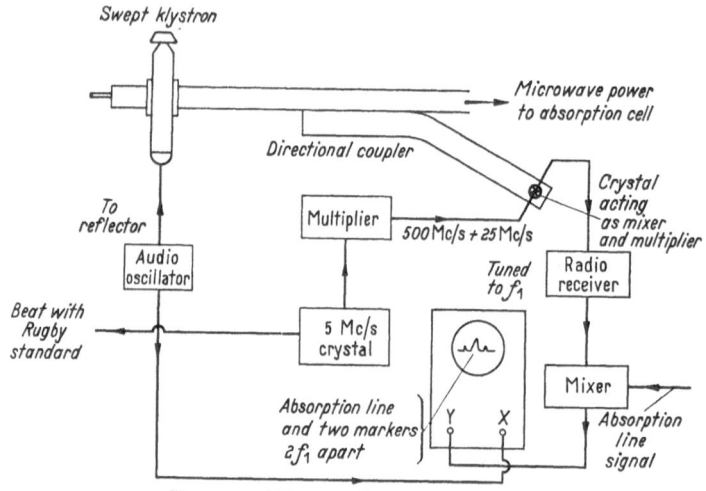

Figure 22. Microwave frequency determination

through the values $f_0 \pm f_1$, f_0 being the frequency of the crystal harmonic. The klystron frequency can be accurately adjusted to be equal to the harmonic frequency by just aligning these two pips to be equidistant from the centre of the screen, the magnitude of the klystron reflector sweep and the frequency of reception of the radio receiver being gradually reduced to minimum values for the highest accuracy.

A block diagram of this form of frequency measurement is shown in *Figure 22*, and from experimental results it would appear that frequencies can be accurately determined to within ± 3 kc/s by these methods (i.e. 1 part in 10^7). Further details of more recent microwave techniques and frequency locking circuits are given in Chapter 11.

REFERENCES

[1] BECK, A. H. W. *Velocity-Modulated Thermionic Tubes*, C.U.P. (1948) p. 46

[2] POUND, R. V. *Rev. sci. Instrum.* 17 (1946) 490

[3] CLEETON, C. E. and WILLIAMS, N. H. *Phys. Rev.* 45 (1934) 234

REFERENCES

[4] KLEIN, J. A., LOUBSER, J. H. N., NETHERCOT, A. H. and TOWNES, C. H. *Rev. sci. Instrum.* 23 (1952) 78

[5] GORDY, W. *Ann. N. Y. Acad. Sci.* 55 (1952) 775; KING, W. C. and GORDY, W. *Phys. Rev.* 90 (1953) 319

[6] BURRUS, C. A. and GORDY, W. *ibid.* 93 (1954) 897

[7] BERINGER, R. and CASTLE, J. G. *ibid.* 78 (1950) 581

[8] ROBINSON, N. and INGRAM, D. J. E. Unpublished

[9] STRUM, P. D. *Proc. Inst. Radio Engrs, N.Y.* 41 (1953) 875

[10] GORDY, W. *Rev. mod. Phys.* 20 (1948) 668

[11] GESCHWIND, S. *Ann. N. Y. Acad. Sci.* 55 (1952) 751

[12] BAGGULEY, D. M. S. and GRIFFITHS, J. H. E. *Proc. phys. Soc.* A. 65 (1952) 594

[13] LAWSON, J. L. and UHLENBECK, G. H. E. *Threshold Signals* McGRAW-HILL (1948) p. 253

[14] PAKE, G. E. and PURCELL, E. M. *Phys. Rev.* 74 (1948) 1184

[15] SCHUSTER, N. A. *Rev. sci. Instrum.* 22 (1951) 254

3

WAVEGUIDE TECHNIQUES

3.1 Propagation of Radiation in Waveguides

EXCEPT for the initial experiment of CLEETON and WILLIAMS[1], all microwave spectroscopes have used a waveguide system to transmit the radiation from the source to the absorption cell and on to the detector, in contrast to optical, ultra-violet, and infra-red work, where free space propagation is employed. A brief summary of the properties of waveguides is therefore necessary in order to understand the design of actual spectroscopes.

The detailed theory follows from an application of Maxwell's four electromagnetic equations, and the substitution of the appropriate boundary conditions. These lead to an equation of the form

$$\frac{\partial^2 \phi}{\partial x^2} + \frac{\partial^2 \phi}{\partial y^2} + k_0^2 \cdot \phi = 0 \qquad \ldots (3.1)$$

where

$$k_0^2 = \gamma^2 + \frac{\omega^2}{v^2} \qquad \ldots (3.2)$$

In this equation ϕ represents either E_z or H_z, γ is the propagation constant, ω is the angular frequency, and v the velocity of the wave in the unbounded medium. There are thus two fundamentally different forms of propagation down a waveguide, (i) where $H_z = 0$ and only the electric field has a component in the direction of propagation. This is known as an E wave (British terminology), or a TM wave (American terminology), and (ii) where $E_z = 0$, and only the magnetic field has a component in the direction of propagation, this being known as an H or TE wave.

Equation (3.1) can be solved by separating the variables; and substitution of the appropriate boundary conditions for a rectangular guide, of cross-section '$a \times b$', then gives

$$E_z = E_0 \cdot \sin \frac{m\pi}{a} x \cdot \sin \frac{n\pi}{b} y \cdot \exp[j\omega t - \gamma z] \quad \text{for } E \text{ waves} \quad \ldots (3.3)$$

and

$$H_z = H_0 \cdot \cos \frac{m\pi}{a} x \cdot \cos \frac{n\pi}{b} y \cdot \exp[j\omega t - \gamma z] \quad \text{for } H \text{ waves} \quad \ldots (3.4)$$

where m and n are integers.

The x and y components of E and H, for the two different cases, can be derived directly from these, by using the differential relations obtained from Maxwell's equations. Unattenuated propagation will only occur if γ is purely imaginary, and this condition is only satisfied if

$$\frac{\omega^2}{v^2} > \left(\frac{m\pi}{a}\right)^2 + \left(\frac{n\pi}{b}\right)^2 \qquad \dots (3.5)$$

Propagation will therefore only take place above a certain critical frequency, or, as more often expressed, below a critical cut-off wavelength given by

$$\lambda_c = 2 \left/ \left[\left(\frac{m}{a}\right)^2 + \left(\frac{n}{b}\right)^2 \right]^{\frac{1}{2}} \right. \qquad \dots (3.6)$$

The wavelength of the radiation in the guide is then given by

$$\frac{1}{\lambda_g^2} = \frac{1}{\lambda_{f.s}^2} - \frac{1}{\lambda_c^2} \qquad \dots (3.7)$$

where $\lambda_{f.s}$ is the wavelength in the free unbounded medium (normally air or free space).

The different values of m and n produce different modes of propagation, that corresponding to the longest cut-off wavelength for any given cross-section being the H_{10} mode ($m = 1$, $n = 0$). For this mode $\lambda_c = 2a$, showing that the longest wavelength that the guide will pass is equal to twice the length of its largest side. The cut-off wavelengths for different modes of a rectangular waveguide of cross-section $a \times b$ are listed in *Table 3.1*, and it can be seen that if the waveguide size is chosen so that $0.706 . \lambda . > a > 0.50 . \lambda$ then only the H_{10} mode can be propagated, as it is then impossible for other modes to be set up, and the points of maximum coupling to the E and H fields are therefore accurately known. Other factors tend to favour a slightly larger a in practice, however.

Table 3.1

Mode	Cut-off Wavelength	
H_{10}	$2.0a$	does not depend
H_{20}	$1.0a$	on b
H_{11} and E_{11}	$1.414a$	if $a = b$,
H_{12} and E_{12}	$0.894a$	smaller if $b < a$

A physical picture of the electric and magnetic field variations down a waveguide carrying an H_{10} mode is shown in *Figure 23*. The phase relation between the electric and magnetic fields depends on whether the guide is matched. For no reflected wave, the maximum values of E_y and H_x occur together at the same point in the guide,

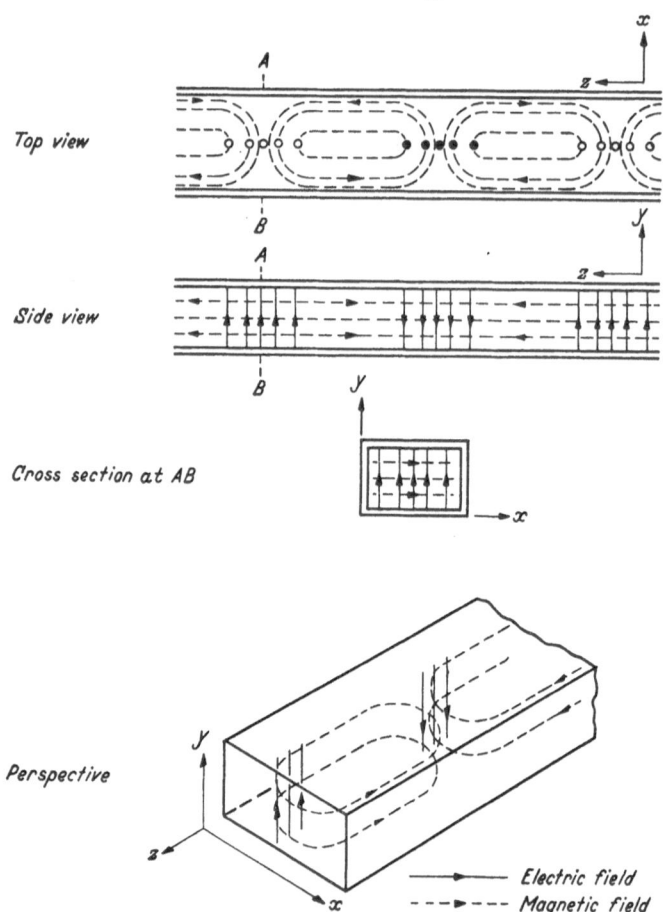

Top view

Side view

Cross section at AB

Perspective

—————▶ Electric field
- - - ▶ - - - Magnetic field

Figure 23. Travelling wave in H_{10} mode

the whole pattern moving down the guide as a travelling wave. If
there is a complete standing wave, however, with no power flow, the
magnetic lines of force loop round the position of maximum electric
field, the two being in both space and time quadrature. A similar set
of diagrams is given in *Figure 24* for the E_{11} mode to illustrate the
difference between E and H waves, though in nearly all practical
applications only the H_{10} mode is used.

3.2 Choice of Waveguide Size

Waveguides are always made as small as possible to reduce weight
and cost, and to ensure that only one mode is propagated, but there

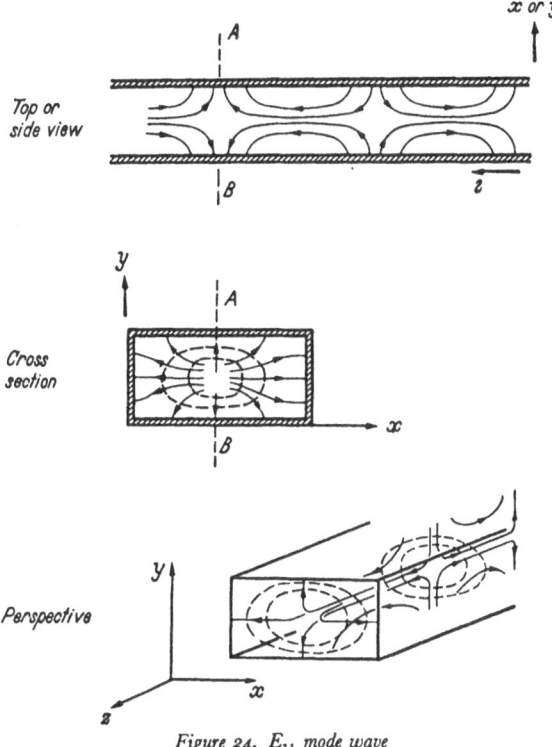

Figure 24. E_{11} mode wave

are two factors which enter to shift the optimum size some way from the cut-off value, these being the attenuation in the walls of the guide, and the group velocity variation.

The attenuation is produced by the flow of radiofrequency currents in the finite resistance of the walls, and, for the H_{10} mode, can be expressed as:

$$\alpha = \frac{R_s}{120\pi b} \cdot \frac{[1 + 2b/a \cdot (f_c/f)^2]}{[1 - (f_c/f)^2]^{\frac{1}{2}}} \text{ nepers/cm} \qquad \ldots \ldots (3.8)$$

The term in the denominator causes the attenuation to rise to infinity as the critical frequency is approached, and *Figure 25* shows the variation of α with frequency for a K band waveguide with a cut-off wavelength of 2·14 cm (14,000 Mc/s). It can be seen that the region for minimum α is some way from the cut-off value.

A consideration of the variation of group velocity with frequency also shows that the guide size must be appreciably greater than the cut-off value, or considerable signal distortion will occur. It can be

47

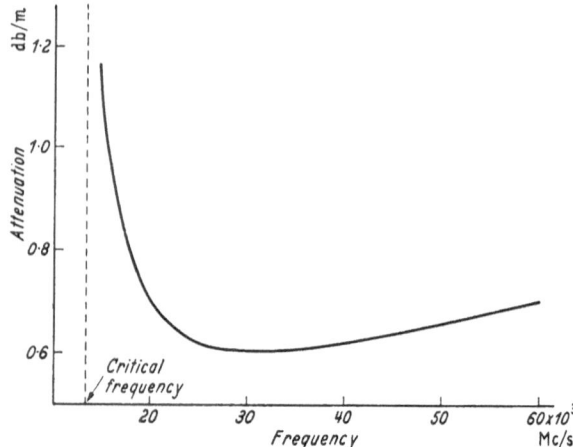

Figure 25. Variation of attenuation with frequency (K-band guide)

shown that if the free-space velocity (V) is put equal to c, then the group velocity (V_g) and phase velocity (V_p) are given by:

$$V_g = \frac{c^2}{V_p} \quad \text{and} \quad V_p = \frac{c}{[1 - (\lambda/\lambda_c)^2]^{\frac{1}{2}}} \qquad \ldots (3.9)$$

The variation of both phase and group velocity with wavelength is plotted in *Figure 26*, and it is seen that there is a very steep fall in V_g

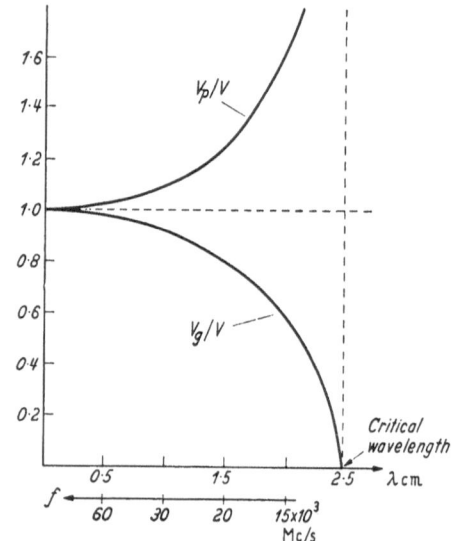

Figure 26. Variation of phase and group velocities, in guide with $\lambda_c = 2\cdot 5$ cm

48

as the critical wavelength is approached. Hence any pulses or sharp absorption lines, which cover a relatively large bandwidth, will be badly distorted if passed as modulation down a guide in this region.

A list of the different sizes of rectangular waveguides commonly used, with the wavelength bands that they cover, and the letters usually employed to denote them, is given in *Table 3.2*.

Table 3.2

Size of Guide Internal Dimensions	Wavelengths Covered in Centimetres	Name of Band*
7·62 × 2·54 cm	8·9 –10·5	S
2·54 × 1·27	3·0 – 3·5	X
1·064 × 0·432	1·2 – 1·5	K
0·702 × 0·315	0·75– 1·2	J (Q)
0·457 × 0·218	0·5 – 0·8	I } H } (V)
0·295 × 0·132	0·3 – 0·5	
0·190 × 0·086	0·2 – 0·3	G
0·124 × 0·056	0·1 – 0·2	F

* The nomenclature of the last five bands is that adopted by Gordy at Duke University, the letters in brackets being standard British nomenclature.

3.3 CIRCULAR GUIDES

The theory of microwave propagation in circular guides is very similar to that of the rectangular case, the general solution being in

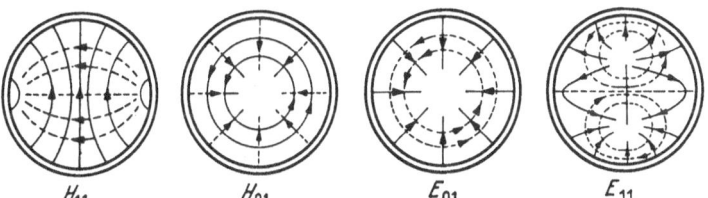

H_{11} $\qquad$ H_{01} $\qquad$ E_{01} $\qquad$ E_{11}

Figure 27. Different modes in circular guides

terms of Bessel functions instead of exponentials, and sines and cosines. Cross-sectional diagrams of the field patterns for some of the lower modes are shown in *Figure 27*, and the subscript zero now indicates circular symmetry and is not necessarily attached to the lowest mode. Values of the cut-off wavelengths, in terms of the guide radius a are given in *Table 3.3* for the four lowest modes, and it is evident that H_{11} is now the dominant mode, its similarity to the rectangular H_{10} mode being seen by a comparison of *Figures 23* and *27*.

Table 3.3

Mode	H_{01}	H_{11}	E_{01}	E_{11}
Cut-off Wavelength	1·64a	3·42a	2·61a	1·64a

D

49

The great disadvantage of circular waveguides is that they are not stable against change in the direction of polarization, and hence the direction of maximum field strength is never certain. On the other hand, their internal dimensions can be made to a much greater accuracy, and hence the use of circular guides is usually confined to such specialized items as wavemeters and resonators.

3.4 COUPLING INTO WAVEGUIDES. DIRECTIONAL COUPLERS

Power can be fed into waveguides by coupling to either the electric or magnetic field, and *Figure 28* illustrates these two methods of

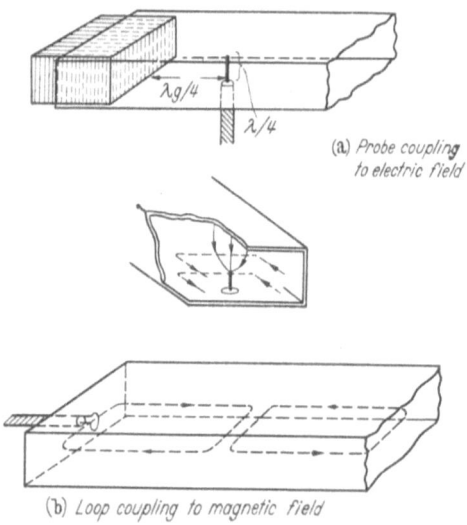

(a) *Probe coupling to electric field*

(b) *Loop coupling to magnetic field*

Figure 28. *Launching of H_{10} mode*

launching an H_{10} mode. In the first case a quarter-wavelength of the inner conductor of the coaxial line projects into the position for maximum electric field, which is set up by voltage variation on the probe. In the second case radiofrequency currents flow in the closed loop, inducing magnetic lines of force which thread through it. It is also possible to couple directly between different waveguides by means of holes or irises. The type and magnitude of the coupling will then depend on both the position and size of the hole, and such coupling systems can be made with directional properties.

A brief consideration of the construction and properties of 'directional couplers' is essential when designing a microwave spectroscope as they are used in a variety of ways for monitoring power, coupling to wavemeters and the like. The name is given to any device which abstracts a certain proportion of the power from the main waveguide run, and feeds it into another waveguide with a preferred direction of propagation. There are two forms of directional couplers commonly used in microwave work, the one employs coupling via two iris holes in the narrow side of the guide, while the other uses just a single hole in the wide side. Since only the longitudinal H component is present along the narrow side of the guide, the coupling in the first case will be entirely magnetic, and each hole will radiate equally in both directions into the second guide. If, however, the two holes are placed $\lambda_g/4$ apart, interference will take place between them, both

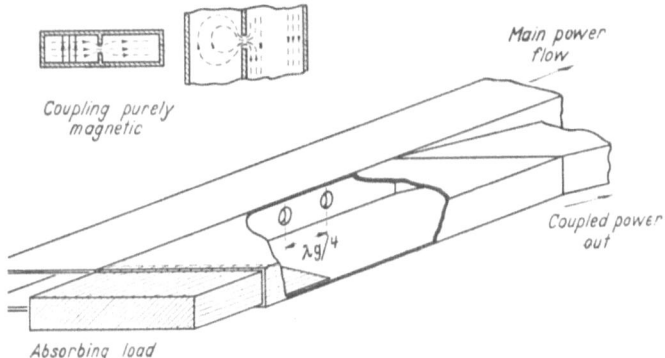

Figure 29. Two-hole coupler

holes adding to a wave travelling in the same direction as in the first guide, but producing a combined phase-change of π for a backward-travelling wave. This type of coupler is illustrated in *Figure 29*, the tightness of the coupling depending on both the size of the holes and the thickness of the guide wall. As an example, for standard K band (1·25 cm) guide with 0·04 in. wall thickness there will be coupled power ratio of 75 dB for a hole diameter of 0·05 in., and a coupling ratio of 30 dB for a hole diameter of 0·14 in.[2] The coupler is frequency-sensitive, as it depends on an interference effect, but can be made broadband by replacing the holes by slots. These are usually $\lambda_g/4$ long, and are staggered in the guide wall, so that they do not overlap; it is also easier to couple more power through by the use of slots, a power ratio of 8 dB can be obtained with the same guide size as above, with slots 0·17 in. long and 0·06 in. wide.

Another very useful and simpler form of directional coupler, which is often used is the Bethe single-hole coupler[3]; illustrated in *Figure 30*. This employs a single hole in the centre of the wide side of the guide,

51

and coupling is by both the electric and magnetic fields. The electric vector induces a field of the same direction in the second guide, but the magnetic field, acting as a dipole, through the coupling hole, induces a field in the opposite direction in the second guide. The wave produced in the second arm will thus travel in the opposite direction to that in the initial arm, but for this directivity to be

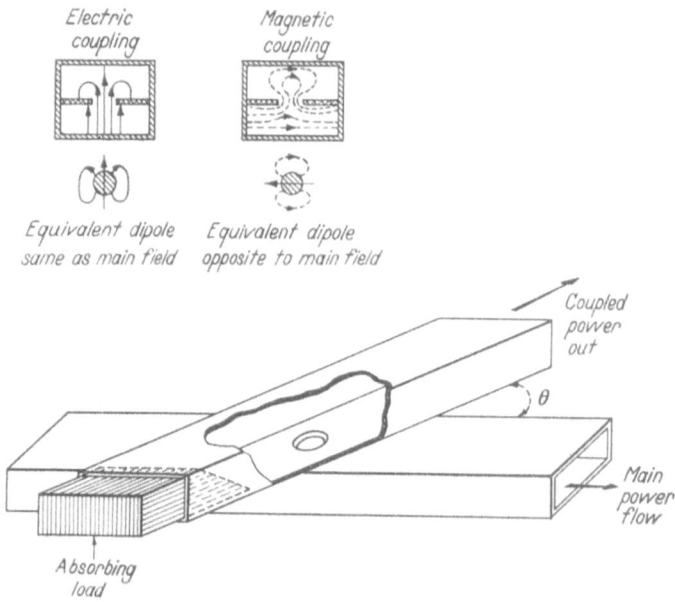

Figure 30.   Bethe single-hole coupler

perfect, the electric and magnetic coupling must be of equal amplitude. This condition is obtained by rotating the two guides relative to each other, which does not alter the electric coupling, but reduces the magnetic coupling by the cosine of the angle turned through, as 't is the transverse component of the magnetic field which is effective. The wall thickness also enters into the calculations as it attenuates the two forms of coupling by different amounts, but by suitably adjusting the angle and hole size it is possible to obtain very good directivity for a required coupling factor. BETHE [4] has calculated the necessary conditions in detail, and the angle required for maximum directivity is given by

$$\cos \theta = \frac{1}{2}\left(\frac{\lambda_g}{\lambda}\right)^2 \frac{F_E}{F_H} \qquad \dots (3.10)$$

52

and the coupling factor by the expression

$$C = 20 \log_{10} \left\{ \frac{2\pi d^3}{3ab \cdot \lambda_g} \cdot \cos \theta \cdot F_H \right\} \qquad \dots (3.11)$$

where d is the diameter of the hole and F_E and F_H are the two attenuation factors of the hole for E and H coupling. As a numerical example, it may be noted that for K-band waveguide of 0·04 in. wall thickness, a coupling ratio of 30 dB is obtained for a hole of 0·15 in. diameter with the angle between the guides equal to 58°. This form of directional coupler is not so frequency-sensitive as the first, and is used a great deal for power monitoring, and coupling to wavemeters.

3.5 MATCHING TECHNIQUES

In any microwave spectroscope it is important to reduce the reflected power and standing waves to as small a value as possible, and this

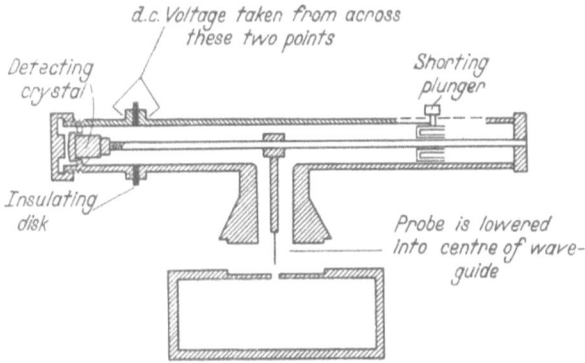

Figure 31. Cross-section of simple standing wave indicator

requires some form of variable matching element that can be used to balance out the effect of internal reflections. The balancing can be accomplished by the use of a standing-wavemeter, the matching element being adjusted until the same power level is detected by the meter at any position of its probe along the narrow slot in the broad side of the guide. A cross-section of a simple standing-wavemeter is shown in Figure 31, the whole being moved along the guide in a direction normal to the paper. In many cases, however, a standing-wavemeter is not used, and the variable matching units are adjusted to give optimum output power at the far end of the run.

Different forms of matching irises are illustrated in Figure 32, together with their equivalent circuits, and such are often used in coupling into wavemeters and resonators. Some form of stub, or screw matching is usually employed, however, if a continuously vari-able adjustment is required. A short screw, inserted in the centre of the broad waveguide face will act as a series combination of capacity and inductance, and by adjusting its position and depth of penetra-tion, it can be made to cancel the standing waves. This single screw is usually replaced by three screws, fixed in position along the centre line of the guide face, spaced about $\frac{3}{4}\lambda_g$ apart. By varying their depths of penetration independently, they can be made equivalent

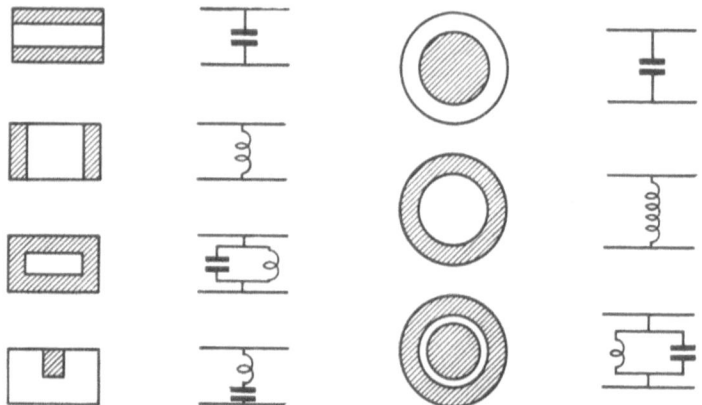

Figure 32. Matching irises with equivalent circuits

to a single screw at any intermediate position and penetration. This 'three screw' matching element is an easily adjustable form of con-tinuously variable matching, and should be incorporated whenever there is any sudden transition in the waveguide, such as at the coupling into a cavity resonator.

It is often necessary to vary the amplitude and phase of the radia-tion independently, especially when balancing microwave bridges. Attenuators, which do not introduce reflexions can be made rela-tively easily from sheets of resistive material, which are gradually inserted into the region of maximum electric field. These may take the form of a thin sheet of graphite-covered plastic, which is lowered as a flap through a slot in the middle of the wide side of the guide, the flap being cut to a shape so that no sudden transition is involved. More recent types of attenuator use a metallized glass strip, which is moved over from the narrow side of the guide to the centre, thus gradually coming into the region of maximum electric field. These films are usually made of nichrome, and are evaporated on to a glass plate of about one sixteenth inch thickness, until it has a resistance of 80 ohms/cm^2, corresponding to a layer about 50 molecules thick. A thin film of magnesium fluoride is then deposited on the surface to give chemical and mechanical protection[5].

Phase-changing devices employ one of two principles, either a dielectric is inserted into the guide so that the wave is slowed up in passing through it, or the size of the guide itself is slightly altered, this changing λ_g and hence the phase velocity. The type employing a dielectric slab is very similar in design to the attenuators, the slab being inserted either through a slot in the centre of the broad side of the guide, or moved across from the narrow side into the centre. In the latter case its ends are tapered to reduce reflexions that occur at the change of impedance which it produces in the guide. To employ the second principle in a phase-shifter, a long slot is cut down the middle of the wide side of the guide, in both faces, and a clamp is used to force the narrow walls together. This works very well at wavelengths of 1·25 cm and below, giving an easily adjustable phase shift, but the guide becomes too rigid for it to work very well above this wavelength.

3.6 CRYSTAL HOLDERS

These have already been discussed briefly in the previous chapter, and here the different devices used to match the crystal into the waveguide so that no reflexions and loss of power are produced, are briefly noted. Above 1 cm wavelength the matching is usually done via a waveguide-coaxial line transition, some form of pick-up stub being used to couple to the E vector of the wave and feed a radiofrequency voltage across the detecting crystal, which is held in a side arm. This type of holder has been illustrated in *Figure 14*, and the main design problems are concerned with the actual stub placed across the guide. These were investigated in great detail during the war, and crystal holders can now be obtained as a standard item for this wavelength band, being designed to hold 1N26 or similar crystals. As already mentioned, the power is matched into the coupling stub by the use of a shorting plunger, approximately $\lambda_g/4$ away.

Below 1 cm wavelength it is normal to use a radically different form of crystal holder in which the crystal is of open construction mounted in the guide itself, the metal-whisker being across the guide in the position of maximum electric field to couple the power straight across the silicon crystal. For the 8-mm wavelength region this type of holder can be bought commercially in Britain from Hilger and Watts Limited, and will take either the BTH or GEC standard 8-mm crystals. For the lowest wavelengths GORDY and KING[6] found it

55

necessary to break the crystals into smaller size, repolish and re-mount these *in situ* in the guide, and use a differential screw mech-anism to advance a 2-thou tungsten-whisker across the guide to make contact with the silicon surface. Such crystal holders, in which none of the discontinuities or elements are greater than a wavelength in size, have worked satisfactorily down to 0·8-mm wavelength[7]. When used as harmonic generators similar types of crystal holders are employed, the crystal itself being across the smaller waveguide

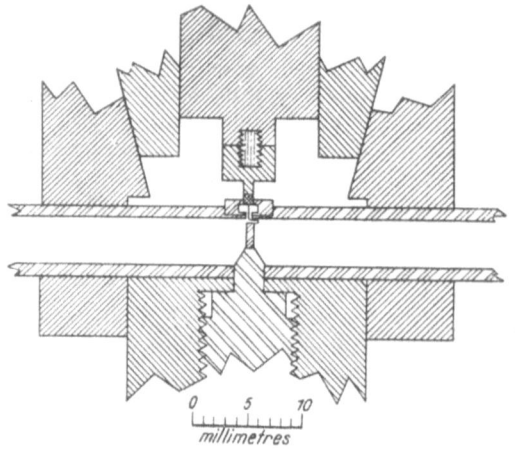

Figure 33. Crystal multiplier from 1·25 cm to 1·0 mm bands

into which it radiates directly, and is fed by a stub which passes across the centre of the larger guide, coupling to its E vector. A multiplier of this kind, designed to multiply from 1·25 cm to 1·0 mm wavelengths is illustrated in *Figure 33*, being of the type developed by Gordy and King for the first successful spectroscopic measurements in the 1 mm wavelength region.

3.7 MICROWAVE BRIDGE BALANCING

In most spectroscopes using superheterodyne detection, a microwave bridge circuit is also incorporated so that a signal is only fed into the detecting arm when the bridge is unbalanced[8]. The essential ele-ment of a microwave bridge is the junction between four waveguide arms, one of which introduces the input power, another acts as the detecting arm, while the other two are the comparison arms of the bridge, no signal being detected when these are balanced against each other. There are two microwave elements which are ideally suited for this, the 'magic-T' and the 'hybrid-ring'. Both of these have the property that when the second and third arms are matched and terminated by equal admittances, then the input admittance at the first arm is independent of the output admittance at the fourth, so that power fed into the first arm will be transmitted equally to the second and third arms, and none to the fourth. The magic-T obtains these properties from the different polarizations existing in its arms, whereas the hybrid ring depends on actual interference effects, the path-length between successive guides being exactly $\lambda_g/4$. They are

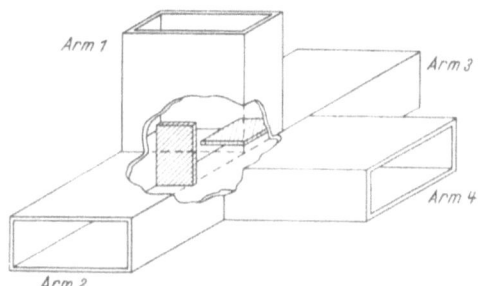

Figure 34(a). Magic-T with compensating irises

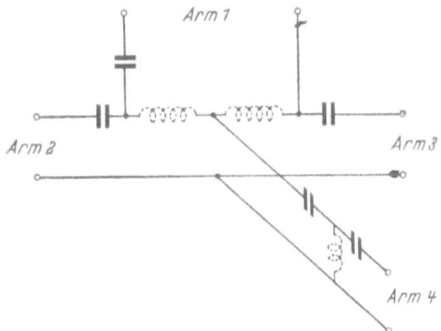

Figure 34(b). Equivalent circuit of magic-T

illustrated in *Figures 34* (*a*) and (*b*) and *35*, the arms are numbered so that the above analysis applies to both, the comparison arms of the hybrid-ring being on either side of the 'fourth' detecting arm.

A magic-T has to be both correctly matched and terminated if it is to have the above properties. The sudden discontinuities at the actual T-junction (which correspond to lumped capacity or inductance in the equivalent circuit representation) will introduce mismatches unless they are compensated. This compensation is effected by irises, or posts, placed a short distance away from the junction, and are adjusted in size so that their susceptance cancels that introduced by the junction discontinuities. The analogous compensation of the equivalent circuit by the dotted elements is also shown in *Figure 34*. The magic-T is usually fabricated complete with matching irises or stubs, and the only balancing required, when used in a bridge circuit, is the correct adjustment of the terminating admittances of the various arms. Theoretically, a perfectly absorbing load would be placed in the guide attached to arm two, this producing no reflexion, and hence would be a perfect termination and match for that arm. Then,

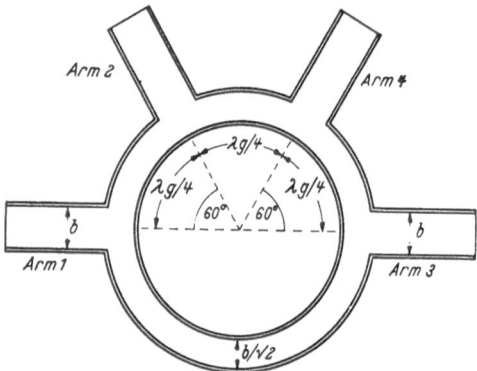

Figure 35. Hybrid ring

in order to balance the bridge, the admittance of arm three would have to be adjusted until it was equal to the admittance of arm two (the characteristic admittance of the waveguide); and as arm three usually contains the waveguide cell or cavity resonator, it is necessary to have both a phase-shifter and attenuator incorporated in it so that the real and imaginary parts of the admittance can be adjusted.

In practice, the T is never perfect and a certain amount of asymmetry or cross-coupling exists, and this has to be balanced out by a slight mismatch introduced purposely into arm two; which may take the form of tuning screws or stubs placed in this arm, or a 'sliding match' placed at the end. The absorbing load which is usually formed of a tapered piece of graphite or polyiron, is never quite reflexionless, and may be used to balance out the effect of the asymmetry of the T and when so used is termed a 'sliding match'. If it is moved up and down along the length of the guide over a distance of $\lambda_g/2$, its impedance will describe a circle centred on the characteristic impedance of the guide, and hence the phase of the reflected wave at the junction can be adjusted to cancel that produced by the

58

cross-coupling. When the T is correctly matched and terminated, no power should be obtained in the fourth arm, and the bridge is then said to be balanced.

The hybrid ring relies on the interference between waves travelling along different paths to produce its balancing properties. The power is fed into arm one (*Figure 35*), the detector is placed in arm numbered four, arms numbered two and three again being used for reference matching. If all the arms are matched it can be seen that no power will be fed from arm one to arm four, as the two path lengths in opposite directions round the ring will have a difference of half a wavelength, and the phases at arm four will cancel. It can be shown that the correct condition for the internal matching of such a hybrid ring is that the characteristic impedance of the ring itself should be $\sqrt{2}$ smaller than that of the waveguide arms. In order to balance a hybrid-ring, the arms are terminated as for the case of the magic-T, any unbalance produced in arm three producing power to be detected in arm four. The hybrid-ring is rather frequency sensitive, as it should only work for the one wavelength where its circumference equals $3\lambda_g/2$, but it can be suitably matched to make it relatively broadband. The use of isolators and circulators is discussed in Chapter 11.

3.8 CAVITY RESONATORS. WAVEMETERS

Cavity resonators are used to a very great extent in microwave spectroscopy, both as actual absorption cells, and also as wavemeters to determine the wavelength to 1 part in 10^4, or better. The fact that their Q values can be made of the order of 20,000 means that they can store a large amount of microwave energy, as well as acting as highly selective tuned circuits. This ability to store energy means that a substance placed inside the cavity will be in radiofrequency magnetic and electric field strengths very many times larger than those existing in the waveguide feeding the cavity. For this reason cavity resonators are used when investigating the absorption spectra of crystalline solids, where the volume of the specimen is limited, the crystal being placed in the region of maximum radiofrequency field. The second advantage of a cavity resonator is that, owing to its small size and compact nature, it can be placed in a homogeneous magnetic field much more easily than a long length of waveguide. Its great disadvantage, from the point of view of microwave spectroscopy, is that the high power density can cause saturation of the molecular absorption, this being especially so in the case of gases, where the electric dipole of a molecule couples very strongly with the electric field of the radiation. Thus, generally, waveguide cells are used for gaseous work, and cavity resonators for solid state studies.

The dimensions of a cavity required to produce resonance for any given wavelength and mode can be very easily derived from previous waveguide theory. A cavity can be regarded as a length of wave-guide, an integral number of half-wavelengths long, with a complete standing wave existing inside it. Each half-wavelength will then contain one full standing-wave pattern, the electric and magnetic fields oscillating in quadrature both in space and time. If the length of the cavity is d, the resonance condition will be

$$d = p \cdot \frac{\lambda_g}{2} \text{ where } p \text{ is an integer.}$$

$$\lambda_g \text{ is given by } \frac{1}{\lambda^2} = \frac{1}{\lambda_g^2} + \frac{1}{\lambda_c^2}$$

Hence
$$\lambda^2{}_{f.s} = \left[\left(\frac{p}{2d} \right)^2 + \frac{1}{\lambda_c^2} \right]^{-1} \qquad \ldots (3.12)$$

This may be expressed in terms of the actual cavity dimensions by substituting for λ_c in the different cases. Thus for a rectangular cavity

$$\lambda_{f.s} = \left[\left(\frac{p}{2d} \right)^2 + \left(\frac{m}{2a} \right)^2 + \left(\frac{n}{2b} \right)^2 \right]^{-\frac{1}{2}} \qquad \ldots (3.13)$$

and for circular cavities, the expressions for the different modes become:

H_{11p} mode $\qquad \lambda_{f.s} = \left[\left(\frac{p}{2d} \right)^2 + \left(\frac{1}{3 \cdot 42a} \right)^2 \right]^{-\frac{1}{2}}$

E_{01p} mode $\qquad \qquad = \left[\left(\frac{p}{2d} \right)^2 + \left(\frac{1}{2 \cdot 61a} \right)^2 \right]^{-\frac{1}{2}}$

H_{01p} and E_{11p} modes $\quad = \left[\left(\frac{p}{2d} \right)^2 + \left(\frac{1}{1 \cdot 64a} \right)^2 \right]^{-\frac{1}{2}} \quad \ldots (3.14)$

A third subscript is now used to denote how many of the half-wave-length standing-wave patterns are included in the cavity, i.e. $p = 1, 2, 3 \ldots$. The dominant mode, being the first one to be excited as the wavelength is reduced, is obtained from the dominant mode of the corresponding waveguide.

i.e. H_{011} for a rectangular cavity
and H_{111} for a circular cavity.

When designing a cavity resonator its Q value must also be considered, especially if it is to be used as a wavemeter. This value is determined by the proportion of energy that is lost as the induced radiofrequency currents flow through the resistance of the cavity walls. Expressions for the Q values of both rectangular and cylindrical cavities can be derived[9], but are rather complicated. In the rectangular case the expression takes the general form

$$Q = \frac{\lambda}{\delta} \cdot \frac{a \cdot b \cdot c}{4} \cdot f(p, m, n).$$

where δ is the skin depth in the walls

a, b and c are the dimensions of the cavity

and $f(p, m, n)$ is a function of the mode numbers and cavity size.

The expression is more involved for the cylindrical case, but it can be shown that for H_0 modes the maximum values of Q are always obtained when the diameter of the cavity is equal to its length. Generally speaking, the Q value rises as the mode possesses a larger number of standing-wave patterns in the cavity (there then being a smaller proportion of the electric field exciting currents in the wall). It follows that an H_{02p} mode has a larger Q value than an H_{01p} mode, which, in turn, has a larger value than an H_{11p} mode. The respective Q values for the H_{021}, H_{011} and H_{111} modes are in fact in the ratio of $4 \cdot 18 : 2 \cdot 4 : 1 \cdot 0$.

For this reason an H_0 mode is often chosen in wavemeter design, although it is not the dominant one, and in these cases care must be taken not to excite the other possible modes. Wavemeters with Q values of over 20,000 can be obtained in the 1 cm wavelength band if H_0 modes are employed[9].

The advantage of using cylindrical cavities in the construction of wavemeters is twofold, they can be made to very precise limits on a lathe, and it is also possible to drive the shorting plunger, which alters their length, directly from a rotating micrometer thread. The choice of the size of cavity for any particular wavelength is determined by two opposing factors, larger cavity size means that variations in the diameter will produce a smaller percentage error, but if the cavity is made too large a great number of different modes will be excited and it will become very difficult to distinguish between them. The input and output feeds of the cavity are usually adjusted so as to eliminate unwanted modes, and the diameter is then made as large as possible consistent with this. *Figure 36* shows a cavity wavemeter designed on this principle by BLEANEY, LOUBSER and PENROSE[10]. The H_{01p} modes are used for determining the wavelength, as although they are not the dominant H_{11p} modes, their Q values are over twice as high, and they also have the advantage that there is no flow of current in the radial direction on the end walls, which means that the shorting plunger does not introduce any sudden discontinuity. The cavity is fed through two holes in the narrow side of the waveguide, only magnetic field coupling being used. These are a half-guide wavelength apart, and spaced equally on either side of the centre of the cavity. They feed magnetic fields π out of phase into the

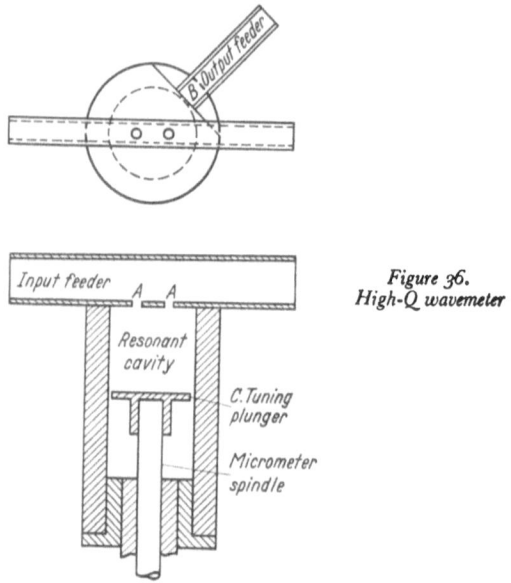

Figure 36.
High-Q wavemeter

cavity, with components in opposite direction; and since, in the cylindrical case, fields are measured radially out from the centre, this corresponds to an excitation of 'even' order modes. Thus no odd-order modes such as H_{11p} or E_{11p}, in which the magnetic field runs continuously across the centre of the end face, will be excited by this type of coupling. The cavity is large enough, however, to allow excitation of the H_{21p} modes, and the output coupling system has to be designed to eliminate them. This is done by feeding out into a guide making 45° with the input guide, the H_{21p} modes having zero components at this point; and hence only resonances due to the H_{01p} modes will couple power out into the detecting arm. Such a cavity, which only gives resonance readings for one mode set, although it is large enough for many others to be excited, can be made with a very high Q value, typical figures being:

Free-space wavelength range 1·15 cm to 1·55 cm
Cavity diameter 30 mm

Power detected from H_{21p} mode resonance $< \dfrac{1}{100}$ that in H_{01p}.

$Q = 20,000$. Determination of λ to 1 part in 10^4, or better.

3.9 ABSORPTION CAVITIES

When designing a cavity resonator to act as an absorption cell, variations in its diameter are not so important as for the case of a wavemeter, since it is only used to obtain a resonance, and not to measure different resonant positions accurately. Cavities are therefore usually operated in their dominant mode, a small size being an added advantage when low-temperature measurements are to be made. The dimensions of the dewar surrounding the cavity can then be made a minimum, and the magnetic field gap reduced as much as possible.

In cylindrical cavities the H_{111} mode is normally employed as this is the dominant one, and therefore no others can possibly be excited. It also has the advantage that the centre of the end walls of the cavity are then regions of maximum radiofrequency magnetic field strength, and the substance being investigated can be very easily mounted in these positions. It is also easier to incorporate variable tuning in a cylindrical cavity, as its size can be altered simply by a circular shorting plunger at the bottom, running in a screw thread. On the other hand rectangular cavities are sometimes preferred, as the direction of polarization is then known very accurately, and, since the probability of transitions induced by magnetic dipole coupling varies with the angle between the d.c. and radiofrequency magnetic fields, this is often of importance; such rectangular cavities are usually tuned to the dominant H_{011} mode. Sometimes modes other than the dominant one are chosen for particular purposes, thus KIKUCHI and COHEN[11] used a cylindrical cavity in the H_{011} mode (not dominant for cylindrical case), with the crystal mounted at the centre of the resonator axis. This mode is also useful if one wishes to mount a long, thin crystal in the direction of the cavity axis, as most of it will then be in the region of maximum radiofrequency magnetic field, and the Q value of this mode is twice that of the dominant H_{111} mode.

For wavelengths below 3 cm, iris holes are used to couple into and out of cavity resonators, the size of the hole and thickness of intervening wall determining the tightness of the coupling[12]. Tight coupling will reduce the effective Q of the resonator, and hence also the absorption produced by the specimen, and a compromise must be made between sufficient transmission of power into the cavity, and the lowering of its Q value. For 3 cm wavelength, and above, coaxial input and output leads are often taken directly to the cavity, the inner conductor being extended into the cavity as a probe to excite the electric field components of the resonant mode. If the cavity is designed for room temperature measurements the normal coupling system consists of two iris coupling holes at diametrically opposite

63

sides of the cavity, fed from the ends of two waveguides, as shown in *Figure 37*.

For low temperature work it is usual to have the input and output coupling at one end, so that nothing extends beyond the external diameter of the resonator, since it has to be surrounded by a cylindrical dewar of as small a size as possible. Such a resonator, designed

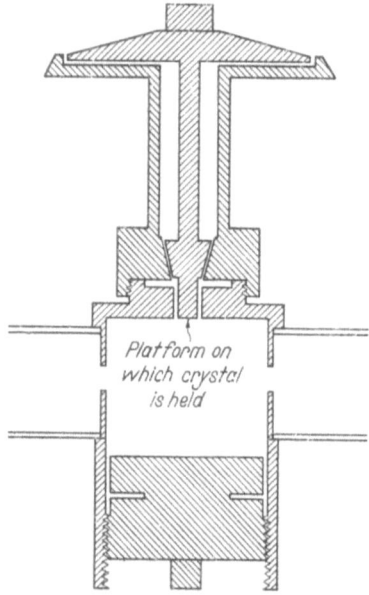

Figure 37. Room temperature 3 cm wavelength resonator

for work at 1·25 cm wavelength is illustrated in *Figure 38*. The cavity is turned from a brass block, and silver plated, having an internal diameter of 12 mm and a length which can be varied between 6 mm and 11 mm. A cylindrical tube of thin-walled German silver leads down to the centre of the top face of the cavity, so that a platform holding the paramagnetic crystal to be studied can be lowered down this until flush with the cavity wall. The crystal can then be rotated from above when it is *in situ* in the cavity, and cooled to liquid air or hydrogen temperatures. The input and output iris holes are on either side of this central tube, being fed by German silver waveguide of internal dimensions 2·5 mm × 6 mm. This size of air-filled waveguide would be beyond cut-off for a wavelength of 1·25 cm and so it is filled with distrene until near the top uncooled end, where the guide tapers out to the normal K band size. Resonators of this design have been used in two or three laboratories with consistently good results.

A photograph of a cross-sectional cut through such a 1·25-cm wavelength cavity is shown in the centre of *Figure 39*, together with general views of 3-cm and 8-mm wavelength low-temperature resonators.

Waveguide size prohibits its use as a direct feed for the low-temperature 3-cm wavelength resonators, and hence coaxial lines having

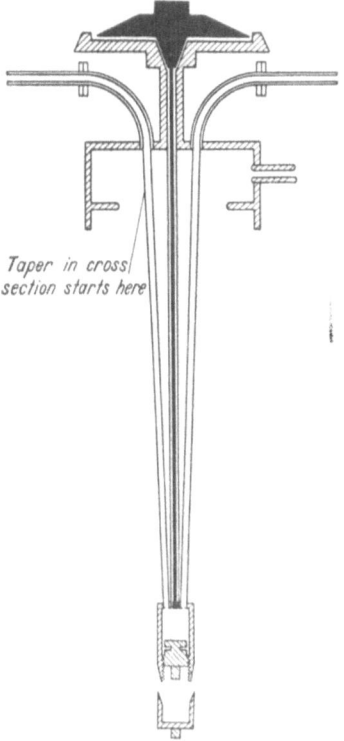

Taper in cross section starts here

Figure 38. Low temperature 1 cm wavelength resonator

German silver tubes as their outer conductors are used instead, the coupling being via a probe to the electric field. The general form of such a resonator can be seen from *Figure 39*.

3.10 WAVEGUIDE CELLS

In gaseous microwave spectroscopy the absorption cell nearly always takes the form of a long length of waveguide, rather than a cavity resonator; the exception to this being when a large magnetic field is used to study the Zeeman effect[13]. The substance being studied no longer has a limited volume, and owing to the possibility of power

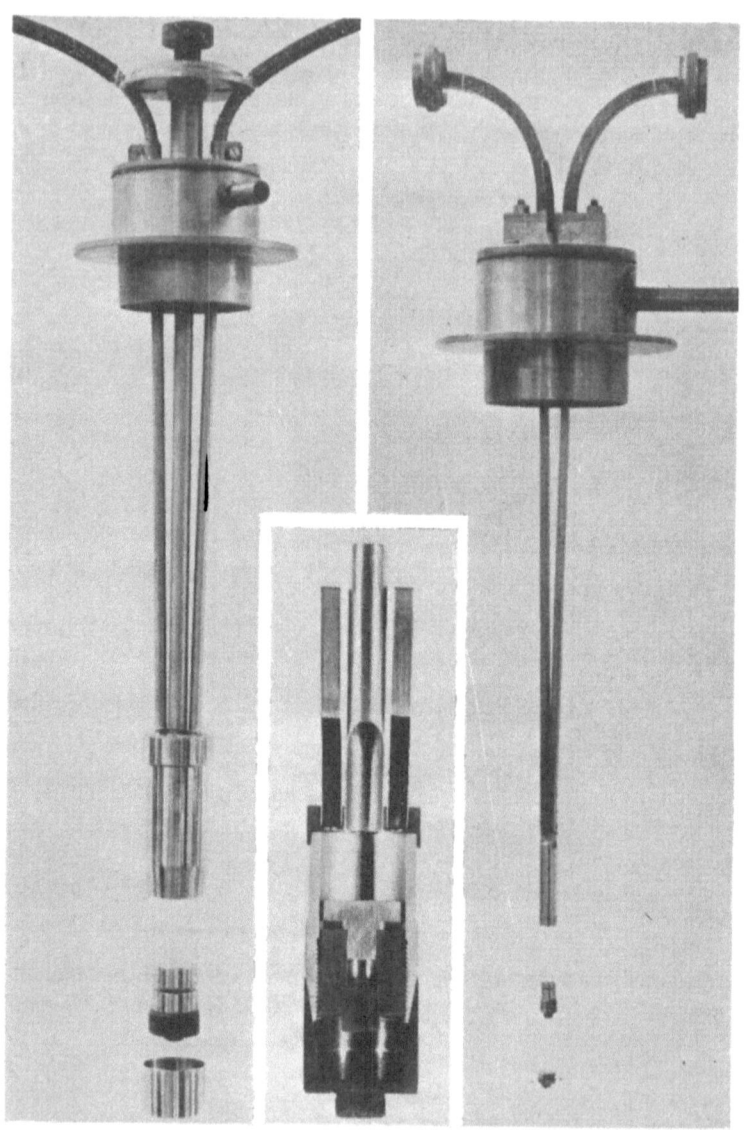

(a) 3 cm wavelength;   (b) cross-section of 1·25   (c) 8 mm wavelength
cm wavelength cavity (3
times scale of (a) and (c));

Figure 39. Low temperature resonators as used for paramagnetic resonance

saturation broadening of the spectral lines, a high concentration of microwave power is in fact a disadvantage.

If the walls of the absorption cell were perfect conductors, its length could be increased without limit to give a better ratio of absorbed to transmitted power for the frequencies of the spectral lines. If the absorption due to the finite resistance of the walls is considered, however, an optimum length for the guide is obtained, because this absorption increases much more rapidly than that due to the gas. Hence, for guide lengths beyond a certain value, the relative absorption at the spectral line frequency (defined as the difference between absorbed power at this frequency and that at other frequencies, with constant incident power will begin to decrease. It may be shown[14] that, with a superheterodyne system of detection, the detected power representing the absorption line is given by

$$(\Delta P) = \tfrac{1}{4}P_{\text{input}} \cdot e^{-\alpha_w \cdot l} \cdot (\alpha_g \cdot l)^2 \qquad \ldots \ (3.15)$$

where l is the length of the waveguide cell, α_w is the attenuation constant of the walls of the waveguide and α_g is the attenuation constant of the gas.

If this expression is differentiated with respect to l, and maximized, the optimum length of the waveguide cell is obtained as

$$l_{\text{opt}} = 2 \cdot \alpha_w^{-1}$$

TOWNES and GESCHWIND[15] have derived the same result by a somewhat different method, and it can be shown to be true for any form of linear detection. If square-law detecting systems are employed (as in low power crystal-video detection) the expression becomes:

$$l_{\text{opt}} = \alpha_w^{-1}$$

In deriving the above expressions, no account has been taken of the variation of conversion loss and noise temperature with the power incident on the crystal. A recent analysis by STRANDBERG, JOHNSON and ESCHBACH[16] has shown that if these variations are considered the expressions for optimum cell length are changed to:

$$l_{\text{opt}} = 8 \cdot \alpha_w^{-1} \text{ for crystal detection}$$
$$l_{\text{opt}} = \alpha_w^{-1} \text{ for bolometer detection}$$

and they also deduce that bolometers should give a better signal-to-noise ratio, if the input power is above 100 microwatts.

The cross-sectional area of a waveguide cell is often made considerably larger than the normal waveband size, for two reasons. First, using a larger size reduces the power density at any given point, and hence reduces power saturation broadening of line widths.

Secondly, the larger size makes it easier to insert an insulated central electrode which is necessary for Stark modulation. As an example, when working in the 1·25-cm wavelength band, standard 3·2-cm wavelength guide is often used for the actual absorption cell, with tapered sections at the input and output. The H_{10} mode will continue to be propagated down the length of the larger guide, provided that there are no major discontinuities or mismatches in it, as these might excite higher order modes. For this reason the mica windows of the absorption cell are placed in the smaller side of the tapers, so that any higher modes excited will be attenuated; it is also possible to use thinner windows in these positions because of their smaller area and closer support, and there will thus be less reflexion of the microwave power. (Normal mica thickness that will stand a pressure of 1 atmosphere, when placed across K-band guide, is about 2 thousandths of an inch.)

As is seen in the next chapter, one very common method of detection in gaseous spectroscopy employs Stark modulation of the spectral lines, which is produced by a strong electric field applied across

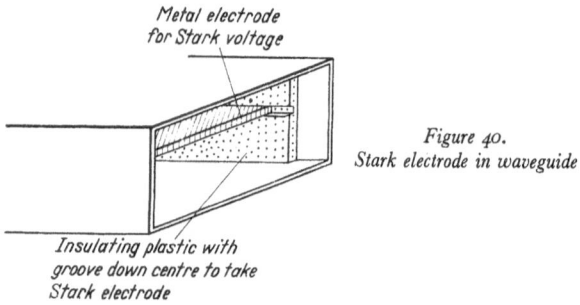

Metal electrode
for Stark voltage

Figure 40.
Stark electrode in waveguide

Insulating plastic with
groove down centre to take
Stark electrode

the waveguide, its amplitude varying at the modulation frequency. The simplest way of doing this is to run an insulated electrode down the centre of the guide, parallel to the wide face, and feed this with the modulated H.T. voltage. This inner electrode must be placed in position very accurately, as any deviation from the centre will produce a different value of electric field in the two halves of the guide, and will thus artificially broaden the Stark components, which can be obtained as separate lines if a square-wave modulation is used. It is sometimes necessary to apply over 1,000 volts modulation to the inner conductor, in order to obtain the required separation of the component lines, and hence it must be well insulated from the rest of the waveguide. These requirements are usually best met by placing insulated strips down both the narrow sides of the guide, these having a recess milled accurately in the centre along their whole length, into which the inner electrode is forced. The method of construction is illustrated in *Figure 40*.

No Stark modulation cells have been constructed for work at the lowest wavelengths, as it is difficult to design the insulating supports when the dimensions become very small. A large number of problems also arise when designing waveguide cells for operation at high temperatures (when many compounds, which are normally liquids or solids, can be studied in the gaseous state). Details of waveguide cells designed for temperatures up to 1,000° C can be found in articles by MAYS [17], and STITCH, HONIG and TOWNES [18]. Typical sizes of waveguide absorption cells vary from 20 cm length and $\frac{1}{3}$ cc volume at 1 mm wavelength [19] to 760 cm length and 600 cc volume at 1 cm wavelength [20]. Recent advances in cavity resonator design, including helices, are discussed in Chapter 11.

REFERENCES

[1] CLEETON, C. E. and WILLIAMS, N. H. *Phys. Rev.* 45 (1934) 234

[2] MONTGOMERY, C. G. *Technique of Microwave Measurements* M.I.T Radiation Laboratory Series No. 11 p. 875(1947)

[3] — *ibid.* p. 858

[4] BETHE, H. A. Radiation Laboratory Reports Nos. 194 and 199 (1943)

[5] MONTGOMERY, C. G. *Technique of Microwave Measurements* p. 774 (1947)

[6] GORDY, W. and KING, W. C. *Phys. Rev.* 93 (1954) 407

[7] BURRUS, C. A. and GORDY, W. *ibid.* 93 (1954) 897

[8] GESCHWIND, S. *Ann. N. Y. Acad. Sci.* 55 (1952) 751

[9] CONDON, E. U. *Rev. mod. Phys.* 14 (1942) 341; HANSEN, W. W. *J. Appl. Phys.* 9 (1938) 654

[10] BLEANEY, B., LOUBSER, J. H. N. and PENROSE, R. P. *Proc. phys. Soc.* 59 (1947) 185

[11] KIKUCHI, C. and COHEN, V. W. *Phys. Rev.* 93 (1954) 394

[12] RAGAN, G. L. *Microwave Transmission Circuits* M.I.T. Radiation Laboratory Series No. 9 p. 655 (1948)

[13] JEN, C. K. *Phys. Rev.* 72 (1947) 986; *ibid.* 74 (1948) 1396; *ibid.* 76 (1949) 1494

[14] GORDY, W. *Rev. mod. Phys.* 20 (1948) 675

[15] TOWNES, C. H. and GESCHWIND, S. *J. Appl. Phys.* 19 (1948) 795

[16] STRANDBERG, M. W. P., JOHNSON, H. R. and ESCHBACH, J. R. *Rev. sci. Instrum.* 25 (1954) 776

[17] MAYS, J. M. *Ann. N. Y. Acad. Sci.* 55 (1952) 789

[18] STITCH, M. L., HONIG, A. and TOWNES, C. H. *Rev. sci. Instrum.* 25 (1954) 759

[19] KING, W. C. and GORDY, W. *Phys. Rev.* 93 (1954) 411

[20] LYONS, H. *Ann. N. Y. Acad. Sci.* 55 (1952) 854

4

MICROWAVE SPECTROSCOPES

4.1 General Considerations

Details of microwave spectroscopes that have been designed and used are now considered, and the particular advantages of each discussed. Those employed for gaseous work are described first, and in these the energy level splittings which give rise to the absorption spectra, are produced by the rotational motion of the molecule. The resonance frequencies are therefore unaffected by external fields, to a first approximation, and the essential experimental requirement is thus a radiation source of continuously variable frequency. In contrast to this, the spectroscopes designed for paramagnetic resonance absorption are usually operated at a fixed frequency, and the applied magnetic field is varied to obtain resonance; since, in this case, the energy level splitting is produced by the direct interaction of the applied magnetic field with the magnetic moment associated with unpaired electrons. The other basic difference between the two types of spectroscope is that, in normal gaseous spectroscopy the absorption of energy is via the interaction of the microwave electric field with the electric dipole moment; whereas, in paramagnetic resonance it is via the microwave magnetic field and the magnetic moment.

The essential elements of a microwave spectroscope for gaseous work are thus (i) a monochromatic source, the wavelength of which can be changed continuously over a given range, (ii) an absorption cell, and (iii) a detector. When searching for new spectra it is advisable to have as few tuned elements in the apparatus as possible, so that a large wavelength sweep can be used without having to reset. In paramagnetic resonance apparatus a fourth item is essential in addition to the above three, namely, a variable magnetic field. New spectra can then be investigated by varying this field instead of the wavelength, and hence tuned circuit elements, such as resonators, have no disadvantage, and superheterodyne systems are also easier to design and operate.

The other practical feature that has to be considered is the question of line width, since even if a molecule or atom produces an absorption line, too large a width will make it very difficult to detect, and will also prevent resolution of any fine or hyperfine structure. The main factors contributing to line width for the two cases of

70

gaseous spectroscopy and paramagnetic resonance are summarized in *Table 4.1.*

Table 4.1. Factors Affecting Line Width

	A. *Gaseous Spectroscopy*		
	Cause	*Method of Reduction*	*Magnitude**
1	Natural Line Width	Unnecessary	10^{-4} c/s
2	Doppler Effect	Select velocities	40 kc/s
3	Pressure Broadening	Reduce pressure	100 kc/s at 10^{-2} mm
4	Wall Collisions	Use larger cell	15 kc/s
5	Power Saturation	Lower input power	200 kc/s at 200 μW
6	Modulation Broadening	Lower modulation frequency	~120 kc/s
	B. *Paramagnetic Resonance*		
1	Natural Line Width	Unnecessary	
2	Spin-Lattice Relaxation	Reduce temperature	
3	Spin-spin Relaxation	Dilute with diamagnetic	
4	Exchange Interaction	Dilute for dissimilar ions	
5	Power Saturation	Lower input power	
6	Inhomogeneous Fields	Polish pole faces	

* Values quoted are very approximate, and apply to OCS at K-band and room temperature[1].

4.2 Line Widths in Gaseous Spectroscopy

The six factors listed in *Table 4.1* are now considered in detail, in order to see what experimental methods are required to reduce the line widths as much as possible.

(a) *Natural line width.* This arises from the finite time that an atom or molecule spends in a given state, and is directly related to Heisenberg's Uncertainty Principle. Thus the energy spread of this state will be given by $\Delta E = h/\tau$ and, as τ is the mean life of the atom in the given state, it follows that the breadth of the line, measured in absolute frequency units, is of the order of $1/\tau$. From the probability distribution of the energy states it is possible to show that [2] the natural line width due to spontaneous emission is given by

$$\Delta \nu = \frac{32\pi^3 \cdot \nu_0^3 \cdot |\mu_{m.n}|^2}{3hc} \cdot \text{cm}^{-1} \qquad \dots \text{(4.1)}$$

This width, which varies as the cube of the frequency, has a value of about 10^{-4} c/s in the microwave region and is completely negligible compared with the other factors.

(b) *The Doppler effect.* This broadening is produced by the spread in the velocities of the molecules, relative to the radiation which they absorb. It can be shown that the intensity of absorption will have a Gaussian distribution curve given by

$$I_\nu = I_{\nu_0} \cdot \exp - \left[\frac{mc^2}{2kT} \cdot \left(\frac{\nu - \nu_0}{\nu_0} \right)^2 \right]. \qquad \dots \text{(4.2)}$$

71

The separation between the two frequencies, for which this expression falls to half its maximum value, is then given by:

$$2 . \Delta\nu = 7\cdot2 . 10^{-7} \left(\frac{T}{M}\right)^{\frac{1}{2}} . \nu_0 \qquad \ldots(4.2a)$$

where M is the molecular weight and T is the absolute temperature. The only way to reduce the Doppler broadening is, therefore, to lower the effective temperature. There are two methods whereby this may be done without actually cooling the gas (which would often condense). The first is by the use of molecular beams, the molecules being fired along a line at right angles to the propagation direction of the radiation. This method has been successfully demonstrated in the case of ammonia, by STRANDBERG and DRIECER[3] and GORDON, ZEIGER and TOWNES[3], who obtained reductions in width by a factor of about six. Another ingenious method of reducing Doppler width has been devised by NEWELL and DICKE[4] in which a travelling wave of electric potential is passed down the waveguide at a velocity in the molecular region. The Stark effect associated with this so acts on the different molecules that only those with velocities very close to that of the wave will constructively interfere, and produce a signal.

The techniques involved in both of these methods are highly complex, however, and in most conditions Doppler broadening is not the main effect, and hence such methods are not necessary.

(c) *Pressure broadening.* This form of broadening is usually by far the most important, and it is necessary to use pressures below 1 mm of Hg if much resolution of the spectra is to be obtained. The theory of pressure broadening is somewhat complicated, but has been approached in general terms by VAN VLECK and WEISSKOPF[5], who considered the random interruption of an assemblage of harmonic oscillators. The frequency before interruption is taken as ν_0 and after collision a Boltzmann distribution of energies is assumed; then, if the duration of the collision is short compared to a period of oscillation, the following expression can be derived for the absorption, due to a single line of resonant frequency ν_0.

$$\alpha = \frac{8\pi^2 . \nu^2 . \mathcal{N}_{JK}}{3ckT} | \mu_{JK} |^2 . \left\{ \frac{(2\pi\tau)^{-1}}{(\nu - \nu_0)^2 + (2\pi\tau)^{-2}} + \frac{(2\pi\tau)^{-1}}{(\nu + \nu_0)^2 + (2\pi\tau)^{-2}} \right\}$$
$$\ldots(4.3)$$

where τ is the average time between collisions

$\mathcal{N}_{JK}$ is the number of molecules per unit volume in the level $J.K.$

and $| \mu_{JK} |^2$ is the component of the dipole moment in the direction of the field.

For frequencies near resonance the second term becomes negligible compared with the first, and the 'shape factor' becomes

$$S(\nu\nu_0) = \frac{1}{(\nu - \nu_0)^2 - (2\pi\tau)^{-2}} \qquad \dots (4.4)$$

which gives a line width between half-power points of

$$2 \cdot \varDelta\nu = \frac{1}{\pi\tau} \cdot \qquad \dots (4.5)$$

This expression thus predicts that the line width is inversely proportional to the time between collisions, or directly proportional to the pressure of the gas. Experimentally this is found to be true over a wide range of pressures, and it can also be seen that, to a first order, the maximum of the absorption curve remains constant, independent of pressure, the change of pressure only affecting the line width. The shape of pressure-broadened lines also differs from the Gaussian shape of Doppler broadening, in having a greater proportion of intensity in the wings. A large amount of experimental work [6, 7, 8, 9] has been carried out on the problem of pressure broadening; the agreement with theory is good, and several more detailed theoretical treatments [10, 11, 12] have been made, since the original work of Van Vleck and Weisskopf.

From the experimental point of view it is seen that as low a pressure as possible is required for high resolution, a width of 100 kc/s being a typical value at a pressure of 10^{-2} mm of Hg. The pressure cannot be reduced indefinitely, however, for two reasons. First, the optimum length of the guide is fixed, and hence reduction in pressure will reduce the actual absorption until this becomes comparable with noise. Secondly, reduction of pressure will raise the power density per molecule and hence saturation broadening will become prominent unless the input power is lowered. Some form of compromise must therefore be made, this depending on both the gas under investigation and the detecting system employed.

(d) *Wall collisions.* Collisions of the molecules with the walls of the absorption cell will produce line broadening in exactly the same way as collisions between the gas molecules themselves, and this will become noticeable when the mean free path is of the same order as the waveguide dimensions (which usually occurs at about 10^{-3} to 10^{-4} mm of Hg). The average number of collisions per second with the walls for each molecule is given by $\frac{S}{V}\left(\frac{RT}{2\pi M}\right)^{\frac{1}{2}}$ where S is the surface area of the waveguide cell and V is its volume. These collisions will produce a line broadening given by

$$2 \cdot \Delta \nu \cdot = 1 \cdot 15 \cdot 10^3 \cdot \frac{S}{V} \left(\frac{T}{M}\right)^{\frac{1}{2}} \text{c/s} \qquad \ldots \ldots (4.6)$$

which, for K-band waveguide and room temperature, gives widths of 15 kc/s and 30 kc/s for OCS and NH_3 respectively.

This value of line width is usually small compared with that produced by the other factors, but may have to be considered when working at very low pressures, and at short wavelengths, with small waveguide size.

(e) *Power saturation broadening.* This arises whenever the incident radiation is sufficiently intense to lift molecules from the lower to the upper energy level, at a rate faster than the relaxation processes can restore thermal equilibrium. It results in a reduction of the peak intensity of the line, while the line width increases. The possibility of this effect was first considered by PURCELL, TORREY and POUND[13] when searching for nuclear resonance absorption at radiofrequencies. It was first investigated in the microwave region by BLEANEY and PENROSE[14], who studied the effect of a change of input power on line breadth for the ammonia spectrum at different pressures. They were able to show that the measured peak absorption started to fall rapidly once the power present per molecule rose above a certain value.

Bleaney and Penrose developed a simple theory of this process in which the rate of increase of the excess molecules in the lower level due to collisions[15] is equated to the number of transitions produced by the absorbed radiation. By substitution of the transition probabilities[16] at the spectral line frequency, they were able to derive an expression for the modified absorption coefficient in the case of a plane wave as:

$$\frac{\alpha}{\alpha_0} = 1 \left/ \left\{ 1 + \frac{32\pi^3}{3h^2c} \cdot \mid \mu_{m,n} \mid^2 \cdot W \cdot \tau^2 \right\} \right. \qquad \ldots \ldots (4.7)$$

where $W = \dfrac{cE^2}{8\pi}$ = energy transmitted across unit area per second.

The corresponding expressions for a waveguide or cavity resonator can be obtained by integrating over the field distributions concerned[17]. In general the absorption coefficient, α, is found to fall quite rapidly as the power initially rises, but flattens out to a nearly constant value for very high power levels.

The disturbance of thermal equilibrium not only reduces the peak absorption but also increases the line width, and more detailed treatments[18, 19, 20], which consider the whole frequency range, give for the general line shape:

$$\alpha = \frac{8\pi^2 \mathcal{N}_{m,n} \nu^2}{ck T} \cdot \sum_{m,n} |\mu_{m,n}|^2 \cdot \left\{ \frac{(2\pi\tau)^{-1}}{(\nu - \nu_0)^2 + (2\pi\tau)^{-2} + [8\pi P|\mu^2_{m,n}|]h^2c} \right\}$$
$$\dots (4.8)$$

where P is the average incident power per unit cross-section of the waveguide, and the other symbols are the same as in equation (4.3). It is seen that saturation broadening, like pressure broadening, also varies from line to line, as the components of the dipole moment change. The initial work of Bleaney and Penrose[14] seemed to indicate that the thermal relaxation time τ was slightly larger than the relaxation time obtained from collision broadening, but later theoretical work[21, 22] has shown that the two are identical, if the degeneracy produced by the different Zeeman levels is allowed for.

The actual broadening produced by this effect can be estimated from the expressions derived above, and it is much larger for a cavity resonator than for a waveguide cell for a given input power. Even for the case of a waveguide cell an input power of 100 microwatts is sufficient to cause noticeable broadening of the (3,3) ammonia line, and the $\mathcal{J} = 1 \rightarrow 2$ transition of OCS is broadened by over 200 kc/s for an incident power of 200 microwatts[23]. This factor often sets a limit on the resolution attainable in a given spectroscope, as it is impossible to decrease the number of transitions per molecule by an increase of pressure, since this will increase the pressure broadening; and any reduction of input power below 100 microwatts usually results in a very poor signal-to-noise ratio on detection, as the conversion loss of the crystal becomes very large for low powers.

(f) *Modulation broadening.* When the resonant frequency of absorption is varied with time, as in the Stark effect, an additional line broadening may be produced. KARPLUS[24] has investigated this problem theoretically and shown that the same effect is produced whether the absorption frequency, or the incident microwave frequency, is modulated, only their difference entering into the expressions. He has shown that sinusoidal modulation should produce a series of extra absorption lines on either side of the true line, separated by the modulation frequency (ν), and hence only resolved, or contributing to the line width, if ν $1/\tau$, or greater, τ being the time between collisions. This effect has been observed experimentally[25].

The case for square-wave modulation is very similar, there is no noticeable broadening if $\nu \ll 1/\tau$, and when $\nu \approx 1/\tau$ the change in

75

peak intensity and half-width of the fundamental component can be expressed as

$$\alpha_{\nu \sim \frac{1}{\tau}} / \alpha_{\nu \ll \frac{1}{\tau}} = 1 - \tfrac{3}{8} \nu^2 . \tau^2. \qquad \dots (4.9)$$

and

$$\Delta\omega_{\nu \sim \frac{1}{\tau}} / \Delta\omega_{\nu \ll \frac{1}{\tau}} = 1 + \tfrac{1}{4} \nu^2 . \tau^2 \qquad \dots (4.10)$$

Since $1/\tau$ corresponds to a frequency of about 50 kc/s at the pressures normally used, it follows that any form of Stark or klystron modulation at a frequency above 100 kc/s will start to broaden the absorption lines appreciably. On the other hand the modulation frequency cannot be reduced too much, or the excess noise that the detecting crystal generates, which is proportional to $1/\nu$, will become large. The combination of these two effects results in 100 kc/s, or thereabouts, being used as the modulation frequency for most microwave spectroscopes using Stark or Source modulation.

4.3 EARLY GASEOUS MICROWAVE SPECTROSCOPES

The importance of the different factors affecting line width can be seen in the results obtained from the earliest measurements. The initial experiment of CLEETON and WILLIAMS[26], already referred to, used ammonia gas at atmospheric pressure, and as a result the pressure broadening obscured all the fine structure of the spectrum, and only a broad hump centred on a wavelength of 1·25 cm was obtained. The next gaseous spectroscope[27], constructed by Bleaney and Penrose (see *Figure 2*), employed the far more sensitive detecting systems developed during the war, and by reducing the pressure of the ammonia gas to 1 mm of Hg and below, they were able to resolve all the fine structure of the spectrum corresponding to the different rotational energy states. They used a cavity resonator as their absorption cell, however, with the result that saturation broadening became greater than pressure broadening at pressures of the order of 0·4 mm of Hg, and they were thus unable to reduce the line width any further, and could not resolve the hyperfine structure of the lines.

The first reported resolution of the hyperfine structure was by GOOD[28], who worked with a waveguide absorption cell, and was able to reduce the pressure of the gas below 10^{-2} mm of Hg before any saturation broadening became noticeable. The first three microwave spectroscopes constructed thus show how the effect of pressure broadening and saturation broadening necessitated the introduction of new techniques.

The basic principle of 'crystal-video' detection, using a waveguide cell, which was the new technique introduced by Good, is illustrated in *Figure 41*. This was an advance on that used by Bleaney and

Penrose in that it allowed direct display of the spectrum on an oscilloscope as well as reducing saturation broadening by the use of a waveguide cell. A saw-tooth sweep at an audiofrequency was applied to the klystron reflector, so that the klystron frequency swept to and fro across the absorption line. The power detected at the far end of the absorption cell was then modulated by the line shape, once every cycle of the sweep (or twice if a sinusoidal voltage sweep is applied). This was then fed into a video-amplifier with a bandwidth sufficient to pass the Fourier components of the line shape. The output of this was applied to the Y plates of an oscilloscope, the time base being in synchronism with the klystron sweep, and in this way the

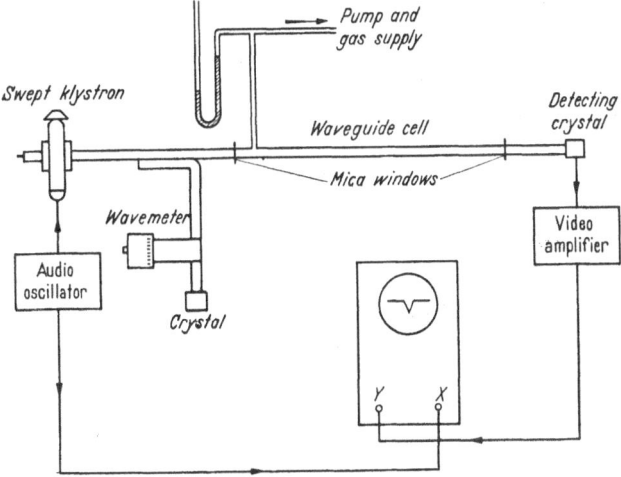

Figure 41. Simple crystal-video detection

actual line shape was displayed directly on the screen. This form of detection is very simple and the spectra can be obtained quickly and easily recorded; it has the one great disadvantage that noise from the detecting crystal is very large, since it is working at low frequencies and with relatively high detected currents.

Following the initial work of Bleaney, Penrose and Good, the study of microwave spectra was taken up rapidly by many workers, and the main advances in technique were concerned with the reduction of the crystal noise, and hence improvement of signal-to-noise ratio, and sensitivity. Two methods were soon introduced whereby the detecting crystal fed into a tuned radio receiver at a frequency of about 100 kc/s, and in this way the large low-frequency crystal noise was eliminated; the first of these was Stark modulation, the second Source modulation.

(a) *Stark modulation.* The Stark effect of an applied d.c. electric field in splitting and shifting the absorption line had been observed by Good[28], but HUGHES and WILSON[29] were the first to introduce the idea of a Stark modulation at a radiofrequency, and use this as a means of overcoming crystal noise and improving sensitivity. They applied the radiofrequency voltage to an electrode in the waveguide, and placed a radio receiver, tuned to this modulating frequency, immediately after the detecting crystal. A diagram of such a spectroscope is given in *Figure 42*. The alternating electric field causes a

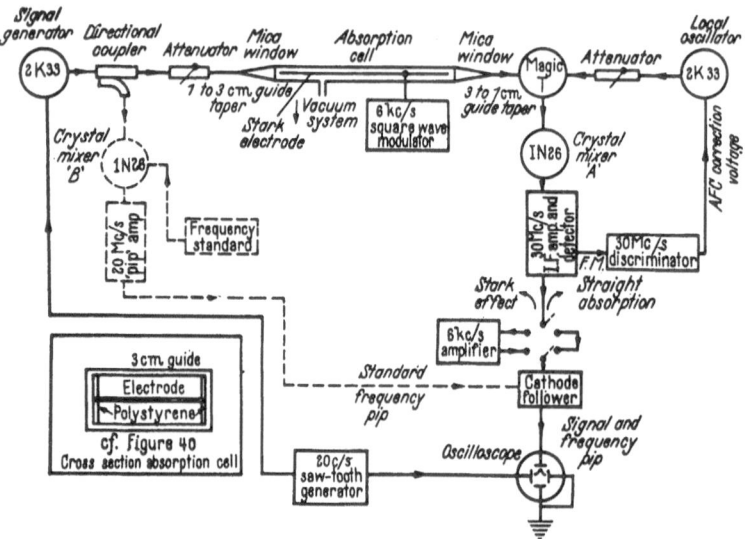

Figure 42. Spectroscope employing Stark modulation with superheterodyne detection

Stark effect in the gas molecules, and hence an intensity modulation of the absorption at the same frequency. The detected microwave power will thus also have an intensity modulation at this frequency, and can be amplified by the tuned radio receiver. The line can still be displayed on an oscilloscope screen by applying an audio-frequency sweep voltage to the klystron reflector as before. The other great advantage of this form of detection is that it differentiates between the signal due to the absorbing gas (which is the only one with a radiofrequency modulation) and that obtained from reflexions in the waveguide run, which cause power variation as the frequency is altered by the klystron sweep.

(b) *Source modulation.* An alternative method of overcoming low-frequency crystal noise is to modulate the frequency of the klystron at a radiofrequency. Gordy and his co-workers have employed this with much success, and in many ways it is equivalent to Stark modulation, although no additional molecular information is obtained. The absorption line acts as a discriminator, which converts the radiofrequency modulation of the klystron into amplitude modulation, this being amplified by a tuned radio-receiver placed after the detecting crystal, as before. A slow sweep, also applied to the klystron reflector, again enables the absorption line to be traced out, since it appears as the envelope of the detected radiofrequency. The great disadvantage of this type of detection is that there is no differentiation between signals from the absorbing gas and those from reflexions in the waveguide run. This is not so important when the lines are very narrow as they can then be easily distinguished from the reflexions, but for wider lines it can be very troublesome. The advantage of this method over Stark modulation is that no central electrode is required in the cell, and because of this it has been used to a very large extent in the mm wavelength region where Stark absorption cells are difficult to construct.

A summary of the three basic methods of detection can be made as follows:

I *Crystal-video*—i.e. only a slow sweep on the klystron. This suffers from bad crystal noise but has no modulation broadening. Its sensitivity can approach that of the other methods when only very small power is available, such as (i) in the millimetre region (ii) when saturation broadening has to be reduced to a minimum.

II *Stark modulation*—This removes the low-frequency noise, but can introduce broadening. Has the advantage of distinguishing between signals and reflexions; disadvantage of complexity of waveguide cell.

III *Source modulation*—Similar to Stark modulation in reducing low-frequency noise, but cannot differentiate between signals and reflexions; has the advantage of being very simple and easy to apply at shorter wavelengths.

Any of these three methods can be combined with a second microwave oscillator to produce a superheterodyne system of detection [30]. This removes the low-frequency crystal noise completely, but has the disadvantage of great complexity, especially when a wide wavelength range has to be searched.

Greater sensitivity can also be obtained with any of these methods by employing phase-sensitive (or lock-in) amplifiers. These will reduce the bandwidth of the detecting system, as discussed in Chapter 2, and for this reason it is normal to use pen-recorders of bandwidths less than 1 c/s with them, rather than oscilloscopes.

4.4 RECENT GASEOUS SPECTROSCOPES

The detailed form of a particular microwave spectroscope will depend mainly on whether high sensitivity or high resolution is required, and also to a very large degree in what wavelength region it is working. In the low-mm and sub-mm bands the only sources of radiation are harmonics from a crystal multiplier, and hence the available power is very low, only of the order of a few microwatts. It is also difficult to construct Stark cells which are not very lossy at these wavelengths, and hence some form of crystal-video detection is usually employed, with or without Source modulation.

(*a*) *Mm-wavelength spectroscope.* An illustration of a typical spectroscope for work in the low-mm wavelength region is given in *Figure 43*, which was designed by GILLIAM, JOHNSON and GORDY[31] for investigations in the 2- to 3-mm wavelength band. In order to obtain

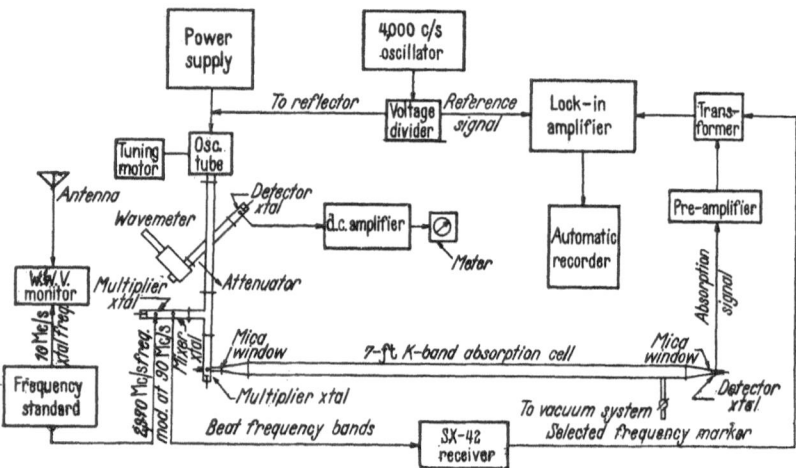

Figure 43. Mm-wavelength spectroscope

the greatest sensitivity Source modulation followed by phase-sensitive detection is employed. The reflector of the klystron generating the fundamental is fed with a 4 kc/s modulation from an oscillator, which feeds the same signal as a reference voltage to the lock-in (or 'phase sensitive') amplifier. The first waveguide run contains a Bethe single hole coupler feeding the wavemeter, and a T-junction, in which the microwave frequency and the harmonics from a frequency standard are mixed to give beat frequencies, which can be added to the detected signal as markers. This run is terminated by the multiplier crystal, radiating into the smaller waveguide, which then tapers out to K-band size for the actual absorption cell. The detecting crystal is mounted at the far end after another taper section, the mica windows being across the smaller G-band guide, and only needing to be 1 thousandth of an inch thick. The detected signal is pre-amplified and then fed into the lock-in amplifier, the output of which is applied to a pen-recorder, and the bandwidth is thus reduced considerably below the 500 c/s needed for oscilloscope presentation. This type of microwave spectroscope or its simpler crystal-video version has been the main one employed in the mm region, where the small powers available produce very little low-frequency crystal noise.

In the longer wavelength region, where Stark cells can be constructed without appreciable attenuation, their use is normally preferred, since they can differentiate against the spurious signals obtained from waveguide reflexions. They can produce line broadening and distortion, however, and for this reason they are not usually employed when the highest resolution is desired.

(*b*) *A high-resolution gaseous spectroscope.* An example of a spectroscope specially designed for high resolution work is that of GESCHWIND[32], which is illustrated in *Figure 44*. The following three steps were taken in order to obtain lines as narrow as possible (i) the gas pressures used were of the order of 10^{-4} mm of Hg, in order to eliminate pressure broadening, (ii) the input power level was kept at 10 microwatts in order to eliminate saturation broadening, and (iii) no Stark or Source modulation was used, in order to prevent broadening or distortion of this type. The signals obtained would be very small if ordinary video-detection were used, as the actual absorption is very low, and the detecting crystal would be working under its worst conditions, having both high conversion loss and large low-frequency noise. Hence superheterodyne detection is employed, this raising the detected signal on to the linear portion of the crystal characteristic, where conversion loss is small, and also eliminating the low-frequency noise.

A slow saw-tooth sweep is applied to the 2K50 klystron, on the left of the figure, which feeds into the balanced bridge microwave circuit containing the absorption and comparison cells. When absorption takes place, the unbalance signal is obtained in the fourth arm of the second 'magic-T', is mixed with the local oscillator signal, and detected by the crystal in the first arm of the T at the top right of the figure. The complexity of the spectroscope arises from the various balancing arms employed to mix the signals, and control the automatic frequency correction circuit.

The details of the balanced bridge are seen in the top left-hand side of the figure. The incident power from the 2K50 klystron is fed

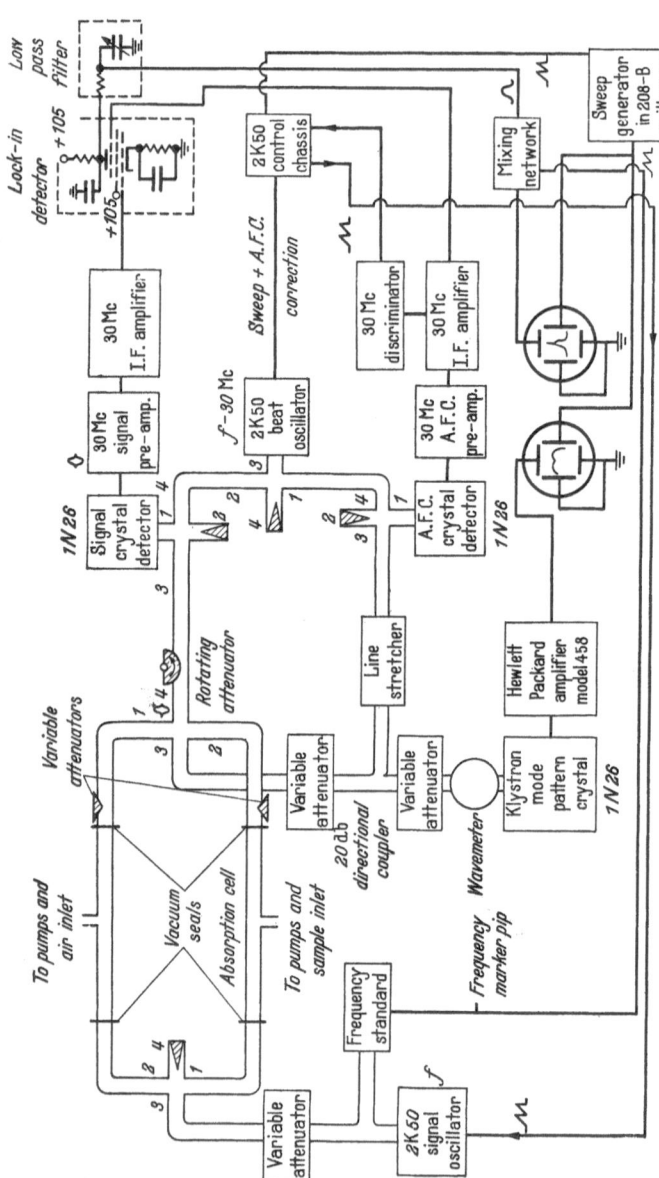

Figure 44. High resolution spectroscope

into a magic-T and equally divided between arms one and two. Arm one contains the absorption cell proper, while arm two contains an identical cell which can be filled with air at a variable pressure, and is used to equalize the phase shift in the two arms. The signals from these two arms recombine at the second 'magic-T', and if their amplitudes and phase are equal they unite to feed down arm three, and no signal is obtained in arm four. The rotating attenuator in arm four is used to obtain initial balance, the attenuators and phase change in the bridge being adjusted until no intensity-modulated signal is obtained at the crystal detector. The 2K50 beat oscillator feeds into the detecting system via another 'magic-T', its signal being equally divided between the detecting T proper, and the 'automatic frequency control and detector T', at the lower right of the figure. This receives the original microwave frequency via a directional coupler, attached to the third arm of the T of the balanced bridge. Although this system appears highly complex it allows independent adjustment of the various mixer units, and the klystron mode pattern and frequency markers can also be obtained, as shown.

The signal from the detecting crystal feeds into a 30 Mc/s inter-mediate frequency amplifier chain, the lock-in detector which follows this obtaining its reference signal from the 30 Mc/s amplifier of the automatic frequency control circuit. The output signal, which cor-responds to the shape of the absorption line (or its differential if the sweep amplitude is smaller than the line width) is then applied to the Y plates of the oscilloscope, the X plates of which are fed from the sweep generator supplying the initial klystron. Absorption lines as close as 55 kc/s [32] could be resolved by this spectroscope, and its sensitivity was found to be limited by spurious reflexions, rather than noise. The sensitivity is being further improved [33] by applying Stark modulation at an audiofrequency—this will not be high enough to cause any broadening, but will eliminate the spurious signals.

4.5 COMPARISON OF SENSITIVITIES

It is of interest to calculate the theoretical sensitivities that can be obtained with the two spectroscopes described above, and compare them with experimentally observed values. Normally, superhetero-dyne systems of detection are orders of magnitude more sensitive than 'straight-through' detectors, but this factor does not hold when applied to microwave spectroscopes, especially when Stark or Source modulation is incorporated in the latter. In order to obtain an ex-pression for the minimum detectable signal, for a given system, the power representing the signal must be equated to the noise power at the input to the detector.

(a) *Superheterodyne detection.* By substitution of the optimum wave-guide length, $2\alpha_w^{-1}$, into the equation already derived[36, 37] for the signal power (equation (3.15)), the maximum detectable signal, representing the absorption line, is obtained as:

$$\delta P = \frac{P_I}{4e^2} \cdot \left(\frac{\alpha_g}{\alpha_w}\right)^2 \qquad \dots (4.11)$$

where P_I is the incident power *on the bridge*. The minimum detectable absorption coefficient can then be derived by equating this to the noise of the crystal and input circuit, to give:

$$\alpha_{g\min} = 2 \cdot e \cdot \alpha_w \sqrt{\frac{2F \cdot kT \cdot \Delta f}{P_I}} \qquad \dots (4.12)$$

where Δf is the bandwidth of the detecting system, and F is the over-all noise-figure of the crystal, including Johnson noise and excess crystal noise varying as $1/f$, the local oscillator and intermediate fre-quency amplifier noise, and also the effect of conversion loss in the crystal[38, 39]. It will have a value of about 50 dB in a typical case, but this can be improved by the use of a balanced mixer to eliminate the local oscillator noise. The value of Δf will depend on the recording system employed, 100 c/s being the minimum possible for oscillo-scope presentation, but well under 1 c/s can be used if phase-sensitive detection with a pen recorder is incorporated. Substitution of these figures into equation (4.12) give values for $\alpha_{g\min}$ of about 10^{-8} cm^{-1} and $5 \cdot 10^{-10}$ cm^{-1}, respectively, for (i) oscilloscope presentation, and (ii) pen-recorder with balanced crystal detector.

(b) *Stark and Source modulation.* The above analysis for a superhetero-dyne system of detection can be made more general and applied to other cases of linear detection by resolving a modulated signal into its carrier and sidebands. TOWNES and GESCHWIND[36] used this prin-ciple, and considered the carrier frequency as the local oscillator power, beating with the signal power in the sidebands. It can then be shown that the useful input power is given by

$$\delta P = \frac{1}{8} \left(\frac{\alpha_g l}{2}\right)^2 \cdot P_I \exp\left[-\alpha_w \cdot l\right] \qquad \dots (4.13)$$

where P_I is here the incident power on the absorption cell itself (a balanced bridge seldom being employed with Source modulation). The minimum detectable gaseous absorption coefficient can then be derived as before, to give:

$$\alpha_{g\min} = 2e \cdot \alpha_w \cdot \sqrt{4F \cdot kT \cdot \Delta f \cdot /P_I} \qquad \dots (4.14)$$

Comparison of this with equation (4.12) shows that the sensitivities of the two forms of detection are very similar, a typical value for $\alpha_{g_{\min}}$ being 8.10^{-9} cm^{-1}, for the case of phase-sensitive detection with a pen-recorder. Experimentally-obtained minimum absorption coefficients are about 5 to 10 times greater than the theoretical values. In deriving the above theoretical expressions, no account has been taken of the variation of conversion loss and noise temperature with the power incident on the crystal. A recent analysis[34] has shown that, if these variations are considered, the expressions for optimum cell lengths are changed to $8 . \alpha_w^{-1}$ for crystal detection, and α_w^{-1} for bolometer detection, and it is also deduced that bolometers should give a better signal-to-noise ratio if input power is above 100 microwatts[34, 35]. Superheterodyne detection is still more sensitive than Stark or Source modulation, mainly because the intermediate frequency of 30 Mc/s is well above the maximum modulation frequency of 100 kc/s, and the excess noise in the crystal is therefore smaller.

(c) *Crystal-video detection.* The sensitivity of a low-power crystal-video detector can be calculated by the method of BERINGER[40], and TORREY and WHITMER[41]. In this, the crystal is represented as a current generator, shunted by its video resistance, R, and the current is a linear function of the microwave power absorbed by the crystal. The signal voltage that is fed on to the first grid of the video-amplifier is then given by:

$$V_g = \tfrac{1}{2}\beta \left(\frac{\alpha_g . l}{4} \right) . 2R . P_I . \exp \left[- \alpha_w . l \right] \quad \ldots . (4.15)$$

This is then equated to $[4 . kT . \Delta f . (R + R_a)]^{\frac{1}{2}}$, which is the noise present at the output terminals of the crystal[41], to give for the minimum detectable absorption coefficient:

$$\alpha_{g_{\min}} = 8e . \alpha_w . \sqrt{kT . \Delta f} . /M . P_I \quad \ldots . (4.16)$$

where M is written in place of $\beta R/(R + R_a)^{\frac{1}{2}}$ and is known as 'The Figure of Merit' for the crystal, being a measure of the detecting efficiency of the crystal and its noise properties. In the case of welded contact germanium crystals, it is possible to improve this parameter by the use of d.c. bias, but this usually produces too much excess noise in the case of silicon crystals. Substitution of typical values into equation (4.16) give a minimum detectable α_g of about 10^{-7} cm^{-1}, showing that this method is not nearly as sensitive as superheterodyne detection, or radiofrequency modulation, but it has the advantage of simplicity, and is thus often used when searching for new spectra, especially at the lower wavelengths.

4.6 Special Spectroscopes—Zeeman Effect

The Zeeman effect in microwave spectroscopy of gases was first noticed by Coles and Good [42], but first studied in detail by Jen [43], who used a cavity resonator as his absorption cell. The advantage of a cavity is that, being much more compact than a coiled length of guide, it is easier to apply a strong homogeneous magnetic field across it. A block diagram of his apparatus is shown in *Figure 45*, and it can be seen that the poles of the electromagnet form the end walls of the

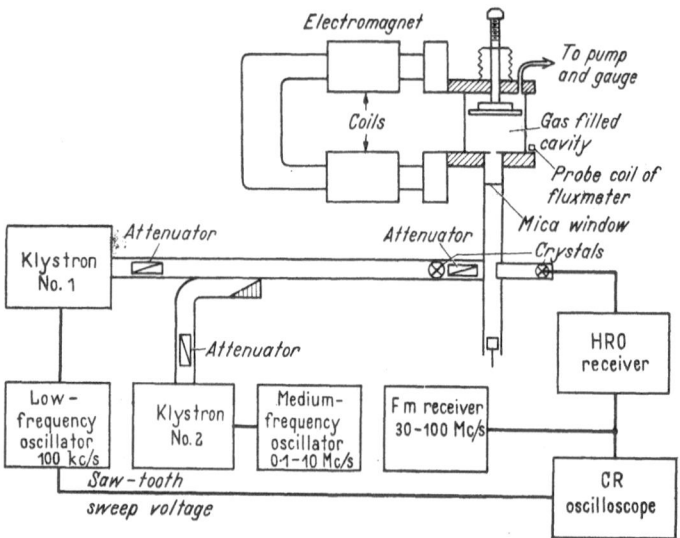

Figure 45. Spectroscope for studying Zeeman effect in gases

cavity, being made of magnet steel and silver-plated on the inside. When the slow saw-tooth sweep voltage is applied to the klystron reflector, the Q curve of the resonant cavity will be traced out by the detecting system, and any absorption by the gas will be displayed as an inverted peak on this curve. Normal spectral line widths correspond to Q values of over 100,000, and hence the whole line is usually obtained right at the centre of the cavity resonance.

Source modulation at 100 kc/s was employed, and a 'magic-T' bridge was used so that change in reflected power at the cavity produced the signal at the detecting crystal. A second klystron, injecting power into the main run via a directional coupler, was used to produce a system of frequency markers. The tuned 30–100 Mc/s receiver selected beat frequencies between the two klystrons and presenting them as pips on the oscilloscope screen, their separation was thus equal to twice the frequency of the tuned 30–100 Mc/s receiver. The sensitivity of this type of spectroscope can be derived in exactly the same way as for the waveguide-cell type, and it can be shown[41] that $(\alpha_{g_{min}})^{-1}$ is directly proportional to the Q factor of the cavity. For this reason a large cavity is employed, which is also effective in reducing Saturation broadening.

Application of a strong magnetic field (about 1,500 gauss) across the cavity produced the Zeeman splitting of the line, which was shown by two inverted peaks on the Q curve, instead of one. The frequency separation between these was measured by the marker pips, and compared with the magnetic field producing the splitting. A second spectroscope, employing large values of magnetic field (up to 10,000 gauss) was also constructed by Jen[44], in which the Paschen-Back effect could be studied as well. The splittings obtained in such cavity resonators will correspond to the π or σ transitions, as the radiofrequency magnetic field is parallel, or perpendicular, to the applied d.c. field. It is in fact easier to obtain more detailed information from the spectra if circularly polarized radiation is used with a parallel d.c. field. With this in mind, Jen also constructed a spectroscope[45] employing a large solenoid round a long run of circular guide, and used phase-changers which could insert mica sheets into the guide at different angles (one set at 45° to the initial plane of polarization of the rectangular guide producing circular polarization). The advantage of this method is that the signs, as well as the magnitudes, of the gyro-magnetic moments can be determined.

4.7 Measurement of Line Shapes and Intensities

When recording a spectrum, the relative intensity of different lines is often very important, and this can usually be determined fairly easily. The crystal detector law must be known, i.e. whether it is working in the square-law or linear portion of its characteristic, and also whether the line itself is being traced out, or its first derivative. If oscilloscope presentation is used, and the bandwidth of the detecting system is sufficient to display the undistorted line shape, then relative intensities can be obtained by a direct comparison of the line heights and widths. If small sweep amplitudes are employed, and the first derivative of the line is obtained, as is normally the case with a pen-recorder, then allowance has to be made for the fact that the recorded signal height is inversely proportional to the line width. A direct method of comparing relative intensities, when using a Stark-modulation spectrograph, has recently been described by Baird and Bird[46]. This relies on the fact that a direct proportionality exists between the absorption coefficient and the component of crystal current at the modulating frequency, provided certain reasonable conditions are fulfilled. The crystal current component at the modulating frequency is separated from the d.c. component by a suitable filter, and then compared with a known standard current, and in this way a direct comparison between absorption at different frequencies is obtained. This method has been further modified[47] to give an estimation of absolute absorption coefficients, by a comparison of the amplitude of the modulated crystal current with that of the d.c. component, when using low microwave power levels.

Determination of absolute absorption coefficients is a more difficult procedure than the comparison of relative values, and is usually best accomplished with the help of a standard calibrated attenuator which is used to restore the signal level, when the gas is absorbing, to the value it had before the absorption[48, 49]. If it is certain that the crystal is working in the square-law region the absolute absorption coefficient can be calculated from the change in detected power when the gas is in the cell, compared with that obtained when the cell is evacuated. BLEANEY and PENROSE[50] employed this method in their original measurements, the absorption coefficient being given by

$$\alpha = \frac{2\pi}{\lambda Q_0} \left(\sqrt{\frac{\delta_0}{\delta_1}} - 1 \right) \qquad \dots (4.17)$$

where Q_0 is the Q value of the empty cavity, and δ_0 and δ_1 are readings of the d.c. crystal current obtained for no absorption and maximum absorption. If the absorption cell is a length of waveguide then the corresponding expression becomes

$$\alpha = \frac{\lambda}{\lambda_g l} \cdot \log_e \left(\frac{\delta_0}{\delta_1} \right) \qquad \dots (4.18)$$

When determining the shape of an absorption line care must be taken to see that the detecting system does not distort the true trace. This applies especially to Stark and Source modulation, where modulation broadening may introduce a spurious line width. In such systems the modulation frequency should be kept less than one quarter of the line width in order to avoid any distortion. The time constants and bandwidth of the amplifiers and recorders should also be such as to pass all the Fourier components of the line shape, and if this is so the lines as traced out on the oscilloscope screen can be photographed and their true shapes directly recorded. If the widths of the lines to be measured are very large (> 10 Mc/s), the line shape is usually best obtained by plotting it out, point by point, from the d.c. value of the detected power at successive small changes of frequency[50]. Such large widths are normally associated with high absorption, and hence d.c. methods of detection are usually sufficiently sensitive. Determination of the actual line shape is often a help in finding the cause of line broadening, since the Gaussian shape produced by Doppler broadening is different from that produced by pure collision broadening, or saturation broadening.

4.8 PARAMAGNETIC RESONANCE*—SOLID STATE SPECTROSCOPY

The construction techniques of microwave spectroscopes used for studying paramagnetic resonance absorption are again largely influenced by the sensitivities required, and the widths of the lines to be observed. The development of paramagnetic resonance spectroscopes followed that of the gaseous ones very closely, the first employing d.c. methods of detection for the observation of intense wide lines[51, 52]. As soon as the line widths were reduced to resolve the hyperfine structure[53] however, a.c. sweep methods were incorporated to increase the sensitivity of the apparatus and allow oscillograph display of the spectrum[54]. A consideration of the various factors influencing the line width is therefore again helpful for an understanding of the different experimental points involved when designing a spectroscope. The six different factors affecting line width, listed in *Table 4.1*, are therefore now dealt with in more detail.

(a) *Natural line width*. This is again quite negligible compared with the other factors in the microwave region, and the remarks made concerning it in section 4.2(a) apply equally to the case of paramagnetic resonance absorption lines.

(b) *Spin-lattice relaxation*. This broadening is produced by the interaction of the paramagnetic ions with the thermal vibrations of the lattice, a 'smearing out' of their energy levels being thus produced. Two types of coupling were initially suggested, the first by WALLER[56], which acted via the magnetic dipole interaction between the spins themselves, this varying periodically with the spatial vibrations of the ions. The second, suggested by KRONIG[55], acted via the effect of the crystalline electric field, the thermal motion of the atoms altering this and thus also the energy level systems of the paramagnetic ions. Waller's theory predicts relaxation times much longer than those observed experimentally, and it would appear that Kronig's suggestion is the correct one. This interaction must act via the spin-orbit coupling, and there are two mechanisms whereby the spins can exchange energy with the lattice. They can either exchange a whole

---

* In later chapters, the now generally accepted term Electron Spin Resonance is used for Paramagnetic Resonance.

quantum directly with a lattice vibration of the appropriate frequency (Direct Process), or transfer energy, by scattering a quantum from the lattice, and changing its value (Raman Process), the latter predominating at the higher temperatures, as all the quanta take part in it. Kronig[55] made an estimate for the relaxation times which might be expected from these two mechanisms, for the case of $S = \frac{1}{2}$, obtaining expressions of the form:

$$\text{Direct Process} \quad \tau_l = \frac{10^4 \cdot \Delta^4}{\lambda^2 \cdot H^4 \cdot T} \text{ secs.} \qquad \dots (4.19)$$

$$\text{Raman Process} \quad \tau_l = \frac{10^4 \cdot \Delta^6}{\lambda^2 \cdot H^2 \cdot T^7} \text{ secs.}$$

where Δ is the height of the next orbital above the ground state, in cm^{-1}, λ is the spin-orbit coupling coefficient[57], and T is the absolute temperature. The term 'relaxation time' is a measure of the time taken for the interaction to restore thermal equilibrium, and the broadening produced by the interaction is given by $\Delta \nu \approx \frac{1}{\tau_l}$.

The main qualitative predictions of this theory are confirmed by experiment, in that ions with small values of Δ have wide absorption lines with the width varying markedly with temperature. Although VAN VLECK[58] has made some more detailed calculations, there is really no quantitative theory at the moment with which to compare experimental results. The line shape produced by spin-lattice broadening has the same form as for collision broadening in gases, but, in practice, the temperature at which measurements are made is always reduced until the line width is due to spin-spin and not spin-lattice interaction.

(c) *Spin-spin interaction.* This interaction produces a broadening of the energy levels by the mutual effect of one paramagnetic ion on another. Each such ion may be regarded as a magnetic dipole which will be precessing in the applied magnetic field. Its component in the direction of the field will have a steady value, and this will produce an extra magnetic field at the neighbouring paramagnetic ions which alters the total field value slightly, and thus shifts the energy levels. It can be seen that this effect will vary markedly with angle of the applied field, and the contribution of each neighbouring dipole will have an angular dependence of the form $(1 - 3\cos^2\theta)$, where θ is the angle between the lines joining the dipoles and the direction of the applied field. The rotating component of the precession will also cause a broadening as it interacts on other spins of the same Larmor frequency and induces transitions, this decreasing the normal lifetime of the energy state.

The theory of spin-spin interaction has been studied by VAN VLECK[59] and PRYCE and STEVENS[60]. Van Vleck derived an expression for the mean-square width of a line, due to dipole-dipole interaction between free spins, of the form

$$(\varDelta H)^2 = \tfrac{3}{4} S (S + 1) g^2 \beta^2 \sum_k \left\{ \frac{1 - 3 \cos^2 \theta}{r^3} \right\}^2_{jk}$$
$$+ \tfrac{1}{3} S' (S' + 1) \frac{g'^4}{g^2} \beta^2 \sum_{k'} \left\{ \frac{1 - 3 \cos^2 \theta}{r^3} \right\}^2_{jk} \quad \dots\dots (4.20)$$

The transition concerned is between the levels of ions with spin S, and spectroscopic splitting factor g, r being the distance apart of the j and kth ion. The second term represents the contribution to the broadening by 'dissimilar' ions of spin S', and spectroscopic splitting factor g'. Thus in some salts (e.g. the double sulphate Tutton salts) there are two ions in each unit cell, differently orientated with respect to the crystalline axes, and the broadening produced by interaction between ions which have 'dissimilar axes' will not be as large as for those with 'similar axes'.

To a first order spin-spin interaction is independent of both magnetic field strength and temperature, and the only way to reduce broadening by this effect is to move the paramagnetic ions farther apart, by diluting the salt with an isomorphous diamagnetic equivalent. In the limit, the line width will be determined, not by the neighbouring paramagnetic ions, but by the magnetic moments of the nuclei of the neighbouring diamagnetic atoms. In hydrated salts the line width can often be further reduced by growing crystals from 'heavy water'[61], since the magnetic moment of the deuteron is less than that of the proton. It can be seen from the terms in r^6, of equation (4.20) that spin-spin interaction falls off rapidly with distance and in calculating the line width it is usually only necessary to consider the immediate neighbours. The line shape produced by this interaction is more or less Gaussian, but not quite so sharp at the peak[59].

(d) *Exchange interaction*. If the paramagnetic ions are close enough together an exchange interaction may occur between them, which can alter the line width considerably. If the ions are similar the effect of exchange is to narrow the lines at the centre, and broaden them in the wings, leaving $(\varDelta H)^2_{Av}$ unchanged in the simplest cases, but reducing the width measured at half-power points. On the other hand, if the exchange is taking place between dissimilar ions the exchange interaction will tend to bring the two different transitions together and hence produce one wider line. This second effect is well illustrated by the measurements of BAGGULEY and GRIFFITHS[62] on $CuSO_4 . 5H_2O$, where the exchange force between the dissimilar ions is so strong that it is impossible to resolve the two lines at 3 cm, wavelengths of 0·85 cm having to be employed before it is possible to do so[62].

The main theoretical treatment of exchange interaction has again been by VAN VLECK[59] and Pryce and Stevens[60], the former dealing with the specific case of free spins and treating it in detail, while the latter consider a more general state of the ion, and only calculate the area and second-moment of the line-shape. Pryce and Stevens[60] show that Van Vleck's prediction that the second-moment of the line shape is not affected by exchange interaction only applies if the ions are similar and precess about parallel axes, otherwise there is a contribution depending on $\sin^4 \theta/2$, if they have the same precessional frequency, and on $\cos^2 \theta$, if they have different frequencies, θ being the angle between their axes. More recent work by ANDERSON and WEISS[63] also attempts to derive a detailed line shape for cases of strong exchange interaction, using the model of 'random frequency modulation'. This predicts a line of 'resonance shape' at the centre, falling off more rapidly in the wings, which seems to be in agreement with the experimental results.

Exchange interaction is found to some extent in most non-diluted paramagnetic salts, and often reduces the expected spin-spin broadening by a large factor. For example, in the case of copper phthalocyanine[64], where the individual copper atoms are only 4 Å apart, the expected width due to spin-spin interaction is over 400 gauss, but in certain directions the measured value is only 35 gauss.

(e) Power saturation broadening. The effect of power saturation broadening is of exactly the same nature in the case of solid state microwave absorption as for the gaseous state (see Section 4.2(e)). That is, when the incident power is so large that the normal relaxation processes can no longer restore thermal equilibrium, the peak absorption will start to fall and the lines become wider. There are two reasons, however, why this effect does not show up in the solid state until much higher power levels are reached. First, the interaction is via the magnetic dipole, not the electric dipole as in the gaseous case, and the transition probability for the former is much smaller. Secondly, even when diluted, the concentration of ions per cc is usually much higher than for gaseous samples, and hence the power density per absorbing atom or molecule is reduced. This is offset by the fact that cavity resonators are employed, and saturation effects will take place at lower input power levels for these[17]; but in practice the power level used hardly ever reaches a value such that

saturation broadening is present in concentrated paramagnetic salts[65]. Dilute crystals and free radicals often do show this effect, especially at low temperatures, and are fully considered in Section 12.3.

(f) *Inhomogeneous fields.* The broadening produced by inhomogeneous magnetic fields is not an intrinsic effect of the paramagnetic ions at all, but is mentioned here as it sometimes contributes a noticeable width to the absorption lines, and must often be considered. In nuclear paramagnetic resonance, absorption lines of the order of a few milligauss width can be obtained[66], and hence special precautions have to be taken in order to obtain fields homogeneous to a milligauss, when of a kilogauss magnitude (i.e. 1 part in 10^6). This involves grinding and polishing the pole faces to optical flats, and if they are not very large in diameter, having small shims at the edge to counteract the end effect of the gap. The requirements for electronic paramagnetic resonance in the microwave region are not usually so severe, although lines with a width of under 0·1 gauss have been observed[67], and hence fields homogeneous to 1 part in 10^5 may be required. (See Chapter 10 for recent developments.)

(g) *Summary.* It can be seen, from the above outline of the different factors producing line broadening, that the position is somewhat simpler than in the gaseous case as the effects are not interdependent on each other like collision and saturation broadening. By taking measurements at different temperatures it may be established whether the line width is due to spin-lattice or spin-spin interaction, the latter being independent of temperature. If the line is broadened by spin-lattice interaction at room temperature good resolution can be obtained by working at a lower temperature where the interaction is smaller than that of the spin-spin. Fortunately the spin-lattice interaction falls off rapidly with temperature decrease, and is usually negligible for most ions by 20° K, although Ti^{3+}, Ce^{3+} and Sm^{3+} are three cases where it is necessary to cool to 4° K before the absorption lines are narrow enough to be observed[68, 69].

The spin-spin interaction is then decreased by diluting the paramagnetic ions with an isomorphous diamagnetic salt. Zinc and magnesium have been mainly used as the dilutant for the iron group salts, and lanthanum for the rare-earth group. If such diluted crystals are to be grown successfully, care must be taken to see that not only the chemical formulae, but also the crystal structure and its dimensions, are the same for the paramagnetic and diamagnetic salts. One or two series of salts have been found very useful in this respect, namely the double-sulphate Tutton salts and the fluosilicates in the iron group, and the ethylsulphates in the rare-earth group. Dilutions employed vary from about 1 in 200 to 1 in 1,000, and for higher dilutions than this the remaining line width is usually due to the magnetic moments of the nuclei of the surrounding diamagnetic atoms, rather than the closest paramagnetic ions. For this reason salts with nitrogens surrounding the paramagnetic ion (e.g. $K_3Cr(CN)_6$) will have narrower line widths when diluted, than those surrounded by the hydrogens of water molecules. As already noted, the line width can be further reduced in the latter case, by growing the crystals from heavy water[61].

4.9 EARLY PARAMAGNETIC RESONANCE EXPERIMENTS

In exactly the same way as the different factors affecting line width called for the introduction of new techniques into microwave gaseous spectroscopy, so the design of paramagnetic absorption spectroscopes altered as higher resolution was required. The initial measurements on paramagnetic absorption lines[52, 62] were all made at room temperature, and hence only those salts with a long spin-lattice relaxation time at 300° K could be investigated. As none of the first crystals studied were diluted, the absorption of power at resonance was quite strong and a simple d.c. method of detection could be employed, the power through the resonant cavity being plotted against successive small changes in field strength; this being very similar to the initial work of Bleaney and Penrose[50] on the ammonia spectrum, except that the magnetic field, and not the frequency, was changed.

The next advance in technique was to reduce the spin-lattice broadening by taking measurements at low temperatures. This necessitated the design of cavity resonators that could be cooled down to liquid air and liquid hydrogen temperatures, similar to those illustrated in *Figure 39*. The absorption of power was still large, however, and d.c. methods of detection could be used as before[70]. It was not until the discovery of hyperfine structure in the solid state by Penrose[53] that very narrow line widths were required, and the dilution of the paramagnetic by a factor of 500 or more meant that much greater sensitivity was needed. This resulted in the a.c. sweep of the magnetic field, with video amplification after detection[54], which provided a much more sensitive method and permitted direct oscilloscope presentation of the spectrum.

4.10 TYPICAL PARAMAGNETIC RESONANCE SPECTROSCOPES

(a) *Crystal-video*. A typical straightforward 'crystal-video' type of paramagnetic absorption spectroscope has already been briefly described, and illustrated in *Figure 16*. The cavity resonator has both an input and output waveguide feed, and the transmitted power is detected, amplified by the video-amplifier ('video-amplifier' is a term taken over from radar usage where the amplifier has to deal with very narrow pulses, whereas here the bandwidth required is only of the order of 10 kc/s and it is really an audio-amplifier), and fed to the Y plates of the oscilloscope. All paramagnetic resonance spectroscopes use variation of magnetic field rather than variation in microwave frequency when searching for new lines, and hence the whole waveguide run need only be correctly matched for one frequency at the beginning of each set of measurements. In order to facilitate this, a 50 c/s voltage modulation is applied to the klystron reflector, so that the particular mode of the klystron is swept across on each forward and backward sweep of the modulation. The signal obtained from the monitor arm will then appear as a trace of the mode pattern (i.e. klystron power versus frequency) when amplified and applied to the oscilloscope. This can be used to check the oscillator power, and its matching. The cavity resonator will only pass a small range of the mode frequency coverage, and hence the trace obtained from the crystal at the end of the waveguide run will be the Q curve of the cavity itself. This signal is used to tune the cavity to the klystron (there being no detected power until the cavity length approaches its resonant value), and is also used to match the rest of the run correctly. Matching stubs are incorporated on either side of the coupling holes to the cavity, and if possible, an attenuator is inserted between the klystron and the cavity so that the frequency of the former is not 'pulled' by a varying load in the latter. This type of spectroscope is very simple to set up and operate, and the spectra can be obtained quickly. It is therefore often used when searching for new spectra, and in measurements where the greatest sensitivity is not required. It suffers from the disadvantage of all low-frequency crystal-video systems, however—i.e. bad low-frequency crystal noise, and low signal-to-noise ratio from the relatively large bandwidth employed.

These disadvantages may be overcome in the same way as is done in the gaseous case, with a few modifications. High-frequency magnetic field modulation has been employed in some cases[71], the amplitude of the sweep being very small, and is used to trace out the derivative of the line which is amplified and detected by phase-sensitive methods. Normally, however, 50 c/s sweep frequency of fairly large magnitude is employed so that several spectral lines can be obtained at once, and the low-frequency crystal noise is eliminated by other means. This may be done by either method employed in gaseous spectroscopy—i.e. superheterodyne detection, or Source modulation. Section 11.1 gives details of recent developments.

(*b*) *Superheterodyne detection.* Superheterodyne detection was first used in paramagnetic resonance absorption by ENGLAND and SCHNEIDER[72], and their apparatus has been illustrated in *Figure 18*. The principles are exactly the same as for the gaseous case, the bridge being balanced so that no power is fed into the fourth arm of the magic-T or hybrid ring when there is no absorption in the cavity, and any absorption then alters the reflexion coefficient of the cavity and unbalances the bridge. The advantages of using a balanced bridge are (i) either absorption, or dispersion, can be selected, (ii)

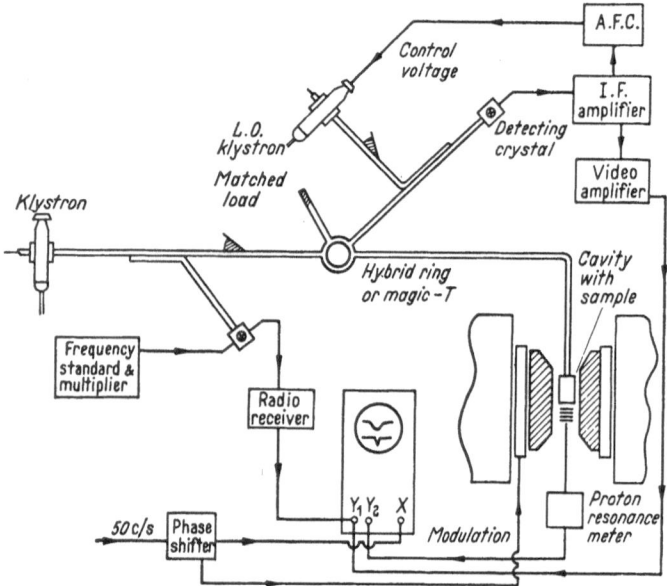

Figure 46. Superheterodyne detection in paramagnetic resonance

the power fed to the detecting crystal can be kept low so as not to saturate the last stages of the intermediate frequency amplifier, and (iii) the use of a 30 Mc/s or 45 Mc/s intermediate frequency eliminates all the low-frequency crystal noise. *Figure 46* shows a diagram of a typical superheterodyne system as employed in paramagnetic resonance absorption together with provision for frequency markers Superheterodyning is much easier in paramagnetic resonance spectroscopes where the microwave frequency is kept constant than in the gaseous case, where continuous change of frequency is required.

(c) *Source modulation.* Source modulation can again be used as an alternative to superheterodyne detection in order to remove the low-frequency crystal noise. The presence of a cavity resonator in the system somewhat complicates the technique, however, as any large change in frequency will cause a drop in power due to the Q curve of the cavity, irrespective of any absorption line. If only narrow lines are to be studied (i.e. width less than one tenth of the width of the cavity Q curve), then a small amplitude modulation of frequency can be employed to obtain the derivative of the line, and a tuned radio-frequency receiver used to amplify this, the derivative of the spectrum being traced out on a cathode-ray tube by the action of the 50 c/s magnetic field sweep.

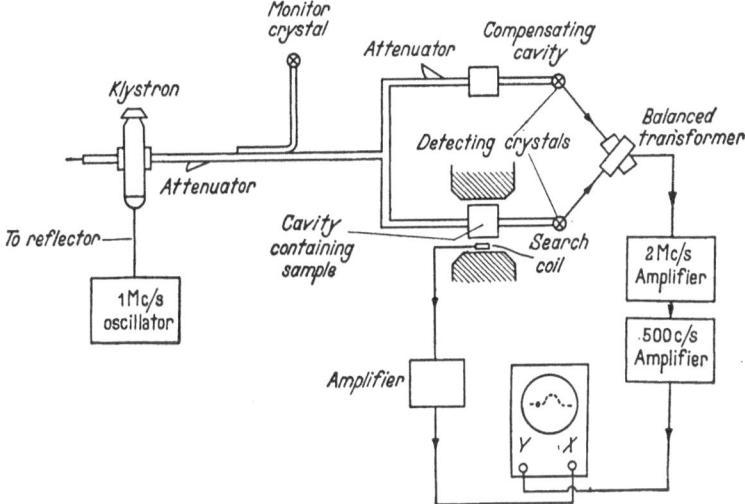

Figure 47. Spectroscope for investigating broad paramagnetic absorption lines

When investigating broad lines, however, it is necessary to balance out the Q curve of the resonator if Source modulation is to be employed. This was done in a spectroscope designed by BAGGULEY[73] which is illustrated in *Figure 47*. Here the effect of change of transmitted power with frequency was compensated by having two identical resonant cavities, and mixing the signals obtained from each in antiphase so that they cancelled. The power from the klystron was divided equally at the T-junction, and after passing through the two resonators and the detecting crystals, was mixed at an out-of-phase transformer. Attenuators were incorporated in the wave-guide arms in order to facilitate initial balancing; and by this means, large modulating voltages ($\sim$40 V) could be applied to the klystron reflector, and yet the only signal passed on to the amplifiers was that due to absorption in the one cavity, its Q curve variation having been eliminated. In practice a 1 Mc/s sweep was applied to the klystron, and a 2 Mc/s amplifier used after the balanced transformer (the absorption line and cavity resonance being swept over twice each cycle). The detected signal was then impressed on a 500 c/s carrier and further amplified, this being incorporated so that a bias could be applied to cancel any signal below a certain level, and was used to compensate for any residual unbalance of the bridge. The magnetic field was not modulated in this case but its value altered by a motor-driven rheostat, changing about 4,000 gauss in 20 seconds. The voltage on the X plates was proportional to the magnetic field strength, so that the spot moved across the screen and traced out the actual absorption line. This was recorded by a camera, the shutter of which was open for the whole period. Such a system appears rather complicated, but is one of the few methods of getting good sensitivity and true line shape when dealing with very broad absorption lines, since it eliminates the low-frequency crystal noise, and the detecting and recording system can have a very small bandwidth.

Another spectroscope designed to study the shapes of wide lines, with high sensitivity, is that used by PORTIS[74]. A narrow bandwidth for the recording system is again obtained by using a long-persistent oscilloscope screen, and a voltage on the X plate which is proportional to the slowly-changing d.c. magnetic field strength. Instead of balancing out the Q variation with frequency, high-frequency *amplitude* modulation is employed. Until recently this was difficult, but the advent of the microwave gyrator [75] has now made it relatively simple. The microwave power is 100 per cent modulated at 6 kc/s, and fed to a balanced bridge, the balancing arm containing a probe, the position and penetration of this being varied to feed either the absorption or dispersion signal onto the detecting crystal, which is followed by a narrow-band 6 kc/s amplifier, and a phase-sensitive detector.

High frequency magnetic field modulation is, at the time of writing, the standard form of high sensitivity detection (see Section 11.1).

4.11 GASEOUS PARAMAGNETIC SPECTROSCOPES

Although most work on paramagnetic resonance has been done with crystalline solids, a few gases are paramagnetic, and some of these have also been studied. BERINGER[35] devised a very sensitive spectroscope for the investigation of oxygen, nitric oxide and nitrogen peroxide which is illustrated in *Figure 48*. This has several features of interest. First a bolometer is used to detect the microwave radiation instead of a crystal; bolometers do not show the large low-frequency noise of crystals, and can also be used at relatively high power levels without introducing excess noise. This means that no balanced bridge is necessary to reduce the average detected power level, and low-frequency modulation can be used, there being no need for superheterodyne detection, or radiofrequency modulation. As can be seen from the figure, the spectroscope incorporates a frequency-

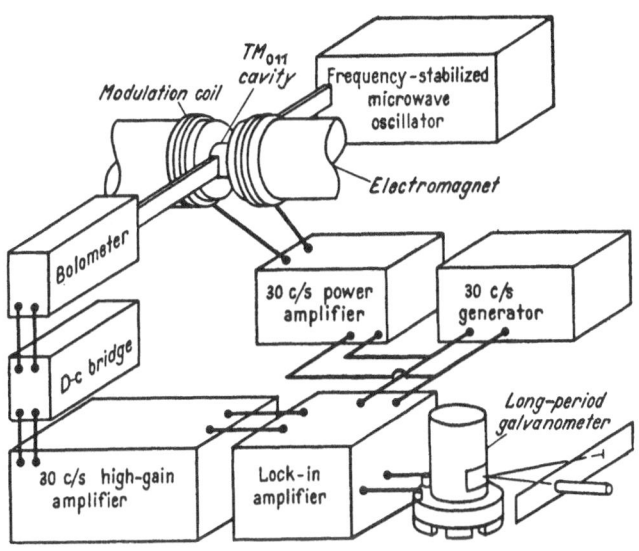

Figure 48. Gaseous paramagnetic resonance spectroscope

stabilized oscillator, and an H_{011} cavity, with 30 c/s magnetic field modulation across it. The bolometer is in a d.c. bridge circuit and feeds into a 30 c/s amplifier, followed by a phase-sensitive detector, the signal being finally applied to a galvanometer of very small bandwidth. The longer the time constant of the recording stage, the more stable must the microwave oscillator be, and it is usually necessary to use Pound stabilization of the frequency if a phase-sensitive amplifier is to be used to its greatest advantage. The results obtained with this spectroscope are discussed in Chapter 6. Until 1954, all gaseous paramagnetic resonance work had been done at Yale University.

4.12 Comparison of Sensitivities

When comparing the sensitivities of the different methods of detection employed in paramagnetic resonance absorption, all the general theory already derived in Sections 4.5(a)–(c) for the gaseous spectroscopes, can be applied. The problem is still one of equating signal strength to input noise power, and the different expressions obtained for superheterodyne detection Source modulation, and simple square-law crystal-video detection still hold. The concept of $\alpha_{g_{min}}$ has to be modified, however, and the minimum susceptibilities that can be detected now replace the actual absorption coefficients. The absorption of microwave power at resonance is due to the imaginary part of the susceptibility, i.e. χ'', where the total susceptibility is written $\chi = \chi' + j \cdot \chi''$.

In order to obtain an expression for the minimum detectable value of χ'', the derivation employed by BLEANEY and STEVENS[76] can be used. The power actually absorbed by the substance, if it is placed in a resonant microwave field $H_1 \cos \omega t$ is equal to $\frac{1}{2}\omega \cdot \chi'' \cdot H_1^2$. The power absorbed in the cavity walls is equal to $\frac{1}{Q_0} \cdot \omega \cdot \frac{H_1^2}{8\pi} \cdot V$, where V would be the total volume if the value of the magnetic field was H_1 throughout the cavity. In practice H varies all round the cavity, and it is thus best to define an 'effective cavity volume', V_{eff}, such that $\int H^2 \, dV = H_1^2 \cdot V_{eff}$. Hence the ratio of power absorbed in the sample to that absorbed by the cavity is $4\pi \, Q_0 \cdot \chi''/V_{eff}$. This ratio is also equal to the fractional change in the resistive component of the cavity impedance, produced by the paramagnetic absorption, i.e. $\left(\frac{\delta r}{r}\right)$. The cavity at resonance can be represented by a tuned circuit of series resistance, r, with one mutual inductance coupling to the oscillator, of power P and internal impedance R, and another to the detector also of input impedance R. When correctly matched, the maximum signal voltage that can be obtained across the detector will be

$$\delta v = (RP)^{\frac{1}{2}} \frac{\delta r}{4r} \qquad \dots (4.21)$$

Which, from the expression for power ratio derived above, gives:

$$\delta v = (RP)^{\frac{1}{2}} \pi Q_0 \chi''/V_{eff} \qquad \dots (4.22)$$

This input signal is then equated to the noise voltage, as before, and by using the expression derived by Beringer[40], as in Section 4.5(c), the following equation is obtained for crystal-video detection:

$$\sqrt{[4\,kT \cdot \Delta f \cdot (R + R_a)]} = (RP)^{\frac{1}{2}} \pi Q_0 \chi''/V_{eff}$$

or $\qquad [4\,N\,kT \cdot \Delta f \cdot R]^{\frac{1}{2}} = (RP)^{\frac{1}{2}} \pi Q_0 \chi''/V_{eff} \qquad \dots (4.23)$

where N is now defined as the noise figure of the receiver embracing the noise sources previously represented by R_a.

Therefore
$$\chi''_{min} = \frac{V_{eff}}{\pi Q_0}\sqrt{\frac{4\,N\,kT.\,\Delta f}{P}} \qquad \ldots(4.24)$$

An effective volume of 3 cc at a wavelength of 3 cm, with a Q_0 of 5,000, gives for χ''_{min}, using oscilloscope presentation

$$\chi''_{min} \approx 2.10^{-12}$$

which can be reduced to 10^{-13} if narrow-band, phase-sensitive detecting systems are employed.

4.13 THE PRODUCTION AND MEASUREMENT OF MAGNETIC FIELDS

For transitions with g values close to the free-electron spin value or 2·0, the magnetic field strengths required for resonance are about 3,500 gauss at 3·2 cm wavelengths, and about 8,000 gauss at 1·25 cm wavelengths. Although the widths of paramagnetic resonance absorption lines are not as small as those encountered in the corresponding nuclear cases, values of less than a gauss, and sometimes less than a tenth of a gauss[77], have been reported, and hence the magnetic field should be homogeneous to a factor of 1 in 10,000. For this reason the largest possible diameter is chosen for the pole faces consistent with the required field strength and size of magnet, and the outer edge of the pole faces has a raised rim in order to compensate for end effects, as was first suggested by ROSE[78]. A typical magnet used for simple resonance work is one supplied by Newport Inst. Ltd., having a U-shaped cast-iron yoke of 62 cm length, which holds two cylindrical steel pole pieces of 10 cm diameter, with the energizing coils placed over these, between the arms. The coils are wound of 16 gauge wire, each having about 3,500 turns, and a current of 8 amps will produce a field of about 5,000 gauss in a gap of 4 cm width between flat pole faces. The pole pieces are interchangeable, conical ones being employed for the shorter wavelengths, although the field homogeneity then drops off more rapidly. These magnets are relatively inexpensive and give a homogeneity better than 1 gauss in most cases. For high precision work on narrow lines a magnet with larger pole faces is required, and ones are used similar to those for nuclear resonance work[66]. See Appendix II for details.

For accurate measurements of spectroscopic splitting factors and line widths, the magnetic field strength needs to be held constant to a high degree. Batteries can sometimes be used for this, although a slow drift is usually still present, and it is generally more satisfactory to stabilize the current by some automatic means, either by the use of an electronic circuit to control a compensating current, or an automatic stabilizing system in the field coils of the generator, which supplies the power to the magnet coils[79]. A system inaugurated by Packard[80] uses a proton resonance signal to control the magnet current, and can hold the field steady to better than 0·02 gauss.

In measurements on single crystals the spectra observed usually vary quite markedly with the angle between the applied magnetic field and the crystallographic axes, and hence the apparatus is usually constructed in such a way that the crystal can be moved relative to the field, when *in situ* in the cavity, so that the whole system does not have to be recooled between each setting. Methods used to do this have been discussed in Section 3.9 and illustrated in *Figures 37* and *38*, the crystal being rotated in the resonator while the direction of the magnetic field remains fixed. In some cases it is better to rotate the magnetic field, however, and this applies especially in work at lower wavelengths where it becomes increasingly difficult to design a cavity with the central tube sufficiently large. The cavity tends to become de-tuned when the crystal is moved and its Q value is also lowered by the large central hole, and hence for work at 8-mm wavelengths and below, it is normal to have the magnet mounted on thrust bearings, so that it can be rotated round the cavity. The crystal then remains fixed in position inside, and there is no central tube down to the cavity.

Since the magnetic field strengths give a direct measure of the g values, and electronic or hyperfine splittings, it is necessary to measure their values with the highest possible accuracy. Proton resonance techniques are always employed for this now, a coil containing some hydrogenous material (waxes, oils or aqueous solutions) being placed in the same magnetic field, and the frequency required for absorption in this, due to the proton spin change, is then measured. The gyromagnetic ratio of the proton has been determined[81] to at least 1 part in 10^5, and the value of the magnetic field can be calculated from the resonant frequency by the equation

$$H = 2·3487 . 10^{-4} . \nu \text{ gauss} \qquad ....(4.25)$$

where ν is the frequency in c/s. This transition is one involving nuclear spin, and not electronic spin, and hence the energy separations are about a thousand times smaller, and the resonant frequencies are now in the 10–50 Mc/s region. A large amount of work[82, 83, 84] has been done on different circuit techniques for investigating nuclear resonance, and these are discussed in more detail in Chapter 8. Since the proton resonance signal is a strong one, quite simple circuits can be employed for its detection[84, 85], and the author and his co-workers have found that a simplified version of POUND and KNIGHT's[85] spectrometer gives very good results, being better than the super-regenerative circuit, incorporating a 6 AK5, previously employed. The circuit diagram is shown in *Figure 49*, and it consists essentially of two cathode-coupled triodes which are kept near the verge of oscillation by adjustment of the potentiometer. The magnetic field strength is swept, and amplitude modulation of the oscillator power is produced as the absorption line is

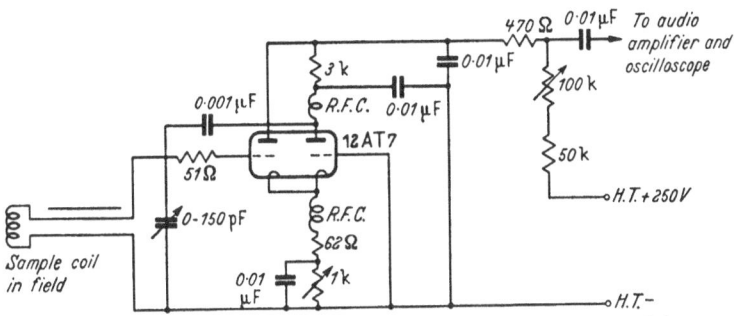

Figure 49. Proton resonance magnetic field meter (after Pound and Knight[85])

passed through. The signal is taken from the H.T. line as shown, only the audio modulation coming through the radiofrequency choke. This signal can then be fed into the same amplifier as the detected microwave signal, via a different input attenuator, so that both signals are obtained on the same trace, or, alternatively, it can be separately amplified and fed onto the second beam of the cathode-ray oscilloscope. Very accurate measurement of the splittings between spectral lines can then be obtained by just moving the proton resonance signal across the screen from coincidence with one line to coincidence with another, this being effected by slowly changing the condenser setting of the proton resonance circuit. The actual frequencies are determined by comparison with a crystal-controlled wavemeter, and hence the magnetic field strength can be obtained to an accuracy of 1 part in 10^5, or better, by this means. Solutions of ferric nitrate or manganese sulphate are suitable samples for use in the coils, the addition of small amounts of paramagnetic ions to the aqueous solutions altering the relaxation times to give narrower lines[85].

For accurate determination of the spectroscopic splitting factors one also needs to know the microwave frequency to 1 part in 10^6, or so; this can be determined by the use of a microwave frequency standard of exactly the same design as those described in Section 2.9, a cavity wavemeter giving values accurate to 1 part in 10^4, and indicating which crystal harmonic is beating with the microwave radiation.

4.14 LOW-TEMPERATURE TECHNIQUE

Low temperatures are used to a very large extent in work on paramagnetic resonance absorption, temperatures of 90° K, or lower, often being required in order to decrease the spin-lattice interaction (i.e. increase the relaxation time) so that the line width is not broadened by this effect. In most cases the apparatus and technique employed have been very simple and direct, a silvered evacuated dewar flask, full of the liquid refrigerant, just being raised up over the cavity resonator so that the whole of it is completely immersed. One of the great advantages of paramagnetic resonance, as compared with other low-temperature measurements giving similar information (e.g. susceptibility and specific heat determinations) is that the thermal capacity of the part of the apparatus to be cooled is very small, and the results are available 'instantaneously', no long run being necessary in order to make corrections for heat losses, or the like. This means that relatively small amounts of the liquid refrigerant are required, and a large number of results can be obtained with simple apparatus. The types of cavity resonator suitable for low-temperature work have already been described and are illustrated in *Figures 38* and *39*. The use of thin-walled nickel silver for the waveguide, or coaxial line, reduces conduction losses considerably, and a dewar flask containing about 150 cc of liquid hydrogen will keep a 1·25 cm wavelength resonator, as illustrated in *Figure 38*, at 20° K for about an hour (the 150 cc including the amount boiled off when initially cooling from liquid air to liquid hydrogen temperatures). For work at 1·25 cm wavelength, or lower, the dewar flask is usually tailed so that the magnet gap can be made as small as possible, the wider diameter at the top of the dewar acting as a storage space, and hence the time between refillings is longer.

Work at liquid helium temperatures usually involves more complex apparatus, although in the U.S.A. the simple direct method, of immersing the resonator directly into a dewar of liquid helium, has been used. In this country, where the helium gas has to be conserved, more elaborate methods have been employed[86, 68], the latest designs[87] arranging that all the waveguide, and other connexions from room temperature, go through a chamber containing liquid hydrogen, thus reducing heat flow to the liquid helium. Most measurements have been made at the fixed temperatures afforded by boiling helium, hydrogen or oxygen (i.e. 4°, 20° and 90° K), but intermediate temperatures can be readily obtained by having a small heating coil wound round the cavity, together with a thermocouple, attached as close to the crystal position as possible. Generally speaking, low-temperature techniques are very simple in experiments on paramagnetic resonance, and temperature is the only parameter that has to be measured in the cold end of the apparatus.

REFERENCES

[1] GESCHWIND, S. *Ann. N.Y. Acad. Sci.* 55 (1952) 752

[2] BLEANEY, B. and PENROSE, R. P. *Proc. Phys. Soc.* 60 (1948) 83

[3] STRANDBERG, M. W. P. and DRIECER, H. *Phys. Rev.* 94 (1954) 1393; and GORDON, J. P., ZEIGER, H. J. and TOWNES, C. H. *ibid.* 95 (1954) 282

[4] NEWELL, G. and DICKE, R. H. *ibid.* 83 (1951) 1064

[5] VAN VLECK, J. H. and WEISSKOPF, V. F. *Rev. mod. Phys.* 17 (1945) 227

[6] BLEANEY, B. and PENROSE, R. P. *Proc. phys. Soc.* 59 (1947) 418; *ibid.* 60 (1948) 540

[7] BLEANEY, B. and LOUBSER, J. H. N. *Nature* 161 (1948) 522

[8] TOWNES, C. H. *Phys. Rev.* 70 (1946) 665

[9] BLEANEY, B. *Rep. Progr. Phys.* 11 (1948) 200

[10] ANDERSON, P. W. *Phys. Rev.* 76 (1949) 647

[11] MARGENAU, H. *ibid.* 76 (1949) 1423

[12] SMITH, W. V. and HOWARD, R. *ibid.* 77 (1950) 840; *ibid.* 79 (1950) 132

[13] PURCELL, E. M., TORREY, H. C. and POUND, R. V. *ibid.* 69 (1946) 37

[14] BLEANEY, B. and PENROSE, R. P. *Proc. phys. Soc.* 60 (1948) 83

[15] FROHLICH, H. *Nature* 157 (1946) 478

[16] PAULING, L. and WILSON, E. B. *Introduction to Quantum Mechanics* McGraw-Hill New York (1935) Ch. II, eqn. 40—12

[17] BLEANEY, B. and PENROSE, R. P. *Proc. phys. Soc.* 60 (1948) 88

[18] KARPLUS, R. and SCHWINGER, J. *Phys. Rev.* 73 (1948) 1020

[19] TOWNES, C. H. *ibid.* 70 (1946) 665

[20] SNYDER, H. S. and RICHARDS, P. I. *ibid.* 73 (1948) 1178

[21] KARPLUS, R. *ibid.* 73 (1948) 1120

[22] — *ibid.* 74 (1948) 223

[23] GESCHWIND, S. *Ann. N.Y. Acad. Sci.* 55 (1952) 753

[24] KARPLUS, R. *Phys. Rev.* 73 (1948) 1027

[25] TOWNES, C. H. and MERRITT, F. R. *ibid.* 72 (1947) 1266

[26] CLEETON, C. E. and WILLIAMS, N. H. *ibid.* 45 (1934) 234

[27] BLEANEY, B. and PENROSE, R. P. *Nature* 157 (1946) 339

[28] GOOD, W. E. *Phys. Rev.* 70 (1946) 109, 213

[29] HUGHES, R. H. and WILSON, E. B. *ibid.* 71 (1947) 562

[30] STRANDBERG, M. W. P., WENTINK, T. and KYHL, R. L. *ibid.* 75 (1949) 270

[31] GILLIAM, O. R., JOHNSON, C. M. and GORDY, W. *ibid.* 78 (1950) 140

[32] GESCHWIND, S. *Ann. N.Y. Acad. Sci.* 55 (1952) 751

[33] GUNTHER-MOHR, G. R., WHITE, R. L., SCHAWLOW, A. L., GOOD, W. E. and COLES, D. K. *Phys. Rev.* 94 (1954) 1184

[34] STRANDBERG, M. W. P., JOHNSON, H. R. and ESHBACH, J. R. *Rev. sci. Instrum.* 25 (1954) 776

[35] BERINGER, R. and CASTLE, J. G. *Phys. Rev.* 78 (1950) 581

[36] TOWNES, C. H. and GESCHWIND, S. *J. appl. Phys.* 19 (1948) 795

[37] GORDY, W. *Rev. mod. Phys.* 20 (1948) 668

[38] MILLER, S. E. *Proc. Inst. Radio Engrs. N.Y.* 35 (1947) 347

[39] VAN VOORHIS, S. N. *Microwave Receivers* (1948) 11

[40] BERINGER, E. R. Radiation Laboratory Report (1944) 638

[41] TORREY, H. C. and WHITMER, C. A. *Crystal Rectifiers* (1948) 344

[42] COLES, D. K. and GOOD, W. E. *Phys. Rev.* 70 (1946) 979

[43] JEN, C. K. *ibid.* 74 (1948) 1396

[44] — *ibid.* 76 (1949) 1494

[45] — *Ann. N.Y. Acad. Sci.* 55 (1952) 822

[46] BAIRD, D. H. and BIRD, G. R. *Rev. sci. Instrum.* 25 (1954) 319

[47] BIRD, G. R. *ibid.* 25 (1954) 324

[48] BERINGER, R. *Phys. Rev.* 70 (1946) 53

[49] STRANDBERG, M. W. P., MENG, C. Y. and INGERSOLL, J. G. *ibid.* 75 (1949) 1524

[50] BLEANEY, B. and PENROSE, R. P. *Proc. Roy. Soc.* A 189 (1948) 358

[51] CUMMEROW, R. L. and HALLIDAY, D. *Phys. Rev.* 70 (1946) 433

[52] BAGGULEY, D. M. S. and GRIFFITHS, J. H. E. *Nature* 160 (1947) 532

[53] PENROSE, R. P. *ibid.* 163 (1949) 992

[54] BLEANEY, B. and INGRAM, D. J. E. *ibid.* 164 (1949) 116

[55] KRONIG, R. de L. *Physica* 6 (1939) 33

[56] WALLER, I. *Z. Phys.* 79 (1932) 370

[57] LAPORTE, O. *ibid.* 47 (1928) 761

[58] VAN VLECK, J. H. *Phys. Rev.* 57 (1940) 426

[59] — *ibid.* 74 (1948) 1168

[60] PRYCE, M. H. L. and STEVENS, K. W. H. *Proc. phys. Soc.* A 63 (1950) 36

[61] BLEANEY, B., BOWERS, K. and INGRAM, D. J. E. *Proc. phys. Soc.* A 64 (1951) 758

[62] BAGGULEY, D. M. S. and GRIFFITHS, J. H. E. *Nature* 162 (1948) 538 and *Proc. Roy. Soc.* A 201 (1950) 366

[63] ANDERSON, P. W. and WEISS, P. R. *Rev. mod. Phys.* 25 (1953) 269

[64] INGRAM, D. J. E. and BENNETT, J. E. *J. chem. Phys.* 22 (1954) 1136

REFERENCES

[65] SCHNEIDER, E. E. and ENGLAND, T. S. *Physica* 17 (1951) 221; and ESCHENFELDER, A. H. and WEIDNER, R. T. *Phys. Rev.* 92 (1953) 869; and PORTIS, A. M. *Phys. Rev.* 91 (1953) 1071

[66] GUTOWSKY, H. S., MEYER, L. H. and McCLURE, R. E. *Rev. sci. Instrum.* 24 (1953) 644

[67] LEVINSTRAL, E. C., ROGERS, E. H. and OGG, R. A. *Phys. Rev.* 83 (1951) 182

[68] BIJL, D. *Proc. phys. Soc.* 63 (1950) 405

[69] BOGLE, G. S., COOKE, A. H. and WHITLEY, S. *Proc. phys. Soc.* A 64 (1951) 931

[70] BAGGULEY, D. M. S. *et al. ibid.* 61 (1948) 542

[71] HAYASHI, I. and ONO, K. *J. phys. Soc. Japan.* 8 (1953) 270

[72] ENGLAND, T. S. and SCHNEIDER, E. E. *Nature* 166 (1950) 437

[73] BAGGULEY, D. M. S. and GRIFFITHS, J. H. E. *Proc. phys. Soc.* A 65 (1952) 594

[74] PORTIS, A. M. *Phys. Rev.* 91 (1953) 1074

[75] HOGAN, C. L. *Bell, S. Tech. J.* 31 (1952) 1

[76] BLEANEY, B. and STEVENS, K. W. H. *Rep. Progr. Phys.* 16 (1953) 107

[77] LEVINSTRAL, E. C., ROGERS, E. H. and OGG, R. A. *Phys. Rev.* 83 (1951) 182

[78] ROSE, M. E. *ibid.* 53 (1938) 715

[79] HUTCHINSON, C. A. *Tech. Report. I. Univ. Chicago* (1951) 64

[80] PACKARD, M. E. *Rev. sci. Instrum.* 19 (1948) 435

[81] THOMAS, H. A., DRISCOLL, R. L. and HIPPLE, J. A. *Phys. Rev.* 75 (1949) 902

[82] BLOCH, F., HANSEN, W. W. and PACKARD, M. E. *ibid.* 70 (1946) 474

[83] BLOEMBERGEN, N., PURCELL, E. M. and POUND, R. V. *ibid.* 73 (1948) 679

[84] HOPKINS, N. J. *Rev. sci. Instrum.* 20 (1949) 401

[85] POUND, R. V. and KNIGHT, W. D. *ibid.* 21 (1950) 219

[86] BOGLE, G. S., COOKE, A. H. and WHITLEY, S. *Proc. phys. Soc.* A 64 (1951) 931

[87] BLEANEY, B. and STEVENS, K. W. H. *Rep. Progr. Phys.* 16 (1953) 123

5

RESULTS AND THEORY OF GASEOUS SPECTROSCOPY

5.1 INTRODUCTION

THE results obtained from normal microwave gaseous spectroscopy are considered in this chapter, and since it is the dipole moment of the molecule as a whole that interacts with the microwave field, the energy states giving rise to the transitions are molecular states and levels, not those associated with just one atom. Molecular spectroscopy covers a wide range of frequencies, since moments of inertia and other molecular parameters vary by several orders of magnitude, and in essence the same kind of spectra are observed in the microwave region as in the infra-red. The former, however, is generally concerned with pure rotational energy levels, whereas the higher energy changes in the latter correspond to vibrational as well as rotational transitions. The great advantage of microwave spectroscopy, compared with normal infra-red work, is that very high resolution is attainable and the frequency of the spectral lines can be measured to a high degree of accuracy with corresponding precision in the determination of such parameters as bond lengths and angles. Microwave spectroscopy is also much more sensitive than corresponding infra-red work, and hence is ideally suited for measurements on very small quantities of gas, such as compounds containing artificial radioactive elements.

The results obtained from microwave spectroscopy can be divided into two main groups—i.e. (i) those which give detailed data of a chemical nature on the molecular binding and structure, and (ii) those which give information on the nuclei of the individual atoms, this being obtained from the hyperfine structure which arises from the interaction of the nuclear quadrupole moment with the electric field produced by the rest of the molecule. One other very practical application of these results is the information that they give on free-space microwave propagation under different atmospheric conditions. During the war this was a major problem, considerable work being done on it, and a brief outline of the absorption spectra obtained from water vapour and oxygen is given below, as it also serves as a good introduction to the other results.

(a) *Water vapour.* The absorption due to water vapour has been of great practical interest, and even before the techniques of microwave spectroscopy had been developed, various measurements at atmospheric pressure were made. One of the first was by means of the DICKE radiometer[1] which was used to study the thermal radiation coming in through the atmosphere. This instrument, which was a form of emission spectroscope, obtained high sensitivity by employing a motor-driven attenuator in the receiving waveguide which modulated the input signal at 30 c/s. A very narrow-band phase-sensitive amplifier was then used after the microwave stage, and the resultant increase in signal-to-noise ratio made it possible to detect the 'thermal radiation', received at the aerial, above the internal noise of the crystal and receiver. The signal intensity obtained was measured throughout the wavelength region in which the water vapour line was expected, and by correlating the results with the measured humidity in the atmosphere above the radiometer the presence of the water vapour absorption was detected. These results indicated that the absorption occurred at about 23,000 Mc/s (1·3 cm wavelength), with a width of 3,000 Mc/s (0·1 cm^{-1}).

Absorption measurements at atmospheric pressure were also made by BECKER and AUTLER[2] using a very large 'multiple-mode cavity'. This was especially designed to be non-resonant, taking the form of an 8-ft. cube containing copper blades which were rotated to 'spread' the input power throughout the cavity, and in this way the modes were mixed to form a continuous band covering several thousand Mc/s. The average energy present in the cavity was then measured for a steady rate of input power, the energy present being inversely proportional to the total losses. The energy in the cavity was determined by over 300 thermocouples distributed throughout its volume, and owing to the high Q of such a cavity (800,000), the $\alpha_{g_{min}}$ detectable was of the order of 10^{-6} per cm. By taking a series of measurements at eleven different frequencies the extra loss produced by the water vapour was found, and a value of 22,320 Mc/s (1·34 cm) was obtained for the centre of the absorption line, with a width of 2,160 Mc/s (0·087 cm^{-1}).

These initial war-time results were confirmed by TOWNES and MERRITT[3], using the standard waveguide techniques of microwave spectroscopy, and they determined the centre of the absorption line as 22,237 ± 5 Mc/s (1·348 cm) with a line breadth of 1·45 Mc/s at a pressure of 0·1 mm of Hg. Recently a second water vapour absorption line has been located by GORDY and KING[4] at 183,311 · 30 ± 0·3 Mc/s (1·64 mm), these two being the only ones expected throughout the microwave region; and the techniques employed in determining this second line are a good example of those used in millimetre-wavelength spectroscopy.

The main interest of these particular results is in the deductions that can be drawn from them concerning microwave propagation through the atmosphere, but they also form a good example of the link-up between infra-red and microwave spectroscopy. The water molecule is an asymmetric top molecule, and its rotational spectrum has been extensively investigated[5], most of the lines being well in the infra-red region. The theory of the energy levels associated with asymmetric top molecules will be considered later, and the two lines occurring in the microwave region are from transitions between energy states corresponding to higher rotational levels and denoted, in the usual nomenclature, by $5_{2,3}$ and $6_{1,6}$ for the 1·35 cm wavelength line, and by $2_{2,0}$ and $3_{1,3}$ for the 1·64 mm wavelength line. Although the infra-red measurements had already given a fairly complete picture of the structure of the water molecule, a more accurate determination of its dipole moment was obtained from Stark effect measurements of microwave spectroscopy[6] and the dipole moments and rotational constants of HDO and D_2O were first obtained by this means[7, 8, 9].

(b) *Oxygen.* The absorption due to water vapour is via the coupling mechanism normally encountered in gaseous microwave spectroscopy—i.e. an interaction between the microwave electric field and the rotating electric dipole of the molecule. The case of oxygen, which is the other gas producing considerable absorption in the atmosphere, is different, however, being one of the few gases possessing a permanent magnetic moment, and it is the interaction of the *magnetic* field component of the microwave radiation with this, that forms the absorption mechanism. Transition probabilities for magnetic coupling are much smaller than the corresponding electric coupling, but, even so, the absorption due to oxygen produces a very opaque region in the 5-mm wavelength band.

The ground state of the oxygen molecule is $^3\Sigma$ and by symmetry it has no electric dipole moment, but its magnetic dipole moment of two Bohr magnetons combines with the rotation of the molecule to produce a number of energy levels which can be denoted by a total quantum number, J, which is compounded of the spin quantum number, S, and the rotational quantum number, K. The selection rules $\Delta J = \pm 1$ and $\Delta K = 0$ allow transitions between the $J = (K+1)$ and $J = (K-1)$ states to the $J = K$ state, these being separated by about 2 cm^{-1}, the exact separation depending on the value of K. A series of absorption lines, all lying close to 5-mm wavelength, is thus produced, together with one peak at 2·5 mm wavelength[10]. The first measurements on the microwave absorption of oxygen were the war-time experiments of BERINGER[11], using oxygen at atmospheric pressure in a waveguide transmission system, and obtaining the 5-mm wavelength radiation from crystal harmonics. At this pressure the individual lines were not separated out, and a broad absorption, stretching from 4-mm to 6-mm wavelength, was obtained. By applying the standard techniques of microwave spectroscopy, and working at low pressures, this absorption has been resolved into 26 different lines[12] corresponding to variation of the rotational quantum number, K, from 1 to 25. The precise determination of the frequencies of these lines has also permitted more accurate calculation of the molecular constants and magnetic dipole moment of the oxygen molecule[12]. Microwave spectroscopy has also determined the structure of the O_3 molecule, measurements by TRAMBARULO et al.[13] in the 42 to 118 kMc/s region showing that it has the form of an isosceles triangle of apex angle $116° 49' \pm 30'$ with internuclear distances of $1·278 \pm 0·001$ Å.

These two cases, of water vapour and oxygen, have been quoted first as examples which have a very practical application in relation to microwave propagation through the atmosphere; a more systematic survey of the results, as applied to the determination of molecular structure, and the derivation of nuclear data, is now given.

5.2 MOLECULAR ENERGY LEVELS

As already mentioned, most of the spectra observed in the microwave region are produced by transitions in the rotational energy of the gas molecules. If the energy states can be expressed in terms of the physical parameters of the molecule (bond lengths and angles, and the atomic masses), then determination of these energy levels, from the very accurate measurements of the spectral line frequencies, enables a correspondingly precise determination of the molecular parameters to be made. It is by this means that bond lengths, mass ratios, and the like, have been determined with accuracies of 1 part in 10^4 or greater.

The first step in applying the results to molecular structure analysis is, therefore, to derive expressions for the rotational energy levels. Since rotational energy is always associated with moments of inertia, the important molecular parameters will be the moments of inertia of the molecule about its different principal axes. In the simplest cases—diatomic molecules—all the masses are concentrated along one line, the internuclear axis, and hence there is no moment of inertia about this axis, and the moments of inertia about the other perpendicular axes through the centre of mass are equal. This is also true for any linear polyatomic molecule. If a molecule is not linear, but still has an axis of over two-fold symmetry, with equal moments of inertia about the two principal axes at right angles to this, then the molecule is called a 'symmetric top' type. Molecules with less symmetry than this, having different moments of inertia about all three axes are called 'asymmetric tops' and have the most complicated system of energy levels and microwave spectra.

5.3 LINEAR MOLECULES

The motion of diatomic and linear molecules has been studied in detail for some time, and is often treated as an example in standard works on Wave Mechanics. It can be readily shown[14] that the allowed energy levels are given by the expression

$$E_J = \frac{h^2}{8\pi^2 I} \cdot J(J+1) \qquad \dots (5.1)$$

where I is the moment of inertia about each of the axes perpendicular to the interatomic axes, and J is the quantum number defining the total angular momentum, P, which is actually equal to $\frac{h}{2\pi}\sqrt{J(J+1)}$.

The total angular momentum can make different angles relative to a fixed direction in space (such as would be provided by a strong external magnetic field), as long as the resolved component along this direction is equal to $\frac{h}{2\pi} \cdot M$, M being an integer which varies from $+J$ to $-J$. In the absence of an external field, these $(2J+1)$ different positions become equivalent, all having the same energy, and thus each of the energy levels defined by equation (5.1), is $(2J+1)$-fold degenerate. The different vectors representing the angular rotation of the molecule and the quantum components are illustrated diagrammatically in *Figure 50*.

The selection rule for a transition in the rotational energy is $\Delta J = \pm 1$, and hence the frequency of the absorption line corresponding to a change from J to $(J+1)$, is given by

$$h\nu = \frac{h^2}{8\pi^2 I} [(J+1)(J+2) - J(J+1)]$$

or

$$\nu = 2\left[\frac{h}{8\pi^2 I}\right](J+1). \qquad \dots (5.2)$$

It can be seen that the only parameter which depends on the particular molecule being investigated is I, and hence the frequencies at which the rotational lines will be observed, depend entirely on the moment of inertia of the molecule about an axis perpendicular to the interatomic axis. As an example of this one can take the two cases of cyanogen bromide (BrCN) and hydrocyanic acid (HCN).

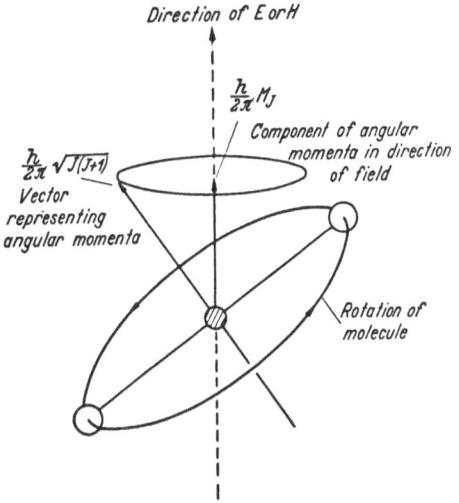

Figure 50. Vector diagram for linear molecule

The BrCN molecule has a moment of inertia about the perpendicular axis of 204 . 10^{-40} gm cm². Therefore, according to equation (5.2), the frequencies of its rotational line spectra should be

$$\nu = 2 . \left[\frac{6 \cdot 624 . 10^{-27}}{8\pi^2 . 204 . 10^{-40}} \right] (J+1)$$
$$= 0 \cdot 8226 . 10^{10} \ (J+1) \ \text{c/s}.$$

The absorption lines corresponding to the first three rotational transitions will therefore occur at 8,226 Mc/s, 16,452 Mc/s and 24,678 Mc/s respectively, and there will be 36 absorption lines between 16,000 Mc/s and 300,000 Mc/s (2 cm to 1 mm wavelengths). In contrast to this, the spectrum of HCN has only three lines in the same frequency range, its moment of inertia being $18 \cdot 9 \times 10^{-40}$ gm cm², and the frequencies of its first two rotational transitions therefore 88,800 Mc/s and 177,600 Mc/s.

It is evident from this that it is the diatomic and linear molecules containing heavier atoms which have most lines in the microwave region, and on which the more complete determinations can be made. If the moment of inertia of the molecule is less than

$15 . 10^{-40}$ gm cm² none of its rotational lines will come into the microwave region (all occurring at wavelengths less than 1 mm).

Equations (5.1) and (5.2) above are only correct to a first order, and have to be modified when making more precise calculations. As the rotational energy of the molecule increases, so its ends tend to move farther apart, and some term to allow for centrifugal distortion must therefore be included. This problem has been treated in detail by the infra-red spectroscopists, and HERZBERG[15] has shown that for the case of diatomic molecules the corrected energy levels can be written as

$$\frac{E}{h} = B . \mathcal{J}(\mathcal{J} + 1) - \frac{4B^3}{\omega^2} . [\mathcal{J}(\mathcal{J} + 1)]^2 \qquad \ldots . (5.3)$$

where B is now written for $h/8\pi^2 I$, as in normal infra-red nomenclature. It is seen that the second term, which represents the correction for centrifugal distortion, varies with the fourth power of $\mathcal{J}$, and also inversely as ω^2, ω being the fundamental vibrational frequency of the molecule—i.e. a parameter governed by its 'elastic restoring forces'. Hence the correction term for centrifugal distortion only becomes appreciable for the higher-order transitions, and if an accurate value of ω is required from these correction factors, it is usually necessary to determine the frequencies of several lines in the millimetre-wavelength region.

In the case of linear polyatomic molecules the correction for centrifugal distortion is not so simple, and the theoretical formulae only show approximate agreement with the experimental results[16, 17]. Another effect which modifies the simple expression for the energy levels is what is known as 'l-type doubling', which arises from the interaction of the rotational energy with the bending vibrational energy. This has been treated theoretically, mainly by NIELSEN[16, 18], and gives rise to a small doublet splitting of the observed lines. Direct transitions between these 'l-type doublets' have been observed by SCHULMAN and TOWNES[19].

The rotational spectra to be expected from diatomic or linear molecules are therefore relatively simple, consisting of a series of nearly equally spaced lines, the separation between each being approximately equal to $h/4\pi^2 I$. Therefore, I can be determined from the measured line frequencies, and for a diatomic molecule the bond length can then be calculated directly from I, and the known nuclear masses. Alternatively, by taking measurements with different isotopes in the same molecule, a very accurate ratio of the isotopic masses can be obtained[20]. The one determination of I is not sufficient to give the internuclear distances in the case of tri- or polyatomic linear molecules, however, and it is necessary to replace some of the nuclei by different isotopes, and thus obtain several different values of I, according as the mass at different points along the molecule is changed. A linear molecule of m atoms requires $(m - 1)$ different isotopic combinations for a complete structure determination. The bond lengths of about twenty linear polyatomic molecules[21, 22] have been determined by this method to date, most of the work on diatomic molecules being on halides[23].

The latest techniques for observing rotational lines are illustrated by the spectrum shown in *Figure 51*. This shows the 8th, 10th, 12th,

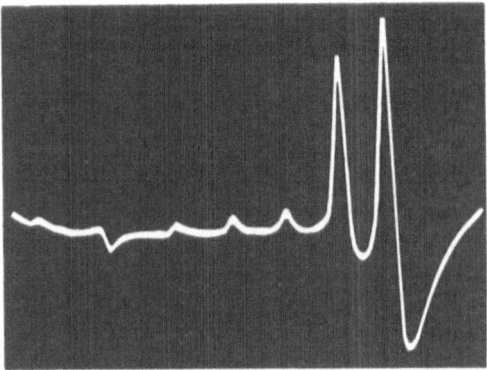

Figure 51. The 8th, 10th, 12th, 14th, 16th, 18th and 20th (from right to left) rotational lines of OCS obtained with the 4th to 10th klystron harmonic. Frequencies range from 97 kMc/s to 243 kMc/s (3·08 mm to 1·23 mm wavelength)

14th, 16th, 18th, and 20th rotational lines of OCS and they are all obtained on one sweep by using the frequencies of the fourth to tenth harmonics of a 2K33 klystron at the same time. This shows the effect of the centrifugal distortion factor very nicely. Thus, if the absorption frequencies were just given by $\nu = 2B(J + 1)$, all the absorption lines would have appeared superimposed, the jump in frequency, corresponding to the next harmonic, being equal to $2B$, which is the right value for the next J transition. The centrifugal distortion term $-4D(J + 1)^3$ separates the different lines, however, as they no longer have a constant difference in frequency between them, and the time-base scale thus corresponds to differences of the absorption line frequency from the nearest $2B(J + 1)$ value, and is not just a linear frequency scale. One result of this method of presentation is that the distortion constant, D, can be determined directly from the splitting between the successive lines, no absolute measurement of frequency being required.

5.4 SYMMETRIC-TOP MOLECULES

As noted already, symmetric molecules are those which have equal moments of inertia about two of their three principal axes. The rotational motion of these molecules is such that, while the whole molecule rotates about its main axis of symmetry, this axis itself precesses about the direction of the total angular momentum vector. Since the electric dipole moment of the molecule lies along its main axis only changes in the precessional energy will cause absorption of the microwave radiation. The quantum mechanical solutions of the rotational states of a symmetric-top molecule are derived in standard works on infra-red spectroscopy[24]. They can be expressed as

$$\frac{E}{h} = \frac{h}{8\pi^2 I_b} \cdot J(J+1) + \frac{h}{8\pi^2}\left[\frac{1}{I_a} - \frac{1}{I_b}\right]K^2 \qquad \ldots .(5.4)$$

where I_a is the moment of inertia along the main symmetry axis, and $I_b (=I_c)$ is the moment of inertia about the other two perpendicular

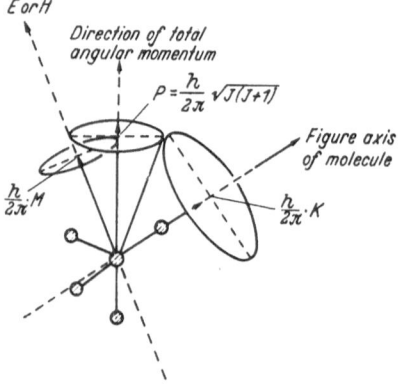

Figure 52. Vector diagram for symmetric tops

principal axes. J is the total angular momentum quantum number as before, and $(h/2\pi) . K$ is the component of angular momentum resolved along the main axis. The angular momentum can still take up different positions relative to a direction fixed in space, the value of the resolved component being $(h/2\pi) . M$, where M is the 'magnetic quantum number'. The relation of these different vectors is illustrated diagrammatically in *Figure 52*.

It can be seen from equation (5.4) that equal positive and negative values of K give the same energy values, and since, in the absence of an external field, each K level is also $(2J+1)$-fold degenerate, due to the different values of M, it follows that the total degeneracy for any level with $K \neq 0$ is $2(2J+1)$.

The above equation only applies exactly to a rigid molecule, and, as in the case of linear molecules, a correction has to be applied for the centrifugal distortion which increases with rotational energy. Theoretical expressions for the distortion constants have been derived[25] in a similar way to those obtained for diatomic and linear molecules, but they have not been confirmed very well by the experimental measurements. It is usual, instead, to represent them by semi-empirical factors D_J and D_K, the former arising from the precessional motion of the whole molecule, and the latter from rotation about its own axis, a cross-term D_{JK} is also included.

The energy levels can then be expressed as[25, 26]

$$\frac{E}{h} = BJ(J+1) + (A-B)K^2 - D_J J^2(J+1)^2 - D_{JK}J(J+1)K^2 - D_K K^4 \quad \dots (5.5)$$

where

$$A = \frac{h}{8\pi^2 I_a}, \quad B = \frac{h}{8\pi^2 I_b}$$

The selection rule for transitions in the rotational energy are $\Delta J = \pm 1$ and $\Delta K = 0$, so that the frequencies of the absorption line for the $J \to (J+1)$ transition are given by

$$\nu = 2B(J+1) - 4D_J \cdot (J+1)^3 - 2D_{JK} \cdot (J+1)K^2 \quad \dots (5.6)$$

The last two terms are very small compared with the first, and hence a series of nearly equally-spaced absorption lines, separated by $h/4\pi^2 I_b$, is to be expected—i.e. exactly the same as for the case of linear molecules. No term involving I_a appears in this expression, because, as already seen, no change in the rotational energy of the molecule about its own axis can absorb energy from the microwave field. The interaction with the radiation is via the electric dipole moment, and hence associated with the precessional motion of the whole molecule.

The main effect of the centrifugal distortion is to remove the degeneracy of the different K states. Without the distortion only one line would appear for each different transition of the principal quantum number, J, because of the $\Delta K = 0$ selection rule. Owing to the cross-term in the centrifugal distortion factor, D_{JK}, the momentum resolved along the axis does show up, however, and as a result each of the main transitions is split into $(J+1)$ components, with a separation between each component of the order of $0\cdot5$ Mc/s[27].

As in the case of linear molecules, a separation of the lines also

117

occurs due to 'l-type doubling', produced by the interaction of the rotational energy with the bending vibrational energy. H. Nielsen[18] first developed the detailed theory of 'l-type doubling', as applied to symmetric tops, in order to explain the extra lines observed by RING, EDWARDS and KESSLER[28] in the microwave spectra of CH_3CN and CH_3NC. His theory predicted that the $J = 1 \rightarrow 2$ transition would be split into two lines separated by about 100 Mc/s, with another two lines at the centre of this doublet, with a separation of 2 Mc/s. In the more general case the splitting of the outer doublet is of the order of $4B^2/\omega\,(J + 1)$, and the number of lines at the centre is equal to $2J$. (J is always taken as the quantum number of the *lowest* state in microwave absorption spectroscopy.)

The absorption spectra of a large number of symmetric-top molecules have now been obtained and analysed, and their molecular structure determined by this means. As in the case of linear molecules, a series of measurements using different isotopes has to be made in order to evaluate the actual internuclear distances from the measured values of I_b. It has already been seen that the value of I_a cannot be determined from the microwave results, but the ratio of I_b/I_a can be derived from infra-red measurements[29] and hence, once I_b is known, I_a can also be evaluated.

The first symmetric-top molecules studied in detail were the methyl halides[30], and a systematic comparison of the different carbon-halogen bond lengths was possible, the very short C–F distance of 1·385 Å obtained in methyl fluoride indicating a considerable amount of 'double-bond' character. These results, together with a large number of similar cases, are an illustration of the very useful new information that microwave spectroscopy can bring to chemical theory. Bond lengths can be determined accurately to better than $\pm 0·005$ Å, bond angles to $\pm 30'$, and detailed data concerning higher vibrational states can often be obtained as well. Symmetric-top molecules that have been analysed in detail include: SiH_3-halides[31]; CF_3-halides[32]; $(CH_3)_3C$-halides[33]; Cs-halides[34]; CH_3Hg-halides[35]; GeH_3-halides[36]; As-halides[37]; P-halides[37] and the CH-halides[38].

The reason why so many halides have been studied is that molecules containing a halide radical often exist in a gaseous form; a fact which illustrates one of the shortcomings of this type of spectroscopy, namely, that the only substances which can be investigated are gases or those having a reasonable vapour pressure at the temperature of the absorption cell. This limitation should be considerably reduced, however, by construction of the high-temperature absorption cells, which is now in progress[39, 72, 73].

5.5 ASYMMETRIC-TOP MOLECULES

Asymmetric-top molecules are those having unequal moments of inertia about all three principal axes, and include the great majority of naturally occurring molecules. Owing to the lack of symmetry, exact derivation of their rotational energy levels becomes very difficult in the general case; there are two main differences from the symmetric-tops: (i) the asymmetry removes the degeneracy of the different K levels, and instead of there only being one line for each J transition, each level is now split into $(2J + 1)$ sub-levels of different energy, and the selection rule corresponding to $\Delta K = 0$ no longer applies, (ii) the effect of centrifugal distortion is much larger, the splitting due to this often being greater than 100 Mc/s.

If exact theoretical determination of the energy levels is attempted, algebraic equations are obtained of a degree which increases with increasing J value[40], and hence the actual energy levels can only be derived for low J transitions. An alternative method of approach is to treat the molecule as a symmetric-top, and the asymmetry as a perturbation on this[41], a method which has had considerable success in some cases. This method has been used by GOLDEN[42] in the case of high value J transitions, which are very difficult to solve exactly, and he was able to show that for high J and low K values, the energy levels of the asymmetric-top can be correlated with solutions of Mathieu's equation. This means that the energy levels can be derived directly from the existing tabulated values of these solutions.

WANG[43] has used the correspondence principle to express the energy levels of an asymmetric-top in a form similar to that for the symmetric-top case

i.e.
$$\frac{E}{h} = \tfrac{1}{2}(B + C) J(J + 1) + [A - \tfrac{1}{2}(B + C)] W_\tau \quad \dots (5.7)$$

where $C = \dfrac{h}{8 . \pi^2 . I_c}$ and W_τ, which corresponds to the K^2 of the symmetric-top case, is now a very complex function of the three moments of inertia, and gives rise to the $(2J + 1)$ sub-levels for each J energy state. Other expressions[44, 45] can be used just as conveniently, and the best analytical method to apply to any particular case will depend on which J-value transitions are involved, and how close the molecule is to a symmetrical type.

The selection rule for asymmetric-top rotational levels is $\Delta J = 0, \pm 1$ and, owing to the possible admixture of sub-levels of different J, all three types of transition can be obtained in the microwave region. The spectra of these molecules is thus extremely

119

complicated, and analysis and allocation of energy levels is often very difficult. The extra information obtained from Stark effect measurements is very useful in this connexion, since different types of splitting are obtained for different transitions, as explained later in the chapter. For a detailed treatment of the energy levels and allowed transitions in the rotational states of asymmetric top molecules, reference should be made to one of the standard works on infra-red spectra, such as Herzberg's *Infra-Red and Raman Spectra*, pp. 42–60[54]. Since the subject rapidly becomes very specialized and complex it will not be treated in any further detail here.

In spite of the complexity of the results, and the difficulty in interpreting them, the structure of a considerable number of asymmetric top molecules have been determined by microwave spectroscopy, including SO_2[46]; HDO[47]; HDS[48]; CH_2O[49]; CH_2CF_2[50]; CH_2Cl_2[51]; $HNCO$[52] and $HNCS$[53].

5.6 Inversion Spectra

Although inversion is not the simplest of transitions in molecular structure, it has been of great importance in microwave spectroscopy

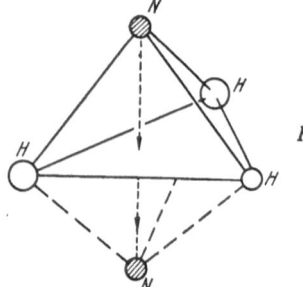

Figure 53. Inversion of ammonia molecule

owing to the relatively intense absorption which it produces in the case of ammonia. The inversion spectrum of ammonia was the first to be observed in the microwave region[55], and it has been studied in detail ever since, to provide information on pressure broadening, quadrupole splitting, Zeeman effects and the like. The most recent examples being the use of molecular beams of ammonia to reduce the Doppler width[56], and also to give emission spectra when fired into a highly tuned cavity[57]. The structure of the ammonia molecule is illustrated in *Figure 53*, in which the two equally-possible positions of the nitrogen atom are shown on either side of the plane formed by the three hydrogens. (In the more general case of any non-planar molecule it can be shown that an identical structure can always be obtained by inverting the position of all the atoms through the centre of mass of the molecule; and the same transformation cannot be obtained by any combination of rotations about the molecular axes.)

It is therefore possible for the nitrogen atom to move from one equilibrium position to the other, but before it can cross the median plane it must overcome the potential hill formed by the closer proximity of the hydrogens. The height of this potential barrier determines the frequency region in which the inversion spectrum will be found, and in some molecules the potential barrier is so high that no transitions occur at room temperature, both the molecular forms existing separately, sometimes with different physical anisotropies. If, however, the potential hill is low the molecule will resonate between the two structures, and two energy levels are obtained from the linear combination of the wave functions corresponding to the two cases. The splitting of these two levels was measured indirectly from the infra-red measurements of WRIGHT and RANDALL[58], who showed, from the doublet splitting obtained in the rotational lines, that an absorption near 0.8 cm^{-1} was to be expected from transitions between the inversion levels. This absorption was that first detected by CLEETON and WILLIAMS[55] and later so thoroughly studied by BLEANEY and PENROSE[59] and other workers. On the simplest theory only one absorption line is to be expected, corresponding to the difference between the two energy levels of the resonant state, but owing to effects of centrifugal distortion a very complex spectrum is obtained as shown in *Figure 3*. The splitting of the energy levels, and hence the frequency of the inversion line, is very sensitive to change of height of the potential barrier, which in turn depends on the position of the atoms. Centrifugal distortion, produced by its molecular rotation, changes the interatomic distances and hence shifts the resonance frequency, and thus a series of absorption lines are obtained corresponding to the different rotational states. Rotation of the molecule about its symmetry axis (K quantum number) will cause the hydrogen atoms to move outwards, and thus reduce the potential barrier, whereas 'end-over-end' rotation of the molecule increases the height of the potential barrier and decreases the inversion frequency.

Over 60 different lines have now been observed in the inversion spectrum of $N^{14}H_3$, corresponding to J and K values up to 15, and these can nearly all be fitted to an equation of the form[60]

$$\nu = \nu_0 \exp\left[a \cdot J(J+1) + b \cdot K^2 + C \cdot J^2(J+1)^2 \right. \\ \left. + d \cdot J(J+1)K^2 + e \cdot K^4\right] \quad \ldots (5.8)$$

A large number of inversion lines have also been obtained for the two asymmetric tops NH_2D and NHD_2[61] and the inversion of ND_3 has also been observed at 2,000 Mc/s[62], a detailed summary and analysis of these results has recently been given by WEISS and STRANDBERG[61].

The inversion spectrum of ammonia has served as a very fruitful field for the detailed investigation of pressure and saturation broadening. These have been discussed briefly in Section 4.2, and it was noticed that the line width should depend markedly on the particular transitions involved.

The agreement between the detailed theory of this effect, and the experimental results of Bleaney and Penrose[59], is very striking, as shown in a paper by ANDERSON[63]. So far ammonia and its isotopic counterparts are the only molecules for which inversion spectra have been obtained in the microwave region. Calculations for such molecules as PH_3 indicate that its inversion is probably at very long wavelengths, and at the moment no other molecules are known which might be expected to show inversion in the wavelength region between 1 mm and 10 cm.

5.7 HINDERED ROTATION

If a molecule contains two or more groups which can rotate against each other about a common molecular axis, a mode of torsional vibration is possible, and this is usually termed 'hindered rotation'. The presence of this type of motion sometimes affects the microwave absorption spectrum to a considerable extent, the two important factors being the height of the potential barrier to the internal rotation, and the presence or absence of a component of the dipole moment at right angles to the axis of the rotating groups.

The theory of hindered rotation in symmetric top molecules has been considered by NIELSEN[64] and by KOEHLER and DENNISON[65], and one of the main effects of the internal rotation is to modify the $\Delta K = 0$ selection rule so that a series of new transitions are allowed. The total rotational energy of the molecule now becomes

$$\frac{E}{h} = BJ(J+1) - BK^2 + \alpha_1 k_1^2 + \alpha_2 k_2^2 \qquad \dots (5.9)$$

where $\alpha_1 = \dfrac{h}{8 \cdot \pi^2 \cdot I_1}$ and $\alpha_2 = \dfrac{h}{8 \cdot \pi^2 \cdot I_2}$ and I_1 and I_2 are the moments of inertia of the two rotating groups about the molecular axis, and k_1 and k_2 are the quantum numbers for the angular momentum of these two groups. The motion of the two groups relative to each other will depend on the ratio of the internal molecular energy to the height of the opposing potential barrier. If this is large, then free rotation will occur, whereas if the ratio is small only torsional oscillation will take place. For the case of free rotation, and if one group possesses a component of dipole moment perpendicular to the axis, it can be shown that $\Delta K = \pm 1$, $\Delta k_1 = \pm 1$ and $\Delta k_2 = 0$ transitions are allowed, where k_1 is the quantum number of the group possessing the dipole moment component. The frequencies of the absorption lines then become:

$$\nu = 2\alpha_1 k_1 + \alpha_1 - 2BK - B \qquad \ldots (5.10)$$

and since α_1 is greater than B, the spectrum should consist of groups of lines separated by $2\alpha_1$, each group containing lines separated by $2B$.

The first spectrum obtained from a molecule possessing hindered rotational levels was that of methyl alcohol[66], and the allocation of some of these lines to such a motion was suggested by their linear Stark effect. Microwave spectra showing the presence of hindered rotation have also been obtained from CH_3CF_3[67], CH_3SiF_3[68] and CH_3SF_5[69], and a detailed analysis of these spectra has given accurate values for the height of the internal potential barrier. Since the actual nature of the interaction between the internal rotating groups is still very uncertain, further determination of such barrier heights should be of considerable help to the theoretical chemists.

5.8 STARK AND ZEEMAN EFFECTS—INTRODUCTION

From a general consideration of the energy levels of gaseous molecules it has been seen that they usually have a $(2J + 1)$ degeneracy, due to the different orientations that the total angular momentum can take up with respect to a direction *fixed in space* (i.e. 'spatial' or 'magnetic' degeneracy). In the absence of any external electric or magnetic field, no axis exists for such a quantization, and the $(2J + 1)$ states are equivalent and identical in energy. The application of such an external field removes this degeneracy, however, each of the $(2J + 1)$ levels, corresponding to a different M value, now having a slightly different energy, and this new series of energy levels can cause a splitting of the observed absorption lines into several different components. An external electric field will act via the electric dipole moment of the molecule, while an external magnetic field will act on the permanent magnetic moment of the molecule.

In paramagnetic resonance, which is concerned with individual ions possessing unsaturated spins, there is a large resultant permanent magnetic moment, and the Zeeman splitting produced by a magnetic field is very large. Such systems do not possess a permanent electric dipole moment, however, and hence no Stark splitting is to be expected. The reverse of this is true in the case of gaseous microwave spectroscopy, as most gas molecules are in a $^1\Sigma$ state, and hence possess no resultant permanent magnetic moment, except for a very small contribution from nuclei and molecular rotation. There are a few exceptions to this, such as oxygen and nitric oxide, but the Zeeman splitting produced by even a large magnetic field is usually very small. On the other hand, the electric dipole moment of many molecules is quite large, and an external electric field of only 50 volts/cm can sometimes produce a large splitting of the absorption lines.

5.9 STARK EFFECT

If the molecule has a component of the electric dipole moment along the direction of the total angular momentum vector, then what is termed a 'first-order Stark effect' will occur, due to the direct interaction of the electric field and dipole moment. Energy splittings of the order of

$$h \cdot \Delta\nu \approx \mu \cdot \epsilon \cdot \quad (\epsilon \text{ is the electric field})$$

are obtained, which produce frequency separations of the absorption lines of about 100 Mc/s for a field of 100 volts/cm, in the case of a $J = 1 \rightarrow 2$ transition.

If there is no component of the dipole moment along the direction of total angular momentum, then no 'first-order' Stark effect will occur, as no interchange of rotational energy can take place via a vector at right angles to it. The external electric field can induce a dipole moment, however, and this will have a component in the direction of J, and hence a small Stark splitting will be obtained. The mechanism of the splitting in this case can be considered as the action of the electric field on the permanent dipole moment to produce a non-uniform rotation. This is termed a 'second-order', or 'quadratic' Stark effect, as it is proportional to ϵ^2, since the induced dipole moment itself is proportional to ϵ. It is very much smaller than the 'first-order' Stark splitting (c.f. diamagnetism and paramagnetism) and is usually only observed when the latter is absent.

(a) *Linear molecules.* Since the electric dipole moment of a linear molecule is always along its internuclear axis, and its angular momentum vector is always perpendicular to this, it follows that no linear molecule can have a 'first-order' Stark effect. A 'second-order' Stark splitting will be produced, however, and a detailed calculation by VAN VLECK[70] shows that the induced moment in the direction of the field is given by

$$\mu_\epsilon = -\frac{4\pi^2 I_a \mu^2}{h^2}\left[\frac{J(J+1) - 3M^2}{J(J+1)(2J-1)(J+3)}\right] \cdot \epsilon \quad \dots (5.11)$$

where μ is the permanent dipole moment.

124

Since the actual energy splitting is given by $-\mu_\epsilon \cdot \epsilon$ it may be written as

$$\Delta E = \frac{\mu^2}{2hB} \left[\frac{J(J+1) - 3M^2}{J(J+1)(2J-1)(2J+3)} \right] \cdot \epsilon^2 \quad \dots (5.12)$$

and it is seen that the splitting of the energy levels is proportional to the square of the applied field, and directly proportional to the moment of inertia of the molecule. The selection rule, for the case of parallel Stark and microwave electric fields, is $\Delta M = 0$ and $\Delta J = \pm 1$, from which the frequencies of the allowed transitions can be obtained.

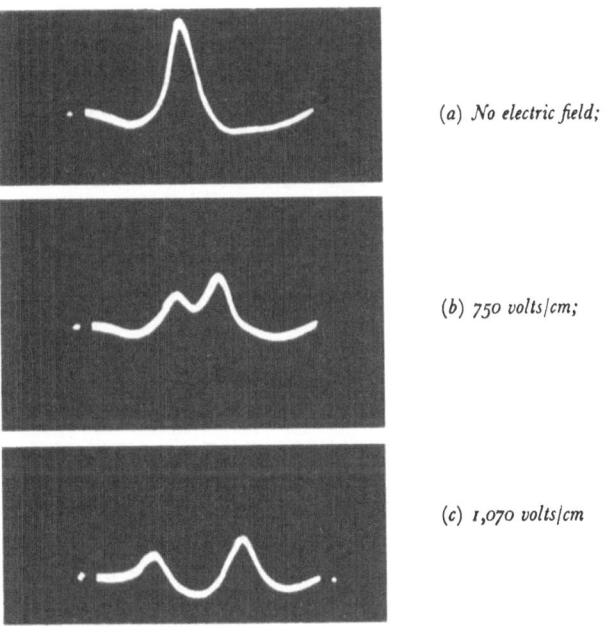

(a) *No electric field;*

(b) *750 volts/cm;*

(c) *1,070 volts/cm*

Figure 54. Stark splitting in OCS

The frequency splitting produced by this 'second-order' Stark effect is very small compared to the 'first-order' case, a field of 100 volts/cm only producing a splitting of about $0 \cdot 2$ Mc/s for a $J = 1 \rightarrow 2$ transition, and this would be even less for higher J values. It is therefore necessary to employ much larger electric fields when studying molecules with 'second-order' Stark effects.

The first Stark effect measurements in microwave spectroscopy were those of DAKIN, GOOD and COLES[71] on the linear triatomic OCS molecule. They observed the splitting of the $J = 1 \rightarrow 2$ rotational line at a field of 1,070 volts/cm and obtained a doublet with a frequency separation of about 5 Mc/s. This splitting is shown in *Figure 54*, which is taken from their original paper. The recent measurements of TATE and STRANDBERG[72] on the Stark effect of KCl and NaCl are a good example of its use in high-temperature microwave spectroscopes for the accurate determination of dipole moments, and shows how these techniques can be extended to include molecules which are not gaseous at room temperature.

(*b*) *Symmetric-top molecules*. Symmetric-top molecules have a component of their electric dipole moment along the direction of the total angular momentum vector equal to $\mu K/\sqrt{J(J+1)}$, and hence, unless $K = 0$, a first-order Stark splitting of the energy levels will occur. The derivation of the energy and frequency splitting produced in this case is very simple. Thus the component of the dipole moment along the direction of the *electric field* will be

$$\frac{\mu K}{\sqrt{J(J+1)}} \cdot \frac{M}{\sqrt{J(J+1)}}$$

and hence the energy level splitting will be

$$\Delta E = -\mu_\epsilon \cdot \epsilon = -\frac{\mu \cdot K \cdot M}{J(J+1)} \cdot \epsilon \qquad \ldots (5.13)$$

which gives a frequency splitting of

$$h \cdot \Delta \nu = \frac{2\mu \cdot K \cdot M}{J(J+1)(J+2)} \epsilon \qquad \ldots (5.14)$$

for the selection rule $\Delta K = 0$, $\Delta M = 0$ and $\Delta J = \pm 1$. This 'first-order' Stark splitting is large, even for small external fields, and it may also be noted that the splitting is independent of the moment of inertia, only the permanent dipole moment μ entering into the expression.

There will, of course, also be a 'second-order' Stark effect present at the same time, the field inducing a dipole moment in its own direction as for the case of the linear molecules. This second-order effect is quite negligible in the presence of the first-order splitting, however, and can only be measured in symmetric-top molecules when the latter effect disappears for the case of $K = 0$. The expression for the energy splittings in this case is identical with that for linear molecules, given in equation (5.12).

(*c*) *Asymmetric-top molecules*. The exact derivation of the energy splittings produced by the Stark effect on asymmetric-top molecules is very difficult as might be expected, and most of the theoretical work has been done by GOLDEN and WILSON[74]. As already seen, the selection rule for these molecules is $\Delta J = 0$, and ± 1, owing to the possible admixture of different J sub-levels, and this tends to make the observed spectra very complicated even without Stark splitting. The Stark effect is often helpful, however, in differentiating between $\Delta J = 0$ and $\Delta J = \pm 1$ transitions. Golden and Wilson[74] show that the intensities of the former are proportional to M^2, whereas those of the latter are proportional to $[(J+1)^2 - M^2]$. An example of the use of the Stark effect in identifying different absorption lines is given by the measurements of Ferguson and Wilson on SOF_2[75]; and its use to determine dipole moments for several different molecules is illustrated by the papers of Townes et al.[76], who also consider in detail the combination of Stark splitting with nuclear quadrupole coupling.

(d) *Inversion spectra.* Although the ammonia molecule is a symmetric top, its Stark splitting[77] is somewhat of a special case as its dipole moment is not 'permanent' but changes with the inversion frequency. This gives rise to a 'second-order' type of splitting rather than the normal linear type of symmetric-tops, and Coles and Good[78] gave as an expression for the splitting

$$\Delta \nu \, (\text{cm}^{-1}) = 1 \cdot 5 \, . \, 10^{-4} \, . \, \mu^2 \left[\frac{KM}{J(J+1)} \right]^2 . \, \epsilon^2 \quad \ldots (5.15)$$

A more detailed theoretical treatment of the Stark splitting produced in ammonia, including the quadrupole interaction term, has been given by Jauch[79].

5.10 The Zeeman Effect

There are three sources which can contribute to the magnetic moment of a gaseous molecule; (i) the moment associated with the orbital and spin momenta of the electrons, (ii) the magnetic moment produced by the rotational moment of the whole molecule, and (iii) the magnetic moments of the nuclei. It has been seen that most molecules are in a $^1\Sigma$ state, which means that there is no contribution from either the orbital or spin momenta of the electrons, and hence only the very much smaller second and third contributions are left.

It can be shown[80] that the rotation of the electron clouds of the molecule gives rise to a magnetic moment which has the direction of J, and a magnitude of approximately $\sqrt{J(J+1)} \, . \, \mu_N$, μ_N being the nuclear magneton, and the quantization of this along the direction of the applied field produces a set of $(2J+1)$ equidistant energy levels separated by $g_J \, . \, \mu_N \, . \, H$ where g_J is the gyromagnetic ratio of the whole molecule, which depends on the particular rotational state, and has values ranging from $0 \cdot 02$ to $0 \cdot 6$[81].

Since μ_N is over one thousand times smaller than the Bohr magneton the splitting of these levels is very small, and relatively large fields have to be applied if the lines are to be resolved in the microwave spectra. Zeeman splitting was first demonstrated in the ammonia spectrum by Coles and Good[78], since when most of the work on the Zeeman effect of gaseous molecules has been carried out by Jen[81, 82]. Such measurements give information on the gyromagnetic ratio of the molecule, and the compounds studied and analysed include NH_3, H_2O, HDO, N_2O, SO_2 and OCS. The sign of the molecular gyromagnetic ratio can be determined by the use of circularly polarized radiation, Jen's apparatus[83] which employs this technique, has been briefly described in Section 4.6. The interaction between the nuclear magnetic moments and the molecular magnetic field is treated in the next section under 'magnetic hyperfine structure'.

It may be noted that transitions are allowed between the Zeeman levels themselves and will give rise to absorption lines in the radio frequency region, some of these having been observed by Ramsey[84] for such molecules as H_2, HD and D_2 by the use of molecular beams. For a field strength of 4,000 gauss the absorption lines occur at about 2 Mc/s. Molecules possessing a permanent electronic magnetic moment (e.g. O_2 and NO) have an energy splitting between the Zeeman levels over one thousand times greater, and the absorption lines due to direct transitions between these occur in the microwave region. This type of absorption is in fact what constitutes gaseous paramagnetic resonance as studied by Beringer and Castle[85].

5.11 NUCLEAR INTERACTION AND HYPERFINE STRUCTURE

So far only the motion of the atoms as a whole has been considered and no attention paid to the fact that they may possess energy individually, which can affect the frequency of the absorbed radiation. Any of the nuclei present in the molecule may be rotating and possess its own spin, magnetic moment and electric quadrupole moment, and it is possible for these to interact with the molecular rotational energy to cause small changes which vary with the nuclear quantum state. There are two ways in which the nucleus can interact with the molecular motion, (i) magnetically—i.e. by the nuclear magnetic moment coupling with the magnetic field produced by the molecular rotation, and (ii) electrically—i.e. by the nuclear electric quadrupole moment interacting with the gradient of the electric field produced by the rest of the molecule. As already noted, the molecular magnetic field is usually very small and hence the splitting of the energy levels by the magnetic interaction is also small. The gradient of the electric field at the nucleus may be quite large, however, and the splitting of the energy levels by the coupling of this to the electric quadrupole moment is often sufficient to produce a well-resolved hyperfine structure. This effect is the one mainly responsible for hyperfine splitting in gaseous spectroscopy and will be considered first.

(a) *Electric quadrupole interaction.* The electric quadrupole moment of a nucleus is usually denoted by Q, and is a measure of the deviation in the charge distribution from spherical symmetry. It may be defined as

$$Q = \frac{1}{e} \int (3z^2 - r^2) \cdot dq \qquad \ldots (5.16)$$

where dq is an element of charge within the nucleus with co-ordinates x, y and z, and $r^2 = x^2 + y^2 + z^2$, and hence Q has dimensions of cm^2. This non-uniform distribution of charge will produce an electric field with a potential of the form[86]

$$V = \frac{Ze}{r} - \frac{eQ}{4r^5}(x^2 + y^2 - 2z^2) \qquad \ldots (5.17)$$

these co-ordinates now extending outside the nucleus. The field thus possesses an axis along the direction of the quadrupole moment, and hence different energies are to be expected for different orientations of this axis, with respect to that of the molecular electric field. The case in which only one nucleus has spin and quadrupole moment is considered initially. CASIMIR[87] was the first to treat this theoretically, and obtained an expression for the interaction energy, in the case of a linear molecule of the form:

$$\Delta E = \left(\frac{\partial^2 V}{\partial z^2}\right)_{Av} \cdot eQ \cdot \frac{\frac{3}{4}G(G+1) - \frac{1}{2}I(I+1)\mathcal{J}(\mathcal{J}+1)}{I(2I-1)(2\mathcal{J}-1)(2\mathcal{J}+3)} \quad \ldots (5.18)$$

where $\qquad G = F(F+1) - I(I+1) - \mathcal{J}(\mathcal{J}+1)$

and F is the quantum number of the total angular momentum inclusive of nuclear spin, and varies from $\mathcal{J} + I$ to $\mathcal{J} - I$. $\left(\frac{\partial^2 V}{\partial z^2}\right)_{Av}$ is the gradient of the electric field and is equal to

$$\int \frac{3z^2 - r^2}{r^5} \cdot dq.$$

This expression has been extended to the case of symmetric-top molecules by COLES and GOOD[88], JAUCH[79] and VAN VLECK[89], who derive an expression for the interaction energy of the form

$$\Delta E = \left(\frac{\partial^2 V}{\partial z^2}\right)_{Av} \cdot eQ \left[\frac{3K^2}{\mathcal{J}(\mathcal{J}+1)} - 1\right] \cdot \frac{\frac{3}{4}G(G+1) - I(I+1)\mathcal{J}(\mathcal{J}+1)}{2(2\mathcal{J}+3)(2\mathcal{J}-1)I(2I-1)}$$
$$\ldots (5.19)$$

which is equivalent to the linear case if $K = 0$. The last term in equations (5.18) and (5.19) is the same and is also common to the energy-level splittings derived for asymmetric-top molecules, only the factor of $\left(\dfrac{3K^2}{J(J+1)} - 1 \right)$ varying between the different molecular types.

The selection rule for transitions between these levels is $\Delta J = \pm 1$, $\Delta K = 0$ and $\Delta F = \pm 1$.

The initial measurements[78, 90] on hyperfine structure produced by quadrupole interaction were found to fit very well with the above expressions, but more accurate determinations[91] on such molecules as ICN and CH_3I soon showed that deviations of the order of a megacycle existed. BARDEEN and TOWNES[92] extended the theory to a second order to account for these deviations, the extra term they introduced being essentially due to an interaction between levels of different J, but having the same total angular momentum F, and the same magnetic quantum number, M. It is only of importance when the splitting of the rotational energy levels is not large compared with the quadrupole splitting.

(*b*) *Two-nuclei quadrupole interaction.* So far it has been assumed that only one nucleus has a quadrupole moment, but the theory can be extended to cover molecules in which there are two interacting nuclei. BARDEEN and TOWNES[93] considered this case, and showed that, if the interaction energy associated with one nucleus is very different from that of the other, then the smaller can be treated as a perturbation on the larger and an expression of the form

$$\Delta E = \Delta E_1 (F_1) + \Sigma_{F_2} c (F_1 F_2)^2 . \Delta E_2 (F_2) \qquad \ldots (5.20)$$

can be used to derive the energy levels.

Here $F_1 = (J + I_1) \ldots | J - I_1 |$ and $F_2 = (J + I_2) | \ldots | J - I_2 |$ and ΔE_1 and ΔE_2 are the expressions obtained by substituting F_1 and F_2 into equation (5.18) or (5.19).

The ClCN[94] molecule is a good example of the two-nuclei interaction case, in which this perturbation approach can be applied. The spectrum under low resolution shows three broadish lines due to the quadrupole interaction with the chlorine nucleus, but higher resolution splits each of these three lines into a collection of smaller lines, which are due to the smaller interaction with the quadrupole moment of the nitrogen nucleus. This spectrum, with the theoretical positions of the various hyperfine lines, is shown in *Figure 55*.

The great importance attached to the measurement and analysis of the hyperfine structure of microwave absorption lines lies in the nuclear information that can be obtained from them. It is worth noting that the hyperfine structure obtained in paramagnetic reson-ance absorption is, in general, very much more simple than that of the gaseous microwave spectra, usually consisting of $(2I + 1)$ equally-spaced lines, and thus giving a direct determination of I. The analysis of the gaseous hyperfine structure is much more complicated, but

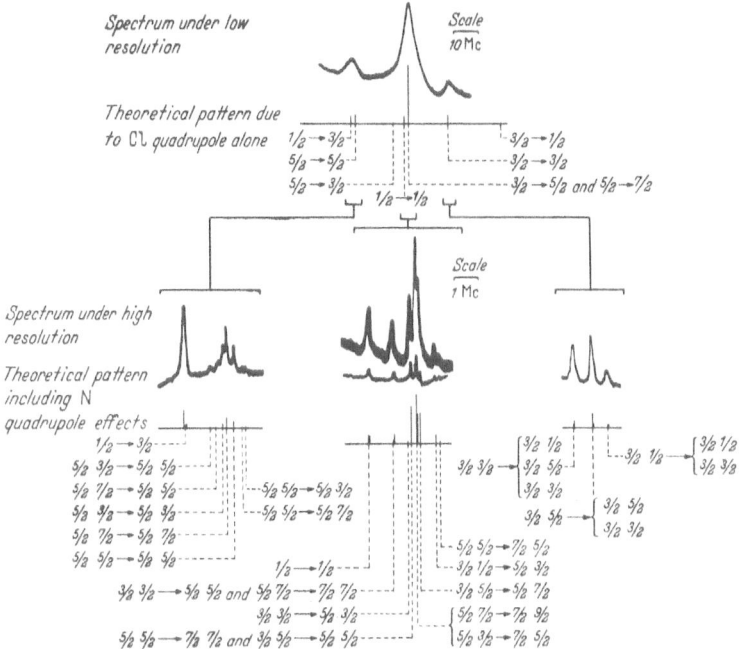

Figure 55. Spectrum of $Cl^{35}CN$ $J = 1 \rightarrow 2$ transition with both low and high dispersion and comparison with theoretical pattern assuming spin of Cl^{35} is 3/2

does yield an accurate value for the quadrupole moments as well as I from the same measurement. Nuclei, the spins of which have been first determined by gaseous microwave spectroscopy, include B^{10} (ref. [95]), B^{11} (ref. [96]), S^{33} (ref. [97]), S^{35} (ref. [97]), Cl^{35}, Cl^{36}, Cl^{37} (ref. [98]), Ge^{73} (ref. [99]), Se^{79} (ref. [100]) and I^{129} (ref. [101]).

(*c*) *Higher nuclear moments.* The interaction of the nucleus with the molecular electric field has been explained and analysed in all the above work by assuming that the charge distribution over the nucleus was accurately quadrupolar. It is possible for the charge distribution to have higher symmetry configurations, however, a hexadecapole distribution being the next permitted state. If some of the charge density is spread in this more complex pattern, slightly different energy levels from the pure quadrupolar case would be expected. The fact that all the measurements to date can be fitted by the simple quadrupole expressions indicates that any higher poles must be very small.

Recent work on atomic beams[102, 103] has shown a deviation from the simple equations for the case of I^{127} and In^{115}, however, and this is taken as the first direct evidence for the existence of a magnetic octopole moment. The frequency shifts are small, but with the increasing resolution now being obtained in microwave spectroscopy the effect of nuclear hexadecapole moments may soon have to be added to the various expressions for interaction energy.

(d) *Magnetic hyperfine structure.* The theory of this interaction, with especial reference to the case of ammonia, has been treated by Jauch[79], and HENDERSON and VAN VLECK[104], who were able to derive expressions for the different interaction energies by relating the molecular magnetic field to the different rotational quantum numbers. In the case of a linear molecule the magnetic field produced is proportional to the total angular momentum, and the interaction is thus directly proportional to $I \wedge J$, and the interaction energies are given by

$$\Delta E \propto [F(F+1) - J(J+1) - I(I+1)] \qquad \ldots (5.21)$$

In the case of symmetric-top molecules the magnetic field can be considered as a summation of two components, one proportional to the angular momentum about the molecular axis—i.e. proportional to quantum number K, and the other normal to this. The general treatment of Henderson and Van Vleck [104], giving the expression of equation (5.21) multiplied by functions of K and J, was shown to be of the following form for the case of ammonia[105].

$$\Delta E = \{aK^2 [J(J+1)]^{-1} + b\} . [F(F+1) - J(J+1) - I(I+1)] \qquad \ldots (5.22)$$

The extra splitting is of course additional to the normal quadrupolar interaction. It was first noticed by SIMMONS and GORDY[106] in ammonia as a shift in the position of the N^{14} quadrupole hyperfine structure, and was explained as due to the interaction of the N^{14} magnetic moment with the molecular magnetic field. More recently[107, 108] additional splitting of the lines has been observed, due to removal of the orientation degeneracy of the hydrogen atoms, thus changing the quadrupole transitions of the nitrogen nucleus by small amounts.

This last experiment of GORDEN, ZEIGER and TOWNES[108], on the magnetic satellites of the ammonia hyperfine structure lines, deserves a brief description as it has introduced a fundamentally new method into the techniques of microwave spectroscopy. They have observed the *emission* spectra of ammonia for the first time, and by the use of molecular beams have reduced the Doppler broadening so that the lines obtained were only 6 kc/s wide. A block diagram of their apparatus is shown in *Figure 56*, and is seen to consist of a source providing a beam of ammonia molecules which pass through a

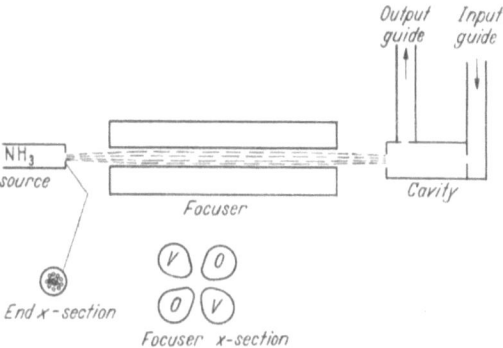

Figure 56. Block diagram of NH_3 *emission spectroscope*

'focuser' into a high-Q resonant cavity. The 'focuser' comprises four electrodes which establish a quadrupolar electrostatic field, with its cylindrical axis in the direction of the beam. This field acts differently on the molecules, according to their inversion level. Those in the upper states experience an inward radial force which focuses them, whereas those in the lower level experience an outward dispersive force. The beam entering the cavity is therefore composed of molecules in the higher inversion level, and these will emit radiation into the cavity at the molecular transition frequency. If the power emitted by the beam is sufficiently high, self-sustained oscillations will be produced in the cavity, and the device will thus act as an extremely well stabilized microwave generator. For lower beam powers radiation of varying frequency is transmitted through the cavity and an emission line appears on the oscilloscope screen when the klystron passes through a transition frequency.

The high resolution is due to the high directivity and small velocity spread of the molecules (as in the case of the other recent molecular beam technique, used by Strandberg et al.[56], to reduce Doppler width). The 3,3 inversion line of ammonia as obtained by this apparatus is shown in *Figure 57*, and the magnetic satellite lines are seen resolved in the main quadrupole hyperfine transitions. Each of the quadrupole levels would be expected to split into four components by the interaction of the hydrogen magnetic moments with the molecular magnetic fields, since the hydrogen spins are in a symmetric state under 120° rotation, and hence $I_H = \frac{3}{2}$.

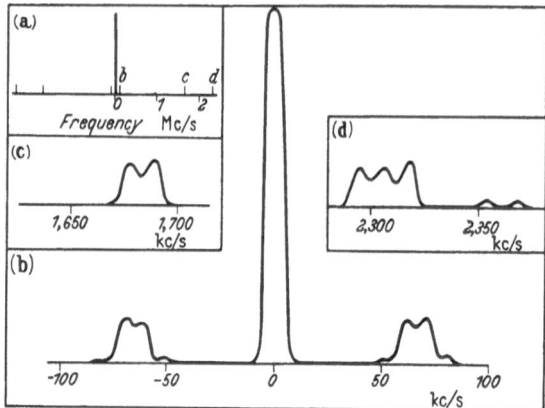

Figure 57. NH$_3$ 3,3 line showing magnetic satellites

This new type of device appears to have considerable potentialities, not only as a high-resolution spectroscope, but also in the practical applications of a microwave frequency standard of great stability, or a very low-noise amplifier. The applications of this method of producing stimulated emission, and the development of masers and lasers, are discussed in Chapter 12.

REFERENCES

[1] DICKE, R. H., BERINGER, R., KYHL, R. L. and VANE, A. B. *Phys. Rev.* 70 (1946) 340

[2] BECKER, G. E. and AUTLER, S. H. *ibid.* 70 (1946) 300

[3] TOWNES, C. H. and MERRITT, F. R. *ibid.* 70 (1946) 558

[4] GORDY, W. and KING, W. C. *ibid.* 93 (1953) 407

[5] DENNISON, D. M. *Rev. mod. Phys.* 12 (1940) 189

[6] GOLDEN, S., WENTINK, T., HILLGER, R. E. and STRANDBERG, M. W. P. *Phys. Rev.* 73 (1948) 92

[7] STRANDBERG, M. W. D., WENTINK, T., HILLGER, R. E., WANNIER, G. H. and DEUTSCH, M. L. *ibid.* 73 (1948) 188

[8] STRANDBERG, M. W. P. *J. chem. Phys.* 17 (1949) 901

[9] BEARD, C. I. and BIANCO, D. R. *ibid.* 20 (1952) 1488

[10] ANDERSON, R. S., JOHNSON, C. M. and GORDY, W. *Phys. Rev.* 83 (1951) 1061

[11] BERINGER, R. *ibid.* 70 (1946) 53

REFERENCES

[12] BURKHALTER, J. H., ANDERSON, R. S., SMITH, W. V. and GORDY, W. *ibid.* 77 (1950) 152 and 79 (1950) 651

[13] TRAMBARULO, R., GHOSH, S. N., BURRUS, C.A. and GORDY, W. *ibid.* 91 (1953) 222

[14] PAULING, L. and WILSON, E. B. *Introduction to Quantum Mechanics* McGraw-Hill (1935)

[15] HERZBERG, G. *Spectra of Diatomic Molecules* p. 104. Van Nostrand (1951)

[16] NIELSEN, A. H. *J. chem. Phys.* 11 (1943) 160

[17] JOHNSON, C. M., TRAMBARULO, R. and GORDY, W. *Phys. Rev.* 84 (1951) 1178

[18] NIELSEN, H. H. *ibid.* 77 (1950) 130; and NIELSEN, H. H. and SHAFFER, W. *J. chem. Phys.* 11 (1943) 140

[19] SCHULMAN, R. G. and TOWNES, C. H. *Phys. Rev.* 77 (1950) 421

[20] TOWNES, C. H., MERRITT, F. R. and WRIGHT, B. D. *ibid.* 73 (1948) 1334

[21] SMITH, A. G., RING, H., SMITH, W. V. and GORDY, W. *ibid.* 74 (1948) 370

[22] WESTENBERG, A. A. and WILSON, E. B. *J. Amer. chem. Soc.* 72 (1950) 199

[23] STITCH, M. L., HONIG, A. and TOWNES, C. H. *Phys. Rev.* 86 (1952) 813; KLEIN, J. A. and NETHERCOT, A. H. *ibid.* 91 (1953) 1018; GORDY, W. and BURRUS, C. A. *ibid.* 93 (1954) 419

[24] DENNISON, D. M. *Rev. mod. Phys.* 3 (1931) 280 and ref. (26) p. 26

[25] SLAWSKY, Z. I. and DENNISON, D. M. *J. chem. Phys.* 7 (1939) 509

[26] HERZBERG, G. *Infra-Red and Raman Spectra* p. 400. Van Nostrand (1945)

[27] ANDERSON, W. E., TRAMBARULO, R., SHERIDAN, J. and GORDY, W. *Phys. Rev.* 82 (1951) 58

[28] RING, H., EDWARDS, H., KESSLER, M. and GORDY, W. *ibid.* 72 (1947) 1262

[29] JOHNSON, M. and DENNISON, D. M. *ibid.* 48 (1935) 868

[30] GORDY, W., SIMMONS, J. W. and SMITH, A. G. *ibid.* 74 (1948) 243

[31] SHARBAUGH, A. H. *ibid.* 74 (1948) 1870; SHARBAUGH, A. H., THOMAS, V. G. and PRITCHARD, B. S. *ibid.* 78 (1950) 64

[32] SHERIDAN, J. and GORDY, W. *J. chem. Phys.* 20 (1952) 591

[33] WILLIAMS, J. Q. and GORDY, W. *ibid.* 18 (1950) 994

[34] HONIG, A., STITCH, M. L. and MANDEL, M. *Phys. Rev.* 92 (1953) 901

[35] GORDY, W. and SHERIDAN, J. *ibid.* 79 (1950) 224

[36] SHARBAUGH, A. H., PRITCHARD, B. S., THOMAS, V. G., MAYS, J. M. and DAILEY, B. P. *ibid.* 79 (1950) 189

[37] KISLIUK, P. and TOWNES, C. H. *J. chem. Phys.* 18 (1950) 1109

[38] GHOSH, S. N., TRAMBARULO, R. and GORDY, W. *ibid.* 20 (1952) 605

[39] TATE, P. and STRANDBERG, M. W. P. *Phys. Rev.* 91 (1953) 464; *Rev. sci. Instrum.* 25 (1954) 956; and STITCH, M. L., HONIG, A. and TOWNES, C. H. *Rev. sci. Instrum.* 25 (1954) 759

[40] RANDALL, H. M., DENNISON, D. M., GINSBURG, N. and WEBER, L. R. *ibid.* 52 (1937) 160

135

[41] KING, G. W., HAINER, R. M. and CROSS, P. C. *J. chem. Phys.* 11 (1943) 27

[42] GOLDEN, S. *ibid.* 16 (1948) 78

[43] WANG, S. C. *Phys. Rev.* 34 (1929) 243

[44] NIELSEN, H. H. *Rev. mod. Phys.* 23 (1951) 90

[45] HAINER, R. M., CROSS, P. C. and KING, G. W. *J. chem. Phys.* 17 (1949) 826

[46] SIRVETZ, M. H. *ibid.* 19 (1951) 938

[47] STRANDBERG, M. W. P. *ibid.* 17 (1949) 901

[48] HILLGER, R. E. and STRANDBERG, M. W. P. *Phys. Rev.* 83 (1951) 575

[49] LAWRANCE, R. B. and STRANDBERG, M. W. P. *ibid.* 83 (1951) 363

[50] ROBERTS, A. and EDGELL, W. F. *J. chem. Phys.* 17 (1949) 742

[51] MYERS, R. J. and GWINN, W. D. *ibid.* 20 (1952) 1420

[52] JONES, L. H., SHODERY, J. N., SHULMAN, R. G. and YOST, D. M. *ibid.* 18 (1950) 990

[53] BEARD, C. I. and DAILEY, B. P. *ibid.* 18 (1950) 1437

[54] HERZBERG, G. *Infra-Red and Raman Spectra of Polyatomic Molecules* Van Nostrand, New York (1945)

[55] CLEETON, C. E. and WILLIAMS, N. H. *Phys. Rev.* 45 (1934) 234

[56] STRANDBERG, M. W. P. and DREICER, H. *ibid.* 94 (1954) 1393

[57] GORDON, J. P., ZEIGER, H. J. and TOWNES, C. H. *ibid.* 95 (1954) 282

[58] WRIGHT, N. and RANDALL, H. M. *ibid.* 44 (1933) 391

[59] BLEANEY, B. and PENROSE, R. P. *Proc. Roy. Soc.* A 189 (1947) 358

[60] COSTAIN, C. C. *Phys. Rev.* 82 (1951) 108

[61] WEISS, M. T. and STRANDBERG, M. W. P. *ibid.* 83 (1951) 567

[62] LYONS, H., RUEGER, L. J., NUCKOLLS, R. G. and KESSLER, M. *ibid.* 81 (1951) 630

[63] ANDERSON, P. W. *ibid.* 76 (1949) 647

[64] NIELSEN, H. H. *ibid.* 40 (1932) 445

[65] KOEHLER, J. S. and DENNISON, D. M. *ibid.* 57 (1940) 1006

[66] HERSHBERGER, W. D. and TURKEVITCH, J. *ibid.* 71 (1947) 554

[67] DAILEY, B. P., MINDEN, H. T. and SHULMAN, R. G. *ibid.* 75 (1949) 1319

[68] SHERIDAN, J. and GORDY, W. *J. chem. Phys.* 19 (1951) 965

[69] KISLIUK, P. and SILVEY, G. A. *ibid.* 20 (1952) 517

[70] VAN VLECK, J. H. *Theory of Electric and Magnetic Susceptibilities* O.U.P. (1932)

[71] DAKIN, T. W., GOOD, W. E. and COLES, D. K. *Phys. Rev.* 70 (1946) 560

[72] TATE, P. A. and STRANDBERG, M. W. P. *J. chem. Phys.* 22 (1954) 1380

[73] STITCH, M. L., HONIG, A. and TOWNES, C. H. *Rev. sci. Instrum.* 25 (1954) 759

[74] GOLDEN, S. and WILSON, E. B. *J. chem. Phys.* 16 (1948) 669

[75] FERGUSON, R. C. and WILSON, E. B. *Phys. Rev.* 90 (1953) 338

[76] LOW, W. and TOWNES, C. H. *ibid.* 76 (1949) 1295; SCHULMAN, R. G. and TOWNES C. H. *ibid.* 77 (1950) 500; SCHULMAN, R. G., DAILEY, B. P. and TOWNES, C. H. *ibid.* 78 (1950) 145

[77] Good, W. E. *ibid.* 70 (1946) 213

[78] Coles, D. K. and Good, W. E. *ibid.* 70 (1946) 979

[79] Jauch, J. M. *ibid.* 72 (1947) 715

[80] Wick, G. C. *ibid.* 73 (1948) 51

[81] Jen, C. K. *Physica* 17 (1951) 379

[82] — *Phys. Rev.* 74 (1948) 1396; and 76 (1949) 1494

[83] — *Ann. N.Y. Acad. Sci.* 55 (1952) 822

[84] Ramsey, N. F. *Phys. Rev.* 58 (1940) 226

[85] Beringer, R. and Castle, J. G. *ibid.* 75 (1949) 1963; and 76 (1949) 868

[86] Born, M. *Optik, Springer,* Berlin (1933)

[87] Casimir, H. B. G. *Physica* 2 (1935) 719

[88] Coles, D. K. and Good, W. E. *Phys. Rev.* 70 (1946) 979

[89] Van Vleck, J. H. *ibid.* 71 (1947) 468

[90] Gordy, W., Simmons, J. W. and Smith A. G. *ibid.* 74 (1948) 243

[91] Gilliam, O. R., Edwards, H. D. and Gordy W. *ibid.* 73 (1948) 635; Townes, C. H., Merritt, F. R. and Wright, B. D. *ibid.* 73 (1948) 1334

[92] Bardeen, J. and Townes, C. H. *ibid.* 73 (1948) 627 and 1204

[93] — — *ibid.* 73 (1948) 97

[94] Townes, C. H., Holden, A. N. and Merritt, F. R. *ibid.* 74 (1948) 1113

[95] Weiss, M. T., Strandberg, M. W. P., Lawrence, R. B. and Loomis, C. C. *ibid.* 78 (1950) 202

[96] Gordy, W., Ring, H. and Bing, A. B. *ibid.* 74 (1948) 1191

[97] Townes, C. H. and Geschwind, S. *ibid.* 74 (1948) 626; Cohen, V. W., Koski, W. S. and Wentink, T. *ibid.* 76 (1949) 703

[98] Townes, C. H., Holden, A. N., Bardeen, J. and Merritt, F. R. *ibid.* 71 (1947) 644

[99] Townes, C. H., Mays, J. M. and Dailey, B. P. *ibid.* 76 (1949) 700

[100] Hardy, W. A., Silvey, G. and Townes, C. H. *ibid.* 86 (1952) 608

[101] Livingstone, R., Gilliam, O. R. and Gordy, W. *ibid.* 76 (1949) 149

[102] Jaccarino, V., King, J. G., Satten, R. A. and Stroke, H. H. *ibid.* 94 (1954) 1798

[103] Kusch, P. and Eck, T. G. *ibid.* 94 (1954) 1799

[104] Henderson, R. S. and Van Vleck, J. H. *ibid.* 74 (1948) 106

[105] Henderson, R. S. *ibid.* 74 (1948) 107

[106] Simmons, J. W. and Gordy, W. *ibid.* 73 (1948) 713

[107] Good, W. E., Coles, D. K., Gunther-Mohr, G. R., Shawlow, A. L. and Townes, C. H. *ibid.* 83 (1951) 880; and 94 (1954) 1184

[108] Gorden, J. P., Zeiger, H. J. and Townes, C. H. *ibid.* 95 (1954) 282

6

RESULTS AND THEORY OF
PARAMAGNETIC RESONANCE*

6.1 INTRODUCTION

IN passing from the study of gaseous microwave spectroscopy to that of paramagnetic resonance, the consideration of individual molecules is changing to that of atoms, or ions, embedded in the crystal lattice, and hence subject to all the various solid-state forces and inter-actions. This means that the theoretical interpretation of the experimental results is no longer so simple, the pure rotational energy of a molecule now being replaced by the action of internal and external electric and magnetic fields which produce splitting of the energy levels. Conversely, paramagnetic resonance can be very useful for the study of solid-state forces. The results already obtained have provided an immense amount of new information for the magnetic theories of solids, detailed data on ionic and covalent bonding, and nuclear information from the hyperfine splittings.

Generally speaking, paramagnetic resonance can take place when-ever there is an electron present in the system which has an unpaired spin. Thus in any full, closed shell of electrons, the orbital and spin angular momenta of the individual electrons will all cancel out to give a zero resultant and produce a diamagnetic substance, no electron being left over with uncompensated momentum. Similarly, when chemical binding takes place between different atoms, the valency electrons are so arranged that closed compensated shells are formed. For example, in the case of ionic binding such as in NaCl, the sodium atom loses one electron to become Na^+ with an outer closed $2p$ shell, and the chlorine atom gains one electron to become Cl^-, and also has a closed outer shell $(3p)$. Hence there are no electrons present in the sodium-chloride crystal that have uncom-pensated spin or orbital momenta. This kind of compensation also occurs in other types of chemical binding. Thus in a covalent bond electrons are shared by the participating atoms, each atom itself having a balanced shell left, and the shared electrons have momenta which cancel each other. It can be seen from this very simple qualita-tive approach that the vast majority of substances in the solid state

* Electron Spin Resonance is now the generally accepted term.

will be diamagnetic and have no unpaired electrons in their structure. There are two major exceptions to this general condition (i) atoms in which the shells inside the valency electrons are not filled, and hence unpaired electrons are present in the atom all the time, and (ii) compounds in which some of the normal bonds have been modified or broken, to leave unpaired electrons scattered throughout the sample. The first of these groups comprises the so-called 'transition elements', which are considered in detail in this chapter. The second group contains such cases as organic free radicals, irradiated crystals, and structures with many dislocations or breaks, and is considered in the next chapter together with the results on ferromagnetic resonance (really a special case of the first group). See also Chapter 11.

6.2 THE TRANSITION ELEMENTS

Transition elements occur in the periodic table when electrons enter an outer s shell before completely filling inner d or f shells, the first ion showing this effect being Ti^{3+}. This has one unpaired electron in its $3d$ shell, inside the valence electrons, and therefore, it and the following ions of V, Cr, Mn, Fe, Co, Ni and Cu, can all possess uncompensated electron momentum, with an associated permanent magnetic moment. These eight elements form the first transition group of the periodic table.

A similar group is formed higher up the periodic table each time the orbital number of a shell is so high that the energy required to put another electron into it is larger than that needed to start a new shell. The higher transition elements occur in successive groups, as follows: the palladium group (unfilled $4d$ shell); rare earth group (unfilled $4f$ shell); the platinum group (unfilled $5d$ shell) and the uranium group (unfilled $5f$ shells). In the absence of a magnetic field, the energy levels of the atom or ion are independent of the direction of the magnetic moment of the uncompensated electron(s); but if an external field is applied, quantisation will take place along this direction and the energy levels of the atom will be split and depend on the resolved component of the magnetic moment in the direction of the applied field. This splitting of the energy levels is an essential feature of paramagnetic resonance, as it is transitions between these component levels which give rise to the observed absorption. It is really a case of Zeeman splitting, but instead of the Zeeman splitting producing fine structure in the main electronic transitions, as in optical spectroscopy, transitions between the Zeeman levels themselves occur in paramagnetic resonance absorption.

The fact that an atom or ion possesses an unfilled shell, and hence a permanent magnetic moment, and will therefore have its energy levels altered by an external field, does not necessarily mean that paramagnetic resonance can be observed in its compounds. It is possible for all the degeneracy of the energy levels to be removed before the application of the external field (e.g. by the action of the very strong electric fields present in the crystal). In this case, although the applied magnetic field will change the energy of the levels, it may never bring them close enough together for transitions to be induced between them by frequencies in the microwave region. The general effect of the internal crystalline electric fields is considered in the next section, but it may be noted that a very general theorem, due to KRAMERS[1], states that if there is an odd number of unpaired electrons in the ion, then no electric field can completely remove the degeneracy and the bottom energy level is always at least two-fold degenerate in spin, and hence paramagnetic resonance is always theoretically possible in such cases.

The other main factor governing the observation of paramagnetic resonance absorption is the question of line width. It has already been seen that very wide lines are difficult to detect, and one of the problems of the spectroscopist is to reduce the width as much as possible. A list of the different causes of line broadening in paramagnetic resonance was given in Sections 4.1 and 4.8 with a brief consideration of their mechanism and possible means of reducing their effect. It was seen that the two major factors are the spread of the energy levels by spin-lattice and spin-spin interactions. The latter is associated with the proximity of the neighbouring paramagnetic ions, and can always be reduced by diluting with an isomorphous diamagnetic salt. The former, on the other hand, is due to the interaction with the lattice as a whole and cannot be reduced by just altering some of its constituents. It is, in fact, mainly governed by the splitting of the orbital levels of the paramagnetic ion concerned, those with relatively close energy levels having short spin-lattice relaxation times and wide absorption lines. The only means of reducing the line width in this case is to cool the specimen to lower temperatures. It is therefore evident that the effect of the crystalline field has to be considered in detail before it is certain that paramagnetic resonance should be observed in any particular case.

6.3 THE EFFECT OF INTERNAL CRYSTALLINE FIELDS

If the paramagnetic ion in free space is first considered by itself, the normal concepts of vector coupling can be applied to its quantum numbers. Thus the total orbital quantum number L compounds with the total spin quantum number S to produce a resultant J, and in the absence of any electric or magnetic fields all the different combinations of L and S states will have the same energy, and the ion will have one energy level, $(2J + 1)$-fold degenerate. If an electric field (such as produced by the negative ions in the surrounding lattice) now acts on the ion quantization will take place about this direction, and the degeneracy of the level will be removed as different quantum states have different energy levels. This quantization can take place in three distinct ways:

(i) The effect of the electric field can be relatively small so that L and S are not uncoupled but still precess about the resultant J. The vector of the total quantum number J will then precess about the direction of the applied electric field, and the energy level will be split into components corresponding to different values of the resolved component of J—i.e. an M_J quantum number. (This state of affairs occurs in the rare earths, where the unpaired electrons are somewhat shielded from the direct effect of the crystalline electric fields, and the LS coupling is not broken.)

(ii) The effect of the electric field can be quite strong and break the coupling between L and S, so that each precesses about the direction of the electric field separately, and J no longer has any meaning. The quantum numbers defining the different energy levels will now be M_L and M_S. It is usually found that the splitting produced between the different M_L levels by such a field is very large (about 10^3 cm^{-1}), and hence only the lowest level is populated at ordinary temperatures. This is usually referred to as 'quenching of the orbital levels' and occurs in nearly all the iron-group transition elements, where the unpaired electrons are more directly affected by the crystalline field. In such cases the lowest level is just $(2S + 1)$-fold degenerate in spin, and some of this degeneracy may also be lifted by the crystalline field to produce small zero-magnetic-field splittings of the energy levels.

(iii) The effect of the electric field can be very strong, so that it breaks down not only the LS coupling, but also the coupling between the angular momenta and spins of the individual electrons. The orbital motion is again quenched, and in this case the spins tend to 'pair off' under the exclusion principle, and an effect corresponding to the odd electron spin leftover is obtained. This kind of interaction is usually produced by covalent bonding and has been studied mainly in the ionic complexes. The effect of the crystalline fields on different ions will be considered in detail later, under the different transition group headings, and it will be seen that the symmetry of the field, as well as its magnitude, is important in determining how the energy levels are split.

6.4 THE RESONANCE CONDITION

The picture can be simplified initially by considering an ion which has all its orbital motion quenched, and just one effective electron spin, so that only this lowest level is populated and is two-fold degenerate in spin. (Cu^{2+} is a good example of this simple case.) This lowest level can be split by the application of an external magnetic field which will act directly on the magnetic moment associated with the electron spin. The condition for paramagnetic resonance to occur is that the splitting, produced by the magnetic field, should be equal to the quantum of energy of the incident radiation, i.e.

$$h\nu = g\beta H \qquad \qquad \dots (6.1)$$

where β is the Bohr magneton, H the strength of the applied magnetic field, g is termed the 'spectroscopic splitting factor' and effectively measures the rate at which the energy levels diverge with the applied

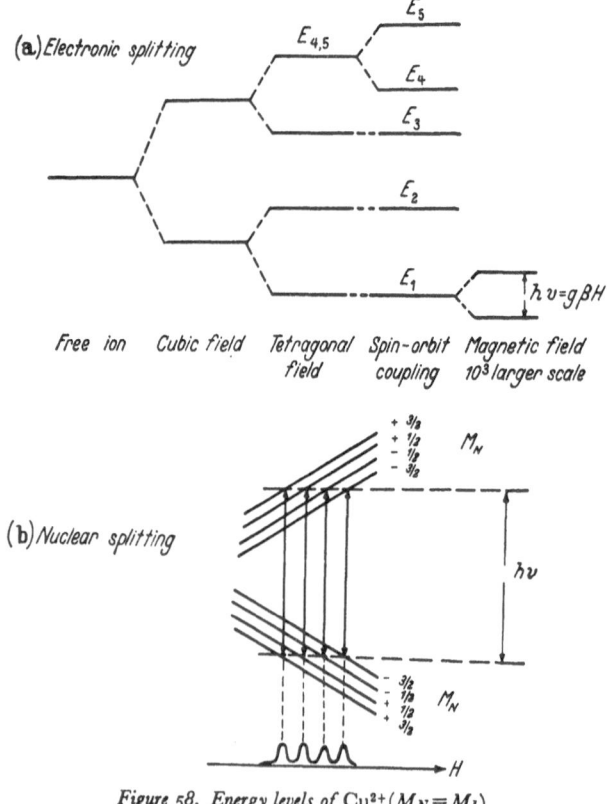

Figure 58. Energy levels of Cu^{2+} ($M_N \equiv M_I$)

142

magnetic field. If the electron spin were absolutely free, and had no coupling to any orbital motion, the spectroscopic splitting factor or 'g value', as it is often called, would be 2·0 (or 2·0023 with the relativistic correction). The fact that the electron is still bound to the paramagnetic atom, and thus has some interaction with the orbital motion via the spin orbit coupling, causes the g value to change from 2·0 by varying amounts, the actual shift depending mainly on the height of the next orbital level. To illustrate this point, the energy level diagram for the Cu^{2+} ion is shown in *Figure 58*, as predicted from POLDER's[2] original theoretical treatment. Copper has one electron missing from its $3d$ shell, and the problem can be treated as a case of one positive hole in an empty shell with a change of sign. The orbital and spin quantum numbers will therefore be $L = 2$ and $S = \frac{1}{2}$, giving a total degeneracy of $(2L + 1) \times (2S + 1) = 10$ for the energy level of the free ion, as represented at the left hand side of *Figure 58*. The effect of the cubic and tetragonal electric fields of the crystalline lattice is to remove most of the orbital degeneracy, the whole of this degeneracy being lifted by the addition of the spin orbit coupling interaction. The splitting between these orbital levels is very large (about 10^4 cm^{-1}), and hence all the copper ions will be in the lowest state, denoted by E_1 in the figure. Each orbital level is still two-fold degenerate in spin, and application of the external magnetic field is required to remove this and produce the splitting, which gives rise to the paramagnetic resonance absorption. Polder's original theory[2] predicted that the g value would be given by the expressions

$$g_{\parallel} = 2\left[1 - \frac{4 \cdot \lambda}{E_3 - E_1}\right] \qquad \ldots (6.2)$$

$$g_{\perp} = 2\left[1 - \frac{\lambda}{E_{4,5} - E_1}\right]$$

where λ is the spin-orbit coupling parameter[3] (equal to -852 cm^{-1} for copper), and E_1 to E_5 denote the energies of the different orbital levels as shown in *Figure 58*. The $g_{\parallel}$ and $g_{\perp}$ values correspond to the cases when the applied magnetic field is parallel, and perpendicular to, the axis of symmetry of the tetragonal crystalline electric field. It is a general feature of paramagnetic resonance that the spectra will change, often markedly, with variation in the angle between the applied magnetic field and the crystalline axes.

It can be seen from equation (6.2) that the larger the orbital splitting the closer will the g value be to the free-electron value, and this applies to the other ions as well (e.g. copper has g values varying from 2·15 to 2·4[4] corresponding to orbital splitting of the order of 10^4 cm^{-1}; whereas cobalt has g values varying from 3·5 to 6·5[5] corresponding to orbital splittings of about 300 cm^{-1}). It is, in fact, from the experimentally obtained g values and their angular variation that most of the information concerning the higher orbital states is derived. This g value is not the same as the Landé splitting factor, as that only applies to free atoms, whereas in the solid state the orbital momentum is subjected to very strong forces, and the theoretical calculation of the spectroscopic splitting factor is often very involved. Covalency effects are discussed in Chapter 11.

If values of the physical constants are substituted into equation (6.1), a numerical relation of the form[6]

$$g = \frac{hc}{\beta} \cdot \frac{1}{\lambda H} = \frac{4\pi mc^2}{e} \cdot \frac{1}{\lambda H} = \frac{21\cdot4184}{\lambda \cdot H} \qquad \cdots (6.3)$$

is obtained where λ is the wavelength of the absorbed radiation in centimetres, and H is the strength of the magnetic field required for resonance in kilogauss.

Measurements are usually made at 1·25 cm or 3 cm wavelengths, with corresponding fields of about 8,000 or 3,000 gauss, but, in principle, paramagnetic resonance can be carried out at any frequency, and recent work has extended the high-frequency measurements[7] to 60,000 Mc/s and the low-frequency measurements[8] to under 1 Mc/s. The advantages of working at high frequencies are: (i) the intensity of absorption goes up, the higher the resonant frequency; (ii) smaller amounts of material are required to fill the same percentage volume of the resonator, a fact which is very important for substances which only form very small crystals, or are only obtainable in very small quantities, such as artificial radioactive elements; and (iii) ions which have a zero-field splitting of their spin levels will only give a resonance if the $h\nu$ of the incident radiation is greater than the resultant splitting (e.g. NiSO$_4$ 7H$_2$O shows a resonance at 0·54 cm wavelength[7] but does not at 1·25 cm, because it has a zero-field splitting of 2 cm^{-1}). The disadvantages of working at high frequencies are purely practical, i.e. (i) the power available at wavelengths below 6 mm is very small, and (ii) very high magnetic field strengths are required to obtain the resonance, and, if work at low temperatures is also necessary, it is very expensive to construct a magnet which will produce a field of over 25,000 gauss across the dewar flask containing the resonant cavity.

A considerable amount of work[8, 9, 10] is now starting on paramagnetic resonance at low frequencies, and the great advantage of this is the extreme simplicity and low cost of the apparatus. It has two fundamental disadvantages however, (i) the intensity of absorption decreases with decreasing frequency, and (ii) the widths of the absorption lines are often much greater than the field strength required for resonance (which is only about 10 gauss at 30 Mc/s). For narrow, intense absorption lines, paramagnetic resonance at radiofrequencies is a very simple technique, however, and details of suitable circuits are discussed in Chapter 8, with the nuclear radiofrequency spectroscopes.

6.5 NUCLEAR INTERACTION AND HYPERFINE STRUCTURE

So far only the electronic spin and magnetic moment have been considered, and no account taken of the spin and moment of the nucleus of the paramagnetic ion. This will interact with the electronic motion to produce a hyperfine splitting of the levels, as occurs in the gaseous case. The nuclear interaction is stronger and more direct for paramagnetic resonance absorption, however, with the result that the hyperfine splitting of the spectra is usually much bigger and more easily interpreted. This splitting is produced by the interaction of the magnetic field due to the nuclear moment, with the electron spin undergoing transition. The $(2I+1)$ different orientations of the nucleus thus produce $(2I+1)$ component levels in each electronic state, the splitting between these being equal and independent of the external field, to a first approximation. The axis of quantization for the nucleus is provided by the very large field produced by the unpaired electrons (about 10^5 gauss), and hence the nuclei do not change their orientation during an electronic transition. Despite this $\Delta M_I = 0$ selection rule, a splitting of the spectra is still obtained, as the order of the $(2I+1)$ component levels is inverted for electron spin states of opposite sign. This is illustrated at the bottom of *Figure 58*, where the nuclear splitting of the two electronic spin levels is shown. Cu^{63} and Cu^{65} both have nuclear spins $I = \frac{3}{2}$, so four component energy levels are produced. The $\Delta M_I = 0$ transitions are shown, and it can be seen how radiation of constant frequency will be absorbed at different magnetic field strengths, and hence a hyperfine splitting of the spectrum is obtained.

The fact that $(2I+1)$ equally-spaced hyperfine lines are produced makes this technique an ideal method for the determination of nuclear spins[11]. Values of the nuclear magnetic moments, and hence the nuclear gyromagnetic ratio, can also be obtained if the magnetic field produced by the electrons at the nucleus can be calculated.

6.6 SECOND-ORDER EFFECTS AND QUADRUPOLE INTERACTION

The above simplified theory, predicting equally-spaced hyperfine splittings, is only strictly true if the applied magnetic field is very much stronger than that produced by the nucleus itself. If the two become comparable, the component hyperfine levels will contain an admixture of the different quantum states M_I, which gives rise to an unequal splitting of the levels. This second-order effect, which shows up as unequal spacings between consecutive hyperfine lines, is more noticeable at 3-cm wavelength than 1·25 cm, as the resonance field for the former is two-and-half times smaller. A very good example of this small change in spacing is observed in the spectra of diluted manganese salts[12], and can be seen clearly in *Figure 61*.

Another effect which gives rise to unequal spacing of the levels and also to the appearance of small intensity transitions normally forbidden by the $\Delta M_I = 0$ selection rule, is the interaction with the nuclear electric quadrupole moment. Since the main interactions are now magnetic, this will only be a very small effect in paramagnetic resonance (instead of the main effect as in gaseous spectroscopy). The quadrupole moment is acted on by the gradient of the electric field at the nucleus, as before, and this interaction is transferred to the energy states via the nuclear magnetic coupling, now being a second-order perturbation rather than a first-order effect. If the applied magnetic field is parallel to the symmetry axis of the crystalline electric field, this interaction will shift all the energy levels by an equal amount and hence produce no change in the observed transitions. If, however, the applied field H makes an angle with the axis of the quadrupole interaction, it is possible for the $\Delta M_I = 0$ selection rule to be broken, and small intensity 'forbidden' transitions occur. The reason for this is that the two types of interaction are now competing, and trying to quantize the nucleus along their two different axes, and this effect will therefore be most pronounced when the applied magnetic field is at right angles to the axis of the crystalline electric field.

The quadrupole interaction also causes a small shifting of the energy levels which will not be the same for each hyperfine component, if H is not parallel to the electric field axis. This interaction has two effects on the spacing of the hyperfine lines (i) a term which produces a constant change in the spacing, and (ii) a term which causes the spacing to be greater at the ends than in the middle when the two fields are at right angles. The quadrupole interaction, therefore, shows up in two ways, (i) by the appearance of normally forbidden transitions, and (ii) by unequal spacings of the hyperfine lines. This last effect can be differentiated from the other second-order shift, discussed above, in that the quadrupole interaction will produce the greatest separations in the centre or at the ends of a hyperfine group, rather than a progressive increase or decrease. A

146

more detailed analysis of these second-order shifts, with mathematical expressions for their angular variation, is given in Section 6.8.

The presence of quadrupole interaction in paramagnetic resonance spectra was first discovered in the case of a diluted copper salt[13], and *Figure 59* shows a series of the spectra as the angle θ

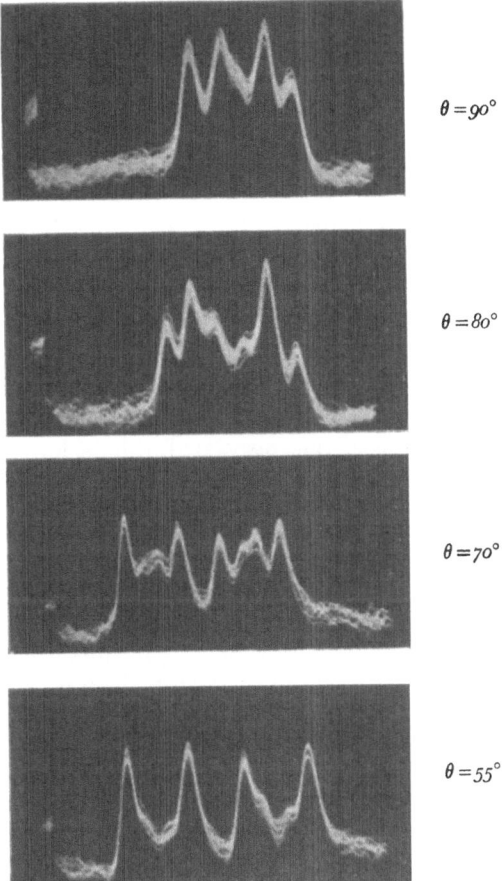

$\theta = 90°$

$\theta = 80°$

$\theta = 70°$

$\theta = 55°$

Figure 59. Quadrupole interaction in $CuK_2(SO_4)_2 \cdot 6H_2O$

between the two axes approaches 90°. At $\theta = 55°$ the four nearly-equally spaced lines due to the magnetic interaction are clearly seen, although two 'forbidden' transitions are already obtained between the two highest-field lines. At $\theta = 70°$ the 'forbidden' lines have risen in intensity to become equal to the 'main' lines, the unequal spacing between which has become apparent. At $\theta = 80°$ the intensity of the second-order transitions is considerably greater than that of the original lines, and they form the main component of the rather complex pattern obtained at $\theta = 90°$. An analysis of this spectrum shows that both $\Delta M_I = \pm 1$ and $\Delta M_I = \pm 2$ transitions occur, the former being prominent between $\theta = 60°$ and $80°$, but die away rapidly between $\theta = 80°$ and $90°$, while the $\Delta M_I = \pm 2$ transitions grow to a maximum at the perpendicular direction. The effect of the nuclear electric quadrupole interaction is not usually as large as in this case of Cu^{2+}, and quite often only very small changes are produced in the normal magnetic splitting.

6.7 THE THEORETICAL HAMILTONIAN

A more exact mathematical form can now be given to the various interaction energies of the paramagnetic ion. The general theory of paramagnetic resonance in crystals is due to ABRAGAM and PRYCE[14], and they express the different factors contributing to total energy of the ion by the Hamiltonian.

$$W = W_F + V + W_{LS} + W_{SS} + \beta \mathbf{H} (\mathbf{L} + 2\,\mathbf{S}) + W_N - \gamma \cdot \beta_N \cdot \mathbf{H} \cdot \mathbf{I}$$

$$\dots\dots (6.4)$$

W_F represents the energies associated with the levels of the free paramagnetic ion due to the configuration of the electrons, and the splittings of these levels is usually of the order 10^5 cm^{-1}. V is the electrostatic energy due to the effect of the crystalline field, and it has been seen that the splittings which it produces are of the order of 10^3–10^4 cm^{-1}. W_{LS} represents the spin-orbit interaction, which causes the energy to deviate from its 'spin-only' value, and is of the order of 10^2 cm^{-1}. W_{SS} is a term representing the magnetic interaction between the electrons themselves and is usually very small but may produce noticeable effects in the case Mn^{2+}, Cr^{2+} and Cr^{3+} of the order of 1 cm^{-1}. The term $\beta \mathbf{H} (\mathbf{L} + 2\mathbf{S})$ represents the effect of the external magnetic field, it produces the splitting of the electronic levels between which the transitions are observed, and is thus about 1 cm^{-1}. W_N is the interaction energy due to the nucleus and contains both the magnetic and electric (i.e. quadrupole) terms, being of the order of 10^{-2} cm^{-1}. The last term, $\gamma \cdot \beta_N \cdot \mathbf{H} \cdot \mathbf{I}$, is to represent the direct effect that the applied magnetic field has on the nuclear magnetic moment. This is very small, but can be observed as a slight shift in the absorption lines in some cases, of about 10^{-3} cm^{-1}.

Since only the lowest energy levels are concerned, the higher levels associated with W_F normally have no effect unless the lowest level is not degenerate when the admixture of higher eigenstates of W_F may

148

be necessary to obtain splitting of the spin degeneracy (as was thought to be the case with Mn^{2+} and Fe^{3+}). By considering the other terms in detail[14, 15], a first-order perturbation Hamiltonian is obtained which gives the shift in energy from the lowest level of the free ion.

i.e.
$$\begin{aligned}
W = V + (\lambda - \tfrac{1}{2}\rho)\,(\mathbf{L}\,.\,\mathbf{S}) &- \rho\,(\mathbf{L}\,.\,\mathbf{S})^2 + \beta\,.\,\mathbf{H}(\mathbf{L} + 2\mathbf{S}) \\
&+ P\{(\mathbf{L}\,.\,\mathbf{I}) + (\xi L\,(L+1) - K)\,(\mathbf{S}\,.\,\mathbf{I}) - \tfrac{3}{2}\xi\,(\mathbf{L}\,.\,\mathbf{S})\,(\mathbf{L}\,.\,\mathbf{I}) \\
&\qquad\qquad - \tfrac{3}{2}\xi(\mathbf{L}\,.\,\mathbf{I})\,(\mathbf{L}\,.\,\mathbf{S})\} \\
&+ q\{(\mathbf{L}\,.\,\mathbf{I})^2 + \tfrac{1}{2}(\mathbf{L}\,.\,\mathbf{I})\} - \gamma\,.\,\beta_N\,.\,(\mathbf{H}\,.\,\mathbf{I}) \qquad\quad \ldots\ldots(6.5)
\end{aligned}$$

where
$$P = 2\gamma\,.\,\beta\beta_N/r^3 \qquad q = \eta e^2 Q/2I(2I-1)r^3$$

γ is the nuclear gyromagnetic ratio in units of $e/2Mc$

r is the radius of the electron orbits

η, ξ and K are constants depending on the electronic configuration, and

λ and ρ are constants for the particular ion

Q is the nuclear electric quadrupole moment

Q, as used here, is the quadrupole moment as normally defined, i.e.

$$Q = \frac{1}{e}\int \rho\,.\,(3z^2 - r^2)\,.\,dv$$

having units of 10^{-24} cm^2 (see MACK[16]). It is therefore the same as used in the equations of gaseous spectroscopy in Chapter 5. In the Spin Hamiltonian described later the quadrupole interaction is usually represented by:

$$\frac{3eQ}{4I(2I-1)}\,.\,\left(\frac{\partial^2 V}{\partial z^2}\right)\left\{I_z^2 - \tfrac{1}{3}I(I+1)\right\}$$

This term is often abbreviated to $Q'\{I_z^2 - \tfrac{1}{3}I(I+1)\}$ and so it should be noted that $Q' = \dfrac{3e}{4I(2I-1)}\,.\,\left(\dfrac{\partial^2 V}{\partial z^2}\right)\,.\,Q$. Unfortunately these two terms have been mixed in the literature, and the quadrupole interaction is often written just as $Q\{I_z^2 - \tfrac{1}{3}I(I+1)\}$. Care must therefore be taken when comparing results from different publications; the term Q', as defined above, will be used in the Spin Hamiltonian throughout this book.

It can be seen that the last two lines of equation (6.5) are due to the nuclear interaction, and although the expression is useful in sorting out the different theoretical contributions to the interaction energies it is usually shortened to the more simple form of

$$\mathscr{H} = \beta\,.\,\mathbf{H}(\mathbf{L} + 2\mathbf{S}) + \lambda\,.\,\mathbf{L}.\mathbf{S}. + A.\mathbf{I}.\mathbf{S}. \qquad\quad \ldots\ldots(6.6)$$

$\mathscr{H}$ now being used to represent the general Hamiltonian, as is done in most of the experimental papers. The method of carrying out the perturbation calculation represented by this equation has been described by PRYCE[17], and the operators referring to spin and nuclear variables are treated as non-commuting algebraic quantities. An expression is therefore obtained which involves the components of **S** and **I**; this is called the 'Spin Hamiltonian', and the required energy levels can then be derived as they are the eigenvalues of this operator. Consideration of[14, 15] the properties of the terms in equation (6.6) enables the 'Spin Hamiltonian' to be written as

$$\mathscr{H} = \mathbf{S}.D.\mathbf{S} + \beta . \mathbf{H}.g.\mathbf{S} + \mathbf{S}.T.\mathbf{I} + \mathbf{I}.P.\mathbf{I} - \gamma \beta_N . \mathbf{H}.\mathbf{I} \quad ....(6.7)$$

This is still in vector form, and the various constants such as D and g are tensors. If the crystalline electric field has axial symmetry the equation may be rewritten in its component form, to become:

The Spin Hamiltonian:
$$\mathscr{H} = D\left[S_z^2 - \tfrac{1}{3}S(S+1)\right] + \beta[g_{\parallel}H_zS_z + g_{\perp}(H_xS_x + H_yS_y)] \\ + AI_zS_z + B(I_xS_x + I_yS_y) + Q'\left[I_z^2 - \tfrac{1}{3}I(I+1)\right] - \gamma\beta_N . \mathbf{H}.\mathbf{I} \\ ....(6.8)$$

The term representing the constant interaction $(-\beta^2 . \mathbf{H}. \Lambda \mathbf{H})$ has been omitted because it is normally very small.

The value of S to be substituted into this equation is determined by putting the multiplicity of the ground state equal to $(2S+1)$. This 'effective electronic spin' may not be the same as the S of the free ion. The equation therefore predicts a g value, and hyperfine structure variation which have the same axial symmetry as the crystalline field; and for cases in which this is not so, a higher order perturbation must be made to include terms in S^4. It can be seen that the first two terms now contain the electronic interaction, the third and fourth are those due to the interaction of the nuclear magnetic moment with the electronic field, the fifth represents the quadrupole interaction, and the last is that due to the external magnetic field acting directly on the nucleus.

The Spin Hamiltonian is an expression that has been derived on entirely theoretical grounds, and can be viewed as a shorthand way of representing the different interaction energies known to be present in the paramagnetic ion. The numerical constants are unspecified, but can be related back to the electronic configurations and, in principle, the process can be quite simply reversed. Thus, if numerical values are given for all the parameters in equation (6.8), it is then possible to use them to calculate the actual electron orbitals and energy levels of the ion. It is this last fact which is the essential link between the theory and experiments of paramagnetic resonance.

It is, in fact, the recognized responsibility of the experimentalist to present his results in the form of equation (6.8), so that his measured parameters can then be used by the theoretician to calculate the energy states and hence types of binding of the paramagnetic ion. The actual expressions used to convert absorption line measurements into values of D, g, A, B and Q' are given in the next section; here it may be noted that: (i) the term in D describes how the levels behave in zero magnetic field in the absence of nuclear interaction, and comes from the second-order effects of the crystal field, spin-orbit coupling and the spin-spin interaction; (ii) $g_{\parallel}$ and $g_{\perp}$ are the g values when the applied magnetic field is respectively parallel, and perpendicular to, the crystalline field axis; (iii) A and B measure the splitting of the hyperfine structure parallel, and perpendicular, to the axis; and (iv) Q' is the parameter measuring the small changes in the spectrum produced by the quadrupole interaction.

It can be seen from this that it is usually possible to express all the experimental measurements in the form of equation (6.8), and this equation can be regarded as one of the most important in paramagnetic resonance of the crystalline state, being the essential link between the measurements of the experimentalist and the general considerations of the theoretician.

6.8 Angular Variation

As seen, equation (6.8) can be regarded as a shorthand summary of the experimental measurements on any particular paramagnetic ion; and although only the parameters along, and at right angles to, the crystalline axis enter into the expression, it is important to know how the spectrum might be expected to behave with variation in angle as this variation is often used to sort out the different effects. The question of angular variation has been treated in detail by BLEANEY[18], who started with the Spin Hamiltonian of equation (6.8), and derived expressions for the field position of the absorption lines with varying θ (θ being the angle between the applied magnetic field and the axis of the crystalline electric field).

(a) *Electronic transitions.* Bleaney[18] derives the following equation for the centres of the different electronic transitions, ignoring the hyperfine structure initially:

$$h\nu = g\beta H + D\left(M - \tfrac{1}{2}\right)\left\{3\frac{g_{\parallel}^2}{g^2}\cos^2\theta - 1\right\}$$
$$- (Dg_{\parallel}g_{\perp}\cos\theta\sin\theta/g^2)^2(2g\beta H_0)^{-1}\{4S(S+1) - 24M(M-1) - 9\}$$
$$+ (Dg_{\perp}^2\sin^2\theta/g^2)^2(8g\beta H_0)^{-1}\{2S(S+1) - 6M(M-1) - 3\} \ldots (6.9)$$
$$\text{where } g^2 = g_{\parallel}^2\cos^2\theta + g_{\perp}^2\sin^2\theta$$

151

Here M is the electronic magnetic quantum number, and S is the effective electron spin as before. It can be seen that there are thus $2S$ different transitions, corresponding to the different values of M, and if the second-order terms in D^2 are neglected these transitions are equally spaced (if $S = \frac{1}{2}$ there is only one transition of course). As measurements are usually made at constant frequency with a varying field strength, it is best to rearrange equation (6.9) to give the position of the absorption lines directly in terms of magnetic field. This may be done by defining H_0 as $\dfrac{h\nu}{g\beta}$, and dividing through by $g\beta$, to give:

$$H = H_0 - \frac{D}{g\beta}(M - \tfrac{1}{2})\left\{3\frac{g_\parallel^2}{g^2}\cos^2\theta - 1\right\}$$
$$+ \left(\frac{D}{g\beta}\right)^2 \frac{1}{2H_0}\left\{\frac{g_\parallel g_\perp \cos\theta \sin\theta}{g^2}\right\}^2 . \{4S(S+1) - 24M(M-1) - 9\}$$
$$- \left(\frac{D}{g\beta}\right)^2 \frac{1}{8H_0}\left(\frac{g_\perp \sin\theta}{g}\right)^4 . \{2S(S+1) - 6M(M-1) - 3\} \quad \ldots (6.10)$$

If $\theta = 0$ and the magnetic field is directed along the crystalline axis, all the second-order terms in D^2 are zero, and hence the $2S$ electronic transitions are equally separated in the magnetic field and the value of D can be determined directly from this separation.

The first-order variation of the spectrum with angle follows a $\left\{3\left(\dfrac{g_\parallel}{g}\right)^2 \cos^2\theta - 1\right\}$ law, and thus the splitting between the lines falls to zero when $\theta = \cos^{-1}\left(\dfrac{1}{\sqrt{3}} . \dfrac{g}{g_\parallel}\right)$, increasing to a subsidiary maximum at $\theta = 90°$, the absorption lines then being in the reverse order compared with their positions at $\theta = 0°$. Variation of this kind is seen very nicely in the case of Mn^{2+}, V^{2+} and Gd^{3+} ions, the lines closing in on each other as θ increases from zero, and crossing over at about $54°$, and then separating out to the subsidiary maxima. The second-order terms will produce unequal spacing in the lines when $\theta \neq 0°$, and the lines will therefore not all cross over together as would otherwise be the case. This second-order effect can also be seen at $\theta = 90°$, the last term in equation (6.10) introducing a shift of the transitions relative to each other. Since H_0 occurs in the denominator of these terms, the second-order effects will not be so pronounced in the shorter wavelength measurements.

(b) *Hyperfine structure.* It has been seen that no change in the nuclear quantum state occurs during an electronic transition, and the complete selection rule is therefore $(M, M_I) \rightarrow (M - 1, M_I)$. Each electronic energy level is split by the nuclear interaction, and thus each electronic transition is also split into hyperfine components. Bleaney[18] shows that this splitting can be represented by adding the following terms to equation (6.9):

$$KM_I + \frac{B^2}{4g.\beta.H_0} \cdot \left(\frac{A^2 + K^2}{K^2}\right) \{I(I+1) - M_I^2\}$$

$$+ \frac{B^2}{2g\beta H_0} \cdot \frac{A}{K} \cdot M_I(2M-1) + \frac{1}{2g\beta H_0}\left(\frac{A^2 - B^2}{K}\right)^2 \left(\frac{g_\| g_\perp}{g^2}\right)^2 \sin^2\theta \cos^2\theta . M_I^2$$

$$+ \frac{Q'^2 \cos^2\theta \sin^2\theta}{2KM(M-1)}\left(\frac{ABg_\| g_\perp}{K^2 g^2}\right) M_I \{4I(I+1) - 8M_I^2 - 1\}$$

$$- \frac{Q'^2 \sin^4\theta}{8KM(M-1)}\left(\frac{Bg_\perp}{Kg}\right)^4 . M_I . \{2I(I+1) - 2M_I^2 - 1\} \qquad \ldots\ldots(6.11)$$

where $\qquad K^2 g^2 = A^2 . g_\|^2 . \cos^2\theta + B^2 . g_\perp^2 . \sin^2\theta$

This equation can be rearranged, as before, to give the position in field directly, as summarized in Appendix I.

If all the second-order terms are neglected, it is seen that the main effect of the nuclear interaction is to produce $(2I+1)$ different transitions, with a splitting between each equal to K. In some cases $(Mn^{2+}$ and $V^{2+})$ [12, 19], $g_\| = g_\perp$ and $A = B$, so that K is constant and there is no variation of the hyperfine splitting with angle. In other cases (e.g. Cu^{2+}) [13] g and the hyperfine splitting vary with angle, both having maxima along the crystalline field axis and minima at right angles to it; and when the anisotropy is very large (e.g. Co^{2+}) [20] the hyperfine splitting, as measured in gauss, may go through its maximum at an angle away from the axis.

The effect of the second-order magnetic terms is to cause a slight successive change in the splitting between the hyperfine components, these shifts are usually small but can be used to determine the relative signs of D, A and B. The effect of the quadrupole interaction on the spectrum has already been considered in Section 6.6, and the shifts that they produce described. Equation (6.11) does not contain the 'forbidden transitions' corresponding to $\Delta M_I = \pm 1$ and ± 2, which the quadrupole interaction also allows. The intensity of these lines is of the order of Q'^2/K, compared with the main hyperfine lines, and Bleaney[18] gives detailed expressions for their variation of intensity and spacing with angle. The analysis of this second-order effect allows Q' to be calculated, and from this, actual values of the nuclear electric quadrupole moment Q can be derived. All the expressions are symmetrical, however, and none of the second-order effects, described above, allows an absolute determination of the signs of D, A, B or Q', only values relative to each other being obtained.

An absolute determination can be made, however, if the direct interaction of the external field on the nucleus is strong enough. This interaction, which is expressed as the last term of the Spin Hamiltonian in equation (6.8), has not been considered in detail yet. Its asymmetrical effect arises from the fact that it acts with the quadrupole interaction to increase the separation of the satellite doublet on one side, and in opposition, to decrease the separation on the other. A careful analysis of how the absolute signs can be determined from these shifts is given by Bleaney[18], and it is found that the absolute sign of the quadrupole moment can then be obtained, but not that of the nuclear magnetic moment. Once the magnitudes and signs of the coefficients D, A, B and Q have been thus determined, they can then be used by the theoretical physicist to derive the actual electronic states and binding, as well as the nuclear moments.

6.9 The Iron Group—Effect of Crystalline Fields

Before summarizing the results obtained for salts of the iron group transition elements, the expected effect of the crystalline electric field may be considered in a general qualitative way. The detailed effects of the crystalline field on any particular ion are obtained by expanding the potential as a series of spherical harmonics and considering the effect of these on the different orbital states (see for example Bleaney and Stevens[15], or ABRAGAM and PRYCE[21]). Considerable information of a qualitative nature can be obtained, however, from a more general approach. The first treatment of this kind was due to BETHE[22] who used group theory methods. He showed that for the orbital momentum to be completely quenched, there can be at the most, rhomboidal symmetry of the crystalline field. A field possessing cubic, tetragonal or hexagonal symmetry always leaves two of the three cartesian potential coefficients equal, and hence does not completely remove the spatial degeneracy. The effect of the higher order terms in the expansion of the potential function, however, usually removes the actual axis of symmetry. Bethe showed, nevertheless, that in such cases a partial degeneracy still remains, and the quenching of all the levels then depends on the relative energy values of the different states. Hence, large fields of rhomboidal or lower symmetry always quench completely, while with those of greater symmetry the quenching is complete only if certain states are lying lowest.

VAN VLECK[23] and SCHLAPP and PENNEY[24] have also considered the effect of fields of different symmetry on various iron group ions, and the general results to be expected are summarized in *Table 6.1*.

Table 6.1. Splitting of Orbital Levels by Crystalline Field

K	Free Ion	Cubic	Trigonal	Tetragonal	Rhombic
0	1	1	1	1	1
$\frac{1}{2}$	2	2	2	2	2
1	3	3	1, 2	1, 2	3×1
$1\frac{1}{2}$	4	4	2×2	2×2	2×2
2	5	2, 3	1, 2×2	2, 3×1	5×1
$2\frac{1}{2}$	6	2, 4	3×2	3×2	3×2
3	7	1, 2×3	$3 \times 1, 2 \times 2$	$3 \times 1, 2 \times 2$	7×1
$3\frac{1}{2}$	8	$2 \times 2, 4$	4×2	4×2	4×2

This table shows how the $(2K + 1)$ degeneracy of an ion is removed by fields of different symmetry. K may represent any general quantum number, and in this case is usually the orbital L value. The numbers show the degeneracies of the levels left, e.g. 2×3 indicates that there are two, triply-degenerate levels present, and $n \times 1$ indicates that there are n separate levels with no degeneracy left.

It is still necessary to determine which of these levels is at the bottom in any particular case. The sign of the cubic field component is helpful in this. Thus the field potential may be represented as

$$V = Ax^2 + By^2 - (A + B)z^2 + D\,(x^4 + y^4 + z^4) \quad \ldots . (6.12)$$

and the sign of the cubic component D will determine which way up the orbital multiplet occurs. GORTER[25] has shown that if the ion is surrounded by an octahedron of water molecules, then the sign of D will be positive, whereas four molecules in a tetrahedron, or eight in a cube, will give a negative sign. Since in nearly all the crystals studied the paramagnetic ion is surrounded by an octahedron of water molecules D is usually positive. The value of λ, the spin-orbit coupling coefficient, is also important and has a positive value for ions with less than half the $3d$ shell completed, and a negative value for those with more than half completed. Equivalent ions at either end of the shell (i.e. Ti^{3+}, with one electron in the $3d$ shell, is equivalent to Cu^{2+}, with one electron missing from the $3d$ shell, and so on) have their levels split into the same multiplet structure, but one is inverted compared with the other.

Apart from the orbital degeneracy of the ionic levels there is also the spin degeneracy to consider. There are two general theorems which help in this. First, Kramers's theorem states that an external field will always leave a system containing an odd number of electrons with an even degeneracy for every level. Secondly, there is the Jahn-Teller effect, which states that any system having a degenerate ground state, will spontaneously distort so that as much degeneracy as possible is removed. Hence it follows that the total (orbit plus spin) degeneracy of the ground state will be a singlet if the number of electrons is even, and will be a doublet if the number is odd. The value of S, the effective electronic spin, that must be substituted into the Spin Hamiltonian can also be obtained by general reasoning in most cases. Thus, if there is an orbital singlet lowest, the Jahn-Teller or other splitting of the spin degeneracy is usually small, and hence the actual electronic spin is substituted into the Spin Hamiltonian. (E.g. V^{2+} has an orbital singlet lying lowest, 10^4 cm^{-1} below the first triplet, and with three electrons in the $3d$ shell it has a spin degeneracy of 4. This is split to form two doublets, with a separation of about 0.3 cm^{-1}[19], and Kramers's theorem prevents any further lifting of the degeneracy. Application of a magnetic field splits both of these levels (see *Figure 60*) and three transitions are to be expected between the four resultant levels. Hence a value $S = \frac{3}{2}$ must be substituted into the Spin Hamiltonian.)

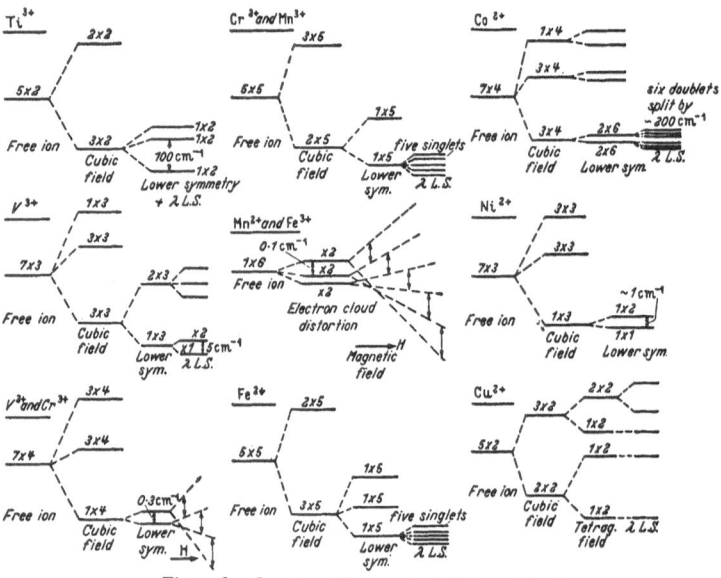

Figure 60. Iron transition group. Splitting of levels
(Not to scale)

If, however, the lowest orbital level is degenerate, the lower symmetry fields and spin-orbit coupling will split the levels by about 100 cm^{-1}, leaving the minimum degeneracy in each. If there is an even number of electrons in the ion, the ground level will therefore be a singlet and no paramagnetic resonance will be observed. If the ion has an odd number of electrons, however, an isolated doublet will be left at the bottom, and the application of a magnetic field will produce two levels between which resonance may occur. Thus, whatever the spin degeneracy of the free ion, the ground state is acting with a degeneracy of only two and hence an effective spin value of $\frac{1}{2}$. Hence $S = \frac{1}{2}$ is substituted into the Spin Hamiltonian. (E.g. Co^{2+}. The ground state of the free ion is ^4F, and hence has a spin degeneracy of 4. The cubic crystalline field leaves an orbital triplet lowest, however, with the result that the total 3×4 degeneracy is lifted to give six doublets[21], the lowest lying about 200 cm^{-1} below the next (see *Figure 60*). This ground level therefore acts with an effective spin of $\frac{1}{2}$, instead of its free-ion value of $\frac{3}{2}$.)

In order to classify the ions of the first transition group, *Table 6.2* lists their different valencies against the number of electrons in the $3d$ shell and the ground state of the free ion.

Table 6.2 .

Ion	Number of electrons in 3d shell	Ground State
Ti^{3+}	1	^2D
V^{3+}	2	^3F
Cr^{3+} and V^{2+}	3	^4F
Mn^{3+} and Cr^{2+}	4	^5D
Fe^{3+} and Mn^{2+}	5	^6S
Fe^{2+}	6	^5D
Co^{2+}	7	^4F
Ni^{2+}	8	^3F
Cu^{2+}	9	^2D

The splittings of the orbital levels of these ions, by crystalline fields of different symmetry, are summarized in *Figure 60*. They are considered in more detail, together with the experimental results on each ion, in Section 6.11, and ligand field methods in Chapter 11.

6.10 General Results from the Iron Transition Group

Before considering the results for each ion separately, the results obtained from the first transition group can be considered together in a general way. Results obtained from paramagnetic resonance can always be divided into two groups as in gaseous microwave work. First, the positions, splittings and intensities of the electronic transitions can be analysed to give information on the higher orbital levels, and hence on the nature of the chemical binding of the particular paramagnetic ion. This information, with its very precise values for the energy levels, is of great importance for the development of magnetic and solid state theory. Then, secondly, the hyperfine structure of the electronic states can be analysed to give data on the nuclear properties. The nuclear spin I can be determined directly, if any hyperfine structure is resolved, and this method has already been used to determine hitherto unknown spins (as in the case of Nd^{143} and Nd^{145})[11], and to confirm others obtained by optical spectroscopy (such as V^{51} and Pu^{239})[19, 26]. The nuclear electric quadrupole moment can also often be obtained, as well as the magnetic moment, if the second-order shifts that it produces are large enough. Apart from these purely nuclear properties, the splitting of the hyperfine structure can also give considerable information on the electronic state of the ion as the hyperfine interaction is very sensitive to the magnetic fields produced by different electron orbits.

(a) *Information from the electronic transitions.* The splitting of the electronic levels in paramagnetic ions has been a subject for theoretical study for some time, and this has been so for salts of the iron transition group[23] especially as a large number of experimental results on such properties as susceptibility and specific heat variations with temperature were available for them. Measurements on susceptibility and specific heat do not give direct values of energy for one level, but rather an 'average' effect over all the levels that are populated at the temperature at which the measurements are made. Various energy level systems can be postulated, however, and then tested by the susceptibilities which they predict. In this way considerable information on the lower energy levels of paramagnetic ions in the crystalline state was obtained before the advent of paramagnetic resonance.

The great advantage of the paramagnetic resonance measurements in this connexion is their ability to select one or two levels only and give very detailed data on their parameters. Thus, instead of having an 'averaged out effect' over several levels, precise information is available about each level separately. This new information from the resonance experiments has sometimes confirmed the previous theories, and sometimes shown them to be quite wrong (as in the case of Mn^{2+}). Nowadays the theoretical procedure is usually reversed. Thus paramagnetic resonance measurements are used to predict the complete set of energy levels for any given salt, and then this information is used in order to choose the best conditions for low-temperature experiments. A classic example of this was the case of cobalt and copper rubidium sulphate. This was the first substance in which alignment of nuclei was obtained at low temperatures[27], and the choice of salt and conditions for the experiment were all based on the paramagnetic resonance measurements previously made[18]. A more detailed discussion of the link-up between paramagnetic resonance and low-temperature physics will be found in Chapter 9.

(b) *Information from the hyperfine structure.* Since most of the nuclear spins and moments are already known, the amount of new nuclear data obtained for the iron group transition elements is relatively small. Two determinations of unknown nuclear spin have been made, however, namely Cr^{53} [28] and V^{50} [29], and many others verified as well as quadrupole moments determined. One of the most striking results obtained, however, from an analysis of the hyperfine structure of the iron group transition elements has been the light thrown on the actual electronic states of the paramagnetic ions. It was shown[14] that the previously accepted idea of the ground states being built up entirely of $(3d)^n$ configurations could not produce the experimentally observed hyperfine splittings. Abragam and Pryce[14] suggested that this discrepancy could be resolved by allowing configurational interaction to admix $(3s\ 3d^n\ 4s)$ states—i.e. where one electron has moved from the $3s$ to the $4s$ state. Only small admixtures of this new configuration are necessary to produce a large hyperfine splitting because of the very great effect of an unpaired $3s$ electron orbital which reaches close in to the nucleus. It was shown that this type of configurational coupling was necessary to explain the hyperfine splittings obtained for nearly all the ions of the first transition group, and it is an interesting point that the g value determinations were not sensitive enough to the s electrons to show up this effect, and it awaited the hyperfine splitting measurements to uncover the detailed electronic states. A comparison[15] of the contributions to the magnetic field at the nucleus, from these unpaired s-electron configurations, shows that they are remarkably constant for all the ions throughout the transition group; only varying by ± 10 per cent for similar types of salts.

Photographs of the hyperfine structures observed for different nuclei of the iron transition group are shown in *Figure 61*. These illustrate the general features met in such hyperfine splittings. Thus *61(a)* is from a diluted cobalt salt and the eight component lines due to the $I = \frac{7}{2}$ nuclear spin of Co^{59} are clearly seen. This photograph was taken when the angle between the applied field and crystalline axis was 86°, and the unequal splitting, due to the large effects of the second-order terms at this angle, are also very evident. *61(b)* is a photograph of the hyperfine structure obtained from a diluted copper salt. This illustrates the effect due to two nuclei, Cu^{63} and Cu^{65}. Both of these have a nuclear spin $I = \frac{3}{2}$ and have nearly equal magnetic moments. Thus each nucleus gives rise to four hyperfine lines, and these two sets do not quite coincide, with the result that the extreme lines, due to the two different isotopes, are clearly resolved. It was necessary to grow the crystals out of heavy water, in this case, before

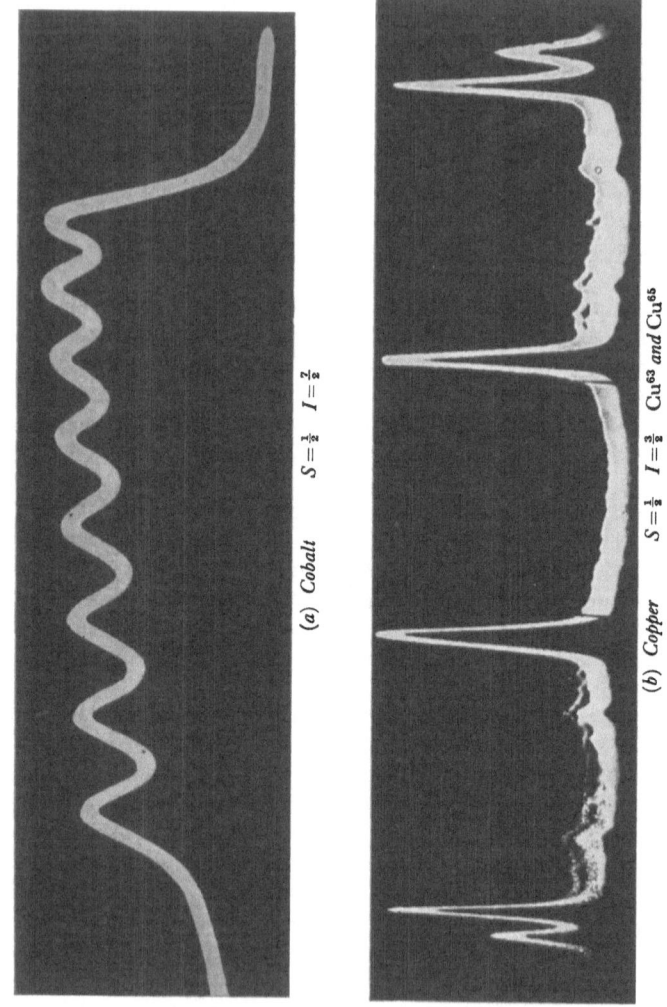

(a) Cobalt $S = \frac{1}{2}$ $I = \frac{7}{2}$

(b) Copper $S = \frac{1}{2}$ $I = \frac{3}{2}$ Cu^{63} and Cu^{65}

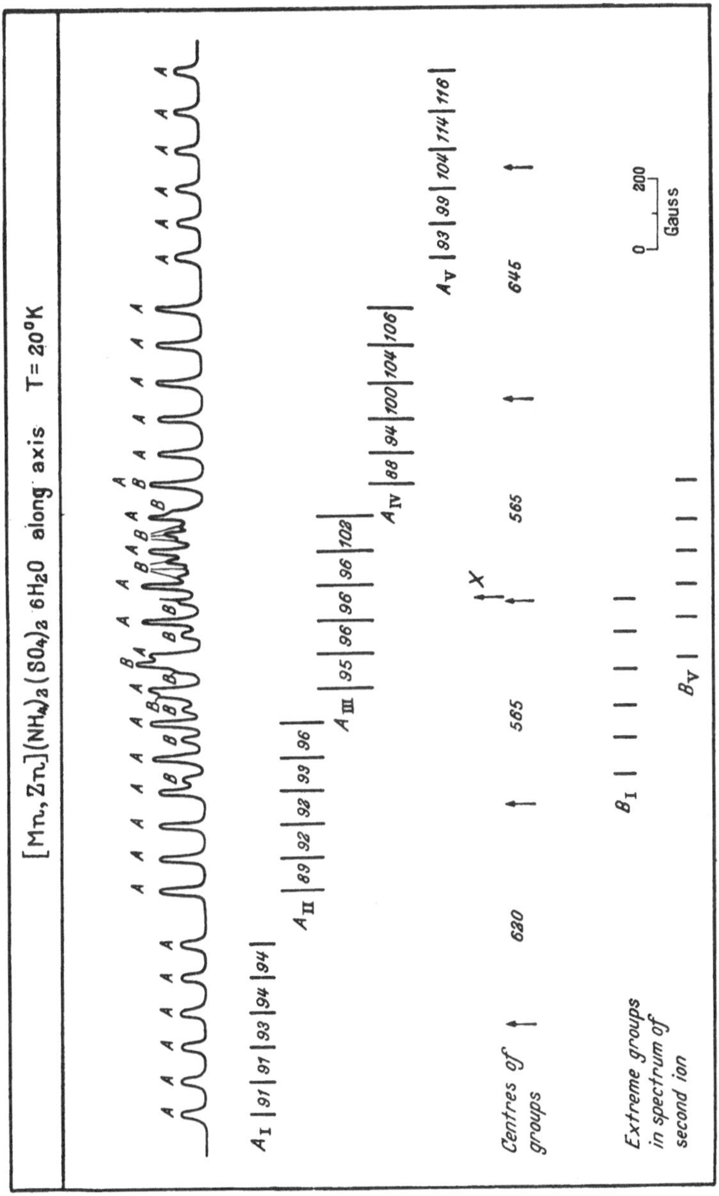

(c) *Manganese* $S = \frac{5}{2}$ $I = \frac{5}{2}$
Figure 61. Hyperfine structure in iron group

the line width could be reduced sufficiently to resolve the outer transition into two separate lines. The very weak lines seen inside the outer pairs are due to transitions induced by the quadrupole interaction. In $6I(c)$ a trace of the complete spectrum obtained from a manganese double sulphate is shown and illustrates the complexity that can be associated with a spectrum. Manganese has an electronic spin of $S = \frac{5}{2}$, and Mn^{55} has a nuclear spin, $I = \frac{5}{2}$. Five different electronic transitions are therefore expected, each being split into six hyperfine components. These thirty lines are further complicated, in the case of the double sulphates, by the fact that there are two dissimilar ions per unit cell, and hence each gives rise to a set of thirty lines, the two sets having different axes for their angular variation. The tracing shown is for the case when the applied magnetic field is along the axis of the crystalline field of one of the ion types, and hence maximum splitting of its electronic transitions is obtained. The five different transitions for this ion are picked out below (labelled A), and the six hyperfine components for each are clearly seen. The slight shift due to second-order terms is also evident from the successive change in spacing between the hyperfine components, the figures shown being the actual separations in gauss.

6.11 RESULTS FOR INDIVIDUAL IONS OF THE FIRST TRANSITION GROUP

A summary of the orbital splittings of the ions in the first transition group has already been given in *Figure 60*. The ions are now listed in the order of the number of their $3d$ electrons, and the spectroscopic state of the free ion is shown following the d orbital configuration.

(a) Ti^{3+} $(3d)^1$ $^2D_{3/2}$. The orbital levels are split to leave one level doubly-degenerate in spin, about 200 cm^{-1} below the next. This is therefore a case in which the splitting to the higher orbital levels is small, and the spin-orbit coupling would be expected to produce a strong spin-lattice interaction, and also shift the 'g values' considerably from the free-spin value. Both of these predictions are confirmed by the experimental results. No absorption could be observed in titanium caesium alum for temperatures down to 20° K, and it was necessary to go down to 8° K[31] before any resonance line was obtained. The measurements of BIJL[31], between 6° and 8° K, were only made on a powdered sample, but more recent experiments have been made by BOGLE and COOKE[32] who obtained $g_\parallel$ as 1·25 and $g_\perp$ as 1·14, when using a single crystal below 4° K.

Apart from an estimation of the spin-lattice relaxation time, these are the only results obtained for Ti^{3+} so far, no hyperfine structure from the low-abundant odd isotopes having been observed. Bijl[31] estimates that the spin-lattice relaxation time changes from $0.48 \cdot 10^{-10}$ sec at $6.3°$ K to $0.25 \cdot 10^{-10}$ sec at $7.9°$ K, and would probably have decreased to 10^{-12} sec at $10°$ K, thus making any absorption unobservable.

BLEANEY[33] developed an approximate theory and by ignoring contributions from the higher levels, showed that

$$g_{\parallel} = \frac{3(\Delta + \lambda/2)}{S} - 1 \quad g_{\perp} = \frac{\left(\Delta - \frac{3\lambda}{2}\right)}{S} + 1 \quad \ldots (6.13)$$

where Δ is the splitting of the orbital triplet due to the trigonal field alone, λ is the spin-orbit coupling ($= + 154 \, \text{cm}^{-1}$), and

$$S = \{(\Delta + \lambda/2)^2 + 2\lambda^2\}^{\frac{1}{2}}.$$

The results of Bogle and Cooke[32] show that these expressions have to be modified, as there is considerable admixture from the upper levels[15], and also some covalent π-bonding.

(b) V^{3+} $(3d)^2$ $^3F_2$. The sevenfold orbital level of this ion is split by the crystalline fields to leave a triplet level at the bottom. Susceptibility measurements[34] indicate that the ground state is a triplet (i.e. $S = 1$) with a splitting of about $5.0 \, \text{cm}^{-1}$. No paramagnetic resonance has yet been reported on any salts of this ion, and if it is detected it will only be at low temperatures.

(c) Cr^{3+} and V^{2+} $(3d)^3$ $^4F_{3/2}$. I' has already been seen that the crystalline fields split the orbital levels of these ions to give a singlet orbital level at the bottom, about $10^4 \, \text{cm}^{-1}$ below the first triplet. The lower symmetry fields and spin-orbit coupling lift the fourfold spin degeneracy to produce two doublets with a small zero-field splitting. Such conditions should allow observation of paramagnetic resonance absorption at quite high temperatures, and produce spectroscopic splitting factors very close to the free-spin value. This is confirmed by the experimental results, as resonance lines can be obtained from the salts of both ions at room temperature[30], and both have g values very close to 2.0.

Most of the work on trivalent chromium has been with the alums. Chrome alum was, in fact, the first salt investigated in detail by paramagnetic resonance methods[35]. These and other[36, 37] measurements resolved the three transitions between the four energy levels that separate out in the magnetic field, and showed that the zero-field splitting of the two Kramers's doublets was about $0.15 \, \text{cm}^{-1}$, varying slightly with the other ion in the alum structure. Low temperature measurements[38, 39] on the chrome alums indicate that

163

the structure of the unit cell changes on cooling, and a sudden jump in the electronic splittings is often obtained.

No hyperfine structure was observed in the initial measurements on the chromic alums, mainly because the one odd chromium isotope present, Cr^{53}, has an abundance of only 9·4 per cent. Bleaney and Bowers[28] succeeded in resolving out the hyperfine splitting, however, by using a sample artificially enriched in Cr^{53}, and growing the crystals from heavy water. The first observations were on potassium chromium selenate and partial resolution of the hyperfine structure was obtained. Further measurements on potassium chromicyanide[87] gave complete resolution of the four hyperfine components, the much narrower width in this case being due to the fact that nitrogens and not protons or deuterons now surround the paramagnetic ion. This determination of $I = \frac{3}{2}$ for Cr^{53} was the first new nuclear spin of an iron-group element discovered by means of paramagnetic absorption.

The first detailed measurements on a vanadous salt were those of Bleaney, Ingram and Scovil[19], who confirmed the somewhat doubtful nuclear spin of V^{51} as $I = \frac{7}{2}$. The splitting of the triplet in this case is of the order of 0·3 cm^{-1}, and hence in order to avoid large second-order effects and zero-field lines, measurements were made at 1·25 cm wavelengths. Eight hyperfine components with isotropic and relatively large splitting were obtained for each electronic transition. The energy level system giving rise to these lines is very similar to that illustrated for the case of manganese in *Figure 62*, except that there are only two, and not three electronic doublets, and each electronic level is split into eight, instead of six hyperfine components.

The Spin Hamiltonian used to describe the spectrum was of the form

$$\mathscr{H} = g \cdot \beta \cdot \textbf{H.S.} + D\left(S_z^2 - \tfrac{1}{3}S(S+1)\right) + E\left(S_x^2 - S_y^2\right)$$
$$+ A \cdot S_z I_z + B\left(S_x I_x + S_y I_y\right) \qquad \dots (6.14)$$

The rhombic component of the crystalline field was found to be very large in this case, and it is really more correct to think of it as a rhombic field with a small tetragonal component. The measured values of the Hamiltonian parameters were $A = B = 0.0088$ cm^{-1}, $D = 0.158$ cm^{-1}, $E = 0.049$ cm^{-1} and $g = 1.951 \pm 0.002$ along the axis.

Measurements[29] on vanadous salts enriched with V^{50} have been used to determine its previously unknown nuclear spin. Baker and Bleaney[29] used a crystal of $K_4V(CN)_6 \cdot 3H_2O$, diluted with the isomorphous diamagnetic ferrous salt, and employed the normal Spin Hamiltonian, plus rhombic correction, to summarize the results. A small set of hyperfine components were observed among the main lines due to the V^{51}, and from these the nuclear spin was determined as $I = 6$. The hyperfine splitting in the complex cyanide is about 30 per cent smaller than in the hydrated sulphates, the value of A and B being 0·0056 cm⁻¹ for V^{51} in the former case.

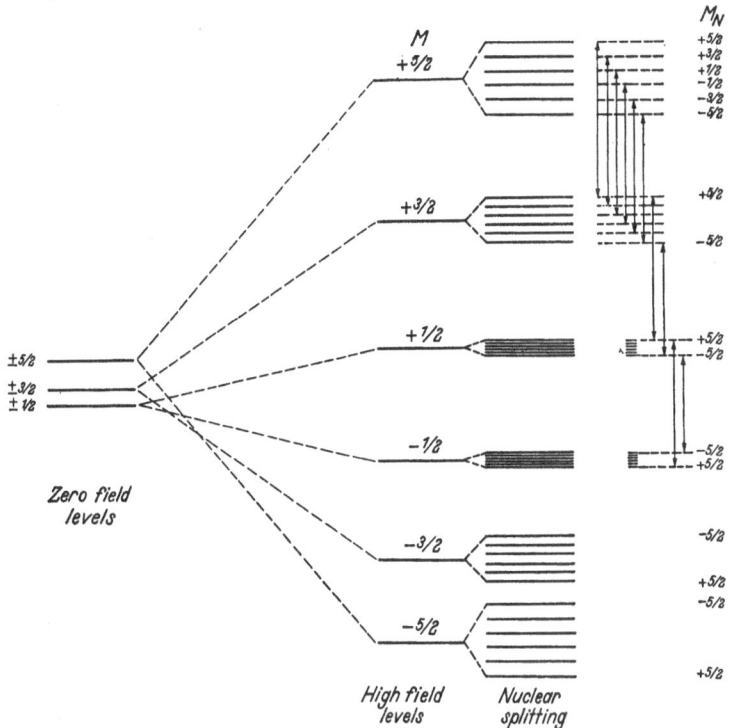

Figure 62. *Energy level splitting of* Mn^{2+}

(d) Cr^{2+} and Mn^{3+} $(3d)^4$ $^5D_0$. The fivefold orbital degeneracy of these ions is lifted by a cubic field to give an upper triplet and lower doublet. A field of tetragonal symmetry will split the doublet to give a singlet orbital level as the ground state with a fivefold spin degeneracy. This spin degeneracy is probably also lifted by the spin-orbit coupling, and since there is an even number of electrons Kramers's theorem no longer applies, so that five singlet levels may be formed by the Jahn-Teller effect. If the splitting between these levels is greater than 1 cm⁻¹ it may not be possible to obtain any paramagnetic resonance between them.

ONO *et al.* [120] have recently made some measurements on $CrSO_4$. $5H_2O$ at wavelengths in the 5 mm region. The results can be fitted to the Hamiltonian of (6.14) with $D = 2·24$ cm⁻¹, $E = 0·10$ cm⁻¹, and show that splittings of the spin quintuplet are large.

(e) Mn^{2+} and Fe^{3+} $(3d)^5$ $^6S_{5/2}$. These two ions are rather a special case, in that the $3d$ shell is now exactly half-full, and the free ion is thus in an S state with no orbital momentum to be quenched. Therefore a very long spin-lattice relaxation time and g values very close to the free-spin value are to be expected. These are ideal conditions for the observation of paramagnetic resonance, and both manganous and ferric compounds have been studied ever since the initial paramagnetic absorption measurements of CUMMEROW and HALLIDAY[40].

The first detailed measurements on the electronic and hyperfine splittings of manganese salts were made by Bleaney and Ingram[12] on diluted crystals of $Mn(NH_4)_2(SO_4)_2 \cdot 6H_2O$ and $MnSiF_6 \cdot 6H_2O$. The sixfold spin-degeneracy is lifted by second-order effects to give three doublets separated by about $0\cdot1$ cm^{-1}. The application of the magnetic field splits these doublets and six different electronic energy levels are obtained, between which transitions can occur, with the selection rule $\Delta M = \pm 1$. The behaviour of these energy levels is shown in Figure 62, which is given in detail as it is typical of the splittings that are produced when the effective spin value to be substituted into the Spin Hamiltonian is greater than $S = \frac{1}{2}$. The splitting of each electronic level into $(2I + 1)$ hyperfine components is also shown, and the actual transitions between the different levels is indicated to show how the spacing between the absorption lines arises. The nuclear spin of Mn^{55} is $\frac{5}{2}$, thus five electronic transitions, each split into six hyperfine components are expected, and the experimental observation of this has already been discussed and is illustrated in Figure 61.

An analysis of the electronic splittings obtained from the paramagnetic resonance results showed that the previous theories on the ground state splitting of the manganous ion were quite wrong. Bethe's[22] group theoretical treatment had shown that the cubic field will split the sixfold spin degeneracy to give a doublet and a quadruplet, and the high-order interactions responsible for this splitting were considered in more detail by VAN VLECK and PENNEY[41]. This cubic field splitting was assumed to be the only one present, until the paramagnetic resonance measurements, when Abragam and Pryce[14] suggested that another mechanism might be mainly responsible for the splitting of the S state. This mechanism depends on the distortion of the otherwise spherical, electron-cloud distribution by the electric field. If the electron-cloud distribution is spherical each of the different sub-states $M = \frac{5}{2} \ldots -\frac{5}{2}$ will have the same energy, but distortion of the distribution causes a splitting of these levels via the dipole-dipole interaction between the spin magnetic moments. This type of interaction is a second-order effect which can be generally represented by $(d_1 . S_z^2 + d_2 . S_x^2 + d_3 . S_y^2)$, and in the case of partial axial symmetry produces the normal second-order terms with D and E as coefficients. The effect of the cubic field via the interactions considered by Van Vleck and Penney[41] is of a fourth order, being proportional to $(S_1^4 + S_2^4 + S_3^4)$, where S_1, S_2 and S_3 are components referred to the cubic axes. The Spin Hamiltonian therefore becomes

$$\mathcal{H} = g\beta . \mathbf{H.S} + D[S_z^2 - \tfrac{1}{3}S(S+1)] + E(S_x^2 - S_y^2) + \tfrac{1}{6}a(S_1^4 + S_2^4 + S_3^4)$$
$$+ A . S_z I_z + B(S_x I_x + S_y I_y) \quad \ldots . (6.15)$$

The quadrupole interaction and direct effect of the external field on the nucleus are left out of this expression as they do not produce a noticeable effect on the manganese spectrum. It is to be noted that the terms in D and E are from the new mechanism suggested by Abragam and Pryce[14], and not from the direct action of the crystalline fields and spin-orbit coupling, as is usually the case. The comparative splittings produced by this mechanism and by the cubic field can be seen from the values of D and a obtained for the manganese double sulphate and fluosilicate[12]. These are: $D = 277$ and -134, and $a = +8$ and $+9$ respectively, all being given in units of 10^{-4} cm^{-1}. Measurements[42-45] have since been made on several other hydrated manganous salts, and show that, although the hyperfine splitting remains remarkably constant, the electronic splitting coefficients D and E are very sensitive to a change in the crystalline field.

The angular variation of the spectra of manganese salts fits in very well with the expressions derived in Section 6.8, and a complete analysis of this has been given in some detail by Bleaney and Ingram[12]. The collapse of the electronic splittings to zero at $\theta = 55°$, and the separation out to a subsidiary maximum at $\theta = 90°$ can be followed easily. The agreement with theory is most marked in the case of the fluosilicate, as there is no term in E for this salt, the crystalline field having complete axial symmetry as is often the case when there is only one ion per unit cell. In sorting out lines belonging to different electronic transitions, the variation of intensity from one transition to another is often very helpful. The intensity of the M to $(M-1)$ transition is given by the expression[18]:

power absorbed by crystal $= \mathcal{Z} . \{S(S+1) - M(M-1)\}$

where $\quad \mathcal{Z} = \dfrac{\pi . \nu^2 . (g\beta H_{\text{R.F.}})^2}{4kT} . \dfrac{\mathcal{N}}{(2S+1)} . \dfrac{1}{\Delta\nu} \quad \ldots . (6.16)$

$\mathcal{N}$ is number of paramagnetic ions in the crystal and $\Delta\nu$ is the half-width at half intensity. In a given resonance spectrum, M is the only variable, and hence the intensities of the lines in the different electronic transitions will be proportional to $\{S(S+1) - M(M-1)\}$, i.e. $5:8:9:8:5$ for the five electronic transitions in Mn^{2+}. Very weak $\Delta M_S = \pm 2$ transitions have also been observed[12], four electronic transitions then being obtained at a field strength half that of the main transition.

Owing to the ease with which the spectrum of Mn^{2+} can be obtained at room temperature, it has been studied in many other compounds[47] including phosphors and dilute solutions[48]; and its hyperfine structure can also be resolved, even when in a completely amorphous solid if it is sufficiently diluted[49]. Schneider and England[48] found a hyperfine separation of $A = B = 60 \times 10^4$ cm^{-1} for manganese impurities in zinc phosphors, which shows that there is a small effect on the s orbital[46], when the crystalline field changes markedly.

Ferric salts. Most of the work on ferric salts has been on the alums. These crystals, like the chrome alums, appear to have rather a complex structure, and inhomogeneity of the crystalline field will widen the absorption lines of diluted iron ammonium alum[50]. The initial measurements on this salt[30, 51, 52] showed that the electronic splittings were not the same as those deduced from low-temperature measurements[53], assuming a cubic field for their interpretation. The experimental results were not resolved completely by the introduction of a second-order interaction, as in manganese, however; and various calculations have been made concerning the effect of crystalline fields of lower symmetry. MEYER[54] has explained the results on the diluted crystals to some extent by considering the combined effect of a trigonal and cubic field, but the values he obtained do not fit the measurements on the undiluted crystals. The interpretation of the results is really hampered by lack of detailed knowledge concerning the crystal field in any given case, and it seems to be extra sensitive to any change in crystal structure. The work of BLEANEY and TRENAM[55, 56] on ferric rubidium alum confirms this, in that the axes of the cubic crystalline field are twisted by $7°$ from those of the cubic crystal cell. However, the contribution of the cubic field to the splitting does seem to be much greater than for the case of manganese, the value of the coefficient a being $134 \cdot 10^{-4}$ cm^{-1}.

Most of the iron isotopes have zero nuclear spin and would not show any hyperfine structure. Bleaney and Trenam[55] managed to reduce the line widths in diluted ferric rubidium alum to 6 gauss, but even so did not observe any hyperfine structure from samples enriched in Fe^{57}.

(f) Fe^{2+} $(3d)^6$ $^5D_4$. A cubic crystalline field, acting on this electronic configuration, will leave an orbital triplet lowest. There is also a fivefold spin degeneracy, and the crystalline fields of lower symmetry together with the spin-orbit coupling, would be expected to split this bottom multiplet into a number of closely spaced levels according to the Jahn-Teller effect. Since there is an even number of spins, Kramers's theorem does not apply, and hence it is not certain that paramagnetic resonance will be possible between the levels. If it is, it will only be observed at low temperatures, owing to the proximity of the next orbital levels to the ground state.

Paramagnetic resonance has been reported[30] in powders of ferrous ammonium and ferrous potassium sulphate, at temperatures of 20° K, and 1·3 cm wavelength. This shows that in these salts at least, the splitting between some of the levels in the bottom multiplet must be less than 0·8 cm^{-1}. No further work appears to have been published on measurements of ferrous compounds, and so in this case very little extra information has been provided by paramagnetic resonance.

(g) Co^{2+} $(3d)^7$ $^4F_{9/2}$. A cubic field splits the sevenfold orbital degeneracy to give a triplet at the bottom. The fourfold spin degeneracy combines with that of the orbital triplet, and a system of six closely lying doublets are formed, Kramers's theorem ensuring that each level is at least twofold degenerate. Paramagnetic resonance is, therefore, to be expected, but only at low temperatures.

The experimental results show that the salts must be cooled to 20° K before spin-lattice broadening becomes small, and under these conditions only the lowest energy level is occupied. The first analysis of the cobalt spectrum was undertaken by Bleaney and Ingram[20] in cobalt ammonium sulphate, who obtained g values varying from 3·0 to 6·5. This wide departure from the free-spin value is to be expected because of the small orbital splittings. Abragam and Pryce[21] have considered the theory of this and other cobalt salts in detail, and can account for the measured g values very satisfactorily by the effect of the interaction with the higher levels. It is also possible to reverse this process and calculate the splitting to the higher levels from the g value determinations. For the case of cobalt ammonium sulphate the estimated heights of the levels in the lowest multiplet are 250 cm^{-1}, 340 cm^{-1}, 1,700 cm^{-1}, 1,720 cm^{-1} and 1,840 cm^{-1} respectively.

As already seen, the results are interpreted by inserting an effective spin of $S = \frac{1}{2}$ into the Spin Hamiltonian (as in equations (11) to (14) of Appendix I). The quadrupole interaction produces no observable effect in the cobalt spectrum, but both the g-values and hyperfine splitting are very anisotropic[5], the former varying from about 2·5 to 6·5 for ionically bound salts. The second-order terms in the hyperfine splitting become very important near the perpendicular direction, and the transitions sometimes actually cross over between $\theta = 86°$ and $90°$.

Three groups of very weak lines were also observed in the fluosilicate spectrum[5], one with an angular variation opposite to that of the main group. These were interpreted[21] as due to cobalt atoms close to crystal imperfections, and hence in very different crystalline fields; it is interesting to note that lattice defects can be observed in this way.

Co^{2+} ions surrounded by a tetrahedron of neighbouring atoms, as in Cs_3CoCl_5, should have the splitting of their orbital levels reversed[25], and recent work by Stevens and Owen suggests that an orbital singlet does, in fact, lie lowest.

(h) Ni^{2+} $(3d)^8$ $^3F_4$. The orbital splitting is the same as that of V^{3+} but with the levels inverted, and hence an orbital singlet lies lowest. Thus paramagnetic resonance should be observable at room temperature, and the g value should be isotropic. There is an even number of electrons, however, and therefore Kramers's theorem does not apply, and it is possible for the spin degeneracy to be lifted, zero-field splittings being produced which prohibit paramagnetic resonance.

The first salt studied in detail was nickel fluosilicate, and HOLDEN, KITTEL and YAGER[57] found a splitting of 0.5 cm^{-1} between the spin-levels at room temperature. PENROSE and STEVENS[58] studied the variation of this splitting with temperature, and obtained a steady decrease in value to 0.12 cm^{-1} at $14°$ K. The g value remained constant and isotropic at 2.29. Higher values of zero-field splitting have been obtained by GRIFFITHS and OWEN[59] for a nickel Tutton salts. The rhombic component of the crystalline field splits the spin degeneracy into three singlets, the smaller splitting usually being the lowest. Values of over-all splitting varying from 2.5 cm^{-1} to 4.0 cm^{-1} were obtained, though all had the same g value of 2.25. Ono[7] obtains an even larger over-all splitting of 5.0 cm^{-1} for the case of $NiSO_4 . 7H_2O$, and a g value of 2.2. The Spin Hamiltonian used to summarize these results is simply of the form

$$\mathscr{H} = g\beta . \mathbf{H.S} + D . S_z^2 + E(S_x^2 - S_y^2) \qquad ....(6.17)$$

and the experimentally determined parameters are summarized in *Table 6.3*.

Table 6.3. Results for Nickel Salts

Salt	g	D cm^{-1}	E cm^{-1}	Reference
$NiSO_4 . 7H_2O$	2.2	−3.56	−1.50	(7)
$Ni(NH_4)_2(SO_4)_2 . 6H_2O$	2.25	−2.24	−0.38	(59)
$NiSiF_6 . 6H_2O$	2.29	0.32	0	(58) (57)

(i) Cu^{2+} $(3d)^9$ $^2D_{5/2}$. The splitting and general theory of the Cu^{2+} energy levels have already been considered in Section 6.4, and it can be seen that this is one of the ideal ions for paramagnetic resonance, as narrow lines can be obtained at room temperature, and there is only one electronic transition. The Spin Hamiltonian reduces to

$$\mathscr{H} = g_{||}\beta H_z S_z + g_{\perp}\beta (H_x S_x + H_y S_y) + A S_z I_z + B (S_x I_x + S_y I_y) \\ + Q'(I_z^2 - \tfrac{1}{3}I(I+1)) - \gamma\beta_N . \mathbf{H.I.} \qquad \dots (6.18)$$

It is therefore very similar to the case of cobalt in its angular variation, with the exception that both the quadrupole interaction, and the direct effect of the magnetic field on the nucleus, produce observable changes in the spectrum.

Copper is very similar to the case of ammonia in gaseous spectroscopy in that most of the new discoveries associated with microwave spectra of the solid state have been first observed in Cu^{2+} salts. Thus the effect of exchange forces was first discovered in $CuSO_4$. $5H_2O$ [60]; the existence of hyperfine structure was first observed in a diluted copper salt [61]; the effect of the quadrupole interaction was first noticed in the Cu^{2+} hyperfine spectrum [13]; heavy water was first used with copper Tutton salt crystals to resolve out the isotope splitting [62]; the concept of a resonating crystal field was first introduced to account for the copper fluosilicate spectrum [63]; and measurements on copper acetate [64] were the first to show that two paramagnetic ions could interact strongly to form a combined system of energy levels. These different effects will be considered in detail, but the normal behaviour of the Cu^{2+} spectrum, as depicted by the above Hamiltonian, is first described.

The g value variation is usually from a $g_{\perp}$ of 2·1 to a $g_{||}$ of 2·4, although recent work [65] indicates that the spread is not so large for Cu^{2+} ions surrounded by nitrogens, rather than the oxygens of water molecules. The anisotropy of the hyperfine splitting due to magnetic interaction is usually quite high, typical values of A and B being 0·013 cm^{-1} and 0·003 cm^{-1}. The original theories on hyperfine splitting predicted that B should be greater than A, and it was only by the configurational interaction admixing some $\{3s\ (3d)^9\ 4s\}$ state, that the correct anisotropy was obtained. The interesting features of the Cu^{2+} hyperfine structure lie in the last two terms of the Hamiltonian, however. The angular variation of the quadrupole interaction has already been discussed in Section 6.6, and photographs of the actual spectra shown in *Figure 59*. It was also shown in Section 6.8 that analysis of the combined effect of the direct action of the applied magnetic field, and the quadrupole interaction, enables the signs of the coefficients of A, B and Q' to be determined. Such a calculation has been carried out[62] for the deuterated potassium double sulphate, and by substitution of the experimentally measured coefficients into the theoretical expressions[14, 66], the nuclear quadrupole moments were determined as $-0·12 \pm 0·01 . 10^{-24}$ cm² for Cu^{63}, and $-0·11 \pm 0·01 . 10^{-24}$ cm² for Cu^{65}; these values being over ten times more accurate than any previously quoted[67].

The unusual results obtained with copper sulphate, fluosilicate and acetate can be briefly described as follows. Early measurements on copper sulphate[68, 69] showed only one absorption line when two were expected from the two dissimilar ions per unit cell. Experiments repeated at shorter wavelengths (8 mm) resolved the two lines, and this effect was interpreted as due to the strong exchange forces in the copper-sulphate crystal. The fact that the two lines are just not resolved at 1·25-cm wavelength, when they would have an energy separation of $0·15$ cm^{-1} suggests that this is the order of the exchange interaction.

The effect of exchange forces is also responsible for the unusual spectrum obtained by BLEANEY and BOWERS[70] from copper acetate. Instead of a single electronic transition with the usual g value, a spectrum corresponding to an effective spin of $S = 1$ was obtained, with a zero-field splitting of about $0·35$ cm^{-1}. The measurements could be fitted to a Spin Hamiltonian of the form:

$$\mathscr{H} = DS_z^2 + E(S_x^2 - S_y^2) + \beta[g_z H_z S_z + g_x H_x S_x + g_y H_y S_y] \quad ....(6.19)$$

and the measured parameters[70] were $D = 0·34$ cm^{-1}, $E = 0·01$ cm^{-1}, $g_\| = g_z = 2·42$, $g_\perp = g_y = g_x = 2·08$. These results can be satisfactorily explained if strong isotropic exchange forces are postulated between isolated pairs of ions whose individual magnetic axes are parallel to one another. The exchange force will be small compared with the effect of the crystalline field, and can be regarded as coupling the two $S = \frac{1}{2}$ states to form a lower level in which the spins are antiparallel with $S = 0$, and a higher level with effective spin $S = 1$. The lower level is diamagnetic, and paramagnetic resonance is only observed for transitions in the upper $S = 1$ triplet, a fact which is confirmed by the disappearance of the spectrum at 20° K when all the pairs are in the $S = 0$ state. The electronic splitting observed is that of the $S = 1$ triplet, and is given by:

$$D = -\tfrac{1}{4}\mathscr{J} . \{\tfrac{1}{8}(g_\| - 2)^2 - (g_\perp - 2)^2\} \quad ....(6.20)$$

where $\mathscr{J}$ is the coefficient of the exchange interaction term $\mathscr{J} . S_1 . S_2$ and $g_\|$ and $g_\perp$ are given by the expressions for normal copper salts.

The observed hyperfine structure also fits in with this explanation, seven components with relative intensities of $1 : 2 : 3 : 4 : 3 : 2 : 1$

being obtained. These can be represented by an additional term of $(S_1 . A . I_1 + S_2 . A . I_2)$ in the Hamiltonian, and each electronic transition is then split into components shifted by $A(M_{I_1} + M_{I_2})$ from the centre, with both M_{I_1} and M_{I_2} varying from $+\frac{3}{2}$ to $-\frac{3}{2}$. This interpretation of the results also fits in with the susceptibility measurements of GUHA[71]; and other paramagnetic resonance experiments by LANCASTER and GORDY[72], and by KUMAGAI, ABE and SHIMADA[73], have confirmed the work of Bleaney and Bowers. ABE[74] has also shown that a similar spectrum is obtained from copper propionate.

The study of these effects in copper salts throws considerable light on the general nature of exchange forces, and recent measurements[75] on copper double chlorides also show that variation of line width with frequency and orientation are due to this type of interaction.

The results observed in copper fluosilicate are another example of an interesting magnetic system. The original measurements of BLEANEY and INGRAM[77] gave an isotropic g value of 2·24, and only very small anisotropy in the hyperfine splitting, B now being greater than A. ABRAGAM and PRYCE[78] interpreted these results on the basis of a crystalline electric field resonating between different possible orientations. The symmetry of the field is trigonal in the case of the fluosilicate, and this does not split the 'non-magnetic' orbital doublet like the tetragonal field of the sulphate and double sulphates. The lowest level therefore has a total degeneracy of four, but this must be partially lifted by the Jahn-Teller effect and the octahedron of water molecules distorts to do this. There are a number of sets of possible distortions which will all give the same energy splitting[79], and a resonance between these states will give the isotropic g value observed. Further measurements by BLEANEY and BOWERS[80] at lower temperatures have shown that the crystalline field gradually changes, and there are eventually three ions per unit cell, each in a field having a tetragonal axis of symmetry which is one of the three edges of a cube, the trigonal axis of the cube being the normal crystal axis.

These measurements on different copper compounds show that paramagnetic absorption measurements are not only of use in giving nuclear data and magnetic information, but can often shed a great deal of light on crystal structure and other solid-state problems.

6.12 COVALENT COMPOUNDS

So far consideration has only been given to salts possessing crystalline electric fields strong enough to break the coupling between L and S, but not strong enough to uncouple the individual $l_1 l_2$---and

$s_1 s_2$ ---. Fields that can break down the coupling between individual orbits and spins usually only occur when the paramagnetic ion is covalently bound as in most complex salts. Relatively few measurements have been made on such salts to date, but those that have are of interest in the radical departure of their spectra from the corresponding ionic cases.

The first salt studied in any detail was potassium ferricyanide. The initial measurements of Bagguley et al.[30] showed that the spectrum was not that of the normal ferric ion, low temperatures being required to observe the absorption and even at 20° K the line was still exceedingly broad. Measurements by BLEANEY and INGRAM[81] on a single crystal showed that only one resonance line was obtained, the g values along the three monoclinic crystal axes being 2·30, 2·18 and 0·94 respectively. The line width was still very large (minimum of 800 gauss along the a axis, as compared with a theoretical dipole-dipole width of 200 gauss) and did not change when the temperature was lowered from 20° K to 12° K.

In order to explain the effective electronic spin of $\frac{1}{2}$, the method of molecular orbitals[82] assumes that there is a set of octahedral bonds to the six cyanide groups of the d^2sp^3 type, which use up two of the 3d orbitals. Only three d-orbitals are therefore available for the electrons which normally go in the 3d shell. The five electrons of Fe^{3+} will therefore occupy the 3d shell leaving one vacancy, and the magnetic properties will therefore correspond to that for $S = \frac{1}{2}$, with an orbital contribution. The same kind of reasoning explains why cobalticyanide and ferrocyanide with six 3d electrons are diamagnetic. VAN VLECK[83] has shown that the crystal field theory approach will give the same result, and HOWARD[84] has considered the effect of lower symmetry fields and spin-orbit coupling. This work has been revised by BERSOHN[85], who shows that the g values obtained by paramagnetic resonance are in good agreement with the low-temperature susceptibility measurements[86]. The large line width has not been explained, however; dilution with isomorphous cobalticyanide reduces the width showing that it is due to spin-spin interaction, but even so very narrow lines cannot be obtained. A similar spectrum is obtained from the isoelectronic

$$K_4 Mn(CN)_6 . 3H_2O,$$

there being a slightly greater anisotropy of the g values in this case. The manganese hyperfine structure which is resolved in diluted crystals is also anisotropic, as would be expected from the effective $S = \frac{1}{2}$ state.

The only other complex salts investigated in any detail are

$$K_3Cr(CN)_6{}^{(87)} \text{ and } K_4V(CN)_6 . 3H_2O^{(29)},$$

and the results from these have been described in detail in Section 6.11(c). Both of these salts have a spectrum similar to that of the ionic compounds, with an effective spin equal to the actual spin degeneracy of $S = \frac{3}{2}$. It seems that, although the electric field associated with the covalent binding is stronger than normal and produces g values closer to the free-spin value, it is not strong enough to break down the coupling between the individual spins in these two ions. Further results on covalent compounds are discussed in Chapter 11.

6.13 THE RARE-EARTH GROUP—GENERAL RESULTS

In the rare-earth salts the crystalline electric field does not affect the electron orbitals so much as in the iron group, and the coupling between L and S remains unchanged. There is, therefore, still a resultant J quantum number, and this can be considered as precessing round the direction of the crystalline electric field to give different quantized energy levels. Transitions will now have $\Delta J_z = \pm 1$ as a selection rule, and it is necessary to determine which levels are lying lowest before the possibility of paramagnetic resonance absorption can be discussed. The situation is thus very much more complex than in the iron transition group, but some general features can be seen in the results.

First, the splitting produced by the crystalline field is usually only of the order of 10^2 cm^{-1} instead of 10^4 cm^{-1}, mainly because the $4f$ shell is better screened than the $3d$ shell. This has two effects: (i) spin-lattice interaction will be strong, and hence the spectra can usually only be observed at low temperatures; and (ii) the g values depart widely from the free-spin value, and are usually very anisotropic. Secondly, it is found that the electronic levels are always doublets ($\pm J_z$) when the ion has an odd number of electrons, and are singlets or doublets when there is an even number of electrons. The Jahn-Teller effect will split these doublets in the latter case, but usually only by very small amounts, and from the point of view of the resonance spectrum they will still act as doublets. Hence, if a paramagnetic absorption does occur it is always with an effective spin of $S = \frac{1}{2}$ in the Spin Hamiltonian, which therefore takes the general form of:

$$\mathscr{H} = g_\parallel \beta H_z S_z + g_\perp \beta [H_x S_x + H_y S_y] + A S_z I_z + B[S_x I_x + S_y I_y]$$
$$+ Q'[I_z^2 - \tfrac{1}{3}I(I+1)] \qquad \dots (6.21)$$

The main theoretical problem is to determine which J_z levels are lying lowest, and nearly all the theoretical work on rare-earth salts

175

has been done by ELLIOTT and STEVENS[88, 89, 90]. The theory of the crystalline field splitting is quite different from the iron group in that the rare-earth ions nearly always have nine nearest neighbours with a vertical threefold axis of symmetry. There is thus no, or very little, cubic component of the field, and a more detailed expression must be used for its potential. The ethyl sulphates have been the only series of salts systematically studied as yet, and for these Elliott and Stevens[88] take a potential of the form:

$$V = A_2^0(3z^2 - r^2) + A_4^0(35z^4 - 30r^2z^2 + 3r^4)$$
$$+ A_6^0(231z^6 - 315r^2z^4 + 105r^4z^2 - 5r^6) + A_6^6(x^6 - 15x^4y^2 + 15x^2y^4 - y^6)$$
$$\dots\dots(6.22)$$

where the coefficients are the standard ones obtained when Laplace's equation is expanded in spherical harmonics and then converted to Cartesian co-ordinates.

The effect of the crystalline field is applied as a perturbation after the spin-orbit coupling has been considered, as the latter is the stronger interaction in the rare-earth group. Because of the field's

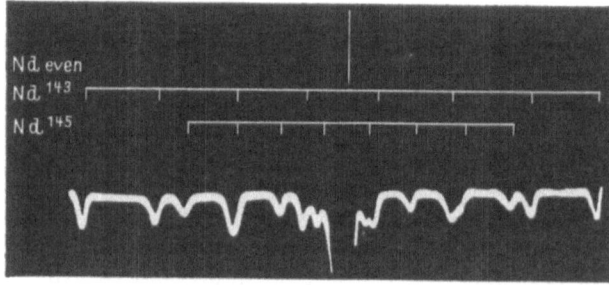

Figure 63. Hyperfine structure of neodymium

smaller effect and its more complex representation, it is now much more difficult to calculate back from the measured parameters of the Spin Hamiltonian to the electronic states of the particular ion. For this reason most of the interesting information obtained from the rare-earth group elements has been from the hyperfine structure rather than the electronic splittings. The separation between hyperfine lines is usually considerably larger than in the iron group, and high resolution is thus obtained. Five new nuclear spins were determined within a year or so of the first measurements on the rare-earth group, and several nuclear magnetic moments have been estimated from the results. *Figure 63* is typical of the hyperfine structure observed in the rare-earth salts, and is taken from BLEANEY and SCOVIL's[91] letter reporting the determination of the nuclear spins of

Nd^{143} and Nd^{145}. The main central line, which goes off the picture, is due to the even isotopes with zero spin, and the two groups of seven lines outside this can be clearly seen. It is interesting to note that this was the first determination of a new nuclear spin by the method of paramagnetic resonance absorption.

6.14 Results for Different Ions of the Rare-earth Group

These results are summarized in a similar way to those of the iron transition group, being listed in order of the number of electrons in the $4f$ shell. Only those with Kramers's degeneracy, i.e. an odd number of electrons, are first considered. Unless otherwise stated, the Spin Hamiltonian used to describe the experimental results is that of equation (6.21), and in very few cases can any signs be given to its coefficients.

(a) Ce^{3+} $(4f)^2$ $^2F_{5/2}$. It is necessary to cool to $4°$ K before paramagnetic absorption is obtained, when two resonances corresponding to the $J_z = \pm\frac{1}{2}$ and $J_z = \pm\frac{5}{2}$ states are observed[92]. Elliott and Stevens[88] found that by taking the perturbation to a second order, coupling elements were obtained between the $J_z = \pm\frac{5}{2}$ and $J_z = \pm\frac{1}{2}$ levels, and hence transitions could be expected from both. The $J_z = \pm\frac{3}{2}$ level, which is also part of the $^2F_{5/2}$ ground state, lies some 100 cm^{-1} above the other two and is therefore not populated at $4°$ K. The measurements showed that the $\pm\frac{1}{2}$ level lies lowest, about 3 ± 1 cm^{-1} below the $\pm\frac{5}{2}$ level in the diluted crystals, but the order is reversed in the concentrated salt. Cerium has no stable odd isotopes and therefore no hyperfine structure is to be expected. The results can thus be summarized by the different g values obtained[18]:

For $\pm\frac{5}{2}$ level: $g_{\parallel} = 3\cdot72 \pm 0\cdot01$; $g_{\perp} = 0\cdot2 \pm 0\cdot05$
For $\pm\frac{1}{2}$ level: $g_{\parallel} = 0\cdot955 \pm 0\cdot005$; $g_{\perp} = 2\cdot185 \pm 0\cdot010$

(b) Nd^{3+} $(4f)^3$ $^4I_{9/2}$. The first detailed analysis of the neodymium spectrum was made by Bleaney and Scovil[91], and has been illustrated in *Figure 63*. Theoretical treatment[88] shows that to a first approximation the $J_z = \pm\frac{7}{2}$ level is lowest, with the $J_z = \pm\frac{5}{2}$ close. The last term in the crystalline field expansion will couple these states so that the lowest level is actually an admixture of both, and hence transitions are allowed between its two components. The g values obtained are in good agreement with this theory, being:

$$g_{\parallel} = 3\cdot535 \pm 0\cdot001 \text{ and } g_{\perp} = 2\cdot073 \pm 0\cdot001.$$

The most interesting feature of the spectrum is its hyperfine structure, however, and estimates of the nuclear magnetic moments as well as determination of their spins can be obtained from the results.

The Spin Hamiltonian parameters were obtained with high accuracy by BLEANEY, SCOVIL and TRENAM[93], and the nuclear moments calculated from these results were[18] $\mu_N^{143} = 1 \cdot 0 \pm 0 \cdot 25$ n.m. and $\mu_N^{145} = 0 \cdot 62 \pm 0 \cdot 15$ n.m.

(c) Sm^{3+} $(4f)^5$ $^6H_{5/2}$. Samarium ethyl sulphate had to be cooled to $4° K$ before showing a resonance[94, 95]. The single electronic transition was then split into sixteen component lines, which enabled the spins of both Sm^{147} and Sm^{149} to be determined as $I = \frac{7}{2}$. The g value is nearly isotropic at 0·6. Theoretical calculations[89] show that the $J_z = \pm \frac{1}{2}$ level lies lowest, but small admixtures from the next multiplet considerably affect the g values. The nuclear magnetic moments were derived from the hyperfine splitting in the usual way (see Section 9.3(b)), and obtained as $\mu_N^{147} = 0 \cdot 83 \pm 0 \cdot 15$ and $\mu_N^{149} = 0 \cdot 68 \pm 0 \cdot 1$.

(d) Gd^{3+} and Eu^{2+} $(4f)^7$ ^8S. These ions are in an S state, and hence are very similar to the case of manganese in the iron group. To a first order the crystalline electric field will produce no splitting, and the mechanism of configurational interaction, suggested by Abragam and Pryce[14], again becomes important. The g values are isotropic and very close to the free spin value, (1·994 and 1·991 respectively)[96], and the spectrum can be observed readily at room temperature. The Spin Hamiltonian has an effective $S = \frac{7}{2}$, and hence seven electronic transitions occur, the spacing between these following a $(3 \cos^2 \theta - 1)$ law, to a first approximation.

The absence of hyperfine structure in the Gd^{3+} spectrum suggests that the admixture of any s configurations is much smaller in the rare-earth than in the iron transition group. Eu^{2+}, as an impurity in a strontium chloride lattice, does give a hyperfine structure, however[96], suggesting a small s admixture in this case.

(e) Dy^{3+} $(4f)^9$ $^6H_{15/2}$. Theoretical treatment[88] indicates that the lowest level is $J_z = \pm \frac{15}{2}$, with an admixture of $\pm \frac{3}{2}$. No transitions are therefore to be expected, and none have so far been observed.

(f) Er^{3+} $(4f)^{11}$ $^4I_{15/2}$. First-order crystalline field theory[88] gives the $\pm \frac{5}{2}$ and $\pm \frac{7}{2}$ levels close together at the bottom, and the A_6^6 term of the field potential admixes these so that transitions are possible for the ground level.

Measurements by BLEANEY and SCOVIL[98] on diluted erbium ethylsulphate showed that the lines were still broadened by spin-lattice relaxation at $13° K$. The width was sufficiently narrow to allow resolution of the hyperfine structure when perpendicular to the hexagonal axis, however, and from this they were able to determine the nuclear spin of ^{167}Er as $I = \frac{7}{2}$. BOGLE, DUFFUS and SCOVIL[99] repeated the measurements at $4° K$, and were able to resolve the spectrum completely in all directions. Their measurements are of interest in being the first in the rare-earth group from which a nuclear quadrupole moment has been determined. The effect of the quadrupole interaction is very similar to that in the Cu^{2+} spectrum, and values of Q' were obtained from both the second-order shifts in the $\Delta M_I = 0$ transitions, and from the splitting of the usually forbidden $\Delta M_I = \pm 1$ and $\Delta M_I = \pm 2$ transitions. The coefficients of the Spin Hamiltonians are as follows:

$$g_{\parallel} = 1 \cdot 47 \pm 0 \cdot 03 \quad g_{\perp} = 8 \cdot 85 \pm 0 \cdot 2$$
$$|A| = 52 \pm 1 \quad |B| = 314 \pm 1 \quad |Q'| = 30 \quad \times 10^{-4} \text{ cm}^{-1}$$

From Elliott and Stevens's[88] estimation of $\dfrac{1}{r^3}$ for the $4f$ orbits, a value of the actual quadrupole moment was obtained as

$$|Q| = 10 \cdot 2 \pm 3 \times 10^{-24} \text{ cm}^2.$$

(g) Yb^{3+} $(4f)^{13}$ $^2F_{7/2}$. This case is very similar to that of Dy^{3+}, the lowest level being a mixture of the $\pm \frac{3}{2}$ and $\pm \frac{15}{2}$ states. No transitions would therefore be possible, and no resonances have been observed experimentally.

(h) Pr^{3+} $(4f)^2$ $^3H_4$. The seven ions listed above are the only ones possessing an odd number of electrons in the $4f$ shell, and hence the only ones to which Kramers's theorem applies. Generally speaking, ions with an even number of electrons will only have singlet levels due to the Jahn-Teller effect, and paramagnetic resonance is therefore not to be expected from them. One exception to this has so far been found, in the case of Pr^{3+} where the Jahn-Teller splitting of the lowest doublet is very small and microwave quanta can still induce transitions between its components. BLEANEY and SCOVIL[100] found a single electronic transition in praesodymium ethylsulphate at 20° K. They used a Spin Hamiltonian of the form

$$\mathscr{H} = g_{\parallel}\beta H_z \cdot S_z + A S_z I_z + \Delta S_x \qquad \ldots \text{(6.23)}$$

to allow for the zero-field splitting. It may be noted that this bottom doublet is not a Kramers's doublet, and since the admixed states do not differ by unity normal transitions are not allowed. Dissymmetry in the crystalline field will mix the two states of the doublet themselves, however, and transitions can then be obtained when the radiofrequency magnetic field is parallel to the crystalline axis. The experimentally determined coefficients are:

$$g_{\parallel} = 1 \cdot 69 \pm 0 \cdot 01 \quad g_{\perp} \to 0$$
$$\Delta = \text{splitting of doublet} = 0 \cdot 04 \text{ cm}^{-1}$$
$$A = 0 \cdot 083 \pm 0 \cdot 001 \text{ cm}^{-1}.$$

179

The hyperfine structure of six equally-spaced lines is due to the $I = \frac{5}{2}$ spin of the only stable isotope Pr^{141}.

Paramagnetic resonance in praesodymium chloride has also been reported by DAVIS, KIP and MALVANO[101].

6.15 RESULTS FROM THE PALLADIUM TRANSITION GROUP

No systematic survey of paramagnetic resonance in any of the other transition groups comparable to that of the iron and rare-earth groups has yet been made. GRIFFITHS, OWEN and WARD[102] have investigated quite a large number of compounds belonging to the 4d and 5d transition groups, however, and have summarized their results in a general survey paper. The palladium group contains ions with an unfilled 4d shell, and resonances have been observed in molybdenum, ruthenium and silver compounds. These results can be summarized as follows:

(a) Mo^{5+} $(4d)^1$. The only compound in which resonance has been observed in this group is $K_3[Mo\,(CN)_8]$[102]. The crystalline field, due to the duodecahedron of cyanides, splits the orbital levels associated with the single 4d electron into a lower singlet and an upper quartic. The energy separation between these is very large, and g values are therefore expected close to 2·0, and the resonance should be observable at room temperature. Both these predictions are confirmed by experiment, resonance being obtained from the powder at all temperatures up to 290° K, with a g value of 2·005 ± 0·005. Octahedral complexes in this group should behave in a similar way to $(3d)^1$ Ti^{3+}, and hence require very low temperatures.

(b) Mo^{4+} $(4d)^2$. The only compound investigated in this group has been $K_4[Mo(CN)_8]$[102] but this has two electrons pairing in the 4d shell and is therefore diamagnetic. Paramagnetic resonance might be expected from ions in an octahedral complex, although small distortions of the octahedron would produce zero-field splittings of the five degenerate levels sufficient to move their transitions out of the microwave region.

(c) Mo^{3+} $(4d)^3$. Paramagnetic resonance of trivalent molybdenum was first observed in K_2MoCl_5, a g value of 1·76 being obtained[103]. Resonances from the three other trivalent salts K_3MoCl_6, $K_2MoCl_5H_2O$ and $KMoF_4$ have since been obtained[102], all with g values of about 1·95. The theoretical interpretation of the results is difficult as only two of these compounds approximate to octahedral complexes.

(d) Ru^{4+} $(4d)^4$. A powdered sample of $K_2(RuCl_6)$ shows a very weak absorption[102] at 20° K, with a g value of 1·92, but this may have been due to trivalent ruthenium present as an impurity. In most $(4d)^4$ salts, the orbital levels may well be singlets split by too large an amount for resonance to be seen.

(e) Ru^{3+} $(4d)^5$. More results have been obtained from this group than any others so far, most of the salts measured having an octahedral complex with axial or rhombic distortions. There is one hole in the otherwise completed subshell and an orbital triplet lies lowest. Trigonal or tetragonal fields, together with the spin-orbit coupling produce three Kramers's doublets, and the behaviour of the lowest doublet can be described by the Spin Hamiltonian of equation (6.21) with $S = \frac{1}{2}$. The magnetic behaviour of these ions is therefore very similar to that of Fe^{3+} in $K_3Fe(CN)_6$.

The results obtained for various ruthenium chlorides and double chlorides are summarized in Griffiths and Owen's[102] survey paper. One of the most interesting features of the spectrum of this ion is its hyperfine structure, which was used to make the first determination[104] of the nuclear spins of Ru^{99} and Ru^{101}. Both of these were found to have $I = \frac{5}{2}$ from the satellite lines, and a ratio of the nuclear magnetic moments was obtained as $\mu_{101}/\mu_{99} = 1 \cdot 09$.

(f) $(4d)^6$ $(4d)^7$ and $(4d)^8$ Palladium ions. If octahedral or square complexes are formed they are diamagnetic, and no resonances have been observed in any of the ions in these three groups.

(g) Ag^{2+} $(4d)^9$. This may have an unpaired electron left in a p_z orbital, and a g value close to $2 \cdot 0$ would be expected with resonance observable up to room temperature. Measurements by GIJSMAN, GERRITSEN and VAN DEN HANDEL[105] on Ag^{2+} persulphate crystals stabilized with pyridine confirm this. They obtain a g value varying from $2 \cdot 06$ to $2 \cdot 09$, and a line width which only changes from 50 to 75 gauss, as the temperature is raised from $4 \cdot 2°$ to $290°$ K. Measurements on crystals diluted with Cd^{2+} give the same results. Bowers[105] also obtained a g value varying from $2 \cdot 04$ to $2 \cdot 18$ from powdered divalent silver compounds.

6.16 Results for the Platinum Group

The theory and results obtained for this group are not summarized in detail because the $(5d)^n$ configuration should behave in a very similar way to the $(4d)^n$ configuration. Thus, quadrivalent ions of Ir, with five $5d$ electrons, should behave in the same way as trivalent ions of Ru, with five $4d$ electrons, and the remarks of Section 6.15(e) therefore apply to them. This is also true for all values of n ranging from 1 to 9. Experimentally, resonance has only been obtained from two different ions in the platinum group.

181

(a) Re^{4+} $(5d)^3$. Six isotropic, equally spaced, broad lines, with a centre at $g = 1·8$, were obtained from a crystal of K_2ReCl_6[102], diluted with the isomorphous diamagnetic $K_2(PtCl_6)$. No resonance was observed in the concentrated salt, and very little information can be drawn from these preliminary measurements.

(b) Ir^{4+} $(5d)^5$. As noted above, this ion is equivalent to Ru^{3+}, and would be expected to behave in a similar way to potassium ferricyanide[81] which has five $3d$ electrons covalently bound, and the g-values obtained from four different chloro- and bromo-iridates[102] confirm this. By far the most interesting feature in the spectrum of this ion, however, is its hyperfine structure, being the first observation of hyperfine splitting caused by surrounding diamagnetic atoms. The first measurements[106] were taken on a crystal of ammonium chloroiridate $(NH_4)_2[Ir Cl_6]$ diluted with the isomorphous diamagnetic platinum salt. This showed a complex hyperfine structure at $20°$ K, which could be resolved into 16 equally spaced lines, with an over-all separation of 15 gauss. The nuclear spins of Ir^{191} and Ir^{193} are both $I = \frac{3}{2}$, and they have very similar magnetic moments, and hence would only be expected to produce four hyperfine lines. The additional lines were shown to be due to Cl^{35} and Cl^{37} (both with $I = \frac{3}{2}$) with the hyperfine structure of the chlorines equal to one-third of that of the iridium.

This contribution from the chlorine nuclei arises from π-bond sharing of electrons. Thus the strong covalent binding of the electrons allows the five electrons of the iridium to pair off in the $5d$ shell leaving one positive hole. This hole may then be considered as spending some of its time on the iridium ion, and some on the chlorines, and hence a contribution to the hyperfine interaction is obtained from both nuclei. A similar and more clearly resolved hyperfine structure[102] has since been observed in $Na_2[(Ir, Pt) Cl_6] . 6H_2O$. This result is very interesting as it gives direct evidence for covalent π-bonding, and other measurements such as this may be a very powerful tool for the theoretical chemist. A more detailed theoretical treatment of π-bonding, and the general theory of covalent XY_6 complexes has recently been given by STEVENS[107].

6.17 RESULTS FROM THE TRANS-URANIC GROUP

Recent work has opened up a very promising field in studies of paramagnetic absorption in the trans-uranic elements. Various workers have tried to obtain resonance from different uranium compounds for some time, but usually with negative results, although GHOSH, GORDY and HILL[112] did obtain a resonance from powdered UF_3

and UF_4. Recently, however, BLEANEY, LLEWELLYN, PRYCE and HALL[108, 109, 110] have obtained and analysed the spectra from single crystals of plutonyl and neptunyl rubidium nitrate, determining or confirming the spins of Pu^{239}, Pu^{241} and Np^{237}. Their results may be summarized as follows:

(a) *Plutonium*. A single crystal of $PuO_2Rb(NO_3)_3$, diluted with the isomorphous diamagnetic $UO_2Rb(NO_3)_3$, was investigated at 3 cm and 1·25 cm wavelength, at 20° K[108]. This contained a 6 per cent ratio of plutonium to uranium and 98 per cent of the plutonium was Pu^{239}, the remainder being Pu^{240}. A single electronic transition was obtained, which was resolved into two components of equal intensity and determined the spin of Pu^{239} as $I = \frac{1}{2}$. The lines had a separation of 340 ±20 gauss, and a half-width of 40 gauss when parallel to the axis, and the observed spectrum could be represented by a Spin Hamiltonian of the form:

$$\mathcal{H} = g_\| \beta H_z S_z + A S_z I_z + \Delta_x S_x + \Delta_y S_y \qquad \dots (6.24)$$

The electronic state of the plutonium ion is thus very similar to that of praesodymium[100] considered in Section 6.14(h). The doublet giving rise to the transition is not a Kramers's doublet, and has a small splitting due to the Jahn-Teller effect, but admixtures in this doublet by the imperfect crystal field are sufficient to permit transitions. The results indicate that the magnetic electrons in the actinide series are in the 5f, and not the 6d shell. A further set of measurements[109] were then made on a similar crystal containing 40 micro-grammes of Pu^{241}. Six equally-spaced, hyperfine lines were obtained from this giving $I = \frac{5}{2}$ for Pu^{241}, and from the two lines of the Pu^{239} also present, μ^{241}/μ^{239} was calculated as 3·53. Estimates of the actual nuclear magnetic moments from these results give

$$\mu_N^{239} = 0·4 \pm 0·2 \text{ n.m.}$$
$$\mu_N^{241} = 1·4 \pm 0·6 \text{ n.m.}$$

HUTCHINSON and LEWIS[113] have also reported measurements on sodium plutonyl acetate.

(b) *Neptunium*. Bleaney, Llewellyn, Pryce, and Hall[110] have also made measurements on a single crystal of $UO_2Rb(NO_3)_3$ containing 400 micro-grammes of Np^{237}. With perpendicular radiofrequency and d.c. magnetic fields, six lines of equal intensity are obtained showing that $I = \frac{5}{2}$ for Np^{237}.

The results can be summarized by the Spin Hamiltonian of equation (6.21), the g value varying from 3·4 to about 1·0. The magnetic moment of Np^{237} was also estimated from the hyperfine splitting as 6 ±2·5 nuclear magnetons.

These preliminary measurements on the trans-uranic elements suggest that there is a large field open for paramagnetic resonance in this transition group.

6.18 Paramagnetic Resonance and Other Low-Temperature Measurements

It has been seen throughout the measurements summarized in this chapter, that results of paramagnetic resonance absorption can be compared directly with other low-temperature measurements such as specific heats and susceptibilities. This is because in both cases the effect of energy level splittings is determined, i.e., the splitting of individual energy levels in the case of paramagnetic resonance absorption, and an average over-all effect of various splittings in the case of specific heat and susceptibility measurements. If the energy level splittings are known from the resonance work their contribution to the specific heat and susceptibility can be quite easily evaluated. BLEANEY[111] gives expressions relating these quantities in terms of the Spin Hamiltonian parameters. His equations may be summarized as follows:

(i) *Specific heats*

Contribution due to electronic splitting

$$D\{S_z^2 - \tfrac{1}{3}S(S+1)\} + E\{S_x^2 - S_y^2\}$$

is given by:

$$\frac{CT^2}{R} = \left(\frac{hc}{k}\right)^2 \{\tfrac{1}{45}(D^2 + 3E)S(S+1)(2S-1)(2S+3)\} \quad \ldots (6.25)$$

Contribution due to nuclear splitting

$$A \cdot S_z I_z + B(S_x I_x + S_y I_y) + Q'(I_z^2 - \tfrac{1}{3}I(I+1))$$

is given by:

$$\frac{CT^2}{R} = \left(\frac{hc}{k}\right)^2 \{\tfrac{1}{9}(A^2 + 2B^2)S(S+1)I(I+1)$$
$$+ \tfrac{1}{45}Q'^2 I(I+1)(2I-1)(2I+3)\} \quad \ldots (6.26)$$

(ii) *Susceptibilities*

Contribution due to an electronic splitting $D(S_z^2 - \tfrac{1}{3}S(S+1))$
is given by:

$$\chi_{\parallel} = (N \cdot g_{\parallel}^2 \cdot \beta^2 S(S+1)/3kT)\left\{1 - \frac{D(2S-1)(2S+3)}{15kT}\right\} \quad \ldots (6.27)$$

$$\chi_{\perp} = (N \cdot g_{\perp}^2 \cdot \beta^2 S(S+1)/3kT)\left\{1 + \frac{D(2S-1)(2S+3)}{30kT}\right\} \quad \ldots (6.28)$$

There is *no* contribution to the susceptibility by the hyperfine splitting since it does not introduce a term in $1/T$ or $1/T^2$ into the susceptibility expressed as a power series in $1/T$, unless the hyperfine splitting becomes of the order of kT, when a complicated variation in χ will be obtained. It therefore follows that the hyperfine splittings cannot be detected by the presence of an effective 'Weiss constant' in the susceptibility measurements. It can be seen that substitution of the paramagnetic resonance data into these expressions will immediately give considerable information concerning the low-temperature behaviour of the given salt.

Zero-field energy levels. Another type of information that can be derived directly from the paramagnetic resonance measurements and is of great importance to low-temperature work, is data on the zero-field energy levels; i.e. the actual splitting of the fine and hyperfine levels in the absence of an applied magnetic field. Bleaney[18] has also considered this problem in detail with special reference to the question of nuclear alignment. For the case of levels with $S = \frac{1}{2}$ the energy levels are given by the expression:

$$W = -\frac{A}{4} + Q'\{k^2 + \tfrac{1}{4} - \tfrac{1}{3}I(I+1)\}$$
$$\pm \tfrac{1}{2}\sqrt{\{k(A - 2Q')\}^2 + B^2\{(I + \tfrac{1}{2})^2 - k^2\}} \quad \ldots (6.29)$$

where $k = M + M_I$ and takes the values $(I + \frac{1}{2})$, $(I - \frac{1}{2})$, ..., $-(I + \frac{1}{2})$. The energy states are all doublets in zero magnetic field, k and $-k$ being degenerate, except for the case of $k = 0$ which gives two singlets of separation $B(I + \frac{1}{2})$. For levels with $S > \frac{1}{2}$, Bleaney[18] obtains the expression

$$W = D\{M^2 - \tfrac{1}{3}S(S+1)\} + A\,M\,.\,M_I + Q'\{M_I^2 - \tfrac{1}{3}I(I+1)\}$$
$$\ldots (6.30)$$

which applies for all levels where $M \neq \pm\frac{1}{2}$, and if $D \gg A$. The $M = \pm\frac{1}{2}$ levels are given by equation (6.29), with B replaced by $B(S + \frac{1}{2})$.

These expressions can be used to determine the energy levels in the absence of the magnetic field, and this information is of the utmost importance in deciding which salts to use for nuclear adiabatic demagnetization experiments, as discussed in Chapter 9.

6.19 PARAMAGNETIC RESONANCE IN GASES

The observation of paramagnetic resonance in gases has already been mentioned in the last chapter, and a brief summary is given here to complete the account of paramagnetic resonance in normal atomic or molecular systems.

All the gaseous paramagnetic resonance work so far reported has been carried out by Beringer and Castle at Yale University, using the spectroscope described in Section 4.11. Three paramagnetic gases have been studied in detail, O_2, NO, and NO_2, and these correspond to three different degrees of coupling between the magnetic moment and the molecular axis.

(a) *Nitric Oxide*, NO. This corresponds to strong coupling between the magnetic moment and the molecular axis, and the magnetic fields required for resonance are not sufficient to produce a Paschen-Back effect. The molecule has one unpaired electron and one unit of angular momentum about its axis. The ground level is therefore $^2\Pi$ and this is split into two doublets of 120 cm^{-1} separation, the lower is diamagnetic, but the upper $^2\Pi_{3/2}$ is paramagnetic. This upper doublet interacts with the rotational energy and a series of levels characterized by the total quantum number J is formed. BERINGER and CASTLE[114] observed the transitions between the components of the $J = \frac{3}{2}$ level, the g value being given by $3/J(J+1)$, and hence equal to $\frac{4}{5}$, for this case. The four components of the $J = \frac{3}{2}$ level will produce three transitions for the normal $\Delta M_J = \pm 1$ selection rule, and hence three absorption lines are to be expected. Each of these was also split into three components of equal intensity, which are hyperfine structure due to the N^{14} nucleus ($I = 1$). The hyperfine splitting arises from both a magnetic and quadrupole interaction, and their different effects can be separated out, to give

$$A = 9\cdot9 \pm 0\cdot1 \times 10^{-4}\ \text{cm}^{-1}$$
and
$$Q' = -0\cdot57 \pm 0\cdot02 \times 10^{-4}\ \text{cm}^{-1}.$$

A similar form of absorption occurs in the alkaline superoxides[119] where the O_2^- ion has one unpaired electron. The energy level system is the same as for NO[115] but the $^2\Pi_{3/2}$ level now lies lowest. The variation of g value from 1·99 to 2·16 can then be used to calculate the distortion of the O_2^- ions in the solid state.

(b) *Nitrogen peroxide*, NO_2. This is the opposite case to NO, as the molecule has one unpaired electron which is only weakly coupled to the molecular axis. A Paschen-Back effect will therefore take place in the field strengths required for resonance, and CASTLE and BERINGER[116] interpreted the spectrum that they observed as due to the three magnetic moments associated with the molecule, precessing separately around the applied field. The energy levels could therefore be represented by the equation

$$E = M_s \cdot g\mu_0 H \cdot + A \cdot M_s \cdot M_I \cdot + B \cdot M_s \cdot M_J \quad \ldots (6.31)$$

the first term being due to the electron spin moment, the second to the moment of the N^{14} nucleus, and the third due to the moment associated with rotation of the molecule as a whole. The observed spectrum[116] consists of three overlapping lines at 10 mm Hg pressure, in agreement with the first term of the above equation, and this is split into a large number of lines at a pressure of 1 mm Hg, these being due to the last term and the large number of J levels which are populated at the temperatures used.

(c) *Oxygen*, O_2. Oxygen represents a case halfway between the other two, in that the magnetic fields used are sufficient to cause appreciable decoupling of the vectors, but not sufficient to break the coupling completely and produce a Paschen-Back effect. The oxygen molecule is in an $^3\Sigma$ state, with zero orbital momentum about its axis but with two unpaired spins. It has already been seen in Section 5.1 that the orbital moment couples with the spin moment to give levels having $J = K$, $K \pm 1$, the latter are nearly degenerate, being separated from $J = K$ by 2 cm^{-1}. Application of a steady magnetic field splits the J levels, and $\Delta J = \pm 1$ transitions can be obtained between them. For small applied field strengths the splitting is linear, and J is still a good quantum number; but at the field strengths used by BERINGER and CASTLE[117], decoupling of the K and S vectors is beginning to occur, and the transitions become very complicated. A large number of the rotational levels which are populated at room temperature contribute to give absorption lines, and the observed spectrum[117] is highly complex. A detailed theoretical treatment has been given by HENRY[118] which accounts for most of the lines, and variation of their relative intensities with temperature can also be used as a means of identification.

(d) *Chlorine Dioxide*. No measurements on gaseous chlorine dioxide have yet been reported, but paramagnetic resonance can be obtained from aqueous solutions of the gas, the chlorine nuclei producing a marked hyperfine structure[119].

6.20 CONCLUSION

It can be seen from the results summarized in this chapter that paramagnetic resonance will give a great deal of information concerning the interactions and binding forces present in the solid state, and this can then be used to test detailed chemical and magnetic theory. Predictions of the low-temperature behaviour of the various salts can also be made, and such form an essential link between low temperature and nuclear physics. The technique can also be used to obtain considerable nuclear data, at least ten new spin values having been determined already, as well as many magnetic moments and electric quadrupole moments.

The more specialized applications of paramagnetic resonance are discussed in the next chapter, and it is seen to be a very powerful method of investigation in these cases also. See also Chapter 11.

REFERENCES

[1] KRAMERS, H. A. *Proc. Acad. Sci. Amst.* 33 (1930) 959

[2] POLDER, D. *Physica* 9 (1942) 709

[3] LAPORTE, O. *Z. Phys.* 47 (1928) 761

[4] BLEANEY, B., PENROSE, R. P. and PLUMPTON, B. I. *Proc. Roy. Soc.* A 198 (1949) 406

[5] BLEANEY, B. and INGRAM, D. J. E. *ibid.* 208 (1951) 143

[6] Numerical values taken from BEARDEN, J. A. and WATTS, H. M. *Phys. Rev.* 81 (1951) 73

[7] ONO, K. *J. phys. Soc. Japan* 8 (1953) 802

[8] CODRINGTON, R. S., OLDS, J. D. and TORREY, H. C. *Phys. Rev.* 95 (1954) 607

[9] CARVER, T. R. and SLICHTER, C. P. *ibid.* 92 (1953) 212

[10] INGRAM, D. J. E. and TAPLEY, J. G. *Phil. Mag.* 45 (1954) 1221

[11] BLEANEY, B. and SCOVIL, H. E. D. *Proc. phys. Soc.* A 63 (1950) 1369

[12] BLEANEY, B. and INGRAM, D. J. E. *Proc. Roy. Soc.* A 205 (1951) 336

[13] INGRAM, D. J. E. *Proc. phys. Soc.* A 62 (1949) 664

[14] ABRAGAM, A. and PRYCE, M. H. L. *Proc. Roy. Soc.* A 205 (1951) 135

[15] BLEANEY, B. and STEVENS, K. W. H. *Rep. Prog. Phys.* 16 (1953) 108

[16] MACK, J. E. *Rev. mod. Phys.* 22 (1950) 64

[17] PRYCE, M. H. L. *Proc. phys. Soc.* A 63 (1950) 25

[18] BLEANEY, B. *Phil. Mag.* 42 (1951) 441

[19] BLEANEY, B., INGRAM, D. J. E. and SCOVIL, H. E. D. *Proc. phys. Soc.* A 64 (1951) 601

[20] BLEANEY, B. and INGRAM, D. J. E. *Nature* 164 (1949) 116

[21] ABRAGAM, A. and PRYCE, M. H. L. *Proc. Roy. Soc.* A 206 (1951) 173

[22] BETHE, H. A. *Ann. Phys., Lpz.* 3 (1929) 133

[23] VAN VLECK, J. H. *Theory of Electric and Magnetic Susceptibilities.* Oxford: Clarendon Press (1935)

[24] SCHLAPP, R. and PENNEY, W. G. *Phys. Rev.* 42 (1932) 666

[25] GORTER, C. J. *ibid.* (1932) 437

[26] BLEANEY, B., LLEWELLYN, P. M., PRYCE, M. H. L. and HALL, G. R. *Phil. Mag.* 45 (1954) 773, 991

[27] DANIELS, J. M., GRACE, M. A. and ROBINSON, F. N. H. *Nature* 168 (1951) 780

[28] BLEANEY, B. and BOWERS, K. D. *Proc. phys. Soc.* A 64 (1951) 1135; BOWERS, K. D. *ibid.* 65 (1952) 860

[29] BAKER, J. M. and BLEANEY, B. *ibid.* 65 (1952) 952; KIKUCHI, C., SIRVETZ, M. H. and COHEN, V. W. *Phys. Rev.* 88 (1952) 142

[30] BAGGULEY, D. M. S., BLEANEY, B., GRIFFITHS, J. H. E., PENROSE, R. P. and PLUMPTON, B. I. *Proc. phys. Soc.* 61 (1948) 542

[31] BIJL, D. *ibid.* A 63 (1950) 405

[32] BOGLE, G. S. and COOKE, A. H. *ibid* A 68 (1955) 57

REFERENCES

[33] BLEANEY, B. *Proc. phys. Soc.* A 63 (1950) 407

[34] VAN DEN HANDEL, J. and SIEGERT A. *Physica* 4 (1937) 871

[35] BAGGULEY, D. M. S. and GRIFFITHS, J. H. E. *Nature* 160 (1947) 532; and *Proc. Roy. Soc.* A 204 (1950) 188

[36] WEISS, P. R., WHITMER, C. A., TORREY, H. C. and HSIANG, J. S. *Phys. Rev.* 72 (1947) 975

[37] HALLIDAY, D. and WHEATLEY, J. *ibid.* 74 (1948) 1712

[38] BLEANEY, B. and PENROSE, R. P. *Proc. phys. Soc.* 60 (1948) 395

[39] BLEANEY, B. *Phys. Rev.* 75 (1949) 1962

[40] CUMMEROW, R. L. and HALLIDAY, D. *ibid.* 70 (1946) 433

[41] VAN VLECK, J. H. and PENNEY, W. G. *Phil. Mag.* 17 (1934) 961

[42] INGRAM, D. J. E. *Phys. Rev.* 90 (1953) 711

[43] — *Proc. phys. Soc.* A 66 (1953) 412

[44] TRENAM, R. S. *ibid.* (1953) 118

[45] KUMAGAI, H., ONO, K., HOYAHI and KAMBE, K. *Phys. Rev.* 87 (1952) 374

[46] ABRAGAM, A. *ibid.* 79 (1950) 534

[47] HURD, K., SACHS, M. and HERSHBERGER, W. D. *ibid.* 93 (1952) 373

[48] SCHNEIDER, E. E. and ENGLAND, T. S. *Nature* 166 (1950) 437

[49] INGRAM, D. J. E. and BENNETT, J. E. *Phil. Mag.* 45 (1954) 545

[50] UBBINK, J., POULIS, J. A. and GORTER, C. J. *Physica* 17 (1951) 213
WHITMER, C. A. and WEIDNER, R. T. *Phys. Rev.* 84 (1951) 159

[51] WEIDNER, R. T., WEISS, P. R., WHITMER, C. A. and BLOSSER, D. R. *ibid.* 76 (1949) 1727

[52] BIJL, D. Thesis, University of Leiden (1950)

[53] COOKE, A. H. *Proc. phys. Soc.* A 62 (1949) 269

[54] MEYER, P. H. E. *Physica* 17 (1951) 899

[55] BLEANEY, B. and TRENAM, R. S. *Proc. phys. Soc.* A 65 (1952) 560

[56] — — *Proc. Roy. Soc.* A 223 (1954) 1

[57] HOLDEN, A. N., KITTEL, C. and YAGER, W. A. *Phys. Rev.* 75 (1949) 1443

[58] PENROSE, R. P. and STEVENS, K. W. H. *Proc. phys. Soc.* A 63 (1949) 29

[59] GRIFFITHS, J. H. E. and OWEN, J. *Proc. Roy. Soc.* A 213 (1952) 459

[60] BAGGULEY, D. M. S. and GRIFFITHS, J. H. E. *Nature* 162 (1948) 538

[61] PENROSE, R. P. *ibid.* 163 (1949) 992

[62] BLEANEY, B., BOWERS, K. D. and INGRAM, D. J. E. *Proc. phys. Soc.* A 64 (1951) 758

[63] BLEANEY, B. and INGRAM, D. J. E. *ibid.* 63 (1950) 408; ABRAGAM, A. and PRYCE, M. H. L. *ibid.* 63 (1950) 409

[64] BLEANEY, B. and BOWERS, K. D. *ibid.* 65 (1952) 667

[65] INGRAM, D. J. E. and BENNETT, J. E. *J. chem. Phys.* 22 (1954) 1136

[66] ABRAGAM, A. and PRYCE, M. H. L. *Proc. Roy. Soc.* A 206 (1951) 164

[67] BRIX, P. *Z. Phys.* 126 (1949) 725

[68] BAGGULEY, D. M. S. and GRIFFITHS, J. H. E. *Nature* 162 (1948) 538

[69] — — *Proc. Roy. Soc.* A 201 (1950) 366

[70] BLEANEY, B. and BOWERS, K. D. *Phil. Mag.* 43 (1952) 372; and *Proc. Roy. Soc.* A 214 (1952) 451

[71] GUHA, B. C. *Proc. Roy. Soc.* A 206 (1951) 353

[72] LANCASTER, F. W. and GORDY, W. *J. chem. Phys.* 19 (1951) 1181

[73] KUMAGAI, H., ABE, H. and SHIMADA, J. *Phys. Rev.* 87 (1952) 385

[74] ABE, H. *ibid.* 92 (1953) 1572

[75] ONO, K., ABE, H. and SHIMADA, J. *ibid.* 92 (1953) 551

[76] KUMAGAI, H., ABE, H., SHIMADA, J., HAYASHI, I., ONO, K. and IBAMATO, H. *J. phys. Soc. Japan* 7 (1952) 535

[77] BLEANEY, B. and INGRAM, D. J. E. *Proc. phys. Soc.* A 63 (1950) 408

[78] ABRAGAM, A. and PRYCE, M. H. L. *ibid.* (1950) 409

[79] VAN VLECK, J. H. *J. chem. Phys.* 7 (1939) 61

[80] BLEANEY, B. and BOWERS, K. D. *Proc. phys. Soc.* A 65 (1952) 667

[81] BLEANEY, B. and INGRAM, D. J. E. *ibid.* (1952) 953

[82] PAULING, L. *The Nature of the Chemical Bond.* New York: Cornell University Press (1939)

[83] VAN VLECK, J. H. *J. chem. Phys.* 3 (1935) 807

[84] HOWARD, J. *J. chem. Phys.* 3 (1935) 813

[85] BERSOHN, R. Thesis, Harvard University (1949)

[86] JACKSON, L. C. *Proc. phys. Soc.* 50 (1938) 707; GUHA, B. C. *Proc. Roy. Soc.* A 206 (1951) 353

[87] BOWERS, K. D. *Proc. phys. Soc.* A 65 (1952) 860

[88] ELLIOTT, R. J. and STEVENS, K. W. H. *ibid.* 64 (1951) 205 and 932

[89] — — *ibid.* 65 (1952) 370

[90] — — *Proc. Roy. Soc.* A 215 (1952) 437; A 218, 553; 219 (1953) 387

[91] BLEANEY, B. and SCOVIL, H. E. D. *Proc. phys. Soc.* A 63 (1950) 1369

[92] BOGLE, G. S., COOKE, A. H. and WHITLEY, S. *ibid.* A 64 (1951) 931

[93] BLEANEY, B., SCOVIL, H. E. D. and TRENAM, R. S. *Phil. Mag.* 43 (1952) 995

[94] BLEANEY, B., ELLIOT, R. J. and SCOVIL, H. E. D. *Proc. phys. Soc.* A 64 (1951) 933

[95] BOGLE, G. S. and SCOVIL, H. E. D. *ibid.* A 65 (1952) 368

[96] BLEANEY, B., ELLIOTT, R. J., SCOVIL, H. E. D. and TRENAM, R. S. *Phil. Mag.* 42 (1951) 1062; BLEANEY, B. and LOW, W. *Proc. phys. Soc.* A 68 (1955) 55

[97] BLEANEY, B. and SCOVIL, H. E. D. *Proc. phys. Soc.* A 63 (1950) 1369

[98] — — *ibid.* A 64 (1951) 204

[99] BOGLE, G. S., DUFFUS, H. J. and SCOVIL, H. E. D. *ibid.* A 65 (1952) 760

[100] BLEANEY, B. and SCOVIL, H. E. D. *Phil. Mag.* 43 (1952) 999

[101] DAVIS, C. F., KIP, A. F. and MALVANO, R. *Atti. Accad. naz. Lincei* 11 (1951) 77

[102] GRIFFITHS, J. H. E., OWEN, J. and WARD, I. M. *Proc. Roy. Soc.* A 219 (1953) 526

[103] RAMASESHAM, S. and SURYAN, G. *Phys. Rev.* 84 (1951) 593

[104] GRIFFITHS, J. H. E. and OWEN, J. *Proc. Roy. Soc.* A 65 (1952) 951

[105] GIJSMAN, H. M., GERRITSEN, H. J. and VAN DEN HANDEL, J. *Physica* 20 (1954) 15; BOWERS, K. D. *Proc. phys. Soc.* A 66 (1953) 666

[106] OWEN, J. and STEVENS, K. W. H. *Nature* 171 (1953) 836

[107] STEVENS, K. W. H. *Proc. Roy. Soc.* A 219 (1953) 542

[108] BLEANEY, B., LLEWELLYN, P. M., PRYCE, M. H. L. and HALL, G. R. *Phil. Mag.* 45 (1954) 773

[109] BLEANEY, B., LLEWELLYN, P. M., PRYCE, M. H. L. and HALL, G. R. *ibid.* (1954) 991

[110] — — — — *ibid.* (1954) 992

[111] BLEANEY, B. *Phys. Rev.* 78 (1950) 214

[112] GHOSH, S. N., GORDY, W. and HILL, D. G. *ibid.* 87 (1952) 229

[113] HUTCHINSON, C. A. and LEWIS, W. B. *ibid.* 95 (1954) 1095

[114] BERINGER, R. and CASTLE, J. G. *ibid.* 78 (1950) 581

[115] MARGENAU, H. and HENRY, A. *ibid.* 78 (1950) 587

[116] CASTLE, J. G. and BERINGER, R. *ibid.* 80 (1950) 114

[117] BERINGER, R. and CASTLE, J. G. *ibid.* 81 (1951) 82

[118] HENRY, A. F. *ibid.* 80 (1950) 396

[119] BENNETT, J. E., INGRAM, D. J. E., SYMONS, M. C. R., GEORGE, P. and GRIFFITHS, J. S. *Phil. Mag.* 46 (1955) 443

[20] ONO, K., SHOICHIRO, K., SEKIYAMA, H. and ABE, H., *Phys. Rev.* 96 (1954) 38

7

FERROMAGNETIC RESONANCE, FREE RADICALS, AND F-CENTRES

7.1 Introduction

In the last chapter paramagnetic resonance absorption was discussed, as it occurred in well-defined systems containing unpaired electron spins; these included ions located in a crystalline lattice and gas molecules existing in certain definite energy states. In most cases normal theoretical treatment was found to give a good explanation of the experimental results, although the paramagnetic resonance measurements sometimes modified the energy-level system that had been previously assumed. In this chapter consideration is given to the rather specialized aspects of paramagnetic resonance, as occurring in more complex and not so easily defined systems. These have been divided into four groups, namely (i) ferromagnetic resonance, (ii) resonance from free radicals, (iii) resonance from conduction electrons in metals and semi-conductors, and (iv) resonance absorption from F-centres and other crystal dislocations.

Although these systems are not so readily analysed theoretically, the information obtained from the resonance measurements can be of the greatest interest, as it often throws considerable light on solid-state interactions which are very difficult to investigate by any other means. The main features of the four groups are as follows.

(a) *Ferromagnetic resonance* is very similar to ordinary paramagnetic resonance, but it is concerned with unpaired electron spins which are extremely close together and hence acted on by very strong exchange forces. The fact that the actual material studied is a ferromagnetic metal and has a large permeability, also radically affects the experimental observations. Thus the field strength required for resonance is altered markedly from that predicted by the simple paramagnetic resonance condition (i.e. equation (6.3)), because the effective internal magnetic field is many times stronger than the applied external field. Also, since the specimen is a conductor, only very small samples or thin sheets can be used as the skin depth at microwave frequencies is so small. The initial theory of KITTEL[1] was able to account qualitatively for the shift in resonant field value, and also for the shape of the absorption line, but could not explain their large widths and the relatively high departure of the g values from 2·0.

Although a considerable amount of further experimental work has been carried out, no theoretical treatment has so far given a satisfactory account of these two outstanding points.

(b) *Paramagnetic resonance in free radicals** is due to the odd unpaired electrons which are associated with these particular structures. Unlike the normal uncompensated electrons of paramagnetic ions, these free radicals are associated with the valence electrons and a modification of the normal chemical binding. The first observations[2, 3] of paramagnetic resonance in free radicals showed that their g values were very close to that of the free electron, and analysis of their line widths and hyperfine structure[3] showed that the odd unpaired electron must spend its time moving between several atoms. Although the g values are all very close to 2·0023, later work has shown that there is an anisotropy in crystalline specimens. There is also an appreciable variation in the line widths obtained from different free radicals, although in the solid state exchange forces usually have a large effect and narrow the lines considerably. If the exchange narrowing is reduced by preparing dilute solutions of the free radical, hyperfine structure is often obtained as a result of the interaction of the electron with the nuclei of neighbouring atoms.

Paramagnetic resonance due to modified chemical binding has been observed in systems other than the organic compounds known to possess unpaired electrons. One of the most striking examples of this was the discovery of high concentrations of broken bonds in coals[4], and carbons formed by low-temperature charring[5], from the strong paramagnetic resonance absorption that they produced. It would seem, in fact, that the whole field of free-radical chemistry is now open for attack by the methods of paramagnetic resonance.

(c) *Electron resonance from metals and semi-conductors.* Recent experiments[6, 7] have shown that resonance absorption can also be obtained from the conduction electrons in metals and semi-conductors. The electrons undergoing transition are thus those normally pictured as forming the 'gas' which is responsible for electrical and thermal conduction. It is possible to obtain both spin and cyclotron resonance for these electrons in semi-conductors, and the results can give very detailed information on the energy bands in such systems.

(d) *Paramagnetic resonance absorption in F-centres* formed by irradiating crystals with neutrons was first observed by Hutchinson[8], and the study of crystal imperfections and deformations by this method has now been taken up by many laboratories. It is often found that the electrons or holes which become trapped in the crystal lattice

---

* This subject has greatly developed over the last ten years, and is discussed in detail in Chapter 11.

move sufficiently close to neighbouring atoms to interact with their nuclei and thus hyperfine structure of the electron resonance absorption line is obtained. The theoretical treatment of this branch of paramagnetic resonance is still in its initial stages, but it is obvious that the methods of paramagnetic and nuclear resonance are going to be of great use in the study of crystal imperfections.

These four different headings are now considered in more detail.

7.2 FERROMAGNETIC RESONANCE—INITIAL WORK

Ferromagnetic resonance was discovered by GRIFFITHS[9] in 1946, when he observed that the product of the permeability and resistance of ferromagnetic metals measured in the microwave region had a

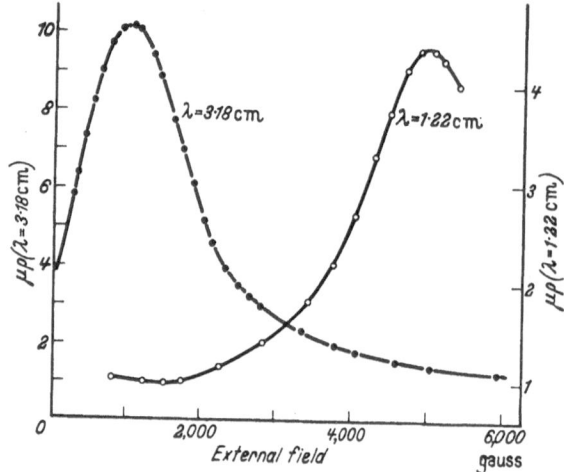

Figure 64. The first observation of ferromagnetic resonance. Absorption lines from nickel

broad maximum as the applied field strength was varied. The original results are shown in *Figure 64*, where the resonance curves for both 3 cm and 1·25 cm wavelengths are given. The experimental technique was very similar to that of the early paramagnetic spectroscopes already described, except that the specimen was formed by electroplating the ferromagnetic metal onto one end wall of the cavity resonator. The variation in Q of the cavity was then measured in the normal way for different field strengths, and the effective value of $(\mu \times \rho)$ calculated from this.

A few months later, Kittel[1] produced a theoretical treatment which accounted very well for the shift of the resonant field from the values expected for a free electron spin. By considering the classical equations of motion for the induced magnetization he derived an expression for the susceptibility component in the direction of the microwave magnetic field (χ_x) as:

$$\chi_x/\chi_0 = \left[1 - \frac{\omega}{\gamma \, (B_z H_z)^{\frac{1}{2}}} \right]^{-1} \qquad \ldots (7.1)$$

The permeability will therefore rise to infinity at the resonance condition

$$\omega = \gamma \, . \, (B_z H_z)^{\frac{1}{2}} \qquad \ldots (7.2)$$

The good agreement between this theoretical equation and the experimental values is illustrated in *Table 7.1* which is taken from Kittel's original paper.

Table 7.1. Experimental and Calculated Resonant Frequencies

Ferromagnetic	d.c. Field H_z	Calculated Larmor Frequencies (i) For H_z	(ii) For $(B_z H_z)^{\frac{1}{2}}$	Exptl. Frequency $\times 10^{10}/2\pi$ c/s
Fe	2,800	5·0	14·5	15·4
	500	0·9	5·8	5·9
Co	510	0·9	5·3	5·9
Ni	5,000	8·8	13·5	15·4
	3,800	6·7	10·9	13·2
	1,030	1·8	4·9	5·9

Equation (7.1) is only a very simple approximation, and Kittel showed that a damping term due to relaxation forces could be added of the form $-\lambda (\mathbf{M} - \chi_1 \mathbf{H})$ to give

$$\frac{\chi}{\chi_0} = \frac{\omega_0^2 + \lambda^2 \mu_0 + j\omega\lambda}{\omega_0^2 + \lambda^2 \mu_0 - \omega^2 + j\omega\lambda (1 + \mu_0)} \qquad \ldots (7.3)$$

where ω_0 is now written for $\gamma \, (B_z H_z)^{\frac{1}{2}}$, and λ is a constant.

So far this treatment has completely ignored the effect of magnetic anisotropy and demagnetizing fields, but even in this simple form it was able to account for the main facts of ferromagnetic resonance. In a more complete treatment[10], Kittel considered these other factors in detail, and was able to derive the following expressions for the resonance condition:

$$\nu = \frac{ge}{4 \, . \, \pi \, . \, mc} \sqrt{(H + N_x M - N_z M + \phi_1)(H + N_y M - N_z M + \phi_2)}$$
$$\ldots (7.4)$$

Here H denotes the externally applied field, assumed along Oz, and M is the component of magnetization along this direction. N_x, N_y and N_z are the demagnetizing factors for fields along the x, y and

z axes, and ϕ_1 and ϕ_2 are correction terms to allow for the effect of the magnetic anisotropy. These are rather complicated in the general case[10, 11], but if the crystal has cubic symmetry, and the xz and 001 planes coincide, they have the simplified form of

$$\phi_1 = \frac{2K_1}{M}\cos 4\theta \quad \text{and} \quad \phi_2 = \frac{K_1}{2M}(3 + \cos 4\theta) \qquad \ldots (7.5)$$

where θ is the angle between H and the 001 direction, and K_1 is the anisotropy energy constant.

The values of N_x, N_y and N_z only have a true meaning if the intensity of magnetization is constant and the sample is ellipsoidal in form. In practice, however, one of three experimental conditions is used, and the demagnetizing factors can be calculated for these. i.e.:

Case (i)—Applied field H and microwave magnetic field both tangential to thin sheet of ferromagnetic (as in Griffiths's original experiment). Then N_x and N_z are zero, and $N_y = 4\pi$; therefore, ignoring the anisotropy corrections, equation (7.4) becomes:

$$\nu = \frac{ge}{4\pi mc}(H.B)^{\frac{1}{2}} \qquad \ldots (7.6)$$

as originally derived by Kittel[1].

Case (ii)—Same as (i), but with the applied field normal to the sheet of ferromagnetic. Then $N_x = N_y = 0$ and

$$\nu = \frac{ge}{4\pi mc}(H - 4\pi M) \qquad \ldots (7.7)$$

It should be noted that the thickness of the ferromagnetic sheet must be less than the skin depth if any of these equations are to apply.

Case (iii)—A small sphere of ferromagnetic material in mutually perpendicular fields. Then $N_x = N_y = N_z$

and
$$\nu = \frac{ge}{4\pi mc} . H. \qquad \ldots (7.8)$$

which is the same as for normal paramagnetic resonance.

Case (i) is that usually employed experimentally, but the validity of the equations for the other two cases has been verified. KITTEL, YAGER and MERRITT[12] showed this strikingly for the first and second cases by observing the shift in the absorption curve of an annealed supermalloy specimen, when the applied field was changed from the parallel to perpendicular direction. At the 1·25 cm wavelength used, the peak was found to shift from 4,880 gauss to 15,550 gauss. Although rather hard to realize in practice, the third case of a small sphere is the ideal experimental specimen as no demagnetizing factors enter into the expression for resonance, and the spectro-scopic splitting factor can therefore be determined directly. The spheres must have diameters less than the skin depth, however, if the equation is to be justifiably applied. BAGGULEY[13] has recently over-come this difficulty by the production of colloidal suspensions in paraffin wax containing spherical particles under 100 Å in diameter; and has shown that the directly measured g values obtained from these are the same as those obtained under the other conditions after correcting for the demagnetizing fields.

7.3 FURTHER EXPERIMENTS IN FERROMAGNETIC RESONANCE

Following Griffiths's original experiments, YAGER and BOZORTH[14] studied ferromagnetic resonance in supermalloy, choosing this be-cause of its low anisotropy. The thin foils extended beyond the cavity walls and the demagnetizing field was reduced to 30 gauss. The absorption line had a much narrower width than those previously observed, and a very good fit between Kittel's theory and the experi-mental curve was obtained.

The effect of the magnetic anisotropy was then studied in detail by KIP and ARNOLD[15], using single crystals of iron. They were able to show that the field strength required for resonance had a 90° period when the crystal was rotated in the 001 plane. Another effect of the anisotropy energy which they observed was the production of a second smaller line at lower field strengths. This occurs when the external field is applied along a direction of difficult magnetization, and arises from the opposing forces of the applied field and the aniso-tropy. If low microwave frequencies are employed (e.g. 9,000 Mc/s) so that the normal resonant-field value is not very high, then the direction of magnetization will not follow that of the applied field initially, and the resonant field value will be reached while its direc-tion is still determined by the anisotropy axis of the crystal. This pro-duces the additional low-field absorption line, whereas if the field is further increased, the direction of magnetization will be 'pulled round' to that of the applied field and absorption will again occur, as predicted by equation (7.4). This low-field line will not occur along directions of easy magnetization, of course, nor when higher micro-wave frequencies are used, as the higher fields required will 'pull the magnetization round' before a resonance value can be reached. It is interesting to note that the agreement between the theoretical treat-ment of anisotropy forces and the measured resonant field values is so good that in certain cases[16] the latter can be used to predict the anisotropy constants of crystals.

197

Measurements on the anisotropy of single crystals of Co-Zn ferrite[30], and on the variation of line width with temperature[17] for nickel and supermalloy, have also been made, and the detailed results in the latter case have still to be explained theoretically. Another recent measurement by RADO and WEERTMAN[54] has shown that the direct effect of exchange interaction in shifting the resonant field position can be observed experimentally if suitable conditions are chosen. They obtained a shift of over 20 per cent by employing polycrystalline specimens, with an effective anisotropy constant of zero and high μ, thus producing a small skin depth which is favourable to a large exchange interaction shift. These experiments indicate that exchange forces do in fact have a large contribution in ferromagnetics, as is to be expected, and hence still leave the problem of why the resonance lines are so wide.

Recent work by KIP, KITTEL and PORTIS[53] on gadolinium metal has shown no change in g values as the metal changes from the paramagnetic state above $16°$ C to the ferromagnetic state below this temperature. They interpret the observed value of $1·95$ as indicating that there must be considerable coupling of the ^8S state to a state including a $5d$ orbit.

7.4 THE THEORETICAL TREATMENT OF FERROMAGNETIC RESONANCE

It has been seen that Kittel's theory[1, 10] was able to account for both the resonant field position and the shape of the absorption lines with considerable success. This theory was developed along classical, macroscopical lines, but Van Vleck[11] has shown that exactly the same results can be derived by a quantum mechanical treatment using the initial Hamiltonian:

$$\mathscr{H} = g\beta H \sum_j S_{z_j} + \sum_{k>j} D_{jk}[\mathbf{S}_j \cdot \mathbf{S}_k - 3r_{jk}^{-2}(\mathbf{r}_{jk} \cdot \mathbf{S}_j)(\mathbf{r}_{jk} \cdot \mathbf{S}_k)] \quad \ldots (7.9)$$

where the coefficients D_{jk} are those associated with the normal interaction of two dipoles. He also showed that part of the anisotropy corrections were probably due to quadrupole-quadrupole interaction of the form

$$\sum_{k>j} K_{jk} \cdot r_{jk}^{-4}(\mathbf{S}_j \cdot \mathbf{r}_{jk})^2(\mathbf{S}_k \cdot \mathbf{r}_{jk})^2 \quad \ldots (7.10)$$

POLDER[18] and others have also studied the theory of ferromagnetic resonance in some detail, but there are still two outstanding points in which the theory and experimental observations are not reconciled. These are (i) the high g values observed, and (ii) the large line widths obtained.

(a) *g-values.* It can be shown[11, 19, 20] that a first-order perturbation treatment of the spin-orbit coupling interaction leads to the general result:

(Spectroscopic splitting Factor – 2) = (2 – Gyromagnetic Ratio)

$$....(7.11)$$

where the term 'Gyromagnetic Ratio' here refers to that obtained from the bulk magnetization-by-rotation experiments. This equation is confirmed qualitatively, but not at all well quantitatively, as can be seen from *Table 7.2.*

Table 7.2. Comparison of Spectroscopic Splitting Factors and Gyromagnetic Ratios

Ferromagnetic	S.S.F.	(S.S.F.–2)	(2–G.R.)	G.R.
Nickel	2·2	0·2	0·08	1·92
Cobalt	2·22	0·22	0·13	1·87
Iron	2·12–2·17	0·15	0·07	1·93
Magnetite	2·20	0·20	0·07	1·93
Permalloy	2·07	0·07	0·09	1·91

This discrepancy between theory and experiment still remains and it is very hard to see what type of interaction will reconcile the two, the only possible effect, which may be capable of producing such high divergencies, seems to be that due to strong anisotropic exchange forces.

(b) *Line widths.* At first sight it might be thought that ferromagnetic resonance absorption lines would be expected to have large widths, as the electron spins are very close together, and hence the ordinary dipole-dipole broadening should be quite large. If it were the only interaction this view would be correct, and the calculated dipole-dipole broadening would produce widths of a thousand gauss or so, as observed. But if exchange interaction is considered[23] (and this must be very much greater in the case of ferromagnetics than for paramagnetics, where it has already been seen to have a large effect) then very much narrower lines are to be expected. Several authors[17, 18, 21, 34] have tried to postulate other interactions, which might cause additional broadening and hence produce the widths observed, but with no great success so far. There are four other mechanisms which might introduce broadening, i.e. (i) spin-lattice interaction, (ii) skin effect, (iii) interaction with the conduction electrons and (iv) anisotropic exchange forces. These may be briefly considered as follows.

(i) AKHEISER[21] suggested that spin-lattice interaction would account for the broadening of the lines, but Polder[18] has criticized his assumptions, and in any case this type of interaction should produce a width which decreases rapidly as the temperature is lowered, and this is not found to be the case[29].

(ii) The skin effect results in a decreasing magnetic moment with depth, and can be regarded as a form of damping effect[11]. This type of interaction will produce a line width which is independent of temperature, but the magnitude is far too small to account for those observed experimentally.

(iii) Interaction between the magnetic and conduction electrons will again only produce a small broadening if calculated on simple first-order theory[22]. If a more detailed treatment is made a larger broadening may be attributed to this effect, but it does not seem very likely.

(iv) The interaction most likely to be responsible for the extra width of the absorption lines is that due to anisotropic exchange forces[11]. This anisotropic exchange coupling arises when spin-orbit interaction is introduced into the normal exchange coupling and can produce short-range anisotropic interaction between different atoms. This type of effect would increase the dipole-dipole broadening, and so might counterbalance the narrowing due to the normal isotropic exchange interaction.

It can be seen from the above summary that the theory of line widths in ferromagnetic resonance is in a very unsatisfactory state, and generally speaking, the understanding and insight into the phenomenon of ferromagnetic resonance is not at all as clear as that into normal paramagnetic resonance.

7.5 ANTIFERROMAGNETIC RESONANCE

A few measurements have been made on antiferromagnetic substances, and the most striking results obtained from all of these has been the sudden broadening and disappearance of the absorption line below the Curie point. This was first noticed in powdered Cr_2O_3 by Maxwell et al.[24], and their results are illustrated in *Figure 65*, where it is seen that the signal intensity falls sharply just below the Curie point at 311° K. Hutchinson has observed an exactly similar effect with single crystals of MnF_2, where the absorption was found to disappear as the Curie point was passed at 64° K, this being independent of the direction of the applied magnetic field. A similar effect has also been observed by MAXWELL and McGUIRE[25] in MnO, MnS, MnSe, and MnTe, which all have resonances with g values of 2·00 in the paramagnetic region, but which disappear on cooling into the antiferromagnetic region.

The disappearance of the signal as the substance enters the anti-ferromagnetic region can be explained by the fact that the internal magnetic fields which are then produced may be strong enough to shift the resonance out of range of the frequencies and applied field strengths available, as can be seen from equation (7.12). If the Curie temperature is high, so is the exchange field, whereas compounds

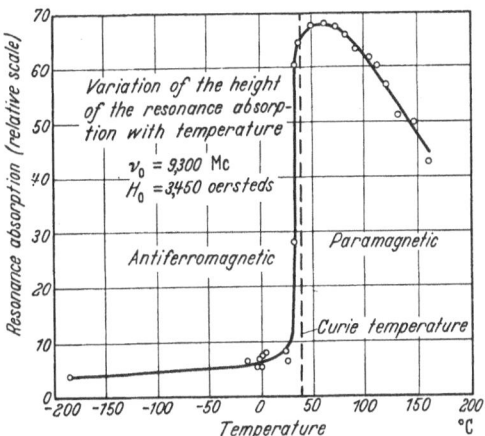

Figure 65. Antiferromagnetic resonance in Cr_2O_3

with low Curie points will only have small internal fields in the anti-ferromagnetic region. This is illustrated by the measurements of UBBINK et al.[25a] on $CuCl_2 . 2H_2O$, which has a Curie point of 4° K, and they were therefore able to observe a resonance doublet in the antiferromagnetic region.

Experiments by ADAM and STANDLEY[35] on MnAs, MnSb and MnBi are also of interest in this connexion. These compounds exist in the ferromagnetic state below the Curie point, and in an anti-ferromagnetic state above the Curie point, later changing to a normal paramagnetic behaviour. For the MnAs, they found a rapid decrease in signal intensity as the Curie point was approached at 45° C, and only a very small signal in the antiferromagnetic region. The g values they obtained in the ferromagnetic region were also of interest, an unusually high value of 3·3 being observed for the MnAs. They explain this as due to the fact that the manganese is not in an S state in this compound, and hence stronger spin-lattice interactions are to be expected than for the manganese ferrites, where the manganese ion is in an S state, and the observed g values are very close to 2·0.

The theory of antiferromagnetic resonance has been treated by STEVENS[26], KEFFER and KITTEL[27], and the position is somewhat

201

similar to that for ferromagnetic resonance, a theoretical explanation being possible in general terms, but not in detail. They derive, as a condition of resonance

$$\omega = g \cdot \frac{e}{2mc}[H_0 \pm \{H_A(2H_E + H_A)\}^{\frac{1}{2}}] \qquad \ldots (7.12)$$

where H_0 is the applied magnetic field, H_E is the Weiss exchange field and H_A is the anisotropy field. It follows from this that it should be possible to obtain antiferromagnetic resonance in zero external field for some substances, as in the case of ferromagnetic resonance. This has not been observed experimentally in antiferromagnetics as yet, but has for the ferromagnetic case by WELCH, NICKS, FAIR-WEATHER and ROBERTS[28] in a ferrite. They varied the radiofrequency from 500 to 2,000 Mc/s, and obtained a marked absorption discontinuity in the permeability.

7.6 PARAMAGNETIC RESONANCE IN SYSTEMS OTHER THAN THE TRANSITION ELEMENTS

It was at first thought that paramagnetic resonance would have limited application, and only elements in the different transition groups could be studied by its methods. Recent work has shown this to be very far from the truth, and a large number of different types of system are now being studied by this means, the only requirement for the possibility of paramagnetic resonance being the presence of free or unpaired electrons in the structure.

The first reported observation of paramagnetic resonance in substances other than the transition elements or their compounds was that of Holden, Kittel, Merritt and Yager[2] on the organic free radical $ON(C_6H_5)C(CH_3)_2CH_2C(CH_3)N(C_6H_5)O$. This was followed very closely by the report of HUTCHINSON[31] on the observation of paramagnetic resonance in LiF and KCl crystals which had been irradiated with neutrons.

An increasingly large number of results have followed these two initial experiments, and are grouped together in the rest of the chapter under (i) resonance from free radicals, or electrons associated with the chemical binding; (ii) resonance from conduction electrons in metals and semi-conductors; and (iii) resonance from electrons trapped in a crystal lattice by natural imperfections, or those induced by irradiation. These subjects are rapidly growing into specialized branches of their own, and only a brief review of them can be given in this chapter.

7.7 RESONANCE IN FREE RADICALS*

The organic salt α,α-diphenyl β-trinitrophenyl hydrazyl

$$(C_6H_5)_2\ N\!-\!NC_6H_2\,(NO_2)_3$$

was the second free radical studied by resonance techniques[3, 36], and has been used and investigated widely ever since. This is mainly due to its narrow resonance, of 3 gauss total half-width, and very strong absorption, such that under 10 micrograms will produce an easily discernable signal at 9,000 Mc/s. It has thus found considerable use as a magnetic field or frequency marker, and can be used to measure g values close to 2·0 with very great accuracy[32]. Its preparation is a relatively simple matter[33], and single crystals up to $\frac{1}{2}$ cm long can be obtained. The initial measurements [3, 36] on diphenyl trinitrophenyl hydrazyl showed that its g value of 2·0036 $\pm$ 0·0002 was very close to the free spin value, and that the narrow width of 3 gauss was due to exchange narrowing.

If the hydrazyl is dissolved in benzene, or other organic solvents, the line broadens out, and in dilute solutions a marked hyperfine structure is obtained, having a central peak with two subsidiary ones on either side [38, 39]. This structure can be explained by assuming that the unpaired electron spends equal time on each of the two nitrogen atoms. The spectrum can then be represented by a Hamiltonian:

$$\mathcal{H} = g\beta \mathbf{S}.\mathbf{H} + A_1\,.\,\mathbf{I}_1\,.\,\mathbf{S}.\,+A_2\,.\,\mathbf{I}_2\,.\,\mathbf{S} \qquad \ldots.(7.13)$$

where A_1 and A_2 are a measure of the times spent by the unpaired electron on each of the two nuclei. If $A_1 = A_2$, five lines are to be expected from the resultant energy levels, with an intensity ratio of $1:2:3:2:1$, as observed. The hyperfine structure of the free-radical 'carbazyl' has also been investigated[38], this being identical in structure to the hydrazyl, but with a valency bond between the two phenyl groups. In this case a double peak is obtained with subsidiary peaks on either side, and the line shape can be analysed[38] and shown to be a combination of seven components of relative intensities $1:1:2:1:2:1:1$. This pattern is to be expected from an A_1/A_2 ratio of $2:1$, and hence it shows that the unpaired electron spends twice as much time on the nitrogen nearest to the amino group. The striking difference in the hyperfine splitting obtained from two compounds with such very similar chemical structures, is a good illustration of the way in which paramagnetic resonance can be used as a very sensitive test of chemical binding.

* For fuller details see Chapter 11.

It would seem that the possible number of resonance structures has a profound influence on both the hyperfine splitting and the line width; the more the unpaired electron can move from one site to another, the narrower the line will become under the process of 'motional narrowing'[40]. This suggestion is also supported by the measurements of SINGER and SPENCER[42] at low temperatures, which show an increase in width of the hydrazyl in cooling from 50° to 20° K, and may be due to the decreased mobility of the unpaired electron.

Another feature of interest in these two free radicals is the anisotropy of the g value. This was first thought to be isotropic, with a value of $2 \cdot 0036 \pm 0 \cdot 0003$ for the hydrazyl, but later measurements showed that there was a change of about ± 4 gauss, at $1 \cdot 25$ cm wavelength, when a single crystal of hydrazyl was rotated about an axis normal to the broad face, this variation following a $(3 \cos^2 \theta - 1)$ law. A similar variation was found for a single crystal of carbazyl, but with about twice the magnitude. In all these measurements[37, 38] a cylindrical cavity operating in the H_{011} mode was used, with the crystals mounted on a quartz rod so that they could be rotated when *in situ* in the centre of the cavity. The magnitude of this anisotropy was found to vary with the applied d.c. field strength in both cases, being smaller at 3 cm wavelength. KIKUCHI and COHEN[38] suggest that the anisotropy is therefore probably associated with the benzene ring, as Wurster's blue, another free radical possessing a benzene ring shows a similar variation.

A brief summary of other free radicals studied by paramagnetic resonance techniques is given in *Table 7.3* but this is far from complete as new observations are continually being reported.

In most cases the line width is found to decrease with decreasing frequency of observation, and this can also be taken as evidence of diamagnetic anisotropy at high field strengths. In many cases the hyperfine structure becomes very complex, owing to the large number of atoms included in the orbit of the unpaired electron, and a good example of this is Wurster's blue which shows over thirty component lines when in aqueous solution[44]. Weissman has also studied 5 other similar derivatives; he obtained highly complex hyperfine structure for each, and has considered the theory of hyperfine structure in polyatomic free radicals at some length[45]. In contrast to the complex structure obtained from the Wurster's derivatives, the peroxylamine disulphonate[43], the anisyl nitric oxide[46] and the phenyl NO-ether[46] all give only three components due to their interaction with the N^{14} nucleus, and thus behave in a very similar way to gaseous NO. Measurements at radiofrequencies[43] have traced the

Table 7.3. Free Radicals Studied by Paramagnetic Resonance

Radical	g value	Line Width in Solid	Hyperfine Structure	References
Diphenyl trinitrophenyl hydrazyl	2·0036	2·9 2·0*	5 lines in benzene	(3, 38)
Picryl-amino carbazyl	2·0036	1·0	7 lines in benzene	(38)
Wurster's blue	2·00	—	39 lines in water	(44, 44a)
Peroxylamine disulphonate	2·0054	—	3 lines in chloro-form	(43)
Naphthalene	2·00	—	19 lines in tetra-hydrofuran	(44)
Difluorenyl nitrogen	2·00	—	10 lines in benzene	(44)
Di-p-anisyl nitric oxide	2·0063	31·4	3 lines in benzene	(36)
β-β-methyl pentane oxime ether	2·0057	18·4	3 lines in solution	(36, 46)
Tri-p-anisyl-aminium perchlorate	2·00	0·68*	—	(41)
Aminophenyl-aminium perchlorate	2·00	0·33*	—	(41)
Triphenyl methyl in hexaphenylethane	2·00	—	100 lines in solution	(94, 95)

* Denotes measured in the radiofrequency region.

zero-field splitting of the transitions back to 54·7 Mc/s for the di-sulphonate. Experiments of BIJL and ROSE-INNES[46] have shown that in some cases the hyperfine structure moves in to form a single central peak as the temperature is lowered, a fact which should throw extra light on the interaction of the unpaired electron with the nitrogen nuclei.

Recent experiments by COMMONER, TOWNSEND and PAKE[47] have shown that paramagnetic resonance can also be used to detect the presence of free radicals in natural organic matter, and this opens up a very wide field of biophysical research. Their preliminary measure-ments proved that free radicals were present among many organic tissues, including those of the brain, muscle, liver and egg; and could also be detected in leaves and roots, showing that a large number of biochemical reactions involve the presence of free radicals in considerable concentration. They also demonstrated that the concentration of the free radicals could be altered by illumination with ultra violet light, and in general, higher concentrations were found among the active organic tissues. See also Sections 11.3 and 11.8.

7.8 RESONANCE FROM BROKEN BONDS

Paramagnetic resonance absorption, due to broken bonds in naturally occurring organic compounds, has recently been discovered by INGRAM and BENNETT[48]. This was first noticed in commercial charcoals, and then found to occur in any natural or artificial carbohydrates that had been charred. It was also found to exist in naturally occurring coal[5], and a systematic survey of different coal samples by Ingram et al.[4] has shown, (i) that the intensity of absorption rises sharply with carbon content once this is greater than 85 per cent, and (ii) that the graph of the intensity, plotted against carbonizing temperature, goes through a maximum at 550° C and then falls sharply to zero. This graph is shown in *Figure 66*, and the peak at

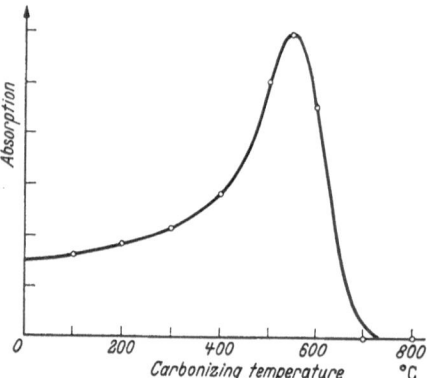

Figure 66. Variation of free radical concentration with carbonizing temperature

550° C may be explained by the large amount of hydrogen that volatilizes off at this temperature, leaving a number of unsaturated bonds which may resonate round the ring structure of the condensed carbons. Then as the aromatic rings become more mobile, these broken edge bonds will join with other rings, and large graphitic planes are formed containing very few broken bonds, and hence the absorption falls to zero. This disappearance of the signal once the substance has been heated above 600° C seems to be a very general feature of carbonaceous chars, and the above hypothesis is supported by the fact that it is just in this temperature region that the electrical conductivity begins to rise, and the substance starts to become gritty or graphitic in nature, showing that the carbon rings have all joined up to form a complete two-dimensional lattice. The increase of absorption with percentage carbon also suggests that the broken bonds are actually associated with the aromatic ring system, as the work of VAN KREVELEN[49] and others has shown that the number of aromatic rings also increases steeply between 85 and 95 per cent carbon content.

The spectroscopic splitting factor of these broken bonds is very close to the free spin value, being $2 \cdot 0030 \pm 0 \cdot 0003$[5], and the line width is the same in nearly all the samples, being 8 ± 2 gauss. This width remains the same when measured in the radiofrequency region[50], which is added confirmation that this type of paramagnetic resonance absorption is different from that observed by CASTLE[51] and HENNING[52] in graphites, due to the conduction electrons.

These measurements are another illustration of the way in which paramagnetic resonance can be used to study chemical and physical systems, and give information which is very hard to obtain by other methods.

7.9 RESONANCE FROM CONDUCTION ELECTRONS

The conduction electrons in metals are really a third type of free electron for which paramagnetic resonance can be observed. They are not the same as the inner unpaired electrons of normal paramagnetic atoms or ions, nor are they associated directly with the chemical binding like organic free radicals or the broken carbon bonds. They can be best visualized qualitatively as forming the 'free electron gas' of metallic conductors, and hence their resonances would be expected to have g values very close to that of a free spin.

The first resonances observed from such a 'free electron gas' were not in metals as such, but from solutions of the alkali metals in ammonia. These will ionize in solution to give positive metal ions with full, closed shells and no unpaired electrons, together with the negative ions which form the free electron gas. HUTCHINSON and PASTOR[55] were the first to observe paramagnetic resonance from such solutions, and obtained a g value of $2 \cdot 0012 \pm 0 \cdot 0002$, and a line width of $0 \cdot 1$ gauss. Measurements were also made in the radiofrequency region[57, 58] on solutions of potassium and sodium in other solvents such as methylamine, as well as in ammonia, and line widths as small as $0 \cdot 08$ gauss were obtained[58]. More recently, HUTCHINSON and PASTOR[56] have reported a whole series of detailed measurements on Na and K in different solvents at different concentrations, and found a minimum width of 0·02 gauss. The g values were independent of concentration, and measured as $2·0012 \pm 0·0002$. KAPLAN and KITTEL[59] have considered the theory of these ammonia-metal solutions in detail, and explain the very narrow lines by motional narrowing effects due to the rotation and diffusion of the ammonia molecules. The electrons are considered as located in cavities, and may be thought of as in molecular orbital states of a proton in an adjacent ammonia molecule.

The first reported observation of resonance due to conduction electrons in metals themselves was by GRISWOLD, KIP and KITTEL[60]. They employed very fine particles of sodium, made by supersonic techniques, suspended in paraffin wax, and obtained a resonance absorption with a g value of $2·000 \pm 0·003$ and a line width of 78 gauss. They were able to prove that this was in fact due to the conduction electrons, and not to a paramagnetic impurity, by showing that the signal intensity did not vary with temperature. The samples used were larger than the skin depth, and the large line width obtained was attributed to this and the damage of the sodium lattice by supersonic treatment. Theoretical calculations of YAFET[62] had predicted a g value of 2·0019; and OVERHAUSER[63] had considered the various factors affecting the relaxation times for electron spins in a metal, and concluded that the largest effect would be due to the translational motion of other electrons, producing a τ of 10^{-6} sec at room temperature, and a variation proportional to the inverse of the temperature. Later measurements by SOLT and STRANDBERG[61] on particles of sodium less than $4\ \mu$ in diameter gave a g value of $2·0014 \pm 0·0002$, which was not in agreement with Yafet's prediction within the experimental error.

Following this initial work on sodium, electron resonance has been observed in several other metals, including lithium[64, 65] and beryllium[66]. In more carefully prepared samples the line width of the sodium absorption was found to be only 10 gauss[61], and measurements by GUTOWSKY and FRANK[67], over a temperature range from 90° to 390° K, showed that this increased linearly with temperature in agreement with Overhauser's prediction. The line width of the lithium resonance was found to remain constant, however, at 4·5 gauss independent of temperature, and therefore another relaxation process may be the most important in this case. If the size of the particles is larger than the skin depth, then an extra broadening is produced which can be represented by a diffusion relaxation time T_D. Both this and the normal electron-spin relaxation time T_2 can be derived from the observed shape and width of the curve, as was done in the case of beryllium[66], where T_2 was determined as about $2 \cdot 10^{-8}$ sec.

Paramagnetic resonance due to conduction electrons has also been observed in graphite by Castle[51], the electrons now moving in the two dimensional graphitic plane. By orientating small particles of graphite so that their planes were all parallel, he was able to obtain an anisotropic line width, which varied from 8 to 40 gauss as the angle between the applied magnetic field and the graphitic plane was varied. In later experiments[68, 69] he also found a variation in width with temperature of preparation, a wide line appearing above 1,400° C, and then narrowing to the width previously observed when prepared at 2,500° C. Soft carbon samples[69] prepared between 1,500 and 2,500° C showed a minimum in their line width at 2,000° C preparation temperature, which Castle correlated to the measured Hall coefficients.

The general theory underlying paramagnetic resonance of conduction electrons has been initiated by ELLIOTT[70], and BROOKS[71] has carried out some detailed calculations on the g values and line widths to be expected for sodium along these lines.

7.10 ELECTRON RESONANCE IN SEMI-CONDUCTORS

Paramagnetic resonance absorption has also been observed in semiconductors in addition to normally conducting metals, and the results are examples of two new types of electron resonance. It was at first thought that the absorption obtained from silicon was due to the conduction electrons[73], as in the metals, but it was later shown[74, 75] that the electron responsible for the resonance was in fact bound to a donor impurity atom. This shows that it is in fact possible to obtain resonance absorption from electrons attached to what are normally diamagnetic atoms. The results obtained from germanium illustrate an entirely different type of absorption mechanism—i.e. 'Cyclotron Resonance' or 'Diamagnetic Resonance', which is due to the spiral motion of the electrons in the applied magnetic field. These different absorptions can be described briefly as follows.

(a) *Resonance from donor atoms in silicon.* Measurements by FLETCHER et al.[74] on arsenic-doped silicon which had been deformed by compression at 1,000° C, showed that four sharp component lines were produced with 73 gauss between each. This was shown to be the hyperfine structure associated with the As[75] nucleus, which has $I = \frac{3}{2}$, and it was found that the intensity of the lines could be increased by compressing the samples by larger amounts. These measurements suggested that the resonance observed was in fact from electrons bound to the As donor atom, rather than those in the conduction band.

These results and their interpretation were confirmed by further measurements on similar samples doped with phosphorus and antimony[75]. The samples containing phosphorus gave two component lines with a separation of 42 gauss, corresponding to an interaction with the P^{31} nucleus of $I = \frac{1}{2}$; and those containing antimony gave a set of six and a set of eight lines, corresponding to the Sb^{121} and Sb^{123} nuclei. The relative intensities and splittings of these two sets of lines agreed exactly with the abundance and nuclear magnetic moment ratios of the two isotopes. They found that if the concentration of donor atoms exceeds about 10^{18} per cc, then the hyperfine structure disappears and is replaced by a single narrow line.

(b) *Cyclotron resonance in germanium.* Cyclotron or diamagnetic resonance is different from normal paramagnetic resonance or electron spin resonance in metals, since it has to do with the curved spiral orbits of the electrons, or holes, as they move under the effect of the applied magnetic field. It is, therefore, not the spin energy which is changing between quantized states, but the 'orbital energy' of the electron, and its associated angular rotational frequency. The condition for resonance is therefore the simple precessional equation

$$\omega = \frac{eH}{m^*c} \qquad \ldots (7.14)$$

where m^* is the effective mass of the electron or hole. It has been seen in the last chapter that the orbital energy of electrons is not coupled directly to the magnetic field, and it will be the microwave electric field, not the magnetic field, which induces transitions in this case. The absorption should thus be correspondingly greater, as the probabilities for electric dipole transitions are many times greater than those for magnetic dipole transitions. The theory of cyclotron resonance in the solid state has been mainly treated by DINGLE[76] and SHOCKLEY[77], and the latter suggested that it might well be observable in the case of germanium.

DRESSELHAUS, KIP and KITTEL[78] were the first to report the observation of such a resonance in germanium crystals for both n and p types. Liquid helium temperatures were used, and with frequencies in the X band they obtained a resonance from the n-type crystals at a field of about 370 gauss, corresponding to an effective mass of $0\cdot11\ m_e$. The line width of 200 gauss gave an estimated relaxation time of $0\cdot7.10^{-10}$ sec. Measurements on p-type germanium with 10^{14} acceptors/cc, showed two resonance lines corresponding to effective masses for the holes of 0.04 m_e and 0.30 m_e. The relaxation times were still estimated as about 10^{-10} secs. The fact that the signal intensity is very dependent on microwave power level suggests that the electrons near resonance gain energy from the microwave field and cause ionization, thus increasing the number of electrons available for the cyclotron resonance. The values for the effective masses of the electrons and holes obtained from these measurements appears to agree with the anomalous infra-red absorption found in germanium[79].

By using infra-red radiation to excite the electrons, and hence only employing low microwave field strengths, DEXTER, ZEIGER and LAX[80] were able to show that there is a small anisotropy in m^* if measurements are made on a single crystal of germanium. They found that the effective mass could be expressed by an equation of the form:

$$m^* = m_0 \left[A + B \left(1 - 3 \cos^2 \theta\right)^2 \right] \qquad \ldots\ldots(7.15)$$

and the actual anisotropy was of the order of 2 per cent.

Recent measurements[81] have also shown that cyclotron resonance can be observed in silicon if optical radiation is used to excite the carriers. The effective masses obtained from such results as these, combined with the theories of Shockley[77] or Kohn and Luttinger[72], should give very interesting additional information on semi-conductor energy states.

7.11 PARAMAGNETIC RESONANCE FROM F-CENTRES

So far all the different types of paramagnetic and electron resonance that have been discussed in both this and the preceding chapter, have been observed in naturally occurring compounds and material. A whole new subject is now springing up concerned with the properties of irradiated substances such as are produced inside atomic piles, or by bombardment with γ-rays. The main interest of this work is centred on crystals, and the dislocations and imperfections that are produced in these by radiation damage. The techniques of paramagnetic resonance are immensely helpful in this field, as most forms of damage have trapped electrons or holes associated with them, and hence will give rise to resonance absorption.

The first experimental observation of paramagnetic resonance in irradiated crystals was by HUTCHINSON[82], who obtained an asymmetrical line about 100 gauss wide from LiF and KCl crystals which had been irradiated with neutrons. Further measurements by SCHNEIDER and ENGLAND[83] on X-ray irradiated KCl crystals, showed that the width of the resonance increases slowly with the number of colour centres, being about 40 gauss wide for 7.10^{16} centres/cc and 50 gauss wide for 3.10^{17} centres/cc (Hutchinson's width of 160 gauss was obtained for about 10^{19} centres/cc). They also found a marked change of width with temperature, and showed that spin-lattice relaxation was at least partially responsible for the width. They obtained a g value of 1·998, which is different from that of the free electron spin by a considerable amount, and suggested that the difference might be due to a resonance between F- and V-centres. F-centres are vacancies in the crystal lattice which have trapped an electron in the place of a negative ion; whereas V-centres are the halogen ions which have had their extra electron removed, and they themselves have become trapped to form 'positive holes'.

More recent measurements by SCHNEIDER [84] have shown that a complex hyperfine structure can be obtained from single LiF crystals irradiated by 180 kV X-rays. A large central peak was obtained of about 90 gauss width, together with 19 weaker and narrower lines of equal intensity, and equally spaced at 60 gauss intervals. Schneider [84] suggests that the broad line is due to the V-centres, while the resolved, narrower lines are due to the interaction of the electrons of the F-centres with the Li nucleus, to produce a hyperfine splitting. A general theory dealing with paramagnetic resonance from F-centres, with special reference to those formed by additional alkali metal concentration, has been described briefly by KAHN and KITTEL [85].

Following the initial measurements on irradiated alkali-halide crystals, paramagnetic resonance has been observed in several other irradiated crystalline structures, including ice and heavy water [86]. The resonance from ice gave a doublet at $g = 2\cdot0$, with separation of 30 gauss, and the width of the individual lines was 6 gauss. Similar absorption in frozen D_2O gave a triplet, with intensity ratio of 1 : 2 : 1. This strongly suggests that electrons have been trapped near H or D atoms, and are interacting with their nuclei to give a hyperfine structure. SMALLER, MATHESON and YASAITIS [86] also found a second resonance on cooling to lower temperatures, and suggested that this might be due to OH radicals.

It is also possible for hyperfine structure to be obtained from impurity atoms in the crystal, electrons often becoming trapped on these during irradiation. A very good example of this is afforded by the work of GRIFFITHS, OWEN and WARD [87] on irradiated quartz. They studied quartz crystals that had been coloured by X-rays, and obtained six electronic absorption lines, each with six hyperfine structure components. The spectroscopic splitting factors for each electron transition are the same, with $g_{\parallel} = 2\cdot06$ and $g_{\perp} = 2\cdot0$, but they have differently orientated axes of magnetic symmetry. The hyperfine structure had an over-all separation of 28 gauss, at both 3·1 cm and 1·2 cm wavelengths, and is ascribed to an impurity atom, probably Al^{27}, to which a trapped electron has become bound. The six different electronic resonances would correspond to the six possible positions of similar atoms in the quartz structure. A photograph of four of the groups of six hyperfine components is shown in *Figure 67*, at an angle such that the two central groups nearly coincide.

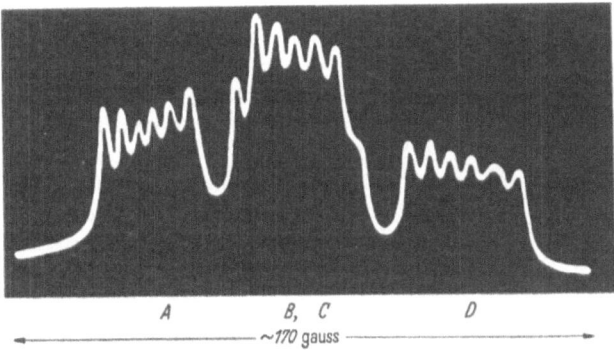

Figure 67. Electron resonance in irradiated quartz. A and D are separate groups showing six hyperfine structure lines. B and C are two groups nearly overlapping, with centres displaced by one hyperfine interval

The measurements of the same workers on diamonds[87] are also of interest in illustrating another type of absorption that can be obtained from crystals damaged by irradiation. They examined neutron-irradiated diamonds over a temperature range from 20° K to 290° K, and were able to distinguish two quite different types of resonance. First, there is an isotropic absorption line with a g value of $2·0028 \pm 0·0006$, which increased in both intensity and width with duration of irradiation, but was reduced after the diamond has been heated above 1,000° C. Secondly, there is an anisotropic spectrum, consisting of lines a hundred times smaller than the central peak, which are not affected by heating to 1,000° C, and have a width of 5 gauss, independent of temperature. They can be explained as arising from paramagnetic units which have an effective spin $S = 1$, and are separated into a doublet and a singlet, with a splitting of $D = 0·010$ cm^{-1}. The axis of crystalline field producing such splitting, is parallel to a side of the fundamental carbon tetrahedron, and as there are six such sides, six groups of two lines are to be expected and are observed experimentally. A preliminary theoretical explanation of these lines is that they are due to displaced interstitial carbon

213

atoms, which have linked together in pairs to give a system with $S = 1$.

Paramagnetic resonance has also been observed from irradiated organic compounds such as perspex[88], sugar and wood[89]. SCHNEIDER, DAY and STEIN[88] obtained sharp resonance lines after the irradiation of several plastics by X-rays, and often a hyperfine structure consisting of three well-resolved lines separated by about 30 gauss was observed. It may be assumed that the irradiation breaks some of the polymer bonds, and the electrons then move to different radicals, and it is thought that the hyperfine structure is due to an interaction of these electrons with the hydrogen nuclei. Other non-crystalline systems have also been studied by this means, as for instance irradiated glass[90], where effective g values of 2, 4 and 6 have been obtained[91]. Since practically any compound can be irradiated with X- or γ-rays, and will then probably show paramagnetic resonance from the electrons or holes trapped in the lattice imperfections, there is obviously a very wide field of research open for exploration in this way. Care must be taken in deducing results from such measurements, however, as radiation damage may take several different forms, and it is also possible for small paramagnetic impurities to be produced if intense sources are used.

A recent result[92] of considerable interest is the spectra obtained from frozen sulphuric and other acids after irradiation by γ-rays. These produce atomic hydrogen, from which intense paramagnetic absorption lines can be obtained, and small satellites have also been observed, spaced at the equivalent proton resonance frequency away from the main line. It would therefore appear that there is an interaction between the nuclear and electron resonances in such systems.

7.12 SUMMARY

In this chapter a very brief outline has been given of systems other than normal paramagnetic salts, which show paramagnetic or electron resonance absorption. It has been seen that ferromagnetic resonance, which is really a specialized case of normal paramagnetic resonance, gives results which are in reasonable agreement with present theory, although there are at least two unresolved problems (i.e. large line widths and high divergence of g value from 2·0). Generally speaking, ferromagnetic resonance has not produced such strikingly new results as paramagnetic resonance, and its applications are somewhat limited.

All the other types of resonance described in this chapter are only in their initial stages, but in most cases it looks as though they will prove to be very useful techniques for studying the properties of different systems. In order to clarify the conditions under which the different types of electron resonance are observed, their main properties and applications are summarized in *Table 7.4*.

Table 7.4. Different Types of Electron Resonance

Type of Resonance	Conditions of Observation	Substances Examined	References
In Free Radicals	Unpaired electrons in molecule	See *Table 7.3*	(2, 3, 36) (37, 43—46)
From Conduction E trons	Small particles or ammoniacal solutions	Na Li Be	(55, 56, 59) (60, 61, 66)
	Graphite	C	(51, 52)
From Donor Atoms	In semiconductors with impurities	Silicon with As, Sb, P	(74, 75)
Cyclotron Resonance	Conduction electrons moving in spirals	Ge and Si *n*- and *p*-type	(78, 81, 93)
From Broken Bonds	Carbon bonds broken by charring	Charred carbonaceous matter Coals, etc.	(4, 5, 48) (50)
Irradiated Crystals	Trapped electrons or holes	Alkali Halides Diamond	(82, 83, 86) (87)
Irradiated Polymers	Broken bonds	Perspex Sugars	(88, 89)

It can be seen from this table that a very large number of different kinds of material can now be studied by means of paramagnetic or electron resonance techniques. Instead of only the few transition group elements to which it was first thought paramagnetic resonance was limited, the different types of substances now studied vary from pure metals, semi-conductors, graphites and irradiated crystals to organic free-radicals and condensed ring systems and even living tissue in its various conditions[47].

It appears that the next ten years or so will show that the techniques of paramagnetic resonance provide an excellent tool for the study of many new problems, ranging from metallurgy to biophysics.

REFERENCES

[1] KITTEL, C. *Phys. Rev.* 71 (1947) 270

[2] HOLDEN, A. N., KITTEL, C., MERRITT, F. R. and YAGER, W. A. *ibid.* 75 (1949) 1614

[3] TOWNES, C. H. and TURKEVITCH, J. *ibid.* 77 (1950) 148

[4] INGRAM, D. J. E., TAPLEY, J. G., JACKSON, R., BOND, R. L. and MURNAGHAN, A. R. *Nature* 174 (1954) 797

[5] BENNETT, J. E., INGRAM, D. J. E. and TAPLEY, J. G. *Report on Bristol Conference July 1954*. Physical Society, London

[6] HUTCHINSON, C. A. and PASTOR, R. C. *Phys. Rev.* 81 (1951) 282

[7] GRISWOLD, T. W., KIP, A. F. and KITTEL, C. *ibid.* 88 (1952) 951

[8] HUTCHINSON, C. A. *ibid.* 75 (1949) 1769

[9] GRIFFITHS, J. H. E. *Nature* 158 (1946) 670

[10] KITTEL, C. *Phys. Rev.* 73 (1948) 155 and 76 (1949) 743

[11] VAN VLECK, J. H. *ibid.* 78 (1950) 266; and *Physica* 17 (1951) 234

[12] KITTEL, C., YAGER, W. A. and MERRITT, F. R. *Physica* 15 (1949) 256

[13] BAGGULEY, D. M. S. *Proc. phys. Soc.* A 66 (1953) 765

[14] YAGER, W. A. and BOZORTH, R. M. *Phys. Rev.* 72 (1947) 80

[15] KIP, A. F. and ARNOLD, R. D. *ibid.* 75 (1949) 1556

[16] BICKFORD, L. R. *ibid* 78 (1950) 449

[17] BLOEMBERGEN, N. *ibid.* 78 (1950) 572

[18] POLDER, D. *Physica* 15 (1949) 253 and *Phil. Mag.* 40 (1949) 99

[19] — *Phys. Rev.* 73 (1948) 1116

[20] KITTEL, C. *ibid.* 76 (1949) 743

[21] AKHEISER, A. *J. Phys. U.S.S.R.* 10 (1946) 217

[22] Section IVc of reference [17]

[23] KITTEL, C. and HERRING, C. *Phys. Rev.* 77 (1950) 725

[24] TROUNSON, E. P., BLEIL, D. F., WANGSNESS, R. K. and MAXWELL, L. R. *ibid.* 79 (1950) 542

[25] MAXWELL, L. R. and McGUIRE, T. R. *Rev. mod. Phys.* 25 (1953) 279

[25a] UBBINK, J., POULIS, J. A., GERRITSEN, H. J. and GORTER, C. J. *Physica* 18 (1952) 361; UBBINK, J. *Physica* 19 (1953) 9

[26] STEVENS, K. W. H. *Phys. Rev.* 81 (1951) 1058

[27] KITTEL, C. *ibid.* 82 (1951) 565; KEFFER, F. and KITTEL, C. *ibid.* 85 (1952) 329

[28] WELCH, A. J. E., NICKS, P. F., FAIRWEATHER, A. and ROBERTS, F. F. *ibid.* 77 (1950) 403

[29] BLOEMBERGEN, N. and DAMON, R. W. *ibid.* 85 (1952) 699

[30] OKAMURA, T. and KOJIMA, Y. *ibid.* 85 (1952) 690

[31] HUTCHINSON, C. A. *ibid.* 75 (1949) 1769

[32] BLEANEY, B. and INGRAM, D. J. E. *Proc. Roy. Soc.* A 205 (1951) 336

[33] GOLDSMICHT and RENN *Berichte* 55 (1922) 636

[34] STEVENS, K. W. H. *Proc. phys. Soc.* A 65 (1952) 149

[35] ADAM, G. D. and STANDLEY, K. J. *ibid.* A 66 (1953) 823

[36] HOLDEN, A. N., KITTEL, C., MERRITT, F. R. and YAGER, W. A. *Phys. Rev.* 77 (1950) 147

[37] COHEN, V. W., KIKUCHI, C. and TURKEVICH, J. *ibid.* 85 (1952) 379

[38] KIKUCHI, C. and COHEN, V. W. *ibid.* 93 (1954) 394

[39] HUTCHINSON, C. A., PASTOR, R. C. and KOWALSKY J. *chem. Phys.* 20 (1952) 534

REFERENCES

[40] BLOEMBERGEN, N., PURCELL, E. M. and POUND, R. V. *Phys. Rev.* 73 (1948) 679

[41] CODRINGTON, R. S., OLDS, J. D. and TORREY, H. C. *ibid.* 95 (1954) 607

[42] SINGER, L. S. and SPENCER, E. G. *J. chem. Phys.* 21 (1953) 939

[43] TOWNSEND, J., WEISSMAN, S. I. and PAKE, G. E. *Phys. Rev.* 85 (1952) 682; and 89 (1953) 606

[44] WEISSMAN, S. I., TOWNSEND, J., PAUL, D. E. and PAKE, G. E. *J. chem. Phys.* 21 (1953) 2227

[44a] WEISSMAN, S. I. *ibid.* 22 (1954) 1135

[45] — *ibid.* 22 (1954) 1378

[46] BIJL, D. and ROSE-INNES, A. C. *Phil. Mag.* 44 (1953) 1187

[47] COMMONER, B., TOWNSEND, J. and PAKE, G. E. *Nature* 174 (1954) 689

[48] INGRAM, D. J. E. and BENNETT, J. E. *Phil. Mag.* 45 (1954) 545

[49] VAN KREVELEN, D. W. and CHERMIN, H. A. G. *Fuel* 33 (1954) 79

[50] INGRAM, D. J. E. and TAPLEY, J. G. *Phil. Mag.* 45 (1954) 1221

[51] CASTLE, J. G. *Phys. Rev.* 92 (1953) 1063

[52] HENNING, F. R., SMALLER, B. and YASAITIS, E. L. *ibid.* 95 (1954) 1088

[53] KIP, A. F., KITTEL, C., PORTIS, A. M., BARTON, R. and SPEDDING, F. H. *ibid.* 89 (1953) 518

[54] RADO, G. T. and WEERTMAN, J. R. *ibid.* 94 (1954) 1386

[55] HUTCHINSON, C. A. and PASTOR, R. C. *ibid.* 81 (1951) 282

[56] — — *J. chem. Phys.* 21 (1953) 1959

[57] GARSTENS, M. A. and RYAN, A. H. *Phys. Rev.* 81 (1951) 888

[58] LEVINTHAL, E. C., ROGERS, E. H. and OGG, R. A. *ibid.* 83 (1951) 182

[59] KAPLAN, J. and KITTEL, C. *J. chem. Phys.* 21 (1953) 1429

[60] GRISWOLD, T. W., KIP, A. F. and KITTEL, C. *Phys. Rev.* 88 (1952) 951

[61] SOLT, I. H. and STRANDBERG, M. W. P. *ibid.* 95 (1954) 607

[62] YAFET, Y. *ibid.* 85 (1952) 478

[63] OVERHAUSER, A. W. *ibid.* 89 (1953) 689

[64] KIP, A. F., GRISWOLD, T. W. and PORTIS, A. M. *ibid.* 92 (1953) 544

[65] CARVER, T. R. and SLICHTER, C. P. *ibid.* 92 (1953) 212

[66] FEHER, G. and KIP, A. F *ibid.* 95 (1954) 1343; 98 (1955) 337

[67] GUTOWSKY, H. S. and FRANK, P. J. *ibid.* 94 (1954) 1067

[68] CASTLE, J. G. *ibid.* 94 (1954) 1410

[69] — *ibid.* 95 (1954) 846

[70] ELLIOTT, R. J. *ibid.* 96 (1954) 266, 280

[71] BROOKS, H. *ibid.* 94 (1954) 1411

[72] KOHN, W. and LUTTINGER, J. M. *ibid.* 96 (1954) 529

[73] PORTIS, A. M., KIP, A. F., KITTEL, C. and BRATTAIN, W. H. *Phys. Rev.* 90 (1953) 988

[74] FLETCHER, R. C., YAGER, W. A., PEARSON, G. L., HOLDEN, A. N., READ, W. T. and MERRITT, F. R. *ibid.* 94 (1954) 1392

[75] FLETCHER, R. C., YAGER, W. A., PEARSON, G. L. and MERRITT, F. R. *ibid.* 95 (1954) 844

[76] DINGLE, R. B. *Proc. Roy. Soc.* A 212 (1952) 38

[77] SHOCKLEY, W. *Phys. Rev.* 90 (1953) 491

[78] DRESSELHAUS, G., KIP, A. F. and KITTEL, C. *ibid.* 92 (1953) 827; 98 (1955) 368

[79] FAN, H. Y. and BECKER, M. *Semi-conducting Materials* 132. Butterworths Scientific Publications, London 1951

[80] DEXTER, R. N., ZEIGER, H. J. and LAX, B. *Phys. Rev.* 95 (1954) 557

[81] DEXTER, R. N., LAX, B., KIP, A. F. and DRESSELHAUS, G. *ibid.* 96 (1954) 222

[82] HUTCHINSON, C. A. *ibid.* 75 (1949) 1769

[83] SCHNEIDER, E. E. and ENGLAND, T. S. *Physica* 17 (1951) 221

[84] — *Phys. Rev.* 93 (1954) 919

[85] KAHN, A. H. and KITTEL, C. *ibid.* 89 (1953) 315

[86] SMALLER, B., MATHESON, M. S. and YASAITIS, E. L. *ibid.* 94 (1954) 202

[87] GRIFFITHS, J. H. E., OWEN, J. and WARD, I. M. *Nature* 173 (1954) 439

[88] SCHNEIDER, E. E., DAY, M. J. and STEIN, G. *ibid.* 168 (1951) 644

[89] COMBRISSON, J. and UEBERSFELD, J. *C.R. Acad. Sci., Paris* 238 (1954) 1397

[90] YASAITIS, E. L. and SMALLER, B. *Phys. Rev.* 92 (1953) 1068

[91] SANDS, R. H., TOWNSEND, J. and HOOD, H. P. *ibid.* 95 (1954) 607

[92] ZELDES, H. and LIVINGSTON, R. *ibid.* 96 (1954) 1702

[93] DRESSELHAUS, G., KIP, A. F., KITTEL, C. and WAGONER, G. *ibid.* 98 (1955) 556

[94] JARRET, H. S. and SLOAN, G. J. *J. chem. Phys.* 22 (1954) 1783

[95] WEISSMAN, S. I. and SOWDEN, J. C. *J. Amer. chem. Soc.* 75 (1953) 503

8

RADIOFREQUENCY SPECTROSCOPY

8.1 THE RESONANCE CONDITION

RADIATION is absorbed in the radiofrequency region by the same fundamental process as at all other wavelengths—i.e. the separation of the energy levels concerned is equal to one quantum of the incident radiation—but the energy of quanta at these frequencies is very small, and is only about a ten-millionth of an electron volt. The small splittings necessary to produce absorption in the radiofrequency region are those normally associated with the hyperfine structure of electronic spectra, and correspond to the different orientations of the nuclear moments in an applied field. In atomic spectra these energy level splittings produce small changes in the total energy of the electronic transition, and are thus responsible for the hyperfine structure; whereas, in the radiofrequency region, it is the transitions between these individual energy levels themselves that produce the absorption spectra. In the last two years electron paramagnetic resonance has also been studied in the radiofrequency region, and this is discussed in Section 11, but by far the greatest amount of research at these frequencies has been concentrated on nuclear resonance, and the initial sections of the chapter are confined to such work.

Nuclei can orientate themselves with respect to either electric or magnetic fields, to produce the small energy level splittings, and the two different cases are analagous to those in microwave absorption spectra. Thus, in the case of gaseous spectroscopy, the nuclear electric quadrupole moment interacts with the gradient of the molecular electric field to produce the hyperfine structure, no external magnetic field being required. Similarly in nuclear quadrupole resonance, as such radiofrequency absorption is called, the electric quadrupole moment can orientate itself in the gradient of the crystalline electric fields, and produce energy level splittings without any external magnetic field being applied. Resonance between these levels is then produced by varying the frequency of the radiation until an absorption line is obtained (e.g. at 38 Mc/s for Cl^{35} in $CHCl_3$).

In a similar way, nuclear paramagnetic resonance is closely analogous to electron magnetic resonance, as in both cases the splitting of the energy levels is produced by the orientation of the magnetic moments in an applied magnetic field; and the magnitude of the splitting is proportional to the strength of the applied field. In both cases the absorption lines are usually obtained by keeping the frequency of the radiation constant, and varying the strength of the magnetic field, while field modulation is used to display the spectrum. The condition for resonance in the nuclear case is therefore very similar to that of equation (6.1) for electron resonance, except that the electronic magnetic moment and g-value are replaced by the corresponding nuclear coefficients.

i.e. $$h\nu = g_I \cdot \beta_N \cdot H \qquad \qquad \ldots (8.1)$$

where g_I is the nuclear g-factor, and β_N is the nuclear magneton. Since the latter is about two thousand times smaller than the Bohr magneton, and the g_I of the proton is about twice that of the electron, it follows that the frequency of the radiation which will produce resonance absorption in a given magnetic field will be about a thousand times smaller than for electron resonance: e.g. in a field of 3,000 gauss, resonance absorption due to a change of spin orientation of a proton will occur at a frequency of 12·8 Mc/s, whereas, in the same field, a frequency of 8,400 Mc/s is required to change the spin orientation of an electron.

The development of these two branches of nuclear spectroscopy took place in the opposite order to that of the microwave case, nuclear paramagnetic resonance being studied in detail for some years before the pure quadrupole resonances, requiring no external field, were detected. The first experiments in radiofrequency spectroscopy were, in fact, carried out on molecular beams by RABI et al.[1] iust before the war, and it was not until 1946 that radiofrequency absorption in solids was first observed[2]. The magnetic resonance condition for absorption of radiation by the nuclei is the same, whether they are in the free molecules of a beam, or embedded in a crystalline lattice, but such factors as relaxation times and line widths vary considerably between the two cases as do the experimental methods.

The factors determining the frequencies of the quadrupole resonances are more complex, depending on the value of the particular nuclear electric quadrupole moment and the strengths of the crystalline electric field gradients. However, when the resonant frequencies have been detected and measured, they can, in turn, be used to give useful information on the solid state forces and interactions. The results obtained from a study of these quadrupole resonances are discussed more fully in Section 8.9, whereas Sections 8.2–8 deal with nuclear paramagnetic absorption in its different branches.

8.2 MOLECULAR BEAM EXPERIMENTS

Work on atomic and molecular beams had been in progress for some time before Rabi *et al.* [1] conceived the idea of using radiation at radio-frequencies to cause transitions in the nuclear magnetic moment orientation, and thus produced a very much more accurate method of measuring nuclear moments. A very brief summary of the development of atomic and molecular beam experiments can be outlined as follows.

(a) *Steady field deflexion methods.* The first work using such beams was the classic experiment of GERLACH and STERN [3] on beams of silver atoms, and from the separation of the two traces, formed by the atoms travelling in a path down an inhomogeneous magnetic field, they were able to show the existence of a quantized electron spin in a very convincing way. The essence of this method was the use of a magnetic field gradient to produce a continuous deflexion of the beam, the direction and magnitude of the deflexion depending on the orientation and magnitude of the electron moment associated with the free silver atoms. The next step was taken by FRISCH and STERN [4], who used a beam of molecules instead of atoms, and were thus able to detect the effect of the field gradient on the *nuclear* magnetic moments. If molecules are used, the electron spin momenta of the two atoms cancel, and the only resultant moment left for the whole molecule is that due to the nuclei; and the deflexion of the beam by the field gradient will therefore be determined by the nuclear magnetic moment. Using this method, they were able to measure the magnetic moments of the proton and deuteron by deflexion of hydrogen molecules, but the experimental difficulties are very much greater as much longer deflecting fields and more sensitive methods of detection have to be employed, since the magnetic moment of the nucleus is so much smaller than that of the electron.

BREIT and RABI [5] then showed that it was not necessary to use molecular beams, in which the electron moments had cancelled, in order to measure nuclear moments. By considering the coupling between the electron and nuclear spin they were able to demonstrate that the deflexion pattern obtained from an atomic beam could be analysed, and the contribution of the nuclear moment and spin deduced from it in an unambiguous way. This removed the disadvantage of the molecular beam deflexion method of Frisch and Stern, in that the presence of the electronic moment produced relatively high deflexions for comparatively short path lengths, as in the original atomic beam experiments. This method was applied by RABI, KUSCH and ZACHARIAS to measure the spins and magnetic moments of the proton [6] and deuteron [7], and by RABI, MILLMAN

221

and Fox to the alkali metals[8], a single deflecting field, of 142 cm length in some cases, being used.

The next advances were in the form of various experimental techniques[9, 10], devised to overcome the effect of beam spreading due to velocity variation among the atoms in the beam. The most powerful method[10] depended on the fact that, for certain strengths of the deflecting field, the effective moments of the atoms would be zero. This would apply to all the atoms whatever their velocities, and hence, for these particular field values, a large maximum at the zero position of the deflexion pattern would be obtained. From the number of such field values, and the spacing between them, the nuclear spin and moment could be determined. The method had the advantage that all the atoms were used to produce the large trace at zero deflexion, and no velocity selection was necessary. This 'zero-moment method' was first used by COHEN[10] to determine the spin and moment of caesium atoms, and other measurements[11] then followed rapidly.

The zero-moment method could not be applied to nuclei with spin $I = \frac{1}{2}$, however, as the only field value giving zero effective moment is then $H = 0$. A new technique was therefore devised to deal with this case, and the principles and design of this 'double field refocusing method' were later taken over into the radiofrequency resonance spectroscopes. The idea, which was due to KELLOG, RABI and ZACHARIAS[12], was to pass the atomic beam, first through a relatively weak deflecting field, and then through a stronger field with the direction of the field gradient reversed. The gradient of the field in the second magnet necessary to cancel the deflexion produced in the first field, and thus bring all the atoms to refocus at the point of zero deflexion at the end, was measured. Then, from this value and the other known physical parameters, the magnitude of the nuclear moment could be calculated. This determination is rather indirect, but it does have the same advantage as the zero-moment method, in that all the atoms, whatever their velocities, are focused to one point, and are all used in obtaining the final result.

These different methods of beam deflexion and detection were employed by various workers to measure the spins and moments of different nuclei, including the proton[12], deuteron[13], indium[14], gallium[15] and the alkali metals. It is to be noted that, in essence, each of these methods depends on the continuous deflexion of the atoms, or molecules, in steady, inhomogeneous magnetic fields.

(b) *Radiofrequency resonance techniques.* Then in 1938, Rabi et al.[1] introduced the entirely new technique of radiofrequency resonance, thus producing a very great increase in accuracy and sensitivity which more or less rendered all the older methods obsolete. Their apparatus was very similar to the two-field refocusing method previously described, but with the addition of a third magnet producing a large homogeneous field, and placed in the centre of the beam path between the two deflecting magnets of opposing field gradients. A coil of wire, carrying a radiofrequency current, was also placed between the pole pieces of the central magnet, so that a radiofrequency magnetic field was produced at right angles to the d.c. field. If the

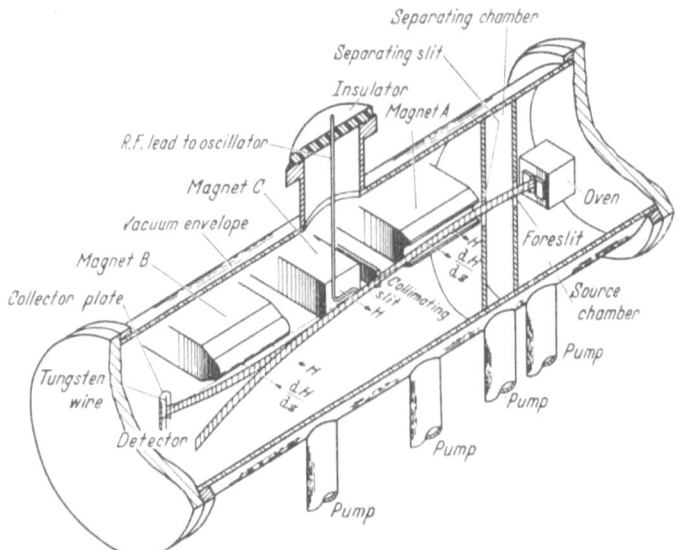

Figure 68. Diagram of typical atomic or molecular beam apparatus

strength of the magnetic field, and the frequency of the current in the coil, were then adjusted to fulfil the resonance condition of equation (8.1), the nuclear spins would change their orientation, absorbing the radiofrequency energy as they did so. The reorientation of the nuclei was detected by the fact that the second deflecting field no longer refocused them on to the collector, which was placed at the end of the run, at the position of zero deflexion. This type of apparatus is illustrated in *Figure 68* (taken from a paper by Kusch), and is actually designed for atomic beams, but exactly the same principles and techniques are used for molecular beams.

It can be seen that the first magnet bends the beam along a curve, so that it crosses the zero-deflexion axis in the centre of its path, and in the region of the strong homogeneous field produced by the central magnet. The coil carrying the radiofrequency power takes the form of one hairpin loop, running the length of the magnet face, and when the field strength is at the resonant value, absorption of energy from this coil causes the nuclei to reorientate themselves. The beam then enters the region of the second inhomogeneous deflecting field, which is adjusted to bend the beam back to pass into the detector, as it crosses the axis again at the end of the run. If nuclear reorientation has occurred, however, the beam is no longer refocused, but deflected to the side of the apparatus, as shown. The experiment therefore consists of : (i) initially adjusting the apparatus, with no radiofrequency power applied, until the detector shows its full, maximum reading; (ii) switching on the radiofrequency power, and varying the strength of the central magnetic field until a dip in the detector output is obtained. The resonance condition $h\nu = g_I \cdot \beta_N \cdot H$ is then fulfilled, and from the measured values of ν and H, the magnetic moment of the nucleus $(I \cdot g_I \cdot \beta_N)$ can be determined. In principle, either the frequency or the magnetic field strength can be varied to search for resonance, but in practice it is usually much easier to keep the frequency constant and vary the magnetic field, as in electron resonance. The molecular beam detector now serves as a null instrument, only being used to indicate when resonance is obtained, and the accuracy of magnetic moment determination is no longer limited by the geometry, or other physical parameters, of the apparatus.

In the first experiment of Rabi *et al.* [1], molecules of lithium chloride were used, the total beam length being 245 cm. The gap of the central magnet was 5 cm long, 6 mm wide, and a field of about 6,000 gauss was applied across the beam. The input radiofrequency was 3·5 Mc/s, and by varying the field strength, sharp resonant lines, due to both the lithium and chlorine nuclei, were obtained. The great advantage of this method is that all the complicated and rather inaccurate calculations on beam deflexion, which require the measurement of path lengths and field gradients, are now quite unnecessary; the only parameters that need to be determined are the frequency of the current in the coil, and the strength of the central homogeneous field.

(c) *Initial results from the resonance method.* The resonance method was very rapidly applied to determine the moments and spin of various nuclei, including [16, 17] Li^6, Li^7, F^{19}, Be^9, N^{14} and the alkali metals [18]. Soon after this, the method was used [19, 20] to show that the large splittings obtained in the spectra of a molecular beam of deuterium could only be explained on the assumption that the deuteron had an electric quadrupole moment. It was thus possible to determine a value for the deuteron quadrupole moment from the fine structure of its radiofrequency spectrum. A detailed survey of these particular results on the proton and deuteron beams has been given by KELLOG and MILLMAN[21], and by RAMSEY et al.[22, 23].

(d) *Recent work on molecular beams.* The method of radiofrequency resonance, in conjunction with molecular beams, has now been applied to a large number of different nuclei to determine both nuclear spins and moments. The basic form of the experimental apparatus has remained surprisingly similar to the original design of Rabi and his co-workers, one of the most important advances in technique being a method introduced by RAMSEY[24] to reduce the width of the resonance lines. The main factor, determining the width of the observed resonance, is the time that any given atom spends in the region of the oscillating field; and the half-width in frequency is given approximately by $t \cdot \Delta v = 1$.

This width was not usually attained in practice, because the field inhomogeneities contributed a greater factor, but the introduction of Ramsey's method not only averaged out the effect of the inhomogeneities, but reduced the theoretical width by about 50 per cent. In this method the long hairpin coil stretching the whole length of the homogeneous field is replaced by two coils, one at the beginning of the field, and one at the end, producing much stronger oscillating fields than before. The idea behind this technique is that nuclei, which are precessing about the applied field at the oscillator frequency, will enter the second field in the correct phase and have high transition probabilities; whereas those for which the frequency is slightly off resonance will acquire a phase shift in their passage down the field between the two coils, which results in a smaller transition probability. Ramsey[24] has considered the theory of this method in detail, and has also shown that, in practice, it does give absorption lines which are considerably narrower than those obtained by the older methods.

8.3 ATOMIC BEAMS AND THE DETERMINATION OF THE FREE ELECTRON GYROMAGNETIC RATIO

So far the radiofrequency resonance technique has only been discussed in relation to molecular beams, but, just as in the case of the d.c. deflexion methods, it was soon found that useful information could also be obtained by applying it to nuclear transitions in beams of free atoms. One striking outcome of the atomic beam resonance experiments has been their contribution to the determination of the precise value of the free electron gyromagnetic ratio, and the proof that the magnitude of this was, in fact, slightly greater than the Dirac value of exactly 2·0.

When dealing with beams of molecules in which the electronic magnetic moments have cancelled out, the resonance condition is quite simple as only the interactions with the nuclear magnetic moment need be considered. Atomic beams possess a resultant moment due to both electron and nuclear contributions, however, and thus the resonance conditions are more complicated, in the same way that the patterns obtained by d.c. deflexion of atomic beams were more involved than those given by molecular beams. In radio-frequency resonance of atomic beams it is the transitions between the hyperfine levels of the electronic ground state that are being observed. If the resonance measurements are made in large magnetic field strengths, the electron and nuclear moments are completely uncoupled, and the splitting between successive hyperfine structure components is then constant and independent of the applied field. The resonance condition for such atomic beams is therefore fundamentally different from the molecular beam case, as it only depends on the physical parameters of the atom concerned and remains constant as the applied magnetic field strength is varied. For this reason resonance absorption in *atomic* beams is always detected by varying the frequency fed to the coil, and not the strength of the central magnetic field.

As the applied magnetic field strength decreases, however, the nuclear and electronic moments begin to couple, and the resulting energy levels are then best described by an 'F' and 'M_F' quantum number, instead of an 'M_J' and 'M_I'. Under these 'intermediate-field' conditions, the splitting between successive hyperfine component levels is no longer constant, and varies with magnetic field. In zero field, all the Zeeman components of the F levels coincide, leaving two levels corresponding to $F = I + \frac{1}{2}$ and $F = I - \frac{1}{2}$ (assuming that $J = S = \frac{1}{2}$ as is usually the case). The splitting between these two zero-field energy levels is normally denoted by '$\Delta\nu$', and called the 'hyperfine structure separation'. The behaviour of the energy levels, as described above, is illustrated qualitatively in *Figure 69*, for the case of an atom with nuclear spin $I = \frac{3}{2}$.

The value of $\Delta\nu$ can be determined from measurements in very strong applied fields, as well as at intermediate or zero-field values. It is evident from *Figure 69* that in strong fields, the over-all splitting of both sets of hyperfine components becomes equal to '$\frac{3}{4}h \cdot \Delta\nu$', and thus transitions between the successive levels will all give rise to an absorption frequency of $\Delta\nu/4$ (or more generally $\Delta\nu/(2I+1)$). This method of determining $\Delta\nu$ from high-field measurements is very useful if the oscillator frequencies available are not high enough to measure $\Delta\nu$ itself in the zero- or low-field region. Determinations of $\Delta\nu$

226

are usually made in the weak or zero-field regions[26], however, and frequency measurements on such transitions as the $(F = 2, M_F = -1)$ to $(F = 2, M_F = -2)$ can also be used to obtain very accurate values of the magnetic field strength, and hence determine other nuclear magnetic moments. Apart from the determination of nuclear spins

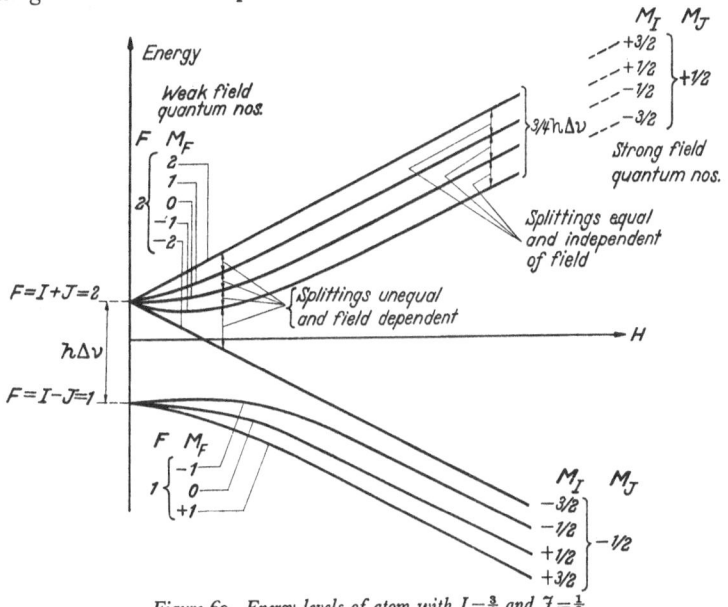

Figure 69. Energy levels of atom with $I = \frac{3}{2}$ and $J = \frac{1}{2}$

and magnetic moments from such measurements on atomic beams, there are three other kinds of data that can be obtained and are of considerable interest. These are summarized briefly under the different headings that follow.

(a) *The hyperfine structure anomaly.* This arises when the hyperfine structure separations, $\Delta\nu$, as measured by the atomic beam method, are compared for atoms of the same atomic number but with nuclei of different mass—i.e. isotopes of the same element. If it is assumed that the nuclei can be represented by charged masses concentrated at a single point, the ratio of their hyperfine separations should then be given by:

$$\frac{\Delta\nu_A}{\Delta\nu_B} = \left(\frac{2I_A+1}{I_A}\right)\left(\frac{I_B}{2I_B+1}\right)\left(\frac{m_A}{m_B}\right)^3\frac{\mu_A}{\mu_B} \qquad \ldots (8.2)$$

A comparison of the ratio predicted by this equation to that obtained from the atomic beam experiments is given in *Table 8.1* for the cases of hydrogen, rubidium and potassium.

Table 8.1. Comparison of Theoretical and Observed Hyperfine Splittings

Isotopes	$\dfrac{\Delta\nu_A}{\Delta\nu_B}$ *from (8.2)*	Observed	References
H^1 and D^2	4·3393876 ±0·0000008	4·3386484 ±0·0000020	(25, 26, 27)
Rb^{87} and Rb^{85}	2·25124 ±0·00035	2·2586 ±0·001	(28, 29)
K^{39} and K^{41}	1·8218 ±0·0002	1·81768 ±0·00001	(30)

It can be seen that the difference between these is well beyond the experimental error in each case, and, in order to explain them, the effects arising from the complex structure of the nucleus have to be considered. The electron is, in fact, not interacting with the magnetic moment of the whole nucleus over all its orbit. BOHR[31] was able to account for nearly all the observed difference by calculating this effect for the case of deuterium, and showed that when close to the nucleus the electron is only interacting with the magnetic moment of the proton. The theoretical treatment has been extended to heavier atoms by BOHR and WEISSKOPF[32], and these measurements on the hyperfine structure anomaly are an interesting example of how the details of nuclear structure itself can be observed by the very precise measurements of radiofrequency spectroscopy.

Recent experiments by BERINGER and RAWSON[33, 34] have shown that the hyperfine structure separation of free atoms can be measured by an entirely different method. In this, paramagnetic resonance techniques at microwave frequencies were employed in magnetic fields of about three thousand gauss. The energy levels thus corresponded to the strong field case of *Figure 69*, and resonance was obtained from transitions in the electron as well as the nuclear spin. The apparatus incorporated a cylindrical H_{011} resonator with a quartz tube passing down its centre, through which was passed a stream of the dissociated atoms, formed in a discharge tube above the resonator[35]. Resonances between the energy levels of the free atoms were then detected by absorption of the microwave radiation, as in a normal paramagnetic resonance spectroscope. The measurements on free atoms at microwave frequencies should be much easier to perform than those using atomic beams, and considerable data on anomalous hyperfine structure separations may be available from them in the near future.

(b) *The determination of quadrupole moments.* If $\mathcal{J} = S = \frac{1}{2}$ the only interaction that will produce the hyperfine separation, $\Delta\nu$, is that between the electron and the magnetic moment of the nucleus. If $J > \frac{1}{2}$, however, the interaction between the nuclear electric quadrupole moment and the electron configuration will also affect the hyperfine splitting. In such cases there may be more than one zero-field transition, as the resultant F will have more than two values. If there were no quadrupole moment all the zero-field $\Delta F = \pm 1$ transitions would have the same energy splitting and resonant frequency, but any quadrupole interaction will cause a variation in these splittings, and more than one resonant frequency will be obtained. It is therefore possible to determine the quadrupole moments of nuclei from the weak- and zero-field transitions in such cases. The method cannot be applied to free atoms with an S ground state, such as the alkali metals, but it has been used to evaluate the quadrupole moments of such nuclei as aluminium [36], chlorine [37], gadolinium [37] and indium [38], the atoms of which can be obtained in a $^2P_{\frac{3}{2}}$ state.

(c) *Determination of the gyromagnetic ratio of the free electron.* Until the very high precision of molecular and atomic beam experiments became available, the g value of the free electron spin had always been taken as exactly 2·0 (i.e. $g_s/g_l = 2 \cdot 0$, where g_l is the g value of the orbital motion and is equal to $e/4\pi mc$). The first suggestion that this was not in fact the case came from the initial determinations of NAFE and NELSON [25] on the hyperfine structure separation of atomic hydrogen. Since the nucleus of the hydrogen atom consists of only one proton, a very accurate expression can be derived relating the hyperfine separation, $\Delta \nu$, to the electronic g value, g_s, and the proton magnetic moment μ_p. When the observed value of $\Delta \nu$ was compared to that calculated from g_s and μ_p, a discrepancy of 0·24 per cent was found, well outside the experimental error for both $\Delta \nu$ and μ_p [39]. BREIT [40] then suggested that the discrepancy was due to the fact that g_s was not exactly 2·0, but had a slightly larger value. This was followed by a new form of quantum electrodynamics proposed by SCHWINGER [41], which predicted a value of g_s equal to $2(1 + \alpha/2\pi) = 2 \cdot 00232$. Later calculations by KARPLUS AND KROLL [42], to a fourth order, predicted a value of $2(1 + \alpha/2\pi - 2 \cdot 973 \ \alpha^2/\pi^2) = 2 \cdot 0022908$, where '$\alpha$' is the 'fine structure constant'.

Several experiments have since been made in order to determine the value of g_s as accurately as possible. The first of these was a series of measurements by KUSCH and FOLEY [43] on atomic beams, using atoms in different ground states. By measuring the hyperfine splitting of Ga in $^2P_{\frac{3}{2}}$ and $^2P_{\frac{1}{2}}$ states; In in a $^2P_{\frac{1}{2}}$ state; and Na in a $^2S_{\frac{1}{2}}$ state, they were able to compare the g_J values of these different states to those calculated by assuming $g_s = 2 \cdot 0 \ g_l$. In every case they found that the results could only be explained by assuming that g_s was slightly larger than 2·0, and the constancy of the factor required over such a range of different atoms confirmed that the anomaly was due to the intrinsic electron magnetic moment, and not to any other perturbing effect.

Two different sets of experiments have since been performed in order to measure g_s more accurately. Both of these are based on the same principle, namely to measure the ratio of the g-factors of the electron and proton as accurately as possible, and then combine this value of g_s/g_p with the value of g_l/g_p determined by GARDNER and PURCELL[44], to obtain an accurate value of g_s/g_l.

In Gardner and Purcell's experiment the cyclotron resonance of a beam of electrons, passed across a 3 cm wavelength guide, was compared to the frequency of proton resonance in the same field, and the ratio of the two frequencies was determined as $657·475 \pm 0·008$. This ratio is also equal to $2\, g_l/g_p$, if the proton g-factor is expressed in terms of Bohr magnetons, instead of the normal nuclear magnetons.

KOENIG, PRODELL and KUSCH[45] were the first to combine this measurement with a determination of g_s/g_p, and thus obtain an accurate value for g_s/g_l. They used an atomic beam method to measure g_s for hydrogen, comparing it directly with g_p by lowering a proton resonance coil into the same magnetic field near the atomic beam. The measurements on the hyperfine splittings of atomic hydrogen actually give g_J but, since the atoms are in a $^2S_{\frac{1}{2}}$ state, this is very nearly equal to g_s, and all the second-order corrections only amount to 1 part in 10^5. They obtained a value of g_s/g_p equal to $658·2288 \pm 0·0006$, which combines with Gardner and Purcell's[44] result to give for the gyromagnetic ratio of the free electron $g_s/g_l = 2·002292 \pm 0·000025$, or, as it can be otherwise stated, for the intrinsic magnetic moment of the free electron, $\mu = 1·001146 \pm 0·000012$ Bohr magnetons. It can be seen that this experimental value is in excellent agreement with that predicted by Karplus and Kroll[42] from fourth-order quantum electrodynamics.

Exactly the same basic method was used by Beringer and Heald[35] to obtain a further check on this value, but they employed paramagnetic resonance techniques instead of atomic beams. Their apparatus was identical to that already described in Section 3(a) for the determination of $\Delta\nu$ of atomic hydrogen, and by comparing the frequencies of the π_1 and π_2 transitions to that of a proton resonance, they were also able to obtain an accurate value of g_s/g_p. Their results agreed with those of Koenig, Prodell and Kusch[45] to within 1 part per million, and this experiment is of interest in showing that the very much simpler and easier techniques of paramagnetic resonance can give as great an accuracy as the highly refined methods employing atomic beams.

These experiments on the determination of the magnetic moment of a free electron have been described in some detail as they are a good illustration of the very high precision now available in the methods of radiofrequency and microwave spectroscopy, and also of the way in which they can be used to determine some of the basic physical constants. Further experimental details on atomic and molecular beam techniques can be found in review articles by ESTERMAN[46], and Kellog and Millman[21].

8.4 INITIAL SOLID STATE EXPERIMENTS

Radiofrequency techniques, as employed in atomic and molecular beam experiments, had become well established before the possibility of radiofrequency spectroscopy of the solid state was realized. This new field of research was opened up by two different groups, namely Purcell, Torrey and Pound at M.I.T., and Bloch, Hansen and Packard at Stanford University, California. They were working independently, and using slightly different methods, but both were essentially observing the resonance of nuclear magnetic moments in the solid state when acted on by a strong applied magnetic field. The first reported observations were those of Purcell, Torrey and Pound[2] on the resonance of protons in solid paraffin. The paraffin was placed in the inductive part of a resonator, formed by a shorted length of coaxial line, with a condenser at one end. This was fed with radiofrequency power from a signal generator, and the output of the resonator was balanced, in phase and amplitude, against power from another tuned circuit fed from the same generator. In this way a signal was only passed on to the detecting system by the unbalance produced in the bridge when the paraffin absorbed some of the radiofrequency power, and lowered the Q of the resonator. The detecting system consisted of an amplifier and detector stage, with a micro-ammeter to read the amplitude of the detected signal. The apparatus was thus similar in principle to the microwave bridges discussed in the preceding chapters, but working at a frequency of 30 Mc/s, instead of in the microwave region.

The resonant condition has already been given in equation (8.1), and the absorption line was obtained by slowly varying the magnetic field, and watching the micro-ammeter reading. A sharp peak with a signal-to-noise ratio of over 20 at a field of 7,100 gauss was obtained and an approximate value for the proton magnetic moment was determined, in agreement with that already established by the molecular beam method. These initial results were of great

231

interest as they showed that nuclear resonance could be observed in the solid state by relatively simple apparatus, and also that the relaxation times were such as to give narrow lines, observable as soon as the magnetic field was applied. Very low radiofrequency input powers were used at first, in case the spin-lattice relaxation time was very long[47] and serious saturation broadening was produced. The initial measurements on paraffin showed that the relaxation time was well under one minute, and later measurements by ROLLIN[48] and others[50] gave an estimate of about 10^{-2} sec for solid paraffin. It was therefore possible in later work to use higher powers for the input radiofrequency, and this, combined with much more sensitive a.c. methods of detection, increased the sensitivity of the method of nuclear resonance by a large factor.

The second reported observation of radiofrequency absorption in the solid state was by BLOCH, HANSEN and PACKARD[49]. Instead of using the decrease in Q of a tuned circuit to detect the resonance absorption, they made use of the fact that at resonance the nuclear magnetization will have a large component at right angles to the direction of the applied d.c. magnetic field. They were able to pick up this component by a coil with its axis at right angles to both the d.c. field and the applied radiofrequency field. The difference between this technique, and that employed by Purcell[2] is best described by the two names 'Nuclear Induction' and 'Nuclear Resonance'. In Bloch's experiments a radiofrequency field is actually *induced* in a separate coil by the Larmor precession of the nuclear magnetization; whereas in Purcell's experiment only one coil is used, and resonance is detected by an absorption of a small amount of power from the coil. The differences between the two methods, with their respective advantages and disadvantages, are discussed more fully in Sections 8.5 and 8.6, where further experiments employing the two techniques are described in detail.

Bloch, Hansen and Packard[49] observed their first proton resonance signals in water at room temperature, using a 60 c/s sweep on the magnetic field, and displaying the signal on an oscilloscope screen. They were also able to show that the addition of small amounts of paramagnetic salt to the water sample had a marked effect on the nuclear spin-lattice relaxation time. Following these initial experiments by the two groups, further observations were soon reported. PURCELL, POUND and BLOEMBERGEN[51] obtained resonance from hydrogen gas at pressures ranging from 10 to 30 atmospheres, and a sharp single line of less than 0·25 gauss width was observed over the whole pressure range. It is therefore not possible to derive as much information from these measurements, as from the corresponding ones on molecular beams of hydrogen, where the six component lines were obtained. The non-appearance of these in the direct radio-frequency absorption method is due to the fact that the molecules suffer collisions at rates much higher than the radiofrequency itself, with the result that the local fields fluctuate rapidly, and very nearly average out.

The next reported observation was by PURCELL et al.[52] on nuclear resonance in a single crystal of CaF_2. They were able to show that the width and peak intensity of the resonance line due to the F^{19} nuclei varied with the direction that the applied magnetic field made with the crystalline axes. This effect can be explained as due to the magnetic dipole-dipole interaction between the nuclei, which is one of the main causes of the line width, and varies with field direction in the manner that was found experimentally. This first measurement of nuclear resonance from a single crystal initiated what has now become a wide field of research, and later work[53, 54] has shown that not only does the line width change with angle, but in some cases the line splits into different components which shift their positions as the crystal is rotated relative to the field.

The next reported experiments were those of BLOCH, GRAVES, PACKARD and SPENCE[55], who used the method of nuclear induction to measure the spin and magnetic moment of tritium. They used a water sample containing 80 per cent H^3, and for a radiofrequency of 41·5 Mc/s obtained resonance lines at 9,160 gauss and 9,770 gauss, the former being due to the tritium and the latter to the protons. In slightly later measurements[55] they obtained a more accurate comparison of the magnetic moments, by feeding two different frequencies to the coil, and adjusting these until the resonance from both isotopes coincided. In this way no measurement of field strength was necessary, and a comparison of moments was obtained directly from the ratio of the two frequencies. This ratio determined the magnetic moment of the triton as $1·066636 \pm 0·00001$ times that of the proton, and they were also able to show, from the ratio of the intensities of the two lines, that the two nuclei had the same value of spin. BLOCH, LEVINTHAL and PACKARD[56] then applied exactly the same technique to measure the magnetic moment of the deuteron, using a water sample containing 50 per cent 'heavy water'. They obtained a value of $3·257195 \pm 0·00002$ for the ratio of the moment of the deuteron to that of the proton.

The success of these initial measurements, which were all made within a year of the discovery of nuclear resonance and induction, encouraged many others to enter this new field of research, and as the apparatus required was relatively simple and inexpensive a large

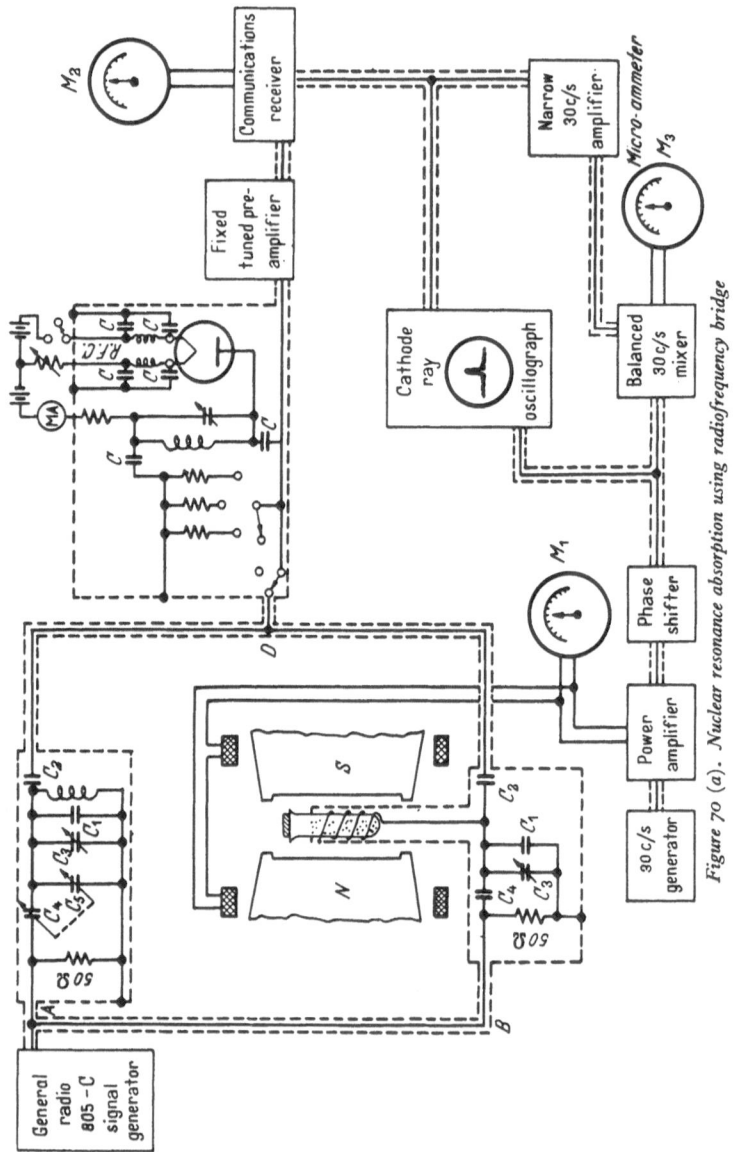

Figure 70 (a). Nuclear resonance absorption using radiofrequency bridge

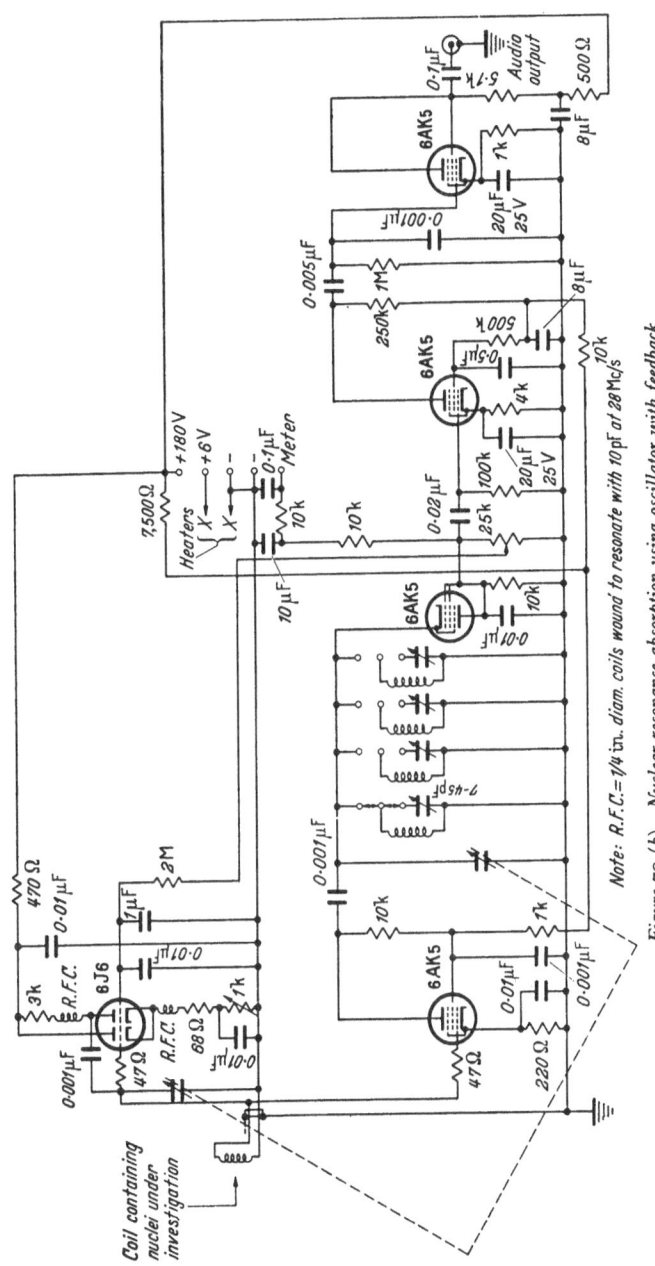

Note: R.F.C.=1/4 in. diam. coils wound to resonate with 10pF at 28Mc/s

Figure 70 (b). Nuclear resonance absorption using oscillator with feedback

number of results were soon being reported. Before summarizing the main points of the later work, a more detailed consideration of the experimental techniques of nuclear resonance and nuclear induction is now given.

8.5 The Techniques of Nuclear Resonance

The first improvement of the experimental apparatus, used in the initial observations of Purcell, Torrey and Pound [2], was the addition of magnetic field modulation, so that a.c. methods of amplification and display of the resonance line were possible. A radiofrequency bridge [57] incorporating such modifications is shown in *Figure 70(a)* and is seen to consist of two arms with a half-wavelength coaxial line connecting them on one side so that the signals should be π out of phase, when mixed at D. Independent amplitude and phase adjustment is incorporated by ganging C_4 and C_5 together in opposition, so that adjustment of C_4 to alter the phase does not disturb the amplitude balance effected by C_3. The general features of the detecting system are thus similar to those of the microwave bridges discussed in Chapters 3 and 4, a balancing system only being necessary to limit the power fed to the amplifying stages of the detecting circuits. The nature of the resultant unbalance is important, however, since a true trace of the nuclear absorption will only be obtained for a pure amplitude unbalance of the bridge, whereas a true trace of the nuclear dispersion signal (which changes the resonant frequency), will only be obtained for a pure phase unbalance.

In contrast to the balanced bridge method of detection, other types of circuit have also been used a considerable amount in nuclear resonance work. An example of one of these is illustrated in *Figure 70(b)* and consists of an oscillator adjusted to work on the curved part of the valve characteristic, so that a relatively large change in the working conditions is produced by a small change in the Q of the oscillator circuit. The sample containing the nuclei is placed in the coil as shown, which with the tuning condenser forms the resonant circuit connected between the cathode and grid of the first triode. The whole system containing the double triode thus forms a cathode-coupled oscillator, the conditions of oscillation being governed by the feed-back from the 25 K potentiometer, which is placed immediately after the detector valve. The coil and sample are placed in the magnet gap, and field modulation is used as before. Each time the field sweeps through the resonant condition the absorption lowers the Q of the resonant circuit, thus producing a change in valve working conditions, and the resultant change in radiofrequency signal amplitude is fed directly, via the 47-ohm resistor, to the grid of the first 6AK5. This acts as a single stage amplifier, and the load in its anode circuit can be switched to match the particular coil being used round the sample. The second 6AK5 acts as a detector, having the envelope of the radiofrequency signal as its output, this corresponding to the shape of the actual absorption line. After two stages of audio amplification, the output can then be fed to an oscilloscope for direct display of the line, or to further stages of amplification and a phase-sensitive detector.

Since this type of circuit only responds to amplitude modulation, only the nuclear absorption will be detected and displayed, the nuclear dispersion having no direct effect on the circuit. It is therefore not quite so flexible as the balanced bridge method. The feedback oscillator type of circuit is very simple to operate, however, requiring none of the balancing adjustments of the bridge, and for this reason is often employed when maximum sensitivity is not required. A very good example of this is its use in magnetic-field measurement, and a proton resonance meter, incorporating a simplified version of *Figure 70(b)*, has already been described and illustrated in *Figure 49*. Various other modifications of these two techniques have been used for the detection of nuclear resonance, both in the way of different bridge circuits [58], and different oscillator-detector systems, such as the autodyne and super-regenerative [59, 61] receivers. A review of some of these latter methods has been made by GUTOWSKY, MEYER and McCLURE [60] who also found that detection of the frequency modulation, produced by the nuclear dispersion, can be more sensitive than the normal amplitude modulation detection, especially for narrow lines. These different techniques are all basically similar to one or other of the two methods described above, however, and, in general, the balanced-bridge circuits are used when accurate setting of the input radiofrequency power is important, and the specialized oscillator circuits are employed when speed of obtaining the results is required.

The factors governing the width of the absorption lines must be considered in detail before it is possible to calculate what sensitivity is needed in order to detect any given absorption, in the same way as for the microwave cases. The main effects contributing to the width of nuclear paramagnetic resonance lines are similar to those of the electron resonance absorption—i.e. spin-lattice interaction and spin-spin interaction (see Section 4.8). In this case the interactions are between the nuclear spins, and not the electron spins, and for this reason the corresponding relaxation times can have quite different orders of magnitude compared to those normally encountered in paramagnetic resonance. Thus the spin-lattice relaxation time for nuclei in crystals is often very long, it is usually of the order of seconds, and can be of the order of hours; whereas values for the electron-spin-lattice relaxation time in similar crystals are usually around 10^{-6} sec. In electron resonance the spin-lattice interaction often broadens the resonance line because it is so strong (i.e. relaxation time so short), and the line width is then given approximately by $\Delta\nu\sim 1/\tau_l$. This case seldom arises in nuclear resonance of the solid state, because of the large values of τ_l or T_1, as the nuclear spin-lattice relaxation time is normally written. On the other hand, it does often contribute to the line width for exactly the opposite reason, i.e. because the interaction is so weak that saturation broadening can easily occur even at the low power levels used. It is important to realize this difference, as it explains why it is necessary to *increase* the spin-lattice relaxation time in order to obtain narrow lines in electron paramagnetic resonance (e.g. from 10^{-8} to 10^{-6} sec); whereas, to obtain narrow lines in nuclear resonance, it is necessary to *decrease* the spin-lattice relaxation time (e.g. from 2 secs to 10^{-2} secs). Because the broadening due to spin-lattice interaction in nuclear resonance is a form of saturation broadening, it follows that the signal strength is not improved indefinitely by increasing the input radiofrequency power level. The condition giving the optimum value for the magnitude of the radiofrequency magnetic field, H_1, was first derived by BLOCH[62] and can be written as:

$$H_1 = \frac{1}{|\gamma| \cdot (T_1 T_2)^{\frac{1}{2}}} \qquad \ldots(8.3)$$

where γ is the gyromagnetic ratio in c.g.s. units

T_1 is the nuclear spin-lattice relaxation time

and T_2 is the nuclear spin-spin relaxation time.

The spin-spin interaction and relaxation times are of the same kind of order in both electron and nuclear absorption, the width due to this interaction varying as $1/\tau$.

The detailed dependence of the line width on these relaxation times was considered theoretically, and observed experimentally, for a variety of compounds by Bloembergen, Purcell and Pound[57], and they also derived an expression for the signal-to-noise ratio that it is theoretically possible to obtain in any given case. This may be expressed as:

$$\frac{A_s}{A_n} = \frac{(V_c Q_0)^{\frac{1}{2}} \cdot \xi \cdot h^2 \cdot \mathcal{N} \cdot \gamma \cdot I(I+1)}{48 (kT)^{\frac{3}{2}} \cdot (B.F.)^{\frac{1}{2}}} \cdot \frac{\nu^{\frac{3}{2}}}{(\pi \cdot \Delta\nu \cdot T_1)^{\frac{1}{2}}} \qquad \ldots(8.4)$$

where V_c is the effective volume of the coil

ξ is the filling factor of the sample in the coil

Q_0 is the Q of the coil in the absence of resonance

N is the number of nuclei present per gramme

B is the bandwidth of the detecting system

F is its over-all noise figure

ν_0 is the resonant frequency

$\Delta\nu$ is the width of the absorption line

and T_1 is the spin-lattice relaxation time.

The actual power absorbed near resonance is, in fact, proportional to:

$$\frac{(\omega\,\omega_0\,H_1)^2/H_0 T_2}{(\omega - \omega_0)^2 + 1/T_2^2 + \left(\frac{\omega_0 H_1}{H_0}\right)^2\left(\frac{T_1}{T_2}\right)} \qquad \dots\dots(8.5)$$

where H_0 and ω_0 are the resonant magnetic field and angular frequency values. It can be seen that the line width is therefore determined by $1/T_2$, $\omega_0 H_1/H_0$ and T_1/T_2, and increases in magnitude the greater the value of each of these quantities.

The widths of absorption lines actually observed varies considerably, from over 20 gauss, in some crystals, down to values of the order of a milligauss, for some liquids. These widths are often determined by the value of the spin-lattice relaxation time, and, whereas in solids $T_1 \gg T_2$, in liquids $T_1 \sim T_2$, and hence very narrow lines are obtained. Further measurements on widths of resonance lines in crystals and liquids, and the other factors which can contribute to the observed magnitude, are considered in Section 8.8.

8.6. The Techniques of Nuclear Induction

The essential experimental difference between nuclear resonance and nuclear induction is that a second coil is employed in the latter case, to pick up the signal induced by the perpendicular component of magnetization at resonance. A block diagram of the apparatus required for nuclear induction is shown in *Figure 71(a)*, and it can be seen that the input coil, which is drawn as in the plane of the paper, is fed by the radiofrequency oscillator, and has no direct connexion with the detecting circuit. The sample containing the nuclei to be investigated is placed at the centre of this coil, and is also surrounded by a second coil which has its axis perpendicular to that of the first, and to the direction of the applied d.c. magnetic field. This receiving coil feeds any induced voltage to a radiofrequency amplifier, the detected output of which is applied to the Y plates of an oscilloscope, via an audio amplifier.

At non-resonant values of the applied magnetic field, the nuclear moments will partially align along its direction, to produce a resultant magnetization, M_z. The radiofrequency field in the input coil will not affect this to any measurable extent, and no signal is picked

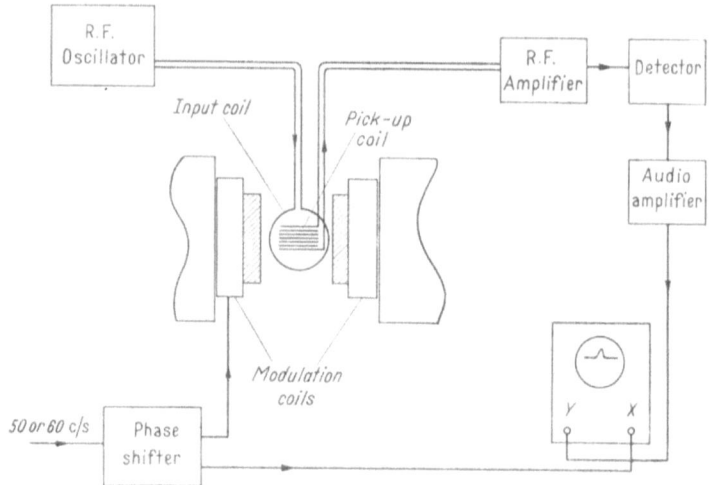

Figure 71 (a). Block diagram of nuclear induction experiment

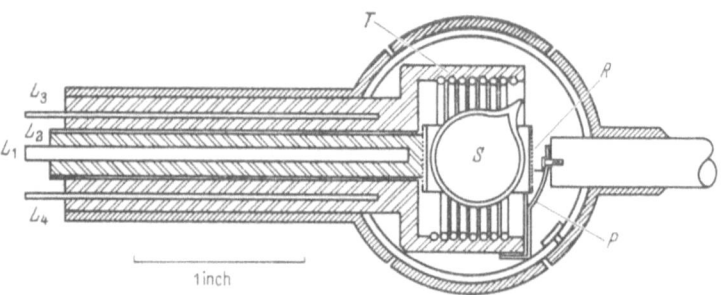

Figure 71 (b). Scale drawing of an xy section of the radiofrequency head. The spherical sample S is surrounded by a receiver coil R, which is in turn surrounded by a transmitter coil T, the whole being encased in a shield. A rotatably mounted paddle P is used to steer the transmitter flux. Leads to the receiver coil are the coaxial leads $L_1 L_2$ while the transmitter leads are L_3 and L_4. The outer shield is split to avoid 60-cycle eddy currents.

up in the receiver coil. If, however, the value of the d.c. magnetic field is adjusted to its resonant value, the magnetization will be markedly affected by the radiofrequency field of the first coil, and will start precessing around the direction of the applied field, with its Larmor frequency. As a result, components of the magnetization will

240

now exist in a plane at right angles to the applied field, and the precession will thus induce a voltage in the receiver coil. Nuclear paramagnetic resonance is therefore detected by the sudden appearance of an induced voltage in the receiving coil and detector circuits.

In order to obtain oscilloscope presentation, magnetic field modulation is employed, as in nuclear resonance, and by feeding the time base from the same source via a phase shifter, direct display of the absorption line is possible. Phase-sensitive methods of detection can be used to give greater sensitivity if required, and it can be seen that the problems of the whole system of detection and display, once the signal has been obtained from the receiving coil, are the same as those in nuclear or electron resonance. The main experimental difficulty that arises in any system employing nuclear induction is the elimination of direct pick-up between the input and receiver coils. This is a very important factor as any sudden variation of the input amplitude, due to valve instability or the like, will produce a signal in the receiving coil identical in appearance to an absorption line. Since the signal produced by the precession of the nuclear moments is usually over a million times smaller than the signal applied to the input coil, it follows that even if the two coils are only coupled by a factor of 1 in 1,000, a variation of the input power greater than 1 part in 10^3 will swamp any signal from the nuclei. It is virtually impossible to reduce the coupling factor to such small amounts by purely geometrical adjustment of the axes to a right angle, and instead, a small amount of coupling is artificially introduced, so that it can be used to compensate that already present. This adjustable coupling usually takes the form of a small semi-circular copper paddle, which can be moved across the end of the input coil, intercepting the flux through it on one side or the other, and thus producing an asymmetry in the field which alters the coupling to the receiver coil. In this way a continuous adjustment is obtained of both positive and negative values. A scale drawing of the radiofrequency head used by Bloch, Hansen and Packard[50] is shown in *Figure 71(b)*, and the paddle used to vary the coupling can be seen on the right-hand side of the diagram. The input, or transmitter, coil consists of seven turns of $\frac{7}{8}$ in. diameter, and has its axis horizontal in the plane of the paper, the receiving coil has 24 turns of $\frac{9}{16}$ in. diameter, with its axis vertical in the plane of the paper, and the d.c. magnetic field would therefore be applied with its direction normal to that of the paper. Further details of the precise experimental techniques that are used in such an experiment can be found in the first paper on the techniques of nuclear radiofrequency spectroscopy by Bloch, Hansen and Packard[50], and in later review articles by PAKE[63]. See also Chapter 10.

The results obtained from nuclear induction experiments are very similar to those observed by the direct resonance technique, and have been used to determine quite a number of nuclear gyromagnetic ratios, quadrupole interactions and spins[64]. Nuclear induction can also be employed to measure relaxation times very accurately, directly from the signals on the oscilloscope. This arises from the fact that the receiver coil can have either positive or negative voltages induced in it, and if the absorption line appears above the time base sweep as the main field is turned up, it will appear below the time base for decreasing values of the main field. A direct measure of the relaxation time can therefore be made by holding the applied field at a value slightly above resonance, but sufficiently close for the line to be still observed on one end of the oscilloscope sweep. The main field is then suddenly reduced to a value an equal amount below resonance, when the line moves across to the other side of the oscilloscope trace.

Initially the signal remains in the same direction, e.g. above the time base, but during the next few seconds its magnitude decreases, goes through zero, and then builds up to an equal value on the opposite side of the time base. The time taken to complete this 'phase reversal' gives a direct determination of the relaxation time.

A detailed analysis of the results to be expected under different conditions from nuclear induction experiments was given very early in the development of solid state radiofrequency spectroscopy by Bloch[62]. In his analysis he used two relaxation times, defined as the 'longitudinal relaxation time' and the 'transversal relaxation time'. The former is a measure of the rate at which the component of magnetization in the direction of the applied field adjusts itself to a new equilibrium value, and the latter measures the rate at which the components of magnetization, in a plane normal to the applied field, readjust themselves. These two times are, in fact, equivalent to the spin-lattice relaxation time, T_1, and spin-spin relaxation time, T_2, already used, as it is only the component of magnetization parallel to the applied field that has to interact with the lattice and change its energy. Bloch[62] was the first to derive the condition for optimum applied radiofrequency field, as quoted in equation (8.3), and the first to consider the variation in signal strength and shape that would be expected for different rates of sweeping through resonance. Generally speaking, the optimum signal strength is obtained when $T_1 = T_2$, and it is for this reason that small amounts of paramagnetic salt are usually added to water samples, in order to reduce the spin-lattice relaxation time of the protons, which is of the order of 3 sec in pure water[65].

8.7 PULSE AND ECHO METHODS

Apart from the different circuits incorporating the autodyne and super-regenerative principles already mentioned[59, 61], the main advance in the experimental techniques of solid state radiofrequency spectroscopy has been the introduction of pulse methods by Torrey and Hahn.

(a) *Nutational resonance.* The first of these was due to TORREY[66, 67], and in it the modulation of the magnetic field is replaced by modulation of the input radiofrequency amplitude. The basic difference between this method and those previously used, is that the nuclear resonance is now observed under transient conditions and not in the steady state. The absorption therefore occurs before saturation sets in, and hence larger signal strengths are obtained. This advantage is offset by the necessity of employing larger bandwidths to pass the pulses through the system, and hence the resultant signal-to-noise ratio is of the same order as that of the previous methods. The larger bandwidths employed allow more rapid search for resonance lines, however, and the method also affords a very direct means of measuring relaxation times. The pulses are produced by applying the output of a gate generator to the signal from the radiofrequency transmitter, and square pulses of power at the nuclear resonance frequency are thus passed on to the balanced bridge circuit. The output of the bridge is passed through a wide-band radiofrequency amplifier, detector and video amplifier, and is then applied to the Y plates of an oscilloscope. The time base of the oscilloscope is triggered by the gating circuit producing the pulses, so that the trace across the screen corresponds to the time during which any one individual pulse of energy is being supplied to the bridge. In the absence of any nuclear absorption the trace will therefore remain horizontal and represent the flat top of each successive pulse.

Theoretical analysis[62, 67] of the equations of motion for the nuclear magnetization under transient conditions shows that if the magnetic field is at the resonant value, the radiofrequency pulse will become amplitude modulated by the nuclear resonance, this modulation having an exponential decay due to the relaxation effects. The modulation is produced by oscillations, which can be regarded as arising from a periodic nutational motion of the magnetic moment vector, and will have an angular frequency given by:

$$\Omega = [\gamma^2 H_1^2 + (\omega_0 - \omega)^2]^{\frac{1}{2}} \qquad \ldots . (8.6)$$

and, at resonance $\Omega \ll \omega$.

When the d.c. field has its resonant value, this effect shows up on the oscilloscope trace as a train of sine waves, of frequency $\Omega = \gamma . H_1$

243

and damped in amplitude by the relaxation processes. Values of the different relaxation times can be determined very directly by this means. Thus, it can be shown[67] that the damping time of the oscilloscope signal is equal to $2T_1 T_2/(T_1 + T_2)$, and also, if the duration of the pulse is long compared with this, that the initial amplitude of the modulating signal is proportional to $(1 - \exp[- T/T_1])$, where T is the time between pulses.

This method of 'nutational resonance' is therefore very useful and direct for the determination of relaxation times, and because of its fast response, also has advantages in searching for unknown spectra. Its resolving power is rather less than the more conventional methods, however, and it is therefore not a good method for studying the structure and shape of resonance lines. The essential difference between this method, due to Torrey, and the 'spin-echo' method due to Hahn, discussed below, is that in this method the resultant signal is observed while the pulse is applied to the sample, whereas in Hahn's spin-echo method, the resultant signal is recorded *after the pulse is switched off*. There is also the difference that Torrey's method used the techniques of nuclear *resonance*, while Hahn's uses those of nuclear *induction*.

(*b*) *The spin-echo method*. At the same time as Torrey was studying the effect of pulses applied to nuclei placed in a resonant bridge circuit, Hahn was also studying pulses as applied to the nuclear induction experiment. His first observations[68] showed that nutational resonance could be detected by this means as well, and he obtained nutational frequencies down to $\Omega = 5$ c/s from protons and F^{19} nuclei.

He then showed[69] that a very direct measure of the relaxation time could be obtained by observing the signal induced immediately *after* the pulse had been switched off. The application of a strong pulse of the resonant radiofrequency can be regarded as orientating the nuclear moments with their axes in a plane perpendicular to the direction of the applied magnetic field. When this pulse is switched off, the moments will precess freely about the direction of the main field. The decay of the induced voltage will therefore be determined by the rate at which these moments lose 'phase memory', and become randomly orientated in the perpendicular plane. They will also start to precess inwards, toward the direction of the applied magnetic field, but this is determined by T_1 and may take much longer; whereas the induced voltage will disappear as soon as random orientation has occurred in the perpendicular plane. If the applied field has zero inhomogeneity, the decay of the signal would be determined only by the spin-spin relaxation time, T_2, and this could therefore be obtained directly from the observed exponential fall on the cathode-ray screen. Unfortunately the field inhomogeneities also effect the rate of re-orientation of the nuclear moments in the perpendicular plane, and in general cause a very much faster decay of the signal than would be produced by T_2 alone. For this reason, accurate values of T_2 cannot be obtained directly by the method, but the same effect facilitates the observation of spin-echoes, as described below, and both T_1 and T_2 can be measured accurately by this latter technique.

HAHN[70] demonstrated the existence of spin-echoes, by applying two short radiofrequency pulses, of duration τ_w, and separated by a time τ, such that $\tau_w < \tau < T_1 \sim T_2$. Each of these pulses will show up on the oscilloscope screen, and the trailing edge of the first will have an exponential decay, depending mainly on the applied field in-homogeneities, as discussed above. This decay is sometimes cancelled out on the second pulse[70], but in every case the signal induced, due to the free Larmor precession, dies out in a time $< \tau$. Then, at a time τ after the leading edge of the second pulse, the nuclei are found to *emit* a pulse of radiation, this occurring in the complete absence of any input radiofrequency. The production of this 'echo-pulse' can be explained by the constructive interference that takes place at this particular instant, between the rotating components of magnetiza-tion produced by the two input pulses. Thus, during the first pulse, all the vectors representing the nuclear moments will orientate into a plane perpendicular to the direction of the d.c. field. Then, during the time between the two pulses, these vectors will continue rotating in this plane but in random directions, reaching an isotropic dis-tribution before the advent of the second pulse, which may be con-sidered as turning the magnetization through π. In the time following the removal of the second pulse, the vectors again rotate in the perpendicular plane, but in the opposite direction to the first case, so that they are in fact 'unwinding their loss of phase', and therefore, at a time τ later, they are all exactly in phase again, and a resultant voltage is induced in the receiver coil. Hahn[70] has carried out a detailed analysis of the general case, and shown that the echo signal lasts for about $(4\pi/\Delta\omega)$ secs, where $\Delta\omega$ is the width of an assumed function, which describes the distribution of magnetization through-out the inhomogeneous external field. From the detailed analysis, it can also be shown that the spin-spin relaxation time is obtained directly, by plotting the logarithm of the maximum echo amplitude at the time $t = 2\tau$ versus different values of 2τ. In a similar way the value of T_1, the spin-lattice relaxation time, can be obtained by the variation in amplitude of 'stimulated echoes'. Stimulated, or second-ary, echoes are produced by a third pulse, applied at a time

$T > 2\tau$ after the first pulse. Under these conditions another echo will be obtained at $(T + \tau)$, and also at $(2T - 2\tau)$, $(2T - \tau)$ and $2T$.

It can be seen that the technique of spin-echoes is very flexible, and also somewhat complicated in the general case. The original paper by Hahn [70] forms a very good introduction to both the theoretical and experimental sides, however, and should be consulted if further details are required. The echo method has several advantages over conventional techniques. It is very suitable as a fast and stable method in searching for unknown resonances, as high power pulses can be fed into the coil, and because of the spread of frequency associated with these, free induction signals can be obtained when the input frequency is away from exact resonance. It is also a very useful method for measuring relaxation times, and has the advantage that, since observations are made when the input radiofrequency field is absent, noise and hum introduced by coupling to the input circuits are eliminated.

8.8 Results of Nuclear Paramagnetic Resonance Experiments

The immediate application of measurements on nuclear resonance is in the determination of nuclear magnetic moments to a high degree of accuracy, from the observed resonant frequency and field. The greatest accuracy is obtained by comparing the resonant frequency for a given nucleus with that for protons in the same magnetic field, and thus eliminating any direct determination of the magnetic field strength. Examples of recent work on these lines are the experiments of Zimmerman, Williams and Chambers [71] to determine over twenty different nuclear gyromagnetic ratios, using the direct resonance method with super-regenerative detection, and the work of Proctor and Yu [64], who measured a large number by induction techniques.

The determination of nuclear magnetic moments by such means has proceeded apace, and a large number of measurements have now been made. This is not the only information that it is possible to obtain from resonance experiments, however, as considerable data can also be deduced about solid state forces and interactions. Methods of measuring the various relaxation effects have already been described, and the variation of these between different solids and liquids, has been studied in detail [57], to test the theories on spin-lattice interactions. Shifts in the resonant frequency due to various causes, such as magnetic shielding and quadrupole interaction, can also be measured, and used to give information on these different effects.

In general, the spin-lattice interactions are much stronger in liquids and metals than in non-conducting solids, and the line widths obtained from them are much narrower, not being broadened by the saturation effects of the long spin-lattice relaxation time of normal solids. The line width of the signals from liquids is, in fact, usually determined by the field inhomogeneities of the main d.c. magnet, and not by any relaxation process. It is for this reason that large magnet pole faces, producing extremely homogeneous fields, are often required for work on nuclear resonance. The main results, obtained from a study of the effects of various interactions on nuclear resonance, are therefore summarized under three different headings, i.e. (a) liquids, (b) metals and (c) crystals.

(a) *Results from liquids.* As mentioned above, liquids approach quite close to the optimum condition of $T_1 = T_2$, having narrow lines with widths often limited by the field inhomogeneity. As a result, very accurate frequency settings can be made on such lines, and the high resolution attainable allows observation of any fine splitting of the resonance. Another simplification that is present in the spectra of liquids, is that the width and splitting of resonance lines, due to non-uniform distribution of the nuclei, is often reduced or eliminated. Thus, in the case of crystals, there may be regions within each molecule where several of the nuclei under investigation are concentrated together in a group, and the magnetic interactions within the group can be much larger than the general spin-spin interaction due to all the other groups. This may result in resolvable structure in the resonance line, as discussed in the next section. Such splitting, or broadening, does not occur to any great extent in liquids, because of the general 'averaging out' effect of the random motion of their molecules.

Nuclear resonance spectra obtained from liquids are therefore of the simplest kind and most easy to interpret, as any splittings produced in the absorption line must be due to interactions which do not vary with orientation. This fact is illustrated in *Figure 72*, which shows the resonance observed in liquid CH_3OPF_2 by GUTOWSKY, McCALL and SLICHTER[72]. The extremely narrow width of the lines can be appreciated from the fact that the total sweep amplitude is only 2 gauss, and the field inhomogeneities are still responsible for the observed width. The high resolution of the triplet is also well illustrated by this photograph, and such well resolved lines are seldom obtained in spectra from crystals. Since the random motion of the molecules in a liquid will average out any splittings due to quadrupole interaction and to direct magnetic interaction between different groups of equivalent nuclear spins, it follows that the splitting observed in this case must be produced by another effect. It is, in fact, due to the magnetic moments of *other nuclei*, which induce an extra magnetic moment in the electron distribution of the molecule, and this then interacts with the nucleus under observation to produce the splitting. Since such an interaction is a property of the molecule

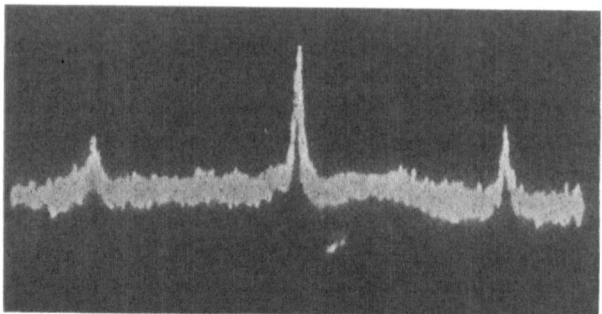

Figure 72. Splitting of phosphorus resonance line in liquid CH_3OPF_2. *Total field modulation is about 2 gauss*

as a unit, and does not depend on the orientation of molecules relative to each other, it will have the same value for each, even when randomly orientated in the liquid state. This type of spectrum, which is a form of 'hyperfine splitting of hyperfine structure', and is often termed 'electron coupled nuclear spin-spin interaction', is a good example of the very precise information that can be obtained by these methods, and leads to detailed calculations of the chemical binding of the molecules concerned (see Sections 10.3 and 10.4).

(*b*) *Results from metals.* The nuclear resonance lines obtained from radiofrequency absorption in metals are similar to those observed in liquids as they have a narrow width and are thus unlike those normally obtained from the solid state. This indicates that the spin-lattice relaxation time is much shorter than usual, and it has been shown by KORRINGA[73], and others, that the relaxation is mainly due to the conduction electrons. Another very noticeable effect produced by the conduction electrons is a shift of the resonant frequency, this was first observed by KNIGHT[74] for the cases of lithium, sodium, aluminium, copper and gallium. The spectra were obtained by using a normal type of resonance spectrometer, with the metal samples in the form of shavings in kerosene, or powders dispersed in paraffin. The shift in resonant frequency between the metal, and that obtained from aqueous solutions of its salts, was first demonstrated by measurements on the two samples separately; and then shown very convincingly by using samples of the metal mixed with its salt, two resonances being observed, with a separation of about 10 kc/s. The shift was found to be greater for the heavier metals, suggesting that it might be linked with the hyperfine structure splitting.

TOWNES[74, 75] showed that this shift was due to the paramagnetism of the conduction electrons when they are in the immediate vicinity of a nucleus, and predicted that the frequency shift would be proportional to the hyperfine splitting, as observed. It follows that such radiofrequency measurements on metals should give detailed information concerning the energy states of the conduction electrons. It is also possible to obtain information about the electron distribution in metals from measurement of the spin-lattice relaxation time, as Korringa[73] has shown that there is a direct relation between the Knight frequency shift and the value of T_1.

Further experiments on radiofrequency absorption in metals, together with simultaneous microwave resonance of the conduction electrons, are discussed in Sections 8.10, 10.8 and 10.9.

(c) *Results from crystals.* Nuclear paramagnetic resonance in crystals has two marked differences from that in liquids and metals. First, the spin-lattice interaction is very much weaker, as there are no conduction electrons, or ions, to assist in the relaxation process, and hence saturation broadening of the resonance line, due to the large value of T_1, can occur. The second striking difference is that the resonance is often split into several components, which vary with the angle that the applied magnetic field makes with the crystalline axes. These splittings are usually due to one of two causes, i.e. (i) the magnetic spin-spin interaction between two or three nuclei close together in a group, and (ii) the interaction between the gradient of the crystalline electric field and any nuclear quadrupole moment.

A summary of the results and theory of relaxation effects in crystals has been given by PURCELL[76], but much more information on the solid state forces and interactions can be obtained from an analysis of the resonance splittings. The first striking example of such splittings was discovered by PAKE[77], who showed that the proton resonance in a single crystal of $CaSO_4 . 2H_2O$ showed a pair of doublets, and that the splitting between these depended on the orientation of the crystal. The splitting of the resonance line in this case can be explained on the assumption that the two protons of the water molecule are acting together as a single group. The interaction 'within this group' is much stronger than the effect of the other groups, and a doublet line is thus produced with a separation corresponding to this energy of interaction, which depends on the angle between the direction of the inter-proton axis and the applied magnetic field. The doublet is itself doubled, because there are two such axes in the crystal. This type of magnetic interaction, between groups of closely-spaced spins, also accounts for the line shapes and splittings that have been observed in more complicated crystals, and an interesting review, comparing predicted and observed line shapes due to this effect, has been made by ANDREW and BERSOHN[78].

As already mentioned, there is another type of interaction which can also produce a splitting of the resonance line, which varies with orientation of the crystal—i.e. that with the electric quadrupole moment. The gradient of the crystalline electric field can interact

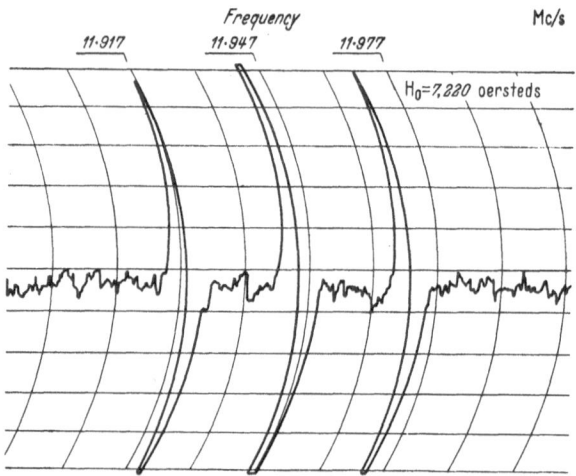

Figure 73. Splitting of Li^7 resonance in crystal of $LiAl(SiO_3)_2$ by quadrupole interaction

with the nuclear electric quadrupole moment to produce a relative shift of the hyperfine levels, as has been seen, and treated quantitatively, in the case of electron resonance in crystals (see Section 6.6). Since transitions between the actual hyperfine structure levels are those observed in radiofrequency spectroscopy, this interaction will produce lines at different resonant frequencies instead of just shifting the resonant frequency of the electronic transitions, which is its main effect in the microwave case. A typical splitting of a nuclear paramagnetic resonance line, due to quadrupole interaction, is shown in *Figure 73*, which was obtained from a single crystal of $LiAl(SiO_3)_2$. In general, a quadrupole multiplet has $2I$ components, and their splitting varies markedly with angle between the applied field and the crystalline axes. It is obvious that a detailed analysis of such splittings not only allows comparison of the quadrupole moments of different isotopes, but, if the value of those moments is known, will give very detailed information on the electric fields existing inside the crystal, and hence also on the chemical binding and structure. In some cases the interaction with the quadrupole moment is so large that the relative shifts of the hyperfine levels so produced have values equal to radiofrequency quanta. It is then possible to observe the transitions between such levels directly, in the absence of any applied magnetic field, and such 'pure quadrupole nuclear resonances' are discussed in Section 8.9.

The above paragraphs give only a very brief summary of the very many measurements that have been made on crystals using nuclear resonance techniques, but it will be evident that such experiments are now becoming a very powerful tool for the investigation of molecular, and crystal, structure.

(d) *Other frequency shifts in nuclear resonance.* All the frequency shifts described so far have been specific to the particular cases considered, e.g. (i) that due to the electron coupled nuclear spin-spin interactions observed in liquids[72, 79], (ii) the Knight shift in metals[74], and (iii) the group magnetic-spin interaction[77, 78] and quadrupole interaction shifts[76] occurring in crystals. There are, however, two other effects which also cause a shift in the resonance frequency, and apply quite generally, irrespective of the physical state of the sample. These are (a) the diamagnetic shielding, and (b) the 'chemical shift effect'.

The effect of diamagnetic shielding is an obvious one, which must occur whenever a nucleus is surrounded by electrons. The shift that it produces in the resonant frequency is small, but with the very high precision now available, it can account for an appreciable correction factor. The first calculations of this effect were by LAMB[80], who represented it as producing an extra field $-\sigma . H_0$ at the nucleus, σ being defined as the magnetic shielding constant. Lamb confined his calculations to the case of a spherically symmetric nuclear electrostatic field, and it was extended to the much more general case of molecules by the work of RAMSEY[81]. He derived a somewhat complicated expression for σ, but also showed that direct substitution could be made for several of its terms from parameters obtained in other experiments.

Apart from this general correction, which should be applied to all nuclear magnetic moment determinations, another effect also occurs, due to the different chemical binding of various nuclei. This was termed the 'chemical shift' and was first noticed by Knight[74], DICKINSON[82], and PROCTOR and YU[83], who found that even in non-metals there was a slight shift in the resonant frequencies obtained from similar nuclei. These shifts can be explained as due to the different electronic environments of the nuclei in different parts of the molecule, which will, therefore, in effect, have different magnetic shielding constants. The sign and magnitude of these shifts can, in fact, be used to give information on the chemical bonding, and to calculate the lowest lying electronic states of the molecule[84].

8.9 PURE QUADRUPOLE NUCLEAR RESONANCE

As already seen in Section 8.8(c) it is possible for the interaction, between the nuclear electric quadrupole moment and the gradient of the crystalline electric fields, to be so large that the splitting produced between the different hyperfine levels will have values equal to radiofrequency quanta, even in zero magnetic field. It is therefore possible to obtain nuclear resonance absorption lines in such cases, without any applied magnetic field, and such absorption lines are referred to as 'pure quadrupole resonance spectra'.

(a) *Initial work.* The first observation of such spectra was by DEHMELT and KRUGER[85], who measured the pure quadrupole frequencies for chlorine and iodine nuclei in crystals containing the quadrupole nuclei in a co-valent p-bond. If the crystalline electric field is rotationally symmetric around the direction of the p-bond (which is taken as the z axis), then the energy levels are given by[86]:

$$E_{M_I} = \frac{eQ}{4} \cdot \left(\frac{\partial^2 V}{\partial z^2}\right) \cdot \frac{3M_I^2 - I(I+1)}{I(2I-1)} \qquad \dots (8.7)$$

where M_I is the component of the nuclear spin along the z axis. It can be seen that, for nuclei with $I = \frac{3}{2}$, as for the two isotopes of chlorine, there will be two energy levels corresponding to $M_I = \pm \frac{3}{2}$ and $\pm \frac{1}{2}$, and therefore one transition frequency for each isotope.

Dehmelt and Kruger observed these transitions in crystals of dichloroethylene, and determined the resonant frequencies as 35·48 Mc/s and 27·96 Mc/s. The ratio of these two frequencies gave the ratio of the chlorine isotope quadrupole moments directly, and with a greater accuracy than any previous method. They also made measurements on iodine in different compounds, and since this has a spin of $I = \frac{5}{2}$, three energy levels result, which give rise to two absorption lines with a frequency ratio of 2 : 1. They also demonstrated that the application of a magnetic field would produce Zeeman splitting of the degenerate levels to give further transitions.

(b) *Later results.* This work was then taken up by others, mainly by Pound and co-workers at Harvard, and Livingston at Oak Ridge, in the first instance. POUND[87] and DEHMELT[88] both obtained pure quadrupole resonances from solid iodine, and were able to show, by this means, that the I_2 molecule is held together by relatively pure covalent p-type bonds. Following these initial measurements, an increasingly large number of results have been reported, for the observation of pure quadrupole resonances of different nuclei, in various crystals. A large number of these experiments have been on molecules containing chlorine, or other halogen nuclei, but resonance has also been detected from those containing nitrogen[89], copper[90], arsenic[91], boron[92], gallium[93] and mercury[94] nuclei, amongst others.

The information derived from such measurements on the pure quadrupole resonance lines can be divided into two groups. First, if more than one isotope of the particular nucleus is present, as is often the case, the ratio of the quadrupole moments for the two isotopes can be measured to a very high order of accuracy. This is because the ratio is determined entirely by the ratio of the resonant frequencies, and hence the accuracy attainable is much greater than by any other method. It is also possible, in a large number of cases, to derive an approximate value for the absolute magnitude of nuclear quadrupole moment. The gradient of the molecular electric field at the nucleus has to be known for this to be done, but in several cases this can be either calculated from the known electronic bonding, or obtained from nuclear paramagnetic resonance measurements on similar crystals.

Then, secondly, as distinct from the nuclear information, a considerable amount of data can be obtained on the solid state crystalline forces and interactions, and the chemical bonding of the molecule. The theoretical calculations for such cases are very similar to those of gaseous microwave spectroscopy, but the relaxation processes are rather different. Determinations of line shape and change with temperature[85, 87] are also of importance in this respect, as they give additional information on the relaxation forces.

(c) Experimental technique. The experimental method and apparatus used for detecting pure quadrupole nuclear resonance is very similar to that used in 'ordinary' nuclear resonance. Since no magnetic fields are applied to the sample, the techniques of nuclear induction cannot be used, and a radiofrequency bridge, or some form of regenerative oscillator or similar circuit, must therefore be employed. The absence of a magnetic field also means that the field modulation must be replaced by frequency modulation, in order that a.c. methods of amplification can be used, and the lines displayed or recorded. It is also necessary to have a wide frequency coverage, since the variation of magnetic field can no longer be used as a substitute for variation of frequency, as it is, to a large extent, in nuclear paramagnetic resonance. Apart from these points, however, the spectrometers used for pure quadrupole resonance are identical with those employed for nuclear paramagnetic resonance, and can usually be adapted for the latter without any alteration, save the addition of the magnetic field.

Thus, in his initial work, LIVINGSTON[95] used a regenerative oscillator incorporating self-detection, which had been developed by HOPKINS[96] for nuclear magnetic resonance. The frequency modulation was effected by means of a vibrating condenser, and a filter peaked at the sweep frequency was also incorporated, in order to remove any amplitude modulation at this frequency that might be produced at the same time. The filtered output was fed, via an audio amplifier, to an oscilloscope, and the resonance lines could thus be displayed in the usual way. The coil containing the sample was constructed on the end of a probe, as before, because, although it was not now necessary to place it in a magnet gap, it was often useful to take measurements over a temperature range, and thus have the coil immersed in a dewar away from the rest of the circuit.

8.10 THE INTERACTION OF ELECTRON AND NUCLEAR RESONANCE —THE OVERHAUSER EFFECT

A consideration of the experiments described in the preceding sections of this chapter, and also of those on electron paramagnetic resonance described in Chapters 6 and 7, will show that, under the right conditions, it should be possible to observe both electron and nuclear resonance from the same substance at the same time. This then raises the interesting question as to whether the two kinds of absorption might affect one another, and what difference this would produce on the observed signal.

OVERHAUSER[97] was the first to consider this question in detail, and he predicted that the nuclear resonance absorption should be enhanced if sufficient input power was used to cause saturation of the electron paramagnetic resonance absorption. This interaction arises essentially from the hyperfine coupling between the magnetic moments of the electron and nuclear spins. Overhauser was able to show that the relaxation processes, which tend to restore equilibrium after absorption of energy by the electron resonance, also induce nuclear transitions via the hyperfine coupling. These induced transitions are predominantly in one direction and, as a result, the normal Boltzmann distribution between the nuclear hyperfine levels is modified, the two energy levels between which nuclear resonance is to be observed now having quite a large difference in population, instead of only a very small one. This increases the transition probability correspondingly, and hence considerably enhanced nuclear resonance lines should be obtained under such conditions. In order to induce appreciable nuclear transitions via this coupling mechanism, the input power to the electron resonance absorption must be sufficient to start saturating the signal. In the limit of complete saturation, Overhauser showed that the steady-state nuclear polarization will be larger, by the same amount that would be expected if the nuclear gyromagnetic ratio was increased to a value equal to that of the electron. An enhancement of the nuclear resonance by a factor of over a thousand was therefore predicted, an effect which could be easily observed experimentally.

The first experimental observation was by CARVER and SLICHTER[98], using particles of suspended lithium. They employed a radiofrequency of 84 Mc/s for the resonance absorption by the conduction electrons in the lithium, and were able to produce input powers of 50 watts and oscillating magnetic field strengths of 4 gauss at this frequency. The strength of the d.c. magnetic field required for the electron resonance was then about 30 gauss, and was provided by an end-corrected solenoid. The frequency of nuclear resonance in this field, for the Li⁷ nuclei, was 50 kc/s; the resonance was detected using a twin-T bridge, and could be displayed on an oscilloscope, or with a phase-sensitive detector. The axis of the 84 Mc/s electron resonance coil, consisting of one turn, was aligned at right angles to both the solenoid axis, and that of the 50 kc/s nuclear resonance coil, of 270 turns; and it was found that, with this arrangement, switching the 84 Mc/s power on or off did not affect the balance of the 50 kc/s bridge. The comparison of oscilloscope traces taken with and without the 84 Mc/s electron resonance input power was very striking. In the absence of electron resonance, the nuclear resonance from the Li⁷ was lost in the noise on the oscilloscope trace, but when saturation of the electron resonance was produced by switching on the 84 Mc/s power, a large nuclear resonance signal was simultaneously obtained. Quantitative measurements showed that it was about 100 times greater than that observed in the absence of electron resonance, and the experiment was thus an excellent confirmation of Overhauser's prediction. The fact that the full enhancement was not realized was probably due to absence of complete saturation.

Overhauser's original theory[97] was limited to the case of metals and their conduction electrons, but Bloch, and others, quickly pointed out that exactly the same kind of interaction could also be expected in other cases of electron resonance, including those from F-centres and free radicals. The second reported case of experimental observation of the Overhauser effect was, in fact, in the hydrazyl free-radical. BELJERS, VAN DER KINT and VAN WIERINGEN[99] were able to show that the proton resonance from the hydrazyl became progressively larger as the input microwave power caused saturation of the paramagnetic resonance signal. They used normal microwave techniques to observe the electron resonance in the hydrazyl, which was placed in the centre of an H_{102} rectangular cavity, and also had a coil wound around it to detect the proton resonance. (Provided some such mode as the H_{102} or H_{011} in a cylindrical cavity is used, where the region of magnetic field stretches down the entire axis with no electric field present, little disturbance of the microwave power is produced by winding an open coil round the sample.) The microwave frequency was about 9,000 Mc/s, and the resonant field had a value of 3,300 gauss, the proton resonance occurring at 14 Mc/s. A 70-watt klystron was necessary to saturate the hydrazyl resonance, and when the power absorbed by the hydrazyl exceeded a value of about 3 watts, enhancement of the proton resonance signal was observed. This case is not quite so direct as that of lithium, as the free electron responsible for the paramagnetic resonance is mainly attached to the nitrogen, not the hydrogen, atoms of the molecule. A much greater change would be expected if the very much weaker nitrogen nuclear resonance were observed.

It seems highly probable that the Overhauser effect will be observed and studied in many other cases in the near future, and may well give useful information on relaxation processes and hyperfine coupling mechanisms, as well as affording a very simple means of nuclear alignment. See also Section 12.5.

8.11 ELECTRON RESONANCE AT RADIOFREQUENCIES

As already seen in Chapters 6 and 7, it is possible, in principle, to obtain electron paramagnetic resonance at any frequency, provided the resonance equation

$$h\nu = g \cdot \beta \cdot H \qquad \ldots (8.8)$$

is satisfied.

Until about 1951, however, most of the experiments on paramagnetic resonance had been performed in the microwave region, and it was not until quite recently that measurements were started at normal radiofrequencies. There are two fundamental disadvantages of radiofrequency electron resonance, as compared with the microwave case. First, the splitting of the electronic energy levels is very much smaller, with the result that the population difference between the two levels at any given temperature also is much smaller, and hence so is the transition probability and the intensity of the observed absorption line. The second disadvantage arises out of the very small magnetic field strengths that are required for resonance at these frequencies. A substitution into equation (8.8) will show that, at a frequency of 30 Mc/s, the resonant magnetic field strength is only about 10 gauss. This means that absorption lines with widths greater than 30 gauss or so are very difficult to observe and study. Moreover, the line width cannot be reduced very much by the usual method of diamagnetic dilution, since the intensity of absorption is small, even in the concentrated salt.

It therefore follows that the only substances which can be studied in detail by electron resonance at radiofrequencies, are those producing narrow and intense absorption lines. Fortunately this limitation is not so severe as might appear at first, as nearly all the organic free radicals as well as the absorption lines obtained from the conduction electrons in metals fall into this class. Some of these experiments have already been described in Chapter 7, and diphenyl picryl hydrazyl[100], the broken carbon bonds in charred organic matter[101], and the conduction electrons in lithium[98] are typical examples of the kind of systems that can be studied at these frequencies.

To date, by far the greatest amount of work has been concentrated on the hydrazyl free radical, as its narrow and intense absorption line is the ideal case for such experiments. It also provides a simple way of measuring magnetic field strengths very accurately in the 0–100 gauss region, which is especially useful as it takes over just where the proton resonance magnetic field determination starts to become difficult. A good example of the measurements on the hydrazyl free-radical is afforded by the recent experiments of GARSTENS, SINGER and RYAN[102], which showed that both the g value and line shape changed markedly with frequency when the resonance was observed in very small magnetic fields. They used a standard radiofrequency oscillator and detector system of the Pound and Knight type, and the absorption line could either be displayed directly on the oscilloscope screen, or mixed in a phase-sensitive detector, and traced out by a pen recorder. The coil containing the sample, which was about 3 gm of the polycrystalline hydrazyl, was placed in the centre of a long uniform field produced by a large solenoid mounted with its axis in the direction of the earth's magnetic field. The magnetic field strengths were determined from a measurement of the current flowing through the solenoid, an absolute calibration being obtained from the resonance measurements at 15 Mc/s, where it was known that the hydrazyl still had a g value of 2·0036.

Their experiments showed that once the resonant frequency falls below 6 Mc/s the g-factor starts to increase rapidly, following an inverse fourth-power law of the frequency, and having a value of about 2·13 at 2·7 Mc/s. A change in line shape, and width, was also observed over the same frequency range, and GARSTENS[103] has shown that the variation of both the g value, and the line shape, can be explained theoretically, using magnetic resonance in gases at low fields as a model for the relaxation process.

These measurements illustrate the general experimental techniques that are used for electron resonance at radiofrequencies. It is evident that the radiofrequency oscillator and detector circuits can be identical to those employed for nuclear paramagnetic resonance, and any of the circuits described in Sections 5 and 6 may be used. The large electromagnet producing the field strengths normally required for nuclear resonance is now dispensed with, and replaced by a pair of Helmholtz coils, or a solenoid, which only has to produce a field of about 20 gauss. In most cases it is not necessary to use a d.c. current or field, but just an a.c. modulation, which will sweep the field through about ±25 gauss, and thus produce two resonance lines on the screen, on either side of the zero[101]. If a free-radical or similar compound is being studied this also serves as a very accurate method of calibrating the magnetic field, and oscilloscope time base, for determination of the resonance line width. Thus, the separation between the centres of the two resonance lines will be equal to $2h\nu/g\beta$ gauss, which, for the hydrazyl g value of 2·0036, gives the expression:

$$\text{Line separation} = 7·132 \cdot 10^{-7} \cdot \nu \text{ gauss} \qquad \ldots (8.9)$$

One striking feature about these experiments is the extreme simplicity of the apparatus, and its very small cost. In these days, when most research in modern physics requires very elaborate and expensive machines, it is most refreshing to come across such a useful experimental apparatus which consists only of a pair of Helmholtz coils, a few radio valves and components, and an oscilloscope—the latter being a large factor in the total cost of the entire spectrometer.

REFERENCES

[1] RABI, I. I., ZACHARIAS, J. R., MILLMAN, S. and KUSCH, P. *Phys. Rev.* 53 (1938) 318

[2] PURCELL, E. M., TORREY, H. C. and POUND, R. V. *ibid.* 69 (1946) 37

[3] GERLACH, W. and STERN, O. *Ann. der. Physik.* 74 (1924) 673

[4] FRISCH, R. and STERN, O. *Z. Physik* 85 (1933) 4

[5] BREIT, G. and RABI, I. I. *Phys. Rev.* 38 (1931) 2082

[6] RABI, I. I., KUSCH, P. and ZACHARIAS, J. R. *ibid.* 46 (1934) 157

[7] RABI, I. I., KUSCH, P. and ZACHARIAS, J. R. *ibid.* 46 (1934) 163

[8] RABI, I. I., MILLMAN, S. and FOX, M. *ibid.* 46 (1934) 320

[9] RABI, I.I., and COHEN, V. W. *ibid.* 46 (1934) 707

[10] COHEN, V. W. *ibid.* 46 (1934) 713

[11] MILLMAN, S. and FOX, M. *ibid.* 49 (1935) 867

[12] KELLOG, J. M. B., RABI, I. I. and ZACHARIAS, J. R. *ibid.* 50 (1936) 472

[13] —— —— *ibid.* 50 (1936) 396

[14] MILLMAN, S., RABI, I. I. and ZACHARIAS, J. R. *ibid.* 53 (1938) 331, 384

[15] RENZETTI, N. *ibid.* 57 (1940) 753

[16] RABI, I. I., MILLMAN, S., KUSCH, P. and ZACHARIAS, J. R.: *bid.* 53 (1938) 495; and 55 (1939) 526

[17] KUSCH, P., MILLMAN, S. and RABI, I. I. *ibid.* 55 (1939) 666

[18] —— —— *ibid.* 55 (1939) 1176

[19] KELLOG, J. M. B., RABI, I. I., RAMSEY, N. F. and ZACHARIAS, J. R. *ibid.* 55 (1939) 318

[20] —— —— —— *ibid.* 56 (1939) 213, 728

[21] KELLOG, J. M. B. and MILLMAN, S. *Rev. mod. Phys.* 18 (1946) 323

[22] KOLSKY, H. G., PHIPPS, T. E., RAMSEY, N. F. and SILSBEE, H. B. *Phys. Rev.* 79 (1950) 883; 80 (1950) 483; and 81 (1951) 1061

[23] RAMSEY, N. F. *ibid.* 85 (1952) 60

[24] — *ibid.* 78 (1950) 695

[25] NAFE, J. E. and NELSON, E. B. *ibid.* 73 (1948) 718

[26] PRODELL, A. G. and KUSCH, P. *ibid.* 79 (1950) 1009; and 88 (1952) 184

[27] SMALLER, B., YASAITIS, E. and ANDERSON, H. L. *ibid.* 80 (1950) 137; and 83 (1951) 813

[28] MILLMAN, S. and KUSCH, P. *ibid.* 58 (1940) 438

[29] BITTER, F. *ibid.* 75 (1949) 1326

[30] OCHS, S. A., LOGAN, R. A. and KUSCH, P. *ibid.* 78 (1950) 184

[31] BOHR, A. *ibid.* 73 (1948) 1109

[32] BOHR, A. and WEISSKOPF, V. F. *ibid.* 77 (1950) 94

[33] BERINGER, R. and RAWSON, E. B. *ibid.* 87 (1952) 228

[34] RAWSON, E. B. and BERINGER, R. *ibid.* 88 (1952) 677

[35] BERINGER, R. and HEALD, M. A. *ibid.* 95 (1954) 1474

[36] LEW, H. *ibid.* 76 (1949) 1086

[37] DAVIS, L., FELD, B. T., ZABEL, C. S. and ZACHARIAS, J. R. *ibid.* 76 (1949) 1076

[38] MANN, A. K. and KUSCH, P. *ibid.* 77 (1950) 427

[39] MILLMAN, S. and KUSCH, P. *ibid.* 60 (1941) 91

[40] BREIT, G. *ibid.* 72 (1947) 984

[41] SCHWINGER, J. *ibid.* 73 (1948) 416; and 74 (1948) 1439

[42] KARPLUS, R. and KROLL, N. M. *ibid.* 77 (1950) 536

[43] KUSCH, P. and FOLEY, H. M. *ibid.* 74 (1948) 250

[44] GARDNER, J. H. and PURCELL, E. M. *ibid.* 76 (1949) 1262; and 83 (1951) 996

[45] KOENIG, S. H., PRODELL, A. G. and KUSCH, P. *ibid.* 88 (1952) 191

[46] ESTERMANN, I. *Rev. mod. Phys.* 18 (1946) 300

[47] WALLER, I. Z. *Phys.* 79 (1932) 370

[48] ROLLIN, B. V. *Nature* 158 (1946) 669

[49] BLOCH, F., HANSEN, W. W. and PACKARD, M. *Phys. Rev.* 69 (1946) 127

[50] —— —— *ibid.* 70 (1946) 474

[51] PURCELL, E. M., POUND, R. V. and BLOEMBERGEN, N. *ibid.* 70 (1946) 986

[52] PURCELL, E. M., BLOEMBERGEN, N. and POUND, R. V. *ibid.* 70 (1946) 988

[53] PURCELL, E. M. *Physica* 17 (1951) 282

[54] POULIS, N. J. *ibid.* 17 (1951) 392

[55] BLOCH, F., GRAVES, A. C., PACKARD, M. and SPENCE, R. W. *Phys. Rev.* 71 (1947); 373; and 71 (1947) 551

[56] BLOCH, F., LEVINTHAL, E. C. and PACKARD, M. *ibid.* 72 (1947) 1125

[57] BLOEMBERGEN, N., PURCELL, E. M. and POUND, R. V. *ibid.* 73 (1948) 679

[58] WARING, C. E., SPENCER, R. H. and CUSTER, R. L. *Rev. sci. Instrum.* 23 (1952) 497; RICHARDS, R. E. and SMITH, J. A. S. *Trans. Faraday Soc.* 47 (1951) 1264

[59] ROBERTS, A. *Rev. sci. Instrum.* 18 (1947) 845 and *Phys. Rev.* 72 (1947) 182; HOPKINS, N. J. *Rev. sci. Instrum.* 20 (1949) 401

[60] GUTOWSKY, H. S., MEYER, L. H. and McCLURE, R. E. *Rev. sci. Instrum.* 24 (1953) 644

[61] ZIMMERMAN, J. R. and WILLIAMS, D. *Phys. Rev.* 76 (1949) 350; GINDSBERG, J. and BEERS, Y. *Rev. sci. Instrum.* 24 (1953) 632

[62] BLOCH, F. *Phys. Rev.* 70 (1946) 460

[63] PAKE, G. *J. chem. Phys.* 16 (1948) 327 and *Amer. J. Phys.* 18 (1950) 473

[64] PROCTOR, W. G. *Phys. Rev.* 75 (1949) 522; 76 (1949) 684; PROCTOR, W. G. and YU, F. C. *ibid.* 77 (1950) 716, 717

[65] CHIAROTTI, G. and GIULOTTO, L. *ibid.* 93 (1954) 1241

[66] TORREY, H. C. *ibid.* 75 (1949) 1326

[67] — *ibid.* 76 (1949) 1059

[68] HAHN, E. L. *ibid.* 76 (1949) 461

[69] — *ibid.* 77 (1950) 297

[70] — *ibid.* 80 (1950) 580

[71] ZIMMERMAN, J. R., WILLIAMS, D. and CHAMBERS, W. H. *ibid.* 76 (1949) 350, 638

[72] GUTOWSKY, H. S., McCALL, D. W. and SLICHTER, C. P. *J. chem. Phys.* 21 (1953) 279

[73] KORRINGA, J. *Physica* 16 (1950) 601

REFERENCES

[74] KNIGHT, W. D. *Phys. Rev.* 76 (1949) 1260

[75] TOWNES, C. H., HERRING, C. and KNIGHT, W. D. *ibid.* 77 (1950) 852

[76] PURCELL, E. M. *Physica* 17 (1951) 282

[77] PAKE, G. E. *J. chem. Phys.* 16 (1948) 327

[78] ANDREW, E. R. and BERSOHN, R. *ibid.* 18 (1950) 159

[79] HAHN, E. L. and MAXWELL, D. E. *Phys. Rev.* 84 (1951) 1246; and 88 (1952) 1070

[80] LAMB, W. *ibid.* 60 (1941) 817

[81] RAMSEY, N. F. *ibid.* 77 (1950) 567; 78 (1950) 699; and 86 (1952) 243

[82] DICKINSON, W. *ibid.* 77 (1950) 736

[83] PROCTOR, W. G. and YU, F. C. *ibid.* 77 (1950) 717

[84] —— *ibid.* 81 (1951) 20

[85] DEHMELT, H. E. and KRUGER, H. *Naturwissenschaften* 37 (1950) 111, 398; and *Z. Phys.* 129 (1951) 401

[86] KOPFERMANN, H. *Physica* 17 (1951) 386

[87] POUND, R. V. *Phys. Rev.* 82 (1951) 343

[88] DEHMELT, H. G. *Naturwissenschaften* 37 (1950) 398

[89] WATKINS, G. D. and POUND, R. V. *Phys. Rev.* 85 (1952) 1062

[90] KRUGER, H. and MEYER-BERKOUT, U. *Z. Phys.* 132 (1952) 171

[91] —— *ibid.* 132 (1952) 221

[92] DEHMELT, H. G. *ibid.* 134 (1953) 642; BASSOMPIERRE, A. *C.R. Acad. Sci., Paris* 237 (1953) 1224

[93] DEHMELT, H. G. *Phys. Rev.* 92 (1953) 1240

[94] DEHMELT, H. G., ROBINSON, H. G. and GORDY, W. *ibid.* 93 (1954) 480, 920

[95] LIVINGSTON, R. *Ann. N.Y. Acad. Sci.* 55 (1952) 800

[96] HOPKINS, N. J. *Rev. sci. Instrum.* 20 (1949) 401

[97] OVERHAUSER, A. W. *Phys. Rev.* 92 (1953) 411

[98] CARVER, T. R. and SLICHTER, C. P. *ibid.* 92 (1953) 212

[99] BELJERS, H. G., VAN DER KINT, L. and VAN WIERINGER, J. S. *ibid.* 95 (1954) 1683

[100] SINGER, L. S. and SPENCER, E. G. *J. chem. Phys.* 21 (1953) 939

[101] INGRAM, D. J. E. and TAPLEY, J. G. *Phil. Mag.* 45 (1954) 1221

[102] GARSTENS, M. A., SINGER, L. S. and RYAN, A. H. *Phys. Rev.* 96 (1954) 53

[103] GARSTENS, M. A. 93 (1954) 1228

9

APPLICATIONS OF RADIOFREQUENCY
AND MICROWAVE SPECTROSCOPY

9.1 INTRODUCTORY REVIEW

IN this chapter a summary is given of the different ways in which spectroscopy at radio and microwave frequencies can be applied to obtain information of various kinds, and correlated with other branches of research in physics and chemistry. Some of the applications have already become clear in the preceding chapters, especially those associated with the evaluation of fundamental physical and chemical data, such as the determination of nuclear spins or moments, and chemical bond lengths or angles. Other applications have been mentioned in passing, such as the possibility of extremely precise frequency standards in the microwave region, and the very accurate determination of magnetic field strengths; whereas other problems, that can also be attacked by these techniques, have not been considered at all as yet—e.g. the detection of very small amounts of paramagnetic impurities, the study of photo-chemical reactions involving free-radicals, and several others.

In order to make the chapter complete in itself, all the applications of gaseous spectroscopy, and electron, or nuclear, paramagnetic resonance are listed and considered briefly, irrespective of any previous mention, although more space is devoted to the practical applications that have not been described in the previous chapters. It seems fairly logical to separate the information that can be obtained from spectroscopy at these frequencies into the two branches of 'fundamental' and 'applied', although the distinction becomes a little arbitrary in some instances. In quite a large number of cases the same kind of data can be obtained either by gaseous microwave spectroscopy, or by electron or nuclear resonance, and hence in the first two summary tables the methods of obtaining similar information by the three techniques are listed side by side. Thus, in *Table 9.1*, which is concerned with the 'fundamental data' that can be derived from measurements in this frequency region, the actual parameters determined are listed in the left-hand column, and the way in which the information is derived from the spectra is noted in the other three columns, for each of the different techniques.

A similar presentation is used in *Table 9.2*, summarizing the information that can be obtained which is of an applied, or practical, nature. The different items listed in the two tables are considered in detail in Sections 9.2 to 9.11, and the reference numbers in the tables relate to the sections in which the particular application is discussed. Other applications, which are of a more specialized nature, and only relate to one of the techniques, are listed in *Table 9.3*, and are discussed in more detail in Section 12.

Part I : Applications to Fundamental Research

9.2 NUCLEAR SPINS

The ways in which nuclear spins and moments can be derived from the hyperfine structure of both gaseous and solid state spectra have been considered in detail in Chapters 5 and 6, and are only briefly summarized here.

(a) *Determinations by paramagnetic resonance.* Paramagnetic resonance absorption is probably the most direct and unambiguous method for the determination of nuclear spins. The interaction of the nuclear magnetic moment with the magnetic field produced by the electron configuration causes a splitting of the energy levels into $(2I+1)$ components. The selection rule, which forbids a change in M_I, then gives rise to $(2I+1)$ hyperfine lines in the observed spectra, as illustrated in *Figure 59*. Hence, providing the spectrum can be observed, the nuclear spin of the paramagnetic atom can be evaluated directly by just counting the number of hyperfine components and equating to $(2I+1)$. The method is limited to nuclei which possess an unpaired electron when bound in the solid state, and are thus able to exhibit electron resonance. It was at first thought that this technique could therefore only be applied to paramagnetic atoms of the various transition groups, but the recent measurements on impurities in semi-conductors[1] suggest that an unpaired electron can be made to associate with diamagnetic atoms also, and hence the spins of a large number of other nuclei may soon be determined by this method.

The most important practical requirement for the determination of nuclear spins by paramagnetic resonance, is that the line width should be small enough for the hyperfine components to be resolved. This nearly always means that some form of magnetic dilution must be used to reduce the spin-spin interaction, and usually a diamagnetic salt with isomorphous crystal lattice must be employed, as, unless single crystals are used, the angular variations of g value, or electronic splittings, will smear out the pattern. A notable exception to this is manganese, where the six hyperfine components can be

Table 9.1. The Determination of Fundamental Data

	By gaseous spectroscopy	By electron resonance	By nuclear resonance
(i) Nuclear spins	Indirectly from h.f.s. components (2(b))	Directly from number of h.f.s. components (2(a))	Indirectly from intensities (2(c))
(ii) Nuclear magnetic moments	Very indirectly via quadrupole coupling (3(c))	Directly from the h.f.s. splitting (3(b))	Directly from the resonance condition (3(a))
(iii) Nuclear quadrupole moments	Directly from the h.f.s. splitting (4)	Indirectly from h.f.s. shifts and second-order transitions (4)	Directly from quadrupole resonance experiments (4)
(iv) Data for nuclear alignment	—	Gives zero-field energy levels and predicts best salts to use for nuclear demagnetization (5(a)–(c))	Combined with electron resonance gives direct alignment, via Overhauser or Sweep effects (5(d))
(v) Chemical parameters	Resonant frequencies give I directly —hence bond lengths, angles and isotopic masses (6(a))	Resonant frequencies give g directly —hence electronic configuration and orbitals (6(b))	Electric field gradients from quadrupole resonance, and electronic states from shifts in frequency (6(c))

Note. Numbers in parentheses relate to sections of this chapter.

obtained readily from small quantities of randomly dispersed Mn^{2+} salts in amorphous powders, or in solution. In this case the g value is very nearly isotropic[2] and the positions of the component lines of the central electronic transition are independent of angle.

Nuclear spins, which have been determined by paramagnetic resonance, include those of Cr^{53}, V^{50}, Nd^{143}, Nd^{145}, Sm^{147}, Sm^{149}, Er^{167}, Ru^{99}, Ru^{101} and Pu^{241}. A much more detailed discussion of the hyperfine structure observed in normal paramagnetic resonance spectra will be found in Sections 6.5 to 6.8, and the study of the hyperfine components, due to diamagnetic atoms in semi-conductors, is summarized in Section 7.10.

(b) *Determination by gaseous spectroscopy.* The interaction producing the hyperfine splitting is via the nuclear electric quadrupole moment in gaseous spectroscopy, and thus is not so direct as in the case of paramagnetic resonance, the number of component lines no longer being given simply by $(2I + 1)$. The equations governing the number and spacing of the hyperfine structure lines in gaseous spectroscopy have been considered in Section 5.11, and it was seen that both are

Table 9.2. Practical Applications

		Gaseous spectroscopy	Electron resonance	Nuclear resonance
(i)	Measurement of frequency	Very accurate standards directly from absorption lines $(7(a))$	—	Standards from hyperfine splittings of atomic beams $(7(b))$
(ii)	Measurement of magnetic field	—	In the 2–100 gauss region $(8(b))$	In the 100–25,000 gauss, and under 2 gauss regions $(8(a), (c))$
(iii)	Chemical analysis	Unambiguous qualitative analysis. Quantitative down to 10^{-9} gm. $(9(a))$	Detection of any paramagnetic. Quantitative down to 10^{-12} gm. $(9(b))$	Quantitative down to 10^3 gm.
(iv)	Detection of free radicals	Detection of OH radical (10)	Direct detection of many free radicals. Concentration can be estimated down to 10^{13}/gm. (10, 12)	—
(v)	Study of irradiation damage	—	By resonance from trapped electrons, or interstitial atoms (11)	From change in line widths and shapes (11)

Note. Numbers in parentheses relate to sections of this chapter.

dependent on the J quantum number of the particular transition, as well as on the value of I. The analysis of the spectra and derivation of I may therefore be quite complicated but, because of the very great precision available, there is usually only one set of values which can be made to fit the observed spectrum.

The method can be applied in principle to any nucleus of a gas molecule which has rotational transitions in the microwave region. It cannot be used for nuclei with spins of $\frac{1}{2}$ or 0, however, as such possess no quadrupole moment, and hence no coupling with the molecular fields and motion. The necessary resolution can be obtained by the simple process of reducing the gas pressure, as collision broadening is usually the main effect above 10^{-3} mm Hg. Nuclei, the spins of which were first determined by gaseous spectroscopy, include the isotopes of boron, sulphur, chlorine, germanium and iodine.

Another nuclear parameter of fundamental importance that can be determined very accurately from gaseous microwave spectroscopy, is the ratio of isotopic masses. The frequency of the observed rotational line depends directly on the moment of inertia of the molecule, and if one of the nuclei is replaced by its isotope, a slight shift in the resonant frequency will occur. Since such frequency shifts can be determined very accurately, this affords a very precise method of measuring the ratio of isotopic masses, and a large number have now been obtained in this way. The details of this kind of determination may be found in a review article by GESCHWIND, GUNTHER-MOHR and TOWNES[49].

(c) *Determination by nuclear resonance.* Nuclear resonance differs from both gaseous spectroscopy and electron resonance in that the observed transitions are between the hyperfine levels themselves. The nuclear g factor (which is equal to the magnetic moment in nuclear magnetons, divided by I) is therefore given directly by the resonance condition for the absorption lines, as in equation (8.1), but the value of the nuclear spin itself cannot be obtained so readily. If there is no quadrupole interaction, or other second-order effect, the frequency corresponding to the transitions between successive nuclear orientations will be the same, and hence only one absorption line will be obtained whatever the value of I. The absolute intensity of the absorption line will depend on how many levels are contributing to it, however, and can thus be used to deduce the value of I indirectly (see equations (24) and (25) of Appendix I). This was first employed for the case of tritium [3], where a comparison of the relative signal intensities from a sample containing a known ratio of H^1 and H^3 showed that I must be the same for both isotopes.

If a quadrupole interaction is also present this will cause an energy change in the different nuclear orientations, so that the spacing between successive levels is no longer equal, and more than one absorption line is obtained. This type of splitting has been illustrated in *Figure 73*, for the case of the Li^7 nucleus in a crystal of $LiAl(SiO_3)_2$ and, in general, $2I$ component lines are produced. Similarly, in pure quadrupole nuclear resonance, the value of I can be determined from the number of resonant frequencies obtained, as predicted by equation (8.7).

A large number of different nuclei have now been studied by nuclear resonance, mainly using paramagnetic absorption techniques, and several new spins have been determined by this method. Detailed descriptions of both the magnetic and quadrupole resonance experiments are to be found in Chapter 8.

9.3 NUCLEAR MAGNETIC MOMENTS

As would be expected, the most direct measurements of nuclear magnetic moments are from the experiments on magnetic resonance, rather than from those on gaseous spectroscopy where the inter-action is via the microwave electric field.

(a) *From nuclear resonance.* The experiments on nuclear para-magnetic resonance give the most accurate and straightforward determination, as the nuclear gyromagnetic ratio can be derived directly from the measured resonant frequency and magnetic field strength, and the magnetic moment can then be calculated if the spin is known,

i.e. $$g_I = \frac{h \cdot \nu}{\beta_N \cdot H} \text{ and } \mu_N = g_I \cdot I \qquad \ldots (9.1)$$

where μ_N is the magnetic moment of the nucleus in nuclear magne-tons, β_N is the nuclear magneton, and g_I is the g-factor for the particular nucleus. It can be seen that no factor determined by the electronic configuration enters into this equation, as it is the external field H which causes the splitting of the levels. A very large number of nuclear magnetic moments have now been determined by this means, and it is very hard to envisage any method more simple or direct.

(b) *From paramagnetic resonance.* The experiments on electron para-magnetic resonance can also be used to determine nuclear magnetic moments in a fairly direct manner, from the splitting between suc-cessive hyperfine component-lines. This splitting is produced by an interaction with the magnetic field associated with the electron con-figuration, however, and hence the detailed electron orbits round the nucleus have to be known, if an absolute value of the nuclear magnetic moment is to be derived. Thus the value of the observed hyperfine splitting is given by:

$$A = \left(\frac{g\beta}{hc}\right) \cdot \Delta H = \frac{2\mu_N \cdot \beta \cdot \beta_N}{I}\left(\overline{\frac{1}{r^3}}\right) \cdot F(\mathbf{l}, \mathbf{s}, \mathbf{j}) \qquad \ldots (9.2)$$

The actual determination of magnetic moments by this means is best illustrated by the cases of erbium and neodymium[4, 5]. In order to evaluate the nuclear magnetic moment $(=\mu_N)$, the value of $(\overline{1/r^3})$ and the form of the function $F(\mathbf{l}, \mathbf{s}, \mathbf{j})$ of equation (9.2) must be found. The evaluation of $(\overline{1/r^3})$ is more direct for the case of the rare earths than for the iron group of elements, because the resonance spectrum of gadolinium[8] has shown that the ions in an S state have no hyperfine structure, and hence an admixture of 's' electrons, of the type suggested by Abragam for the iron group atoms, need not be considered. It can therefore be assumed that the magnetic field at the nucleus is produced by a $4f$ electron with a hydrogenic wave function, and that the nuclear charge is $(\mathcal{Z} - \sigma)$, where σ is the screening constant. The value of $(\overline{1/r^3})$ and σ can then be obtained from the two equations:

$$\left(\overline{\frac{1}{r^3}}\right) = \frac{(\mathcal{Z}-\sigma)^3}{n^3 l\,(l+\tfrac{1}{2})\,(l+1)\,.\,a_0^3} \qquad \ldots (9.3)$$

$$\lambda = 2\,(\mathcal{Z}-\sigma)\,.\,\beta^2\,.\,\left(\overline{\frac{1}{r^3}}\right) \qquad \ldots (9.4)$$

where λ is the spin-orbit coupling coefficient.

There is also an alternative way of deriving $(\overline{1/r^3})$, using a method introduced by STERNHEIMER [9], which is based on the Fermi–Thomas model.

The only other factor now required, in order to calculate the magnetic moment, is the value of the expression $F(l, s, j)$. This depends on which electronic level is lying lowest, and has to be derived for each particular ion. Thus, for the two cases of erbium and neodymium, the following parameters are obtained.

(i) *Erbium.* The function $F(l, s, j)$ is calculated by ELLIOTT and STEVENS [5] as $(176/225)\,(7\cos^2\theta - 5\sin^2\theta)$. Here θ is a parameter which is determined by the admixture of different levels in the ground state, and its value is obtained from the best fit with the experimentally observed g values. The value of $(\overline{1/r^3})$ obtained from equations (9.3) and (9.4) is 80 ± 20 Å^{-3}, and substitution of the observed [7] A, of 52.10^{-4} cm^{-1}, into equation (9.2) then gives $\mu_N = 0.50 \pm 0.12$ nuclear magnetons.

(ii) *Neodymium.* The function $F(l, s, j)$ is of the same form as that for erbium, but rather more complicated (see Elliott and Stevens [5]). Its value, combined with the experimental [6] value of

$$A = 380.3 \times 10^{-4} \text{ cm}^{-1}, \qquad \ldots (9.5)$$

gives $(\mu_N/I)\,.\,(\overline{1/r^3}) = 1.156\,.\,10^{25}$ cm^{-3}, and $(\overline{1/r^3})$ can be estimated as 40 ± 10 Å^{-3}, to give $\mu_N = 1.0 \pm 0.25$ nuclear magnetons for Nd^{143}. A similar calculation gives $\mu_N = 0.62 \pm 0.25$ n.m. for Nd^{145}.

Since evaluation of $(\overline{1/r^3})$ is never very accurate, this method cannot be compared in its precision to that of the nuclear resonance determinations. In fact, the initial measurements on hyperfine splittings in paramagnetic resonance were used to calculate $(\overline{1/r^3})$ from the nuclear magnetic moments, rather than vice versa, and thus obtain information on the electron orbitals. It was such work which first showed that the ground state of the iron group transition elements had a considerable admixture of a level containing $(4s)$ electrons. Because of this, the determination of nuclear magnetic moments by paramagnetic resonance has been limited to the rare-earth group of elements, in work so far reported.

(c) *Determinations from gaseous spectroscopy.* It is possible to evaluate nuclear magnetic moments from gaseous spectra, if the Zeeman splitting of the hyperfine structure can be observed. This interaction is only a second-order effect, however, as it acts via the main quadrupole interaction, and it also depends on the electronic wave-functions associated with the molecular binding. It is, therefore, normal to compare the spectra of different isotopes in the same compound, rather than attempt a calculation of the absolute energy splittings, when gaseous spectroscopy is used to determine unknown magnetic moments.

9.4 NUCLEAR QUADRUPOLE MOMENTS

In the same way that accurate values of magnetic moments can only be obtained from magnetic resonance experiments, so accurate values of nuclear quadrupole couplings can only be obtained from experiments in which the interaction is via the electric field—i.e. gaseous spectroscopy, or pure quadrupole nuclear resonance. Estimates of quadrupole moments can be made from the second-order shifts and 'forbidden' transitions of paramagnetic resonance spectra[10], in the same way that magnetic moments can be determined from gaseous spectra; but by far the largest amount of information on quadrupole couplings and moments has been derived directly from work on gaseous spectroscopy, and more recently from pure quadrupole nuclear resonance.

As discussed in detail in Section 5.11, the spacing of the hyperfine components of the resonance lines in gaseous spectroscopy depends directly on the nuclear electric quadrupole moment, the energy levels producing the hyperfine pattern being given by:

$$E_Q = e \cdot Q \cdot \left(\frac{\partial^2 V}{\partial z^2}\right)_{Av:} \cdot \left[\frac{\frac{3}{8}G(G+1) - \frac{1}{2}I(I+1)\mathcal{J}(\mathcal{J}+1)}{I(2I-1)(2\mathcal{J}-1)(2\mathcal{J}+3)}\right] \quad \dots (9.6)$$

where
$$G = F(F+1) - I(I+1) - \mathcal{J}(\mathcal{J}+1)$$
$$F = \mathcal{J}+I, (\mathcal{J}+I-1), \dots \quad |\mathcal{J}-I|$$

and $(\partial^2 V/\partial z^2)_{Av:}$ is the gradient of the electric field, which interacts with the quadrupole moment. This varies for different types of molecule, but can generally be expressed in a simple form and is given by

$$\left(\frac{\partial^2 V}{\partial z^2}\right) \cdot \left(\frac{3K^2}{\mathcal{J}(\mathcal{J}+1)} - 1\right)$$

for a symmetric-top molecule, where Oz is the symmetry axis.

It follows that the field gradients inside the molecule must be evaluated before an absolute determination of the nuclear quadrupole moment can be made, in a similar way that $(\overline{1/r^3})$ had to be evaluated for a determination of the magnetic moment from electron resonance measurements. This limitation still applies in the case of pure quadrupole nuclear resonance, the energy levels then being given by:

$$E_Q = \frac{e \cdot Q}{4} \cdot \left(\frac{\partial^2 V}{\partial z^2}\right) \cdot \frac{3M_I^2 - I(I+1)}{I(2I-1)} \qquad \ldots (9.7)$$

The actual parameter, calculated from the observed frequencies, is therefore always of the form '$Q . (\partial^2 V/\partial z^2)$', and no direct evaluation of Q itself is possible. This is unlike nuclear paramagnetic resonance, where the magnetic moment can be obtained directly from the resonance condition for the magnetic absorption. Because of this limitation, results on the quadrupole moments obtained from both gaseous spectroscopy and nuclear quadrupole resonance are usually tabulated as 'quadrupole couplings', defined as $eQ . (\partial^2 V/\partial z^2)$, and having the dimensions of a frequency. Such quadrupole couplings have been determined for a very large number of nuclei in gaseous molecules, and in crystals, and can be quoted to a high degree of accuracy, but, because of the uncertainty in $(\partial^2 V/\partial z^2)$, very few values of the actual moment Q can be given to more than two figures.

9.5 NUCLEAR ALIGNMENT

Nuclear alignment is a branch of research that has come very much into the foreground during the last few years. It is usually considered as part of low-temperature or nuclear physics, but paramagnetic resonance experiments have provided a vital role in bridging the gap between these two subjects, and have supplied essential information on the kinds of compounds that are best to use for the experiments. It has also been shown quite recently, that resonance techniques alone can be employed to produce a high degree of nuclear alignment without the use of very low temperatures.

(a) *Low temperature methods.* The basic principle behind any method of nuclear alignment employing very low temperatures is to split the energy levels, corresponding to the different orientations of the nuclei, by an amount greater than kT_F, where T_F is the final low temperature reached. Nearly all the nuclei will then be in the lowest level, and have their spins aligned in the one direction. There are, therefore, two essential requirements, (i) a magnetic or electric field to act as an axis of quantization, and (ii) a strong enough interaction to split the nuclear energy levels by an amount greater than kT_F.

The most direct and generally-applicable method is the 'brute-force' technique[11, 12], in which a very large magnetic field is applied across the specimen. This interacts directly with the nuclear magnetic moments, providing an axis of quantization and splitting the levels by the required amount. In order to reach the low temperatures, however, some form of adiabatic demagnetization[17] must be employed; and the great experimental difficulty of this method is the high initial field required, as, after demagnetization to the low temperature T_F, the field strength left must still be large enough to produce energy level splittings greater than kT_F.

There are, however, three other methods which can be employed to align nuclei at low temperatures, and these all make use of internal interactions already present in the solid state to produce the required splitting of the nuclear energy levels. The first was suggested independently by GORTER and ROSE[13] and uses the magnetic field produced by the unpaired electrons in a paramagnetic salt to interact with the nuclei and split the energy levels. The axis of quantization is still supplied by an external field, but this need only be strong enough to split the *electronic* states by an amount large compared to kT_F. Under such conditions all the electron spins will be in the lowest level, and hence all produce a magnetic field in the same direction at the nuclei. This field will have a value of about 10^5 gauss, however, instead of the 100 gauss or so applied externally to align the electron spins. The experimental requirements are therefore not nearly so severe as for the 'brute-force' method, but the polarizing field needed for the alignment of electron spins does limit the final temperature to some extent.

Another method was suggested by BLEANEY[15, 16], which relies on the same basic principle as the Gorter–Rose method but does not require the application of any external field at the low temperatures. Bleaney suggested that the crystalline electric field of a paramagnetic crystal could replace the applied magnetic field as an axis of quantization. Detailed consideration of this interaction is given in the next section, but it can be seen that the required splitting of the nuclear levels is again produced by the large internal magnetic field of the unpaired electrons. The external field is now only used for cooling-by-demagnetization, and since no field is required at the end, lower temperatures, and hence greater percentage alignment, can be obtained than by the Gorter–Rose method, for the same samples.

Another method was also suggested by POUND[14], in which the interaction between the gradient of the crystalline electric field and the nuclear quadrupole moment is used to produce the splitting of the nuclear energy levels, and the axis of quantization is provided by the direction of the electric field gradient. Both this and Bleaney's

271

method produce systems of levels which are degenerate in $\pm M_I$, so that when all the nuclei are in the ground state there are equal numbers pointing in opposite directions. This is usually immaterial as the angular variation of γ-rays emitted from radioactive nuclei involves even powers of $\cos \theta$. The advantage of Bleaney's method over that of Pound is that the hyperfine splitting produced by the magnetic interaction is usually much greater than that produced by the electric quadrupole interaction, and hence appreciable nuclear alignment will occur at higher temperatures.

It is to be noted that all three methods require the use of magnetically diluted single crystals, the dilution being necessary in order to reduce the dipole-dipole interaction, which would otherwise 'smear out' the separate hyperfine levels. It is also evident that unless single crystals are employed in the Pound or Bleaney method, the variation in the orientation of the crystalline electric fields would average out any preferential direction acting as the axis for alignment. At first sight a single crystal does not appear to be necessary for the Gorter–Rose method, but this is not in fact true, because for the small residual fields employed, the crystalline field will compete with the magnetic field to act as an axis of quantization, and unless single crystals are used, alignment will take place about axes orientated at random. It follows from this, that the specimen possessing optimum properties for the Bleaney method will also possess optimum properties for the Gorter–Rose method, and since the former is much easier to carry out experimentally, it was the first used to produce actual alignment[18].

The choice of paramagnetic salt depends mainly on the zero field hyperfine splittings, and it is for this reason that paramagnetic resonance measurements play such an important part in experiments on nuclear alignment. This fact is illustrated in the next paragraph where the details of the Bleaney method are considered more quantitatively.

(b) *Paramagnetic resonance and nuclear alignment.* Since the method due to Bleaney has been that most widely used in aligning nuclei at low temperatures, a slightly more detailed consideration of its principles is worth while, especially as it illustrates the use of paramagnetic resonance determinations very well. The method works best for salts in which the hyperfine structure is very anisotropic, i.e. $A \gg B$ in the usual Spin Hamiltonian. If B can be neglected, compared with A, then the pattern of the zero-field energy levels is that of $(2I + 1)$ doublets, equally spaced by $A/2$ in energy. Each of these states is an equal admixture of the states $\pm M_I$, and they are spaced in the order $M_I = \pm I; \pm (I - 1); \ldots; \mp I$. Hence if $T_F{}^\circ$ is the final temperature reached, after the adiabatic demagnetization, the fraction of nuclei in the state $\pm M_I$ will be $\cosh\left(\dfrac{M_I \cdot A}{2k \cdot T_F}\right)$.

As numerical examples illustrating this, Bleaney[15, 16] quotes the cases of copper and cobalt. In a typical copper tutton salt the ratio of number of nuclei in the $\pm \frac{3}{2}$ levels to those in the $\pm \frac{1}{2}$ levels would be 1·5 to 1 at $kT = 3A/4 \approx 0.015°$ K. In cobalt ammonium sulphate the population of the nuclear levels $\pm \frac{7}{2} : \pm \frac{5}{2} : \pm \frac{3}{2}$ and $\pm \frac{1}{2}$ would be 1 : 0·53 : 0·32 : 0·21 at a temperature of $kT = 0.7A \approx 0.025°$ K (which is the lowest that can be reached by adiabatic demagnetization of the cobalt salt by itself). It can be seen that in both cases there is considerable excess of nuclei in the states of high M_I at temperatures which are relatively easy to reach by normal adiabatic demagnetization methods.

An analysis of the opposite case, for which $B \gg A$, shows that the preponderance of nuclei is now in the states with low M_I values, but the alignment is not so large, because the energy levels tend to concentrate near the bottom. The case of $A = B$ results in two groups of levels, each group containing an equal admixture of all states of M_I, and hence no nuclear alignment will be obtained. The above calculations only apply directly to salts which have an effective spin of $S = \frac{1}{2}$ in their Hamiltonian, such as copper and cobalt. It can also be shown that salts, which have an isotropic hyperfine structure $(A = B)$, may also have a set of doublets as zero-field energy levels, which can produce nuclear alignment provided that they also have a large electronic splitting, i.e. $D \gg A, B$.

The detailed calculation of the zero-field energy levels from the measured values of the Spin Hamiltonian coefficients A, B, D and Q, has already been considered in Section 6.18, and it is quite evident that the determination of these zero-field levels, from paramagnetic resonance measurements, is an essential preliminary before low temperature nuclear alignment is attempted with any particular salt.

(c) *Results of low temperature experiments.* The first definite alignment of nuclei was accomplished by the group working at Oxford[18], using Bleaney's method with a radioactive cobalt salt. In order to obtain greater cooling than could be produced by adiabatic demagnetization of the cobalt salt alone, some of the isomorphous copper salt was added to the specimen. The actual composition of the mixed tutton salt used was (1 per cent Co, 12 per cent Cu, 87 per cent Zn) $SO_4 \cdot Rb_2 SO_4 \cdot 6H_2O$, the zinc acting as the diamagnetic diluent. With this they were able to reach a final temperature of 0·01° K, after adiabatic demagnetization from an initial condition of H/T equal to 30,000 gauss per degree. The total weight of the crystals was 4 gm, and the cobalt contained a small amount of Co^{60} corresponding to 70 microcuries. The γ-rays emitted by the Ni^{60}, which is formed by β decay of the Co^{60}, were detected by four Geiger counters, placed round the specimen at 10·5 cm distance. Initial measurements showed that an anisotropy, in the angular variation of the radiation, of up to 44 per cent was produced, and that this anisotropy disappeared rapidly as the specimen warmed up. From a more detailed quantitative analysis of their results[19], an estimate of the nuclear magnetic moment of Co^{60} was obtained. This can be calculated from the dependence of the anisotropy of the emitted γ-rays on $e^{\mu H/kTI}$; where H is the magnetic field produced by the electrons at the nucleus, and can be estimated from the paramagnetic resonance hyperfine splitting observed for Co^{59}. By assuming a value of $I = 5$, they obtained $\mu = 3\cdot0 \pm 0\cdot3$ nuclear magnetons.

Further experiments were then made by the Oxford group, using the Gorter–Rose method to produce nuclear polarization and alignment. The first of these[20] was again with Co^{60}, but the salt $3 . (Co, Mg) (NO_3)_2 . 2 Ce (NO_3)_3 . 24 HO_2$ was used, as cerium has a very anisotropic g value[22, 23], and the field necessary for quantization can be applied along the low g value direction, after demagnetization to zero by a field applied along the high g value axis. In this way temperatures of 0·004° K could be reached from initial conditions of 25,000 gauss and 1° K. A similar experiment[24] was also performed with Mn^{54} replacing the Co^{60} in the double nitrate, and nuclear information concerning the spin and decay of the Mn^{54} nucleus was deduced from the results. The theoretical work on emission of radiation by aligned nuclei has been mainly carried out by SPIERS[26] and STEENBERG[27, 28].

These experiments, performed at Oxford, were all with radioactive nuclei, and the production of nuclear alignment was detected by the anisotropy of the γ-rays emitted by the specimen, which reached a value of 90 per cent in some cases[24]. Another technique for studying nuclear alignment has been developed in the States by BERNSTEIN et al.[25] using non-radioactive paramagnetic nuclei, and detecting their alignment by passing a beam of polarized neutrons through the specimen. The first experiments[25] were performed on Mn^{55}, using a Gorter–Rose method of alignment. The beam of polarized neutrons was passed through an analyser after transit through the manganese double sulphate[2], and they were able to observe a dependence of the capture cross section of the aligned Mn^{55} nuclei, on the orientation of the incident neutrons. A similar experiment[21] has also been performed on aligned samarium nuclei, and these results are of interest in showing that non-radioactive nuclei can also be studied by this means.

These experiments on nuclear alignment at very low temperatures have been described in some detail as they are a very good example of the combination of nuclear and low-temperature physics, and of the essential role that paramagnetic resonance determinations play in this new line of research. The measurements of COOKE, DUFFUS and WOLF[23] on the cerium double nitrates are very interesting in this respect, as it is the very large anisotropy of the g values that makes this salt an ideal one for such work. In fact, a suggestion by DABBS and ROBERTS[29] has recently been made, that such a salt could be used to produce nuclear alignment of protons by the 'brute force' method; this may be accomplished by just rotating a large field from the direction of large g-axis to that of the small g-axis. In this way considerable cooling by adiabatic demagnetization would be produced, without reducing the strength of the large applied field, and they calculate that a proton polarization factor of 0·006 could be obtained in $Mg_3(NO_3)_6 . Ce_2(NO_3)_6 . 24H_2O$ by using an initial H/T of $2 . 10^5$ gauss per degree.

(d) *Nuclear alignment by direct resonance techniques.* Recently some quite different methods of producing nuclear alignment have been discovered, which require much simpler techniques without the use of very low temperatures and yet attain a high degree of alignment. The first of these[30] has already been discussed in some detail in Chapter 8, and uses the OVERHAUSER[31] effect to produce the change in population of the nuclear energy levels. Another due to HONIG[32], attempts to employ a sweeping process through the hyperfine components of an electron resonance line, which was thought might alter the relative populations of the different hyperfine energy levels.

The essential mechanism in both of these methods is an interaction between the relaxation processes, associated with the electron resonance, and the nuclear spin transitions. This interaction acts via the coupling between the magnetic field produced by the unpaired electrons and the magnetic moments of the nuclei. It has the effect that, when electron resonance absorption is taking place, the relaxation processes which tend to restore equilibrium in the electron spin levels will also act on the nuclear spins and modify their normal Boltzmann distribution. This interaction was first considered by Overhauser[31] and, as discussed in detail in Section 8.10, he showed that the difference in population between the nuclear energy levels could be increased by a factor of over a thousand if sufficient power was used to cause saturation of the electron resonance. The first experimental demonstration, that this could in fact be used to produce nuclear polarization, was shown strikingly by CARVER and SLICHTER[30], who obtained a hundredfold increase in the nuclear resonance from Li^7,

275

when the conduction electrons of the lithium were taking part in electron resonance absorption.

In the second method, first attempted by Honig[32], the change in the population of the nuclear energy levels would be seen from the change in relative intensity of the hyperfine component lines. The first experiments were performed with a sample of silicon containing about 10^{17} As atoms per cc. The work of Fletcher et al.[1] had already shown that in such a case, four component lines are obtained, due to the interaction between the magnetic moment of the As nucleus and the unpaired electron trapped at the arsenic atom. Using a frequency of 9,000 Mc/s, and at 4° K, Honig obtained these lines with a width of 3 gauss and a separation of 73 gauss.

The interaction between the electron resonance and the nuclear energy levels was demonstrated in two ways. First, it was shown that if just one of the hyperfine components was observed, and the magnetic field was swept twice through this one resonance, with a time t elapsing between the two traversals, then the amplitude of the line on the second traversal was reduced, relative to the first, by an amount given by:

$$A_2/A_1 = (1 - \exp[-t/T]) \qquad \qquad \ldots\ldots(9.8)$$

where T is the nuclear spin-lattice relaxation time, and was equal to 16 seconds in this case. The second experimental observation demonstrated the fact that if two neighbouring hyperfine components were swept through sufficiently rapidly, an enhancement of the second, relative to the first, was produced. These results can be explained if it is assumed that on sweeping through a hyperfine component of the electron resonance, the arsenic nuclei corresponding to that particular orientation undergo $\Delta M_I = \pm 1$ transitions, thus depopulating the M_I level associated with the particular line. This explanation requires a mechanism which will cause the number of nuclear spin transitions to be greatly increased when electron resonance takes place. Initial theoretical treatment[33] suggested that the electron spin relaxation interaction would act as such a mechanism, but more recent work by Abragam has cast doubt on this. It would seem that a more thorough analysis of the whole method is necessary before its potentialities are certain.

Recent experiments[99] have in fact shown that these results can be explained more satisfactorily if it is assumed that they are due to a very long *electronic* spin relaxation time, without any nuclear alignment or polarization. An electronic spin relaxation time of 16 seconds is most unusual, and the initial interpretation of the results in terms of a nuclear polarization is thus not surprising. It therefore now appears that this method cannot be used for nuclear alignment, but there are some other very interesting fundamental experiments that can be performed by making use of the very long relaxation time. It is in fact possible to use such samples to produce a solid state analogue of the ammonia microwave amplifier and oscillator described on pages 133–134. This was first suggested by Combrisson, Honig and Townes[100] and has recently been realised in practice by several workers.

There seems little doubt that other similar experiments involving the interactions of spin resonance systems will be devised in the near future, and the production of aligned nuclei, or the generation and amplification of microwaves are only the initial experiments in this field.

9.6 The Determination of Chemical Parameters

In the preceding sections, the use of radiofrequency and microwave spectroscopy for the determination of nuclear parameters has been discussed, and this forms one main type of fundamental data that can be obtained from such measurements. The other main type is that of an atomic or chemical nature, related to the molecular or ionic binding, rather than the properties of the nuclei. The derivation of this kind of information from the microwave or radiofrequency spectra has already been considered in some detail in the preceding chapters, and only the main points are summarized here. In order to clarify the various applications in this field the results are summarized under the headings of the three different techniques.

(a) *Determinations from gaseous spectroscopy*. There are two main groups of chemical parameters that can be obtained from gaseous microwave spectroscopy—i.e. (i) bond lengths and angles, and (ii) molecular dipole moments. Bond lengths and angles are evaluated from the measured frequencies of the rotational absorption lines via the determination of the moments of inertia of the molecule. The way in which the moments of inertia can be calculated from the resonant frequencies, and the equations relating the two, has been considered in Chapter 5. The bond lengths and angles are derived from the moments of inertia, by taking a series of measurements with different isotopes substituted into the molecule. In this way the mass at certain points in the molecule is changed by a known amount, and hence the actual internuclear distances and bond angles can be calculated from the set of simultaneous equations, relating the measured moments of inertia to the known isotopic masses. The great advantage of this technique in determining such parameters is the very high accuracy available, and as described in Chapter 5, bond lengths can be measured to ± 0.001 Å, and bond angles to $\pm 1'$.

The other important chemical parameter that can be determined from gaseous microwave spectra is the molecular dipole moment. This is obtained from the Stark splitting of the rotational absorption lines, the separation in frequency between the different Stark components being related directly to the strength of the applied electric field, and the molecular dipole moment. The theory and detailed expressions for this interaction have been given in Section 5.9, and from the equations listed there or in Appendix I it is evident that the dipole moment can be evaluated directly from such splittings.

The determination of very accurate values for bond lengths and angles, and molecular dipole moments, is of great importance to theories of chemical binding, and at the moment it is probably fair to say that the precision of the available experimental data is well ahead of the detailed chemical theory required to explain it.

(b) *Determinations from paramagnetic resonance.* Information on chemical binding and solid state interactions can be obtained from both the g values and hyperfine splittings, which are observed in paramagnetic resonance spectra. The dependence of these on the precise electron orbitals and wave-functions has been considered at some length in Chapter 6, and two examples will probably suffice to show how such information can be obtained from the experimental observations. The first example is that of copper in an 'ionic' crystal, and the second that of iridium in a 'covalent' complex.

A brief summary of the theory of the copper ion in a tetragonal crystalline field has been given in Section 6.4, and the relations between the g values and orbital level splittings, as derived by Polder, have been quoted. The theory of the copper ion has been considered in much more detail by ABRAGAM and PRYCE[34], who derived more accurate expressions for the g values, and showed that the energy level system originally proposed not only fitted the observed g values, but also agreed tolerably well with the optical absorption data. It thus appeared that a good theoretical explanation of the observed g values could be made on the basis of Polder's model, which assumes that the copper ion can be treated as in a ^2D state, with just one electron missing from the 3d shell.

However, when this theoretical treatment was applied to calculate the hyperfine structure splitting, a complete disagreement with experiment[35] was found. Thus the values of the spin coefficients A and B were predicted as:

$$A = [g_\parallel - 2 + \tfrac{3}{7}(g_\perp - 2) - \tfrac{4}{7}] \cdot P = -0.13 \cdot P$$
$$B = [g_\perp - 2 - \tfrac{3}{14}(g_\parallel - 2) + \tfrac{2}{7}] \cdot P = +0.36 \cdot P \quad \dots (9.9)$$

where $P = 2\gamma \cdot B \cdot B_N \cdot (1/r^3)$, and is a measure of the magnetic interaction between the electrons and the nucleus. The value of P was estimated from the optical hyperfine structure parameters, and the Hartree wave functions of the Cu^+ ion. Substitution into equation (9.9) then gave:

$$A = -0.0046 \text{ cm}^{-1}, \quad B = 0.0131 \text{ cm}^{-1}$$

These values are the right order of magnitude, but are in complete disagreement with experiment, in that the observed values give $A = 3B$ instead of vice versa! This discrepancy, together with a similar failure to explain the large hyperfine splitting in manganese salts, lead ABRAGAM [36] to suggest that the lowest level of the ions was not just $(3d)^n$, but had an admixture of a state, in which one $3s$ electron has been lifted to a $4s$ orbit. Only a small admixture of the unpaired s electrons is required, in order to make a large change in the hyperfine structure splitting, as they pass very close to the nucleus. The modified values of the Spin Hamiltonian coefficients then become

$$A = (-K - 0.13)P$$
$$B = (-K + 0.365)P \qquad \dots (9.10)$$

where K is a constant proportional to the fraction of admixed $3s(3d)^9 4s$. level. By taking a value of $K = 0.25$, very good agreement between the theoretically derived and experimentally observed values of A and B could then be obtained. These results are a typical example of how measurements on paramagnetic resonance spectra can give detailed quantitative information on the admixture of different orbitals into the ground state. See also Chapter 11.

The paramagnetic resonance spectra of iridium [37] in ammonium chloroiridate is an interesting comparison with that of copper, since the binding is covalent instead of mainly ionic, and a very different type of hyperfine structure is obtained. The iridium nuclei Ir^{191} and Ir^{193} have a spin of $I = \frac{3}{2}$, like the copper nuclei, and hence four hyperfine component lines might be expected. Instead, a large number of closely spaced components are observed, the whole pattern appearing somewhat like a Christmas tree. OWEN and STEVENS [37] were able to explain this spectrum by assuming an admixture of hyperfine splitting, due to the chlorine nuclei, Cl^{35} and Cl^{37}, which also have $I = \frac{3}{2}$. The mechanism responsible for this can be pictured as one whereby, for part of the time, an electron is transferred from a Cl^- to Ir^{4+}, cancelling the unpaired spin on the iridium, and leaving an unpaired spin on the chlorine atom. Each iridium atom lies at the centre of a regular octahedron of chlorine ions, and on a purely ionic model, these would be Ir^{4+} $(5d)^5$ and Cl^-, giving one hole in the $5d$ shell. The covalent bonding, however, causes the magnetic electrons (or magnetic hole) to be distributed between the $5d_{xy,yz,zx}$ orbitals of

279

the iridium, and the $3p_\pi$ orbitals of the chlorines. For some of the time there is thus a magnetic interaction with the nuclear moment of the chlorine and hence extra components are obtained in the hyperfine pattern. This admixture also has the effect of reducing the g value from 2·0, and a value of 1·8 was obtained for the ammonium chloroiridate. By a detailed analysis of the hyperfine structure Owen and Stevens[37] were able to show that the electron hole is on any one chlorine for about 3 per cent of its time, the hyperfine structure interaction for all the chlorines being about one-third of that for the iridium. A paper, surveying the results obtained with similar salts, has recently been published by GRIFFITHS and OWEN[38], containing some very good photographs of this type of spectra, as well as a detailed analysis of the bonding from them.

A careful correlation between the optical absorption coefficients and the paramagnetic resonance results of the iron group hydrated salts has recently been made by OWEN[50], and suggests that, even in these cases, there is a small amount of covalent bonding too. He showed that normal crystalline field treatment (as for copper, above) does not give quite the right values for the orbital splittings, and that the systematic discrepancies can be accounted for by introducing weak covalent bonds into the $[M(H_2O)_6]$ complex. This produces a charge transfer between the paramagnetic ion and the attached water molecules, and the absence of any small additional hyperfine splitting is due to the zero nuclear magnetic moment of O^{16}.

One of the advantages of using paramagnetic resonance as a method of determining detailed bonding parameters, is its ability to ignore all the other diamagnetic atoms in the molecule. In this it is very different from such techniques as infra-red spectroscopy, where absorption lines and bands are obtained from the different states of the whole molecule, and it is often very difficult to obtain detailed information on the chemical binding of any one atom. In contrast to this, the paramagnetic resonance spectrum is determined entirely by the energy states of the one paramagnetic atom, and the binding to its nearest neighbours. It is therefore possible to obtain detailed information about the paramagnetic atom, with no obscuring spectra due to the rest of the molecule. This advantage shows up strikingly in the case of large organic molecules containing paramagnetic atoms, and the work of BENNETT and INGRAM[39, 40] on phthalocyanine, chlorophyll and haemoglobin derivatives is a good example of such a case. All three of these organic compounds have a metal atom at the centre of a square of four nitrogens, this atom is iron in the case of haemoglobin, magnesium in the case of chlorophyll, and a whole series of different paramagnetic atoms can be substituted into the centre of the phthalocyanine structure. Any detailed information that can be obtained on the binding of these central atoms to the four nitrogens will obviously be of great biophysical importance, and initial measurements on haemin, and the acid- and fluoro-haemoglobin derivatives, have already shown that the binding to the central iron atom is not as simple as had been previously supposed. It seems that paramagnetic resonance will be a powerful tool in analysing such structures, and detailed information on the binding in copper phthalocyanine is already available from an analysis of the hyperfine splitting variation in different directions relative to the molecular planes[40]. Further details are given in Sections 11.6 and 11.8.

(c) *Determinations from nuclear resonance.* The most direct form of chemical information that can be obtained by nuclear resonance techniques is that from the pure quadrupole experiments. In these, the frequencies of the absorption lines are determined directly by the gradient of the crystalline electric field at the nucleus in question, according to the equation

$$E_Q = e \cdot Q \left(\frac{\partial^2 V}{\partial z^2} \right) \cdot \frac{[3M_I^2 - I(I+1)]}{4I(2I-1)} \qquad \ldots (9.11)$$

where $\Delta M_I = \pm 1$ gives the allowed transitions.

It is evident that if the quadrupole moment of the nucleus (Q) is known, the value of the field gradient $(\partial^2 V / \partial z^2)$ can be calculated immediately, and used to give information on ionic distances and chemical binding. Such measurements were used by GOLDSTEIN and BRAGG[41] and DEAN and POUND[42] to evaluate the percentage double-bond character of the C—Cl bond in various compounds. Even if the absolute value of the nuclear quadrupole moment is not known, comparison between the different frequencies obtained for quadrupole resonance of the same nuclei in different compounds can provide considerable information on the chemical bindings.

In contrast to the pure quadrupole spectra, the resonance condition for nuclear paramagnetic absorption lines is independent of any chemical binding to a first order, and is determined solely by the nuclear magnetic moment and spin. As noted in Section 9.8, however, there are some effects which do shift the resonant frequency by very small amounts, and because of the extremely high precision with which the frequencies can be measured, these shifts can be used to give information on the electronic states of the atoms concerned. There are two such effects which are quite general and apply to nuclear resonance in any compound, namely the spin-spin coupling[43], and the chemical shift[44]. These are both concerned with the screening effect of the electronic motion round the nucleus, and the magnitude of the latter can be used to supply data on the actual orbitals involved in chemical binding (see Sections 10.3, 10.4 and 10.5).

There are also other effects specific to certain cases, which produce small shifts or splittings of the resonant frequency, and can be used to give information on the magnetic or electric fields associated with the binding forces. These include the 'group magnetic-dipole'[45] and the 'quadrupole' interactions[46], which occur in crystals; the 'Knight shift'[47], due to the conduction electrons in metals; and the 'electron-coupled nuclear spin-spin interaction'[48], found in certain liquids. They have been considered in some detail in Sections 8.8(a)–(d), together with the information that can be derived from them.

Part II : Applications of a Practical Nature

The applications of spectroscopy at radio and microwave frequencies, so far considered, have been those which provide information of a fundamental nature concerning nuclear and atomic, or chemical, properties. This branch of spectroscopy is now being used in an increasingly large measure, however, to make determinations of practical value, and the rest of the chapter is concerned with such applications. These are listed in *Tables 9.2* and *9.3*, and are now considered in more detail, starting with those summarized in *Table 9.2*, which are applications common to two or more of the different methods; while *Table 9.3* lists those of a more specialized type.

9.7 Frequency Standards and the Measurement of Time

Since the resonant frequencies of gaseous molecular spectra are determined entirely by the interatomic binding forces and not by any external conditions, they offer the possibility of a very accurate frequency standard that can be reproduced at any place desired, on or off the earth's surface. This does not apply to microwave or radio-frequency spectral lines that are obtained by magnetic resonance methods, as their resonant conditions require the determination of a magnetic field strength as well, and this in turn is dependent on a measurement of frequency. On the other hand, nuclear quadrupole resonance and the hyperfine splitting of atomic beams are two types of radiofrequency spectra which are independent of magnetic field, and hence could also be used as frequency standards.

There are thus three types of spectra that offer possibilities as frequency standards in the microwave or radiofrequency region, and these are now compared in relation to both their inherent Q factors and also the signal-to-noise ratio of their observed absorption lines. Both of these factors will affect the ultimate resolving power and accuracy of the frequency standard. Thus the resolving power of a given standard can be defined as $f_0/\delta f$ where f_0 is the resonant frequency, and δf is the minimum detectable frequency shift of the oscillator locked to the absorption line. The dependence of δf on the signal-to-noise ratio and sensitivity of the recording circuits becomes evident if a small change in the frequency of the oscillator, feeding the absorption cell, is assumed to take place. The output power, detected at the far end, will then change, because the frequency corresponds to a different part of the absorption line, but this change in intensity will only be observable if it is greater than the noise intensity. Hence the noise present at the input to the recording circuits sets an inherent limitation on the resolving power of the standard. It is also evident that greater resolving power can be obtained if the reference frequency is chosen at the point of maximum slope of the absorption curve rather than the point of maximum intensity, as there is a much greater variation of signal magnitude with frequency in the former case. This method requires more complicated apparatus, but the resolving power can be increased by factors of up to a thousand if it is employed, and any variation in line width can be counterbalanced by using both points of maximum slope and taking the mean frequency.

A comparison of the Q's of the three different types of absorption line shows that the values which can be obtained for atomic beams are considerably larger than those for gaseous absorption spectra, or pure quadrupole resonances. This is because the collision and Doppler broadenings are reduced to a negligible value by the use of a beam of atoms all having the same velocity, and the constant supply of fresh atoms also prevents any saturation broadening. The Q factor and width of these absorption lines are therefore determined entirely by the time that the atoms spend in the oscillating field, according to the relation $\Delta E \cdot \Delta t \sim h$, or more exactly

$$\Delta f = \frac{0 \cdot 42}{L}\left[\frac{2kT}{m}\right]^{\frac{1}{2}} \qquad \dots(9.12)$$

if Ramsey's method of excitation[51] is employed. By using a path length, L, of 50 cm, between the two oscillating coils, a Q factor of over thirty million can be obtained for the field-insensitive σ-line of the caesium hyperfine structure. In comparison with this, the Q factors for gaseous absorption lines are only about 10^5, Doppler broadening being the main limitation, although recent work, using molecular beams in gaseous spectroscopy, may increase this factor appreciably. Nuclear quadrupole resonance lines also have a comparatively low Q of about ten thousand, this being limited by the nuclear spin-lattice relaxation time.

The reference frequency can be set to a much greater accuracy than one line-width, however, and hence the Q values must be multiplied by a factor depending on the sensitivity and signal-to-noise ratio of the detecting apparatus, in order to obtain an estimate of the resolving power. LYONS[52] has considered these factors in some detail, and calculated the maximum possible resolving power of a microwave gaseous standard, as 5.10^{12}, if superheterodyne detection is used with the maximum slope of the line as reference point. Experience with experimental standards so far constructed indicates that this theoretical limit may be very hard to achieve, however, and a similar situation is true of the nuclear quadrupole resonances, where the low signal-to-noise ratio at present available, forms the main limitation on the resolving power. On the other hand, initial measurements on atomic beams of caesium[52] suggest that a resolving power of 1 in 10^{10} should be quite easily attainable.

Most of the work on the use of microwave absorption lines as frequency standards has been carried out by a group under LYONS[53, 54] at the National Bureau of Standards in Washington, and is summarized briefly below. See also Section 12.7.

(a) *Gaseous microwave standards*. If an absorption line is to be used as a frequency standard, and also as a fundamental comparison for a clock or other timing device, then two items of apparatus are essential —i.e. (i) a separate oscillator which is used to drive the clock, and (ii) a discriminator system which will compare the frequency of the absorption line with that of the oscillator (or one of its harmonics) and compensate for any change. It can be shown that the inherent resolving power of the system will depend to some extent on the stability of the oscillator, and hence it is usually best to employ a quartz crystal oscillator, and multiply up from this to the frequency of the absorption line, rather than use a klystron centred on the absorption line frequency. The final stages in the multiplier chain from the quartz crystal usually take the form of velocity-modulated frequency multipliers, as illustrated in *Figure 20*, but sufficient power to detect and use the ammonia lines can be obtained from the eighth harmonic of a 1N26 crystal, acting as a microwave source in the waveguide.

Details of the different discriminator circuits that can be used to lock the oscillator to the absorption line may be found in the original papers by LYONS[52, 53]. One of the most effective methods is to use a slow frequency-sweep of the multiplier chain, about the centre frequency of the absorption line, and of total amplitude equal to the line width; the line itself is then used as a demodulator, converting the frequency modulation to amplitude modulation. For correct frequency setting the output signal will be a series of equal peaks, and any shift in the oscillator frequency shows up as a change of their relative amplitudes. This system uses the points of maximum sensitivity on the absorption line, and also balances out any change in pressure broadening.

All the initial work on such atomic clocks has been with ammonia, as this has the most intense absorption lines in the microwave region. It may be possible to obtain higher final resolving powers by the use of the oxygen lines, however, as these have higher inherent Q factors, smaller collision and saturation broadening coefficients, and can still be observed at 50° K where the Doppler width is very much smaller.

(b) *Atomic beam frequency standards.* If the hyperfine separation in atomic beam experiments is to be used as a frequency standard, transitions must be chosen which are as field-independent as possible, and also with as high a frequency as can be obtained, so that any change in frequency by disturbing magnetic fields is only a small fraction of the total. The atoms of caesium and thallium therefore offer the greatest possibilities in this work, as their hyperfine splittings are in the region of 9,000 Mc/s and 21,000 Mc/s respectively, and the σ-line of caesium requires a magnetic field change of 10 per cent to change its frequency by 1 part in 10^{10}. Since it is relatively simple to keep the transition region of the beam in a field constant to this amount, the frequency stability of such lines should be extremely high. It has already been seen that the inherent Q factors are also much larger than those of the gaseous absorption lines, and it appears that clocks controlled by these transitions may well prove more accurate than any others.

The construction of a clock based on such absorption lines is identical, in principle, to that of the ammonia case discussed above. The atomic beam apparatus itself is very similar to those already decribed in Chapter 8, the Ramsey[51] method of separate regions of excitation being employed. The oscillating fields are provided by two X-band cavity resonators, which are placed at either end of the central homogeneous field, and are fed from a multiplier chain actuated by a quartz oscillator as before. The beam of caesium atoms passes down the centre of the system, through the first inhomogeneous field, then through the holes in the microwave cavities, and via the second inhomogeneous field to the detector. On resonance, the output of the detector will be a minimum, and any shift of frequency in the multiplier chain will cause an increase in detected current which can be used to derive an error signal to reset the quartz oscillator.

The problems and characteristics of the control circuits are thus very similar to those of the ammonia case, and detailed accounts and analysis of both forms of atomic clock are to be found in the review paper by Lyons[52].

If the microwave absorption lines are only required as frequency standards, to be used in such applications as locking the frequency of a transmitter, very much simpler circuits than those mentioned above can be used. The only essential requirement then is a discriminator, which will correct the frequency of the transmitter oscillator if it should drift from that of the absorption line. The nuclear quadrupole resonance lines are potentially very useful as frequency standards for this purpose in the normal radiofrequency region, as none of the cumbersome apparatus that goes with atomic beams is required in their case. Frequencies that can be accurately and quickly changed and controlled are also provided by the magnetic resonance absorptions; and if electron resonance of free radicals is used, the magnetic field required can be produced by a small solenoid for frequencies in the normal radio wavebands.

It is apparent from this brief summary, that because of their high Q values, the absorption lines of radio and microwave spectroscopy may prove useful in many applications that require highly stable frequency standards, very highly selective filters, or magnetically modulated frequencies.

9.8 The Measurement of Magnetic Fields

Since the frequency of absorption of electron and nuclear paramagnetic resonance lines depends linearly on the strength of the applied magnetic field, they offer a very direct method of magnetic field calibration, converting the determination of field strength into one of frequency. If a resonance absorption line is to be used for magnetic field measurements it should have as high an intensity as possible so that only small quantities need be employed, and hence allow small regions of the field to be measured separately; it should also have as small a line width as possible, so that it can be set on with the highest accuracy. The two compounds fulfilling these conditions best are (i) for nuclear resonance, the protons in water, with the spin-lattice relaxation time reduced by the addition of a paramagnetic salt, and (ii) for electron resonance, crystals of the free radical α,α-diphenyl β-trinitrophenyl hydrazyl, which has a very intense signal with strong exchange narrowing. These two compounds are, therefore, the ones most commonly employed for magnetic field measurement or stabilization.

Whether nuclear or electron resonance is used depends on both the strength of the magnetic field to be measured, and the frequencies available for the determination. Accurately calibrated wavemeters are usually readily available in the region of 1 to 100 Mc/s, and frequencies within this range can be measured to 1 part in 10^6 quite easily.

The resonance equations for the two cases are given by:

(i) *Nuclear Resonance*

$h\nu = g_I \cdot \beta_N \cdot H$

$h\nu = 5\cdot58552 \cdot 0\cdot50504 \cdot 10^{-23} \cdot H$

(for protons)

$\therefore H = 2\cdot3487 \cdot 10^{-4} \cdot \nu$

(ii) *Electron Resonance*

$h\nu = g \cdot \beta \cdot H$

$h\nu = 2\cdot0036 \cdot 0\cdot92712 \cdot 10^{-20} \cdot H$

(for hydrazyl)

$\therefore H = 3\cdot566 \cdot 10^{-7} \cdot \nu$

(H in gauss; ν in cycles per second)

It can be seen that the values of magnetic field strength corresponding to frequencies between 1 and 100 Mc/s are from 200 to 20,000 gauss for nuclear resonance, and from 0·5 to 30 gauss for electron resonance. The region in between 30 and 200 gauss can be covered by either method, though the signal strength is falling off rapidly for the nuclear resonance, and the oscillator construction and frequency determination is somewhat complicated in the 300 Mc/s region, which is required for the electron resonance.

The intensity of the absorption line is an important factor, as it determines the minimum size of sample that can be used, and very small specimens are normally required if accurate mapping of field homogeneities is to be attempted. Other things being equal, the signal intensity is determined by the difference in population between the two energy levels and is proportional to the square of the frequency of the absorption. Thus, to a first approximation, the same signal intensity will be obtained from $\frac{1}{2}$ cc of hydrazyl at 10 gauss field strength (30 Mc/s frequency), as from $\frac{1}{2}$ cc of water at 6,000 gauss field strength (30 Mc/s frequency); and whereas about 30 gm of water are needed to produce a detectable signal at 80 gauss, only about 1 mg of hydrazyl is necessary. For this reason calibration by proton resonance is usually confined to field strengths above 100 gauss, where good signals can be obtained from small samples, and the standard techniques of nuclear paramagnetic resonance are used to produce and detect the absorption line.

(a) *Nuclear resonance (100 to 25,000 gauss)*. The experimental techniques required for the measurement of magnetic fields by proton resonance have already been described in some detail in Section 4.13, and the circuit diagram of a proton resonance meter is shown in *Figure 49*. This has been used by the author and his colleagues, and gives good signals from coils of 2 mm diameter, or less, in magnetic fields from 2,000–16,000 gauss. If much weaker fields are to be measured by this means, the more sensitive circuits described in Chapter 8 should be used. In most cases the proton resonance line is displayed on an oscilloscope screen, and the magnetic field corresponding to any absorption line or other phenomena is measured by bringing the proton resonance pip into coincidence with it, and then determining the frequency of the proton meter oscillator by an accurately calibrated wavemeter. The main limitation on the accuracy of such determinations is usually the precision with which the two lines can be made to coincide.

(b) *Electron resonance (2–100 gauss)*. The circuits employed for the determination of magnetic field strengths by electron resonance depend largely on the frequencies required. Exactly the same circuit as shown in *Figure 49* can be used to obtain electron resonance in the 30 Mc/s region, and determine magnetic field strengths over the 2–20 gauss range[55], the only alteration being the replacement of the proton sample by a tube containing about $\frac{1}{2}$ cc of hydrazyl. For higher field strengths, the oscillator circuits must be modified and the techniques applicable to the 200–300 Mc/s band substituted. It is interesting to note that if two such oscillators are built, one covering the 0·5 to 100 Mc/s band, and the other covering the 100–350 Mc/s band, then complete coverage of magnetic field is possible, from 2 to 80,000 gauss, the change from electron to nuclear resonance being made at a field of 120 gauss (350 Mc/s electron resonance to 0·5 Mc/s nuclear resonance).

Magnetic field determination by electron resonance is limited on the low field side, not by the intensity of the observed signals, but by the width of the absorption line. The hydrazyl resonance is 2 gauss wide at these frequencies, and it will be very hard to set accurately on an absorption line when its resonant field value is equal to its line width. The work of GARSTENS, SINGER and RYAN[56] has also shown that the g value changes rapidly in fields below 3 gauss, and hence hydrazyl cannot be used as an accurate means of calibration below this value.

There are other free-radicals that have narrower lines than hydrazyl[57], however, and the electron resonance from metal-ammonia solutions[58], where the line width can be as low as 0·02 gauss, also offers possibilities in increased accuracy of magnetic field determination and stabilization by electron resonance. For field strengths below 2 gauss it is best to revert to nuclear resonance, however, and use the precession method of PACKARD and VARIAN[59] described below.

288

(c) *Nuclear precession (0–2 gauss)*. A new and powerful method of measuring magnetic field strengths has recently been introduced by Packard and Varian [59], which utilizes the decaying precession of the magnetized nuclei about the axis of the field to be measured. This was first demonstrated for nuclei precessing in the earth's magnetic field. The principle of the method is to apply a strong magnetizing field, H_0', of about 100 gauss, at right angles to the earth's field (or field to be measured), and thus align all the nuclei in this direction. This field is then reduced to a small value of a few gauss, H_0'', in a time short compared with the spin-lattice relaxation time, so that during this process the magnetization still retains its initial value of $\chi_0 \cdot H_0'$, and is still orientated very nearly in the perpendicular plane. The field, H_0'', is then reduced quickly to zero in a time short compared with $1/(\gamma H_0'')$, so that the magnetization has no time to reorient. There is, therefore, a magnetization of $\chi_0 \cdot H_0'$ left, precessing in a plane perpendicular to the earth's field, at a frequency given by the normal Larmor or nuclear resonance equation, which is about 2 kc/s for the earth's field of 0·5 gauss. This precession dies away in the corresponding relaxation time, but during this period it can be picked up by a coil placed with its axis at right angles to both the earth's field and that of the main magnetizing coil, the decaying sine wave being fed to an audio amplifier and displayed on an oscilloscope.

In their initial experiments, Packard and Varian [59] found that a signal-to-noise ratio of better than 20 could be obtained from a 500 cc water sample, and that the value of the earth's magnetic field could be determined to 1 part in 15,000. The signal was found to last for more than a second, and BLOEMBERGEN and POUND [60] have shown, in a recent analysis, that this is one of the few cases where the line width or relaxation time is determined by 'pure radiation damping', i.e. an example of a radiofrequency absorption line with a 'natural width'. Further details are given in Section 10.10.

(d) *Stabilization of magnetic fields by resonance methods*. The stabilization of magnetic field strengths by resonance methods follows straightforwardly from the techniques used to determine their value. The only extra requirement is a discriminator circuit, which derives an error voltage from any shift in the absorption line, and uses this to control a compensating current, either in the windings of the magnetic field generator or in the coils of the magnet itself.

The easiest way to derive such a voltage from the absorption line is to trace out its differential so that the centre of the line corresponds to zero output, and a shift in frequency one way produces a positive output whereas a shift in the opposite direction produces a negative output. The differential of the line can always be obtained by using a field modulation with amplitude considerably less than the line width, as in normal phase-sensitive detection. Packard was the first to describe such a system in detail, using a 500 c/s field modulation with an induction technique to produce the absorption line, and a standard phase-sensitive method of detection. The same method can be employed with normal resonance circuits; all that is necessary is to produce the differential of the line so that positive or negative error voltages can be derived from it. The strength of the magnetic field can be changed to any other stabilized value by altering the frequency of the proton oscillator circuit correspondingly.

Exactly the same technique can be employed to stabilize magnetic fields by electron resonance, although so far as the author is aware no detailed account of such a system has yet been published. In both cases it is sometimes necessary to incorporate a form of 'rough stabilization' first, so that large changes of field do not take the proton stabilizer right off resonance. Such rough stabilization can be effected by comparison of the potential drop across a resistance, in series with the magnet windings, with a standard reference voltage. Any resulting unbalance is then amplified, and used to alter the current in the field windings of the generator which feeds the magnet. Alternatively, if the magnet is fed from rectifiers, a series of low resistance batteries can be connected in parallel across the input and used as a 'buffer stabilizer'[61]. These initial stabilizing systems, together with the proton resonance control circuits, enable the magnetic field to be kept constant at any desired value to 1 part in 10^5 or better.

9.9 CHEMICAL ANALYSIS

The advantages of chemical analysis by spectroscopy at these frequencies become apparent when only very small quantities of the compounds are available, as in the case of artificially radioactive nuclei, or impurities in crystals. The limit of detection for both gaseous spectroscopy and paramagnetic resonance is below a millimicrogramme; but for resonance at radiofrequencies the limit is much higher, at about 10^{-3} gm, because of the lower energy level splittings and transition probabilities.

(a) *Analysis by gaseous spectroscopy.* The main advantages of microwave spectroscopy in chemical analysis are (i) the ease with which substances can be identified by the frequencies of their absorption lines, (ii) the very small concentration which will still give a detectable signal, and (iii) the very small size of sample required. These features enable rapid and accurate qualitative analysis of any microwave spectrum, as the extremely high precision with which frequencies can be measured will differentiate between any spectra. The tables of microwave absorption line frequencies, now compiled by the National Bureau of Standards[62], can be used as a standard reference for such analysis.

Quantitative analysis by gaseous microwave spectroscopy is not quite so straightforward, however, because of the different broadening factors, which affect the shape and height of the absorption lines. If the pressure is below about 10^{-2} mm Hg, the other broadening factors will predominate, and hence the width of the absorption line will be independent of the amount of sample present, and the concentration can be measured by the height of the line. If the pressure is greater than 10^{-2} mm Hg, however, pressure broadening will be the main effect, and the intensity of the line will remain constant while the width varies with amount of the sample present. Care must therefore be taken to determine in which pressure region the apparatus is working, before peak intensities are taken as proportional to the amount present. If possible, it is probably best to work at the higher pressures, as greater amounts of the total mixture can then be introduced into the apparatus, which is a help if the substance being investigated is only present as a small fraction. The concentration in the mixture can then be obtained from the measured width and intensity of the lines. Care must still be taken in interpreting the results, as foreign gases may have a different collision broadening effect, with respect to the molecules of the material being studied, from that of the molecules of the material themselves.

An estimate of the total amount of gas present can be made from the measured spectra in the ways described in Section 4.7, but it is usually best to employ some form of comparison method, or calibrate the spectrograph with samples of known constitution. Very little work on quantitative analysis by microwave spectroscopy has so far been reported, mainly because of the difficulties outlined above, but it should prove a very useful method for qualitative analysis of small samples. Further information can be found in a review article by HUGHES[63].

(b) *Analysis by paramagnetic resonance.* Chemical analysis by paramagnetic resonance suffers from the drawback that it is only applicable to elements or compounds possessing an unpaired electron; but, for such, it offers a more sensitive method than any other. The minimum quantities of paramagnetic material necessary to produce a detectable absorption can be estimated from the expressions already derived for the sensitivity of such spectroscopes in Section 4.12. By substituting appropriate values for the parameters into equation (4.24), the magnitude of the minimum detectable χ'' was shown to

291

be 10^{-13}. In order to calculate the amount of paramagnetic equivalent to this, the value of the complex component of the susceptibility χ'' must be related to the particular paramagnetic ion. It can be shown[64] that, for a sample containing N paramagnetic ions of spin S, the value of χ'' for the M to $(M-1)$ transition is given by:

$$\chi'' = \frac{\pi \cdot \nu \cdot g^2 \cdot \beta^2 \cdot N[S(S+1) - M(M-1)]}{8k \cdot T(2S+1)} \cdot f(\nu) \quad \ldots (9.13)$$

where S is the effective spin value, as used in the Spin Hamiltonian. The function $f(\nu)$ represents the shape of the absorption line, and has a value of approximately $1/\Delta\nu$ at its centre. The absorption lines will be easier to detect if they are well resolved, with narrow width and hence higher peak intensities. This fact can be seen by expressing χ'' in terms of $(\nu/\Delta\nu)$, and for the case of $S = \frac{1}{2}$, and an isotropic g value, most of the right-hand side of equation (9.13) can be written in terms of the isothermal susceptibility χ_0 to give

$$\chi'' = \chi_0 \left(\frac{\nu}{\Delta\nu}\right) \quad \ldots (9.14)$$

as shown by BLEANEY and STEVENS[64] in their review paper.

The value of $(\nu/\Delta\nu)$ is equal to the resonant field value, divided by the line width in gauss, which is of the order of $(8,000/8) \approx 10^3$, for a diluted salt. Hence the minimum detectable χ'' of 10^{-13} corresponds to a value of $\chi_0 = 10^{-16}$. The isothermal susceptibility, χ_0, of a salt with $S = \frac{1}{2}$ and $g \approx 2 \cdot 0$ (such as copper) is given by

$$\chi_M = \frac{N^2 \cdot \beta^2}{3R \cdot T}[4S(S+1)]$$
$$= \frac{0 \cdot 38}{T} \text{ per gramme-mole} \quad \ldots (9.15)$$

Thus the amount of material, corresponding to a χ_0 of 10^{-16} at $20°$K, is $10^{-16} (20/0 \cdot 38) = 5.10^{-15}$ gramme-moles, which, for a gramme-ionic weight of about 100, gives an equivalent mass of 5.10^{-13} gm.

This is the theoretically attainable minimum, assuming that noise only enters in the 1 c/s bandwidth employed for phase-sensitive detection, and that all the coupling networks are optimized. The experimentally attainable figure may, therefore, be somewhat above this, but even so, the sensitivity is far beyond that of other chemical or physical methods of quantitative analysis, and the figure can be improved by cooling to helium temperatures. The techniques of paramagnetic resonance, therefore, offer a very powerful means of detecting and measuring very small quantities of paramagnetic material, with the added advantage that no chemical preparation or destruction of the sample is required.

The samples used must be such that narrow, well-resolved absorption lines can be obtained, however, and the conditions necessary for this will depend on the actual paramagnetic ion being studied. For ions with isotropic g values, and isotropic hyperfine splitting, very few conditions are required, and their spectra can be observed from powders, glasses or any amorphous mixture. Such examples are provided by divalent manganese, trivalent iron, and trivalent gadolinium, which all possess an isotropic g value of 2·0, and a central electronic transition with an isotropic, or negligible hyperfine structure.

If the ion to be detected has an anisotropic g value, or hyperfine structure, however, it must be incorporated in a crystal lattice before small quantities can be observed, otherwise the random angular variation will smear out any resolved lines. If the ion has strong spin-lattice interaction, it will also be necessary to cool the specimen until the line width produced by this effect is less than that of the spin-spin interaction. The temperatures required vary from ion to ion, but as a rough indication, liquid hydrogen temperatures are sufficient for all the iron group transition elements (except titanium), whereas liquid helium temperatures are required for most of the rare-earth group.

As in the gaseous case, the best way to make quantitative measurements is by comparison with samples containing a known amount of paramagnetic, and if possible, of the same order of magnitude. The power absorbed by the paramagnetic crystal can be derived directly from equation (9.13) and is related to the number of ions, $\mathcal{N}$, by[65]:

Power absorbed

$$= \frac{\pi \cdot \nu^2 \, (g \cdot \beta \cdot H_M)^2}{4k \cdot T} \cdot \frac{[S(S+1) - M(M-1)]}{(2S+1)} \cdot f(\nu)\mathcal{N} \quad \ldots (9.16)$$

where H_M is the magnitude of the microwave magnetic field inducing the transitions. It is therefore important to know the strength of this field if an absolute intensity measurement is to be attempted, and important to keep its strength constant when making comparisons between the absorptions produced by different samples. The function $f(\nu)$ can again be written as the inverse of the line width, when on resonance, and the dependence of the absorption on this factor must also be remembered. If standard samples are available which have a similar line width and intensity (and these can usually be made by a process of magnetic dilution), quantitative measurements of small quantities should be relatively simple. Different paramagnetic spectra can usually be distinguished quite simply, from g values, hyperfine splitting and angular variation, and if needed, variation of line width with temperature.

It is evident from the above summaries that both gaseous microwave spectroscopy and paramagnetic resonance offer considerable possibilities for chemical analysis, especially when dealing with very small amounts, or concentrations.

9.10 THE DETECTION OF FREE RADICALS

The detection of free radicals by electron resonance techniques has been described in some detail in Chapter 7, and it is evident that such experiments are one of the most direct and powerful means for their observation and measurement. The application of paramagnetic or electron resonance in this connexion arises from the fact that all such free radicals have unpaired electrons associated with them. Moreover, the widths of the absorption lines that they produce are usually quite small, and hence signals of strong intensity are obtained from small amounts (e.g. diphenyl picryl hydrazyl gives an easily detectable signal from a tenth of a microgramme in a simple crystal-video spectroscope).

In theory it should be possible to detect free radicals by gaseous microwave spectroscopy as well, since the observed absorption lines will be quite different from those of the undissociated molecules. This has been done for the OH radical by SANDERS et al.[66], using Zeeman modulation at K-band frequencies. The radicals were produced by a radiofrequency gas discharge, and the vapour flowed down the absorption cell at a pressure of 0·1 mm Hg. Four absorption lines were observed, at frequencies between 23,000 Mc/s and 37,000 Mc/s, and were shown to be due to transitions between Λ type doublets of the paramagnetic ground state. This appears to be the only reported observation of free radicals by gaseous spectroscopy, and it would seem, from a review article by MAYS[67], that the experimental difficulties involved in detecting others are quite considerable.

There seems to be no doubt, therefore, that electron resonance is the best technique to apply when searching for the presence of free radicals. Their detection in solids and liquids has been described in some detail in Chapter 7, and it has also been shown recently that their production by ultra-violet irradiation[68] can be detected in this way. The work of COMMONER, TOWNSEND and PAKE[69] is also of great interest in opening up the study of free radical chemistry in living matter by the method of electron resonance, and this may well prove one of the major practical applications of these techniques.

The best way to obtain quantitative estimation of free radical concentrations is by comparison with the signal obtained from weighed amounts of hydrazyl. Its 3-gauss line width is sufficiently close to that of most other free radicals to make direct comparison relatively easy, and an example of such an application is the work of INGRAM *et al.*[70] on the variation of free radical concentration in low temperature carbons. Experiments show that concentrations of 10^{16} per gramme can be detected by simple crystal-video techniques at room temperature, so that about 10^{12} per gramme should be detectable when using 1 c/s bandwidths at low temperatures, or for the 5 mg samples normally employed, a minimum detection limit of about 10^9–10^{11} free radicals should be possible. (See also Section 9.12(*c*).) Many other free radical studies are discussed in Chapter 11.

9.11 THE STUDY OF IRRADIATION DAMAGE

The application of electron and nuclear resonance to investigate irradiation damage is a field of research which has only recently started, but is growing rapidly. The subject has already been discussed at some length in Section 7.11, and only the main points are summarized here.

The possibility of electron resonance, from substances damaged by irradiation, arises from the trapping of electrons in the vacancies or imperfections produced in the crystal or glass. Such trapped electrons may move close to other nuclei in the crystal, and will then interact with any nuclear magnetic moment to produce a hyperfine structure. A very good example of such a case is seen in *Figure 67*. This is the electron resonance spectra obtained from irradiated quartz, and the six hyperfine lines, produced by the interaction of a trapped electron and an aluminium impurity atom, are clearly seen. Measurements of this kind can therefore be used to give qualitative information on the nature of crystal imperfections, as well as a quantitative estimate on the number of damage centres. Different types of spectra are also observed when atoms are knocked bodily out of the crystal lattice and made to occupy interstitial positions; a good example of this being the resonances obtained from irradiated diamonds, by GRIFFITHS, OWEN and WARD[71], and discussed in more detail in Chapter 7.

Nuclear resonance can also be used to study irradiation damage, but in a more indirect way, as the information is obtained from changes in the line shape and width, not from the observation of entirely new spectra. The whole subject of crystal imperfections and their study by resonance techniques was the subject of a recent International Conference, and the report of this[72] should be consulted for further details of the kind of data that can be obtained by this means.

9.12 OTHER APPLICATIONS

The other more specialized applications of spectroscopy at these frequencies are now considered briefly. These are listed in *Table 9.3*.

Table 9.3. Other Practical Applications

Gaseous spectroscopy:	Radio astronomy.
Electron resonance:	(i) The study of conduction levels in metals and semi-conductors.
	(ii) Photo-chemical reactions.
	(iii) The study of gas discharges.

(*a*) *Radio astronomy.* The subject of radio astronomy is really a separate study in its own right, but inasmuch as discrete microwave line spectra have now been found in the radiation received from outer space, its applications and interpretations can be justifiably related to those of microwave spectroscopy in general. Most of the work on radio astronomy has been concerned with the detection and analysis of noise in different wavelength regions. A considerable amount of data can be derived from such measurements, as illustrated by the experiments on the radio and microwave frequencies emitted by the sun[73–76].

Detailed plots[77] of the distribution of radio and microwave noise intensity from outer space as a whole have revealed the existence of point sources of intense radio noise, or 'radio stars' as they are now called. In some cases, these can be identified with large magnitude stars, or nebulae[78], but, in other cases[79], they appear to be in a region containing no outstanding visible objects. This whole subject of galactic radio noise, and the nature of radio stars, is under very active investigation at the present moment.

Studies of noise distribution at these frequencies can hardly be classified as 'spectroscopy', however, as no determinations of discrete frequencies are made. The only example of line spectra that has been found to date is the 1420 Mc/s hydrogen emission line. The ground state of the hydrogen atom is a hyperfine doublet, and its splitting, as measured by the atomic beam resonance experiments described in Section 8.3, is known to be 1420·25 Mc/s[80]. This corresponds to the $F = 1 \rightarrow 0$ transition, and will only be observed as an emission line if there is sufficient population of the upper level at the temperature of the hydrogen atoms in outer space. The possibility of detecting such radiation, as a discrete spectral line above the general level of incoming cosmic radiowaves, was first suggested by VAN DER HULST[81], and the first experimental observations were made, more or less simultaneously, by EWEN and PURCELL[82] in the United States, and

MULLER and OORT[83] in Holland. Both groups used a double super-heterodyne form of detection, shifting the pass band at 30 c/s. In Ewen and Purcell's apparatus, the signal was fed finally to a phase-sensitive detector of 0·016 c/s bandwidth, and the sensitivity was such that hydrogen emission corresponding to a temperature of 3·8° K could be detected. They employed a fixed aerial of 12° half-power beam width, and scanned the sky by using the earth's rotation. The initial measurements showed that the hydrogen line was emitted from an extended source, centred about the galactic plane, and with an effective temperature of 35° K.

Muller and Oort employed a mobile paraboloid of 2·8° beam width as their antenna, and were able to make a more systematic survey of the sky, and use their measurements to obtain an estimate of the galactic rotation. In slightly later work[84] they were able to delineate the galactic arms by this means. Measurements were also made very shortly afterwards by CHRISTIANSEN and HINDMAN[85] in Australia, who showed that the intensity of the spectral line varied along the galactic equator, having a maximum effective temperature of 100° K. Some of the theoretical implications of these results, and estimates of the galactic thickness from them, have been given by WILD[86].

This is a very specialized branch of microwave spectroscopy, but is also a very good illustration of the great variety of information that can be obtained from the study of discrete spectral lines in this frequency region.

(b) *The study of conduction bands in metals and semi-conductors.* The detection of electron resonance, in both metals and semi-conductors, has already been considered in Chapter 7, and it would appear that such measurements will give some very useful information on the details of energy levels and conduction bands in such solids. Apart from the absorption lines obtained from electrons associated with impurity atoms in semi-conductors[87], there are two other types of electron resonance occurring in these compounds—(i) spin reson-ance of the conduction electrons, and (ii) cyclotron resonance of electrons raised into the conduction levels of semi-conductors.

Electron spin resonance has been observed for the conduction electrons in sodium[88], lithium[89] and beryllium[90], and some detailed analyses made of the factors affecting both the g values and line widths. The theory of such interactions[91, 31] is somewhat com-plicated, but detailed information on the energy levels and conduc-tion bands can be derived; and such measurements should prove to be a very powerful addition to the experimental side of metal physics. The results of the cyclotron resonance experiments on germanium[92]

297

and silicon[93] have been used in a similar way, via the determination of the effective electron masses, and these small values result in resonance curves which are often widely separated and in much lower magnetic field strengths than usual. Both the classical[94] and quantum[95] treatment of cyclotron resonance in the solid state has now been given in some detail, and such measurements should prove extremely useful in the elucidation of semi-conductor theory.

(c) *Photo-chemical reactions.* Although very little experimental work has been done as yet, the study of photo-chemical reactions by electron resonance may well prove one of the major applications of this technique in the future. It is known that irradiation of certain chemical systems will produce free radicals, and in principle, it should be possible to detect these by electron resonance in a very straightforward manner. That it is, in fact, experimentally possible to do so has been shown by the work of Commoner, Townsend and Pake[69], and of BIJL and ROSE-INNES[96]. A brief description of the latter experiment affords a good illustration of the kind of technique that may be used. In order to trap the free radicals for a reasonable length of time, solid solutions of the chemicals were made, and then recombination of the radicals formed by irradiation is prevented by the internal viscosity of the glass-like material. Their initial experiments were performed with a solid solution of Wurster's base, $(CH_3)_2 N . C_6H_4 . N(CH_3)_2$, in perspex, both were first dissolved in chloroform, the two solutions mixed and then poured on to a glass plate. A thin plastic film was left behind, which was irradiated for about five minutes by a high pressure mercury lamp, and the formation of free radicals was shown by change in coloration. This film was then folded into a pellet and inserted into the cavity resonator when a line at $g = 2 \cdot 001 \pm 0 \cdot 002$ was obtained. That this was due to the free radicals, so formed, was shown by absence of signal from non-irradiated specimens and the correct variation of signal intensity with temperature.

It is quite certain that this experiment is only the beginning, and that many others will follow, as a large number of free radical reactions can be studied by a similar 'trapping process' in glasses, as may other photo-chemical processes. The ultimate aim, in the application of electron resonance to the study of free radical reactions, is to design an apparatus in which dynamic systems can be studied, so that the concentration of free radicals can be measured as a function of time and chemical conditions. The first step in this direction has been the study of free radical decay in carbon samples, where the radical concentration was recorded continuously[98] (see Section 11.2 for recent developments in rapid recording techniques).

(d) *The study of gas discharges.* Another rather different application of electron resonance may be mentioned in closing, and that is the study of low pressure gas discharges. Such discharges possess a large number of positive ions and electrons, and cyclotron resonance of these will occur under the appropriate field and frequency conditions. The maximum concentration of such ions is of the order of 10^{11} per cc, but since cyclotron resonance acts via the microwave or radio-frequency electric field, its transition probabilities will be very much larger than those associated with the normal magnetic interaction. Cyclotron resonance of the electrons in such discharges has been experimentally observed[97], and the variation of its intensity and line width with gas pressure and ion current determined. A detailed analysis of such factors should give very useful information on the numbers and energies of the ions present without the use of any disturbing electrodes or probes.

Finally it may be remarked that several other potential applications, such as the study of excited phosphor levels, have been omitted because no detailed experimental results have so far been published. The whole field of research is widening rapidly, however, and new and different problems are being studied by electron resonance at an ever increasing rate. It would seem a fair and just conclusion to suggest that, if any physical or chemical system possesses unpaired electron spins, then sooner or later the techniques of electron resonance will be applied to obtain exact and detailed information, which is usually impossible to derive by any other means.

REFERENCES

[1] FLETCHER, R. C., YAGER, W. A., PEARSON, G. L., HOLDEN, A. N., READ, W. T. and MERRITT, F. R. *Phys. Rev.* 94 (1954) 1392 and 95 (1954) 844

[2] BLEANEY, B. and INGRAM, D. J. E. *Proc. Roy. Soc.* A 205 (1951) 336

[3] BLOCH, F., GRAVES, A. C., PACKARD, M. and SPENCE, R. W. *Phys. Rev.* 71 (1947) 373; and 71 (1947) 551

[4] ELLIOTT, R. J. and STEVENS, K. W. H. *Proc. phys. Soc.* A 64 (1951) 205

[5] — — *Proc. Roy. Soc.* A 218 (1953) 553; and 219 (1953) 387

[6] BLEANEY, B. and SCOVIL, H. E. D. *Proc. phys. Soc.* A 63 (1950) 1369 ; BLEANEY, B., SCOVIL, H. E. D. and TRENAM, R. S. *Phil. Mag.* 43 (1952) 995

[7] BOGLE, G. S., DUFFUS, H. J. and SCOVIL, H. E. D. *Proc. phys. Soc.* A 65 (1952) 760

[8] BLEANEY, B., ELLIOTT, R. J., SCOVIL, H. E. D. and TRENAM, R. S. *Phil. Mag.* 42 (1951) 1062

[9] STERNHEIMER, R. *Phys. Rev.* 80 (1950) 102

[10] BLEANEY, B., BOWERS, K. D. and INGRAM, D. J. E. *Proc. phys. Soc.* A 64 (1951) 758

[11] GORTER, C. J. *Phys. Z.* 35 (1934) 923

[12] KURTI, N. and SIMON, F. *Proc. Roy. Soc.* 149 (1935) 152

[13] GORTER, C. J. *Physica* 14 (1948) 504 ; ROSE, M. E. *Phys. Rev.* 75 (1949) 213

[14] POUND, R. V. *Phys. Rev.* 76 (1949) 1410

[15] BLEANEY, B. *Proc. phys. Soc.* A 64 (1951) 315

[16] — *Phil. Mag.* 42 (1951) 441

[17] ROBERTS, J. K. *Heat and Thermodynamics.* Blackie & Son, Ltd., 3rd edn. (1945) 117-121. See also reference [12]

[18] DANIELS, J. M., GRACE, M. A. and ROBINSON, F. N. H. *Nature* 168 (1951) 780

[19] BLEANEY, B., DANIELS, J. M., GRACE, M. A., HALBAN, H., KURTI, N. and ROBINSON, F. N. H. *Phys. Rev.* 85 (1952) 688

[20] AMBLER, E., GRACE, M. A., HALBAN, H., KURTI, N., DURAND, H., JOHNSON, C. E. and LEMMER, H. R. *Phil. Mag.* 44 (1953) 216

[21] ROBERTS, L. D., BERNSTEIN, S., DABBS, J. W. T. and STANFORD, C. P. *Phys. Rev.* 95 (1954) 105

[22] BOGLE, G. S., COOKE, A. H. and WHITLEY, S. *Proc. phys. Soc.* 64 (1951) 931

[23] COOKE, A. H., DUFFUS, H. J. and WOLF, W. P. *Phil. Mag.* 44 (1953) 623

[24] GRACE, M. A., JOHNSON, C. E., KURTI, N., LEMMER, H. R. and ROBINSON, F. N. H. *ibid.* 45 (1954) 1192

[25] BERNSTEIN, S., ROBERTS, L. D., STANFORD, C. P., DABBS, J. W. T. and STEPHENSON, T. E. *Phys. Rev.* 94 (1954) 1243

[26] SPIERS, J. A. *Nature* 161 (1948) 807

[27] STEENBERG, N. R. *Proc. phys. Soc.* A 65 (1952) 791; and A 66 (1953) 399

[28] — *Phys. Rev.* 93 (1954) 678

[29] DABBS, J. W. T. and ROBERTS, L. D. *ibid.* 95 (1954) 307

[30] CARVER, T. R. and SLICHTER, C. P. *ibid.* 92 (1953) 212

[31] OVERHAUSER, A. W. *ibid.* 92 (1953) 411

[32] HONIG, A. *ibid.* 96 (1954) 234

[33] KAPLAN, J. I. *ibid.* 96 (1954) 238

[34] ABRAGAM, A. and PRYCE, M. H. L. *Proc. Roy. Soc.* A 206 (1951) 164

[35] INGRAM, D. J. E. *Proc. phys. Soc.* A 62 (1949) 664

[36] ABRAGAM, A. *Phys. Rev.* 79 (1950) 534; *Proc. Roy. Soc.* A 230 (1955) 169

[37] OWEN, J. and STEVENS, K. W. H. *Nature* 171 (1953) 836

[38] GRIFFITHS, J. H. E. and OWEN, J. *Proc. Roy. Soc.* A 226 (1954) 96

[39] INGRAM, D. J. E. and BENNETT, J. E. *J. chem. Phys.* 22 (1954) 1136

[40] BENNETT, J. E. and INGRAM, D. J. E. *Nature* 175 (1955) 130

[41] GOLDSTEIN, J. H. and BRAGG, J. K. *Phys. Rev.* 78 (1950) 347

[42] DEAN, C. *ibid.* 86 (1952) 607 ; DEAN, C. and POUND, R. V. *J. chem. Phys.* 20 (1952) 195

[43] LAMB, W. *Phys. Rev.* 60 (1941) 817 ; RAMSEY, N. F. *ibid.* 77 (1950) 567; and 78 (1950) 699; and 86 (1952) 243

[44] DICKINSON, W. *ibid.* 77 (1950) 736 ; PROCTOR, W. G. and YU, F. C. *ibid.* 77 (1950) 717

[45] PAKE, G. E. *J. chem. Phys.* 16 (1948) 327

[46] POUND, R. V. *Phys. Rev.* 76 (1949) 1410; and 79 (1950) 685

[47] KNIGHT, W. D. *ibid.* 76 (1949) 1260

[48] GUTOWSKY, H. S., McCALL, D. W. and SLICHTER, C. P. *J. chem. Phys.* 21 (1953) 279

[49] GESCHWIND, S., GUNTHER-MOHR, G. R. and TOWNES, C. H. *Rev. mod. Phys.* 26 (1954) 444

[50] OWEN, J. *Proc. Roy. Soc.* A 227 (1955) 183

[51] RAMSEY, N. F. *Phys. Rev.* 78 (1950) 695

[52] LYONS, H. *Ann. N.Y. Acad. Sci.* 55 (1952) 831

[53] — *Phys. Rev.* 74 (1948) 1203; *Elect. Engng. N.Y.* 68 (1949) 251; and *Proc. Inst. Radio Engrs, N.Y.* 39 (1951) 208

[54] HUSTEN, B. F. *Proc. Inst. Radio Engrs, N.Y.* 39 (1951) 208

[55] INGRAM, D. J. E. and TAPLEY, J. G. *Phil. Mag.* 45 (1954) 1221

[56] GARSTENS, M. A., SINGER, L. S. and RYAN, A. H. *Phys. Rev.* 96 (1954) 53

[57] CODRINGTON, R. S., OLDS, J. D. and TORREY, H. C. *ibid.* 95 (1954) 607

[58] HUTCHINSON, C. A. and PASTOR, R. C. *J. chem. Phys.* 21 (1953) 1959

[59] PACKARD, M. and VARIAN, R. *Phys. Rev.* 93 (1954) 941

[60] BLOEMBERGEN, N. and POUND, R. V. *ibid.* 95 (1954) 8

[61] BATES, L. F. *Modern Magnetism*, 3rd Edn. C.U.P. (1951) 108

[62] *Molecular Microwave Tables.* National Bureau of Standards, Washington

[63] HUGHES, R. H. *Ann. N.Y. Acad. Sci.* 55 (1952) 872

[64] BLEANEY, B. and STEVENS, K. W. H. *Rep. Progr. Phys.* 16 (1953) 108

[65] — *Phil. Mag.* 42 (1951) 444

[66] SANDERS, T. M., SCHAWLOW, A. L., DOUSMANIS, G. C. and TOWNES, C. H. *Phys. Rev.* 89 (1953) 1158

[67] MAYS, J. M. *Ann. N.Y. Acad. Sci.* 55 (1952) 789

[68] BIJL, D. and ROSE-INNES, A. C. *Nature* 175 (1955) 82

[69] COMMONER, B., TOWNSEND, J. and PAKE, G. E. *ibid.* 174 (1954) 689

[70] INGRAM, D. J. E., TAPLEY, J. G., JACKSON, R., BOND, R. L. and MURNAGHAN, A. R. *ibid.* 174 (1954) 797

[71] GRIFFITHS, J. H. E., OWEN, J. and WARD, I. M. *ibid.* 173 (1954) 439

[72] *Report on the Conference on Defects in Crystalline Solids, Bristol, July, 1954.* The Physical Society (1955)

[73] REBER, G. *Proc. Inst. Radio Engrs.* 28 (1940) 68; 30 (1942) 367

[74] SOUTHWORTH, G. C. *J. Franklin Inst.* 239 (1945) 285

[75] RYLE, M. *Rep. Progr. Phys.* 13 (1950) 184

[76] Hey, J. S. *Mon. Not. R. astr. Soc.* 109 (1949) 179 ; Hagen, J. P. *Astrophys. J.* 113 (1951) 547

[77] Reber, G. *ibid.* 100 (1944) 279

[78] Mills, B. Y. *Aust. J. Sci. Res.* 5 (1952) 456

[79] Bolton, J. G. and Stanley, G. J. *Nature* 161 (1948) 312

[80] Prodell, A. G. and Kusch, P. *Phys. Rev.* 88 (1952) 184

[81] Van der Hulst, H. C. *Nederl. Natur Kunde* 11 (1945) 201

[82] Ewen, H. I. and Purcell, E. M. *Nature* 168 (1951) 356

[83] Muller, C. A. and Oort, J. H. *ibid.* 168 (1951) 357

[84] Oort, J. F. H. and Muller, C. A. *S. Afr. J. Sci.* 49 (1952) 87

[85] Christiansen, W. N. and Hindman, J. V. *Aust. J. Sci. Res.* 5 (1952) 437

[86] Wild, J. P. *Astrophys. J.* 115 (1952) 206

[87] Fletcher, R. C., Yager, W. A., Pearson, G. L. and Merritt, F. R. *Phys. Rev.* 95 (1954) 844

[88] Griswold, T. W., Kip, A. F. and Kittel, C. *ibid.* 88 (1952) 951

[89] Carver, T. R. and Slichter, C. P. *ibid.* 92 (1953) 212

[90] Feher, G. and Kip, A. F. *ibid.* 95 (1954) 1343; 98 (1955) 337

[91] Elliott, R. J. *ibid.* 96 (1954) 266, 280

[92] Dresselhaus, G., Kip, A. F. and Kittel, C. *ibid.* 92 (1953) 827; and 95 (1954) 568; 98 (1955) 368

[93] Dexter, R. N., Lax, B., Kip, A. F. and Dresselhaus, G. *ibid.* 96 (1954) 222

[94] Shockley, W. *ibid.* 90 (1953) 491

[95] Kohn, W. and Luttinger, J. M. *ibid.* 96 (1954) 529

[96] Bijl, D. and Rose-Innes, A. C. *Nature* 175 (1955) 82

[97] Ingram, D. J. E. and Tapley, J. G. *Research, Lond.* 7 (1954) S 63; and *Phys. Rev.* 97 (1955) 238

[98] —— *Chem. & Ind. (Rev.)* (1955) 568

[99] Honig, A. and Combrisson, J. *Phys. Rev.* 102 (1956) 917

[100] Combrisson, J., Honig, A. and Townes, C. H. *C. R. Acad. Sci., Paris,* 242 (1956) 2451

10

HIGH RESOLUTION NUCLEAR RESONANCE

10.1 Developments in Experimental Technique

During the last ten years most of the work on nuclear magnetic resonance has been concerned with the high resolution spectra that can be obtained from samples in the liquid state. The very narrow width of these absorption lines has made it possible to distinguish very slight shifts in their resonance frequency or field position, and this in turn, has allowed very detailed chemical analyses of the compounds to be made. It is probably fair to say that the major contribution of 'High resolution NMR' has been to the fields of qualitative and quantitative organic analysis, but it has also proved of great value in the investigation of reaction kinetics, solvation studies and similar subjects.

In the field of organic analysis it has proved possible not only to determine the actual quantity of the particular atom in the specimen, but also to determine how much of it is in a particular chemical group, and moreover to deduce what other chemical groups are in close proximity to it. The wealth of detailed structural analysis that has resulted from the application of this technique has revolutionized this aspect of organic chemistry, and it would appear that 'High resolution NMR' is rapidly becoming a standard technique in most large chemical laboratories.

The essential requirement for high resolution spectra is the production of absorption lines of very narrow width, and the various factors affecting this will be outlined first, before the spectra themselves are discussed in detail. These factors can be divided into two groups, (i) internal interactions, i.e. those arising within the specimen, and (ii) external effects, i.e. those produced outside the specimen. It is shown in Section 10.2 that it is possible for the internal interactions to be averaged to a vanishingly small net effect if the molecular tumbling motion in the liquid state is rapid enough, and in these cases the observed line width is then determined entirely by external effects.

The most obvious and direct external effect on the line width is the inhomogeneity of the applied magnetic field over the specimen.

Any slight variation in the field strength across the specimen will produce a corresponding spread of the absorption line along the magnetic field display axis, and hence an artificial broadening of the spectra. Since in most cases all other forms of broadening can be reduced effectively to zero, it is the actual inhomogeneity of the applied magnetic field which determines the final resolution of the nuclear magnetic resonance spectrometer, and hence the precision with which a detailed analysis can be made from the spectra. It is for this reason that the magnets required for such work are large and expensive.

The inhomogeneity of the magnetic field over the specimen must be less than a milligauss if accurate interpretation of the high resolution spectra is to be achieved. This corresponds to a field homogeneity of 1 part in 10^7 or 10^8 across a specimen of about 4 mm diameter, and such extremely high orders of uniformity can only be achieved in magnets with pole face diameters of 10 in. or more, after careful geometrical alignment and optical polishing. For the optimum conditions it is nearly always best to hand figure the faces as a final adjustment, or to incorporate small coils on the poles themselves, through which 'balancing currents' can be passed. This latter technique was developed by GOLAY[1] in conjunction with RICHARDS[2] and would appear to be a very effective and yet relatively simple method of optimizing the field homogeneity. Thin flat stacks consisting of several coil-pairs, one coil in each stack, are placed on the opposing pole faces of the magnet. The coils are designed so that they produce correcting magnetic fields which can be described by different sets of spherical harmonics. In this way current adjustments in the different coils can be made independent of each other and the resolving power can thus readily be optimized. In practice it is found best to fabricate the actual coils as photo-etched circuits.

In 1954 an experimental technique was introduced by BLOCH[3] which increased the available resolution by a considerable factor. The sample is spun rapidly about a vertical axis, and an averaging of the field inhomogeneities is produced. Thus if the variation of the magnetic field over the sample is given by ΔH, and the sample is rotated at such a speed that each part is exposed to the variation in time t, then the nuclei will effectively see an 'averaged field', provided

$$t < \frac{2\pi}{\gamma \cdot \Delta H} \qquad \dots (10.1)$$

where γ is the nuclear gyromagnetic ratio.

For example, if $\Delta H = 1$ milligauss, then this homogeneity can be averaged out for a proton resonance if $t < \frac{1}{20}$ sec, i.e. if the specimen tube is spun faster than 20 rev/sec, which is quite simple experimentally. It should be noted that the averaging takes place around circular paths, and although the inhomogeneities in the plane perpendicular to the axis of rotation are effectively removed, no averaging of field gradients takes place along the vertical axis. On the other hand it is a much easier matter to locate the region of magnetic field with maximum homogeneity in the vertical direction, than to find the point of optimum homogeneity for all three directions simultaneously. This spinning technique has in fact increased the effective resolution to better than 1 part in 10^8, although the actual field homogeneity over the sample is only 1 part in 10^7.

It is also necessary to have a 'time stability' of the magnetic field of the same order as its space homogeneity. Thus, if it takes 10 sec to sweep through a resonance line, it is imperative that the magnitude of the field strength should remain constant to 1 part in 10^8 during this time, if the line is to have its minimum width. This is a very stringent requirement in practice, and necessitates highly stabilized power supplies with automatic correcting circuits if electromagnets are employed. Permanent magnets have some advantage in this connection, but require very precise temperature control to obtain the required constancy of field. In the present state of development there seems little to choose between electromagnets and permanent magnets, in so far as the ultimate homogeneity of the field is concerned; in either case very precise engineering is required and the total cost of the stabilized magnet is of the order of £5,000.

The sensitivity of any NMR spectrometer is increased by increasing the difference in population between the energy levels, and this can normally only be achieved by increasing the value of the applied magnetic field. It therefore follows that the applied field should have as large a magnitude as possible consistent with the required homogeneity. In the case of both permanent and electro-magnets the limitation on the magnitude of the field is effectively determined by the properties of the magnetic material of the pole pieces. Permanent magnetic materials do not operate well above 12,000 gauss and conical pole tips must be used to obtain fields greater than this. The upper limit to the field in the magnetic gap is then determined by the point at which the material of the pole faces begins to saturate, and this cannot be made much greater than about 25,000 gauss. Nuclear resonance spectrometers operating at 100 Mc/s for protons, and thus requiring a field of about 23,000 gauss, therefore represent, more or less, the limit so far as NMR spectrometers employing conventional electromagnets or permanent magnets are concerned.

The possible use of superconducting magnets should be mentioned in this connection. At the moment these are in the early stages of development, but it would appear that fields of up to 60,000 gauss can be produced by such means. One major problem with superconducting magnets will be the attainment of sufficient field homogeneity, but recent designs have produced fields homogeneous to 1 part in 10^6 or better, and no doubt these figures will be steadily improved. It is therefore quite possible that NMR spectrometers operating at much higher field strengths and frequencies will become a practical proposition in a few years time.

Since the resonance condition is essentially a relation between magnetic field strength and frequency, it is also necessary to have equal stability in the frequency of the applied radiation. For this reason most high resolution NMR spectrometers employ separate fixed-frequency units at such frequencies as 20, 40, 60 and 100 Mc/s, rather than provide an oscillator with a continuously variable frequency. Such fixed-frequency oscillators can be locked to quartz crystals, held at at constant temperature in thermostatically controlled ovens, and the required frequency stability is then relatively easily achieved. Should they be required, it would appear that the more modern frequency standards, such as those derived from the ammonia maser, with two orders of magnitude increase in stability, could be adapted to control the frequency of the resonance absorption. It seems likely, however, that the inhomogeneity of the magnetic field will remain the major limitation on the maximum resolution for some time to come.

A more recent development in this field has been the introduction of a self-correcting feedback link between the magnetic field strength and the applied radiofrequency. This acts in such a way that the resonance condition between the field and the frequency is maintained for a standard absorption line, the radiofrequency being automatically adjusted to compensate for any change in the magnetic field. A small, compact and relatively inexpensive high resolution spectrometer can then be designed around such a system, and this has found wide application as a standard analytical tool in many chemical laboratories.

Any of the basic methods of detection described in Chapter 8 can be used in the high resolution spectrometers, although the Nuclear Induction Balanced Bridge techniques seem to have been most generally favoured. Complete commercial high resolution nuclear resonance spectrometers are now available from several firms in Europe and America and, although somewhat expensive, good electronic facilities would be required by any laboratory which contemplated the construction of its own spectrometer. Details of high resolution nuclear resonance spectrometers employing permanent magnets have been published by ARNOLD [4], PRIMAS and GUNTHARD [5, 6], and Leane, Richards and Schaefer [2]; and of those employing electromagnets by BAKER and BURD [7], and BLOOM and PACKARD [8]. Brief descriptions of High resolution NMR spectrometers now available commercially are found in Appendix II.

10.2 MOTIONAL NARROWING IN LIQUIDS

Very narrow absorption lines are obtained in solution because the rapid tumbling motion of the solvent molecules averages out the anisotropic contributions to the line width. The high resolution spectra thus obtained, which have proved of such very great value in chemical analysis, are entirely dependent on this motional narrowing and a more detailed consideration of the effect is therefore now undertaken.

There are, in fact, several different sources of line-broadening that may exist in any specific case, including those due to interactions of the nuclear quadrupole moment with internal electric fields, or the interaction with electrons that can occur in metals and magnetic material. The most common interaction, however, and one that must always be present, is the direct dipole-dipole interaction between nuclei in which one experiences an additional component of magnetic field due to the magnetic dipole possessed by the other. In a solid, the nuclei usually have definite locations and orientations and this dipole-dipole interaction produces an additional field with an angular variation of $(3 \cos^2 \theta - 1)$, where θ is the angle between the applied field and the internuclear axis. The presence of neighbouring nuclei can thus cause splittings or, more generally, alter the shape of the simple resonance absorption expected.

In the early years of nuclear resonance, before high resolution spectra were discovered, a considerable amount of information was obtained by studying the line shapes of proton resonance signals obtained from single crystals [9]. The magnitude, shape and angular dependence of the line width could be used to determine the position and orientation of neighbouring protons from a careful analysis of this dipole-dipole broadening.

The change towards a liquid may be considered initially in terms of the onset of molecular rotation within the solid. It had been observed in the very early studies on nuclear resonance that the absorption lines would often suddenly narrow when the solid was heated. This was taken as evidence that some groups within the molecule had started to rotate, and the nuclei were then only seeing an average field from the rotating group.

The first detailed theoretical and experimental treatment of this subject was undertaken by GUTOWSKY and PAKE[10]. They showed that, when the frequency of the molecular tumbling is greater than the width of the absorption line measured in frequency units, the magnitude of the mean local field determines the width of the spectral line, and this is usually much smaller than the static local field which exists in the absence of motion. Further theoretical considerations by PAKE[11], however, showed that molecular rotation should not cause a reduction in the second moment of the nuclear resonance absorption, and that, in fact, the fourth moment should be increased. In other words, the molecular motion should not only produce a narrowing of the central region of the line, but this would be compensated by an increase of intensity in the wings.

The fact that the second moment should remain invariant might appear to contradict the experimental fact that the absorption lines do appear to narrow very considerably when molecular motion takes place. In order to obtain a direct experimental check on this point ANDREW, BRADBURY and EADES[12] carried out some very interesting experiments in which solid crystals were rotated at high speed.

The crystals employed were those of sodium chloride: there was thus no appreciable motion of the nuclei within the crystal, and it could be assumed that when the crystal was spun all the nuclei would rotate with the same uniform frequency with respect to their neighbours. In order to test the theoretical predictions it was necessary to spin the crystals at a frequency comparable with the frequency width of the spectral line. The line width of the Na^{23} resonance in the static crystal was 0·74 gauss which is equivalent to a rotation frequency of 833 c/s. By using an air-driven rotor, they found that it was possible to spin the crystal of sodium chloride at speeds up to 50,000 rev/min. In the initial experiments the crystal was mounted so that it spun about its [001] axis and the magnetic field of 6,000 gauss was applied normal to this axis. Their theoretical analysis of the effects to be expected in such a case showed that the central absorption line should narrow, and that sideband absorption lines should be obtained at even multiples of the rotation frequency, on either side of the narrowed central spectrum.

The experimental results completely confirmed these predictions, and at the highest rate of rotation of 800 c/s, the width of the central line was found to be approximately halved, while two first-order side spectra could be clearly seen at 1,600 c/s on either side of the central line. They also carefully measured the over-all second moment for each of the spectra obtained for different rates of rotation, and confirmed that this remained effectively constant at 0·57 gauss² although the general appearance of the spectrum changed radically.

Further measurements[13] were then made with the axis of rotation orientated at different angles to the applied magnetic field. These were carried out to check the more detailed predictions of the theory[14, 15] which can be summarized as follows.

The angular factor in the interaction can be expressed as the sum of the two factors (i) its mean value

$$\tfrac{1}{2}(3 \cos^2 \alpha - 1)(3 \cos^2 \gamma - 1)$$

(ii) the two time-dependent terms

$$\tfrac{3}{2} \sin 2\alpha \sin 2\gamma \cos \omega_r t$$
$$\tfrac{3}{2} \sin^2\alpha \sin^2\gamma \cos 2\omega_r t$$

where ω_r is the angular velocity of rotation, α is the angle between the axis of rotation and the direction of the applied field, and γ is the angle between the internuclear vector and the axis of rotation.

The two time-dependent terms lead to the formation of the side spectra, and it can be seen that if $\alpha = 90°$, as was the case in the initial measurements, only the second harmonic frequencies will be obtained. In general, however, sideband lines are to be expected at both $\pm \omega_r$ and $\pm 2\omega_r$ from the central line.

These expressions also predict that the central absorption should have the shape found for the static crystal but with a width reduced by the factor of $\tfrac{1}{2}(3 \cos^2 \alpha - 1)$. It therefore follows that if α has the value $\cos^{-1}(1/\sqrt{3}) = 54° 44'$ the dipolar broadening of the central line should be reduced to zero by rotation at this angle.

The experiments were therefore repeated for different values of α, and in particular for a value of α equal to 54° 44'. The spectra actually obtained at this angle are shown in *Figure 74*, (a) is from the stationary crystal, (b) is for the crystal rotated at 810 c/s, and (c) is for rotation at 1620 c/s. The vertical lines are frequency markers spaced at 790 c/s intervals. The very great reduction in line width of the central absorption can be clearly seen, together with the sideband absorptions at ±810 c/s and ±1,620 c/s respectively. The simple theory predicts, of course, that the width of the central line should be zero at this angle of rotation instead of the 200 c/s

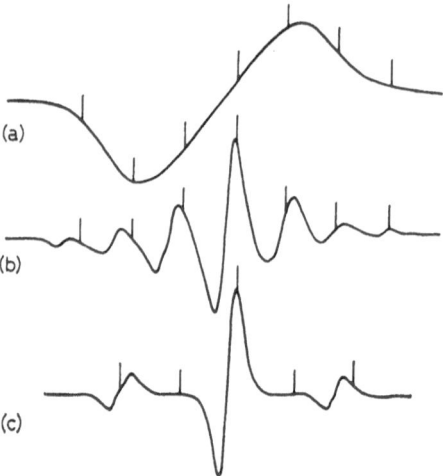

Figure 74. Spectra from Na²³ of sodium chloride: (a) stationary crystal, (b) crystal rotated at 810 c/s, (c) rotated at 1620 c/s.

observed. The theory has only considered magnetic dipolar broadening between the nuclei, however, and the residual width probably arises from interactions between the electric quadrupole moment of the Na^{23} nuclei with internal electric fields caused by crystal imperfections.

The general agreement of the theory with experiment can be clearly seen in *Figure 75* where the width of the central line is plotted against the value of α. The line indicates the plot of the $(3 \cos^2 \alpha - 1)$ variation predicted, while the points represent the actual experimental observations. These particular measurements are not only of inherent interest in confirming the detailed theory of the mechanisms of dipolar broadening under rotation, but also open up several interesting applications, such as the determination of chemical shifts in the solid state.

The above experiments have been reported in some detail since they show very clearly what actually happens to the nuclear absorption signal as molecular rotation starts to occur. If different pairs of nuclei are rotating at different angular velocities, as in the case of a solid approaching the melting point, they will produce sideband absorption signals at a variety of frequency shifts from the central line. In fact if there is any noticeable spread of rotational velocities, as will certainly be so in the hindered rotation of the solid state, these sideband absorptions will all be spread out along the frequency axis and be too weak to detect. This explains why only the narrowed central line is actually observed in such cases—the

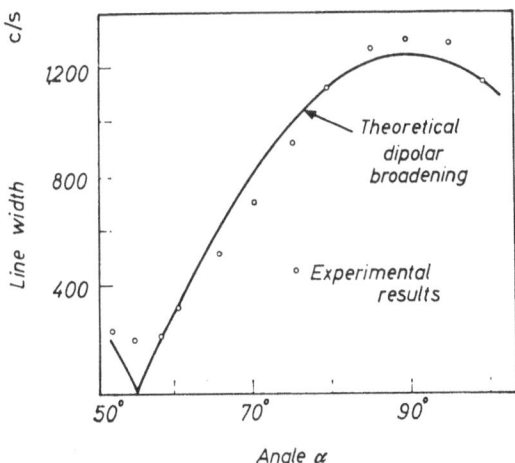

Figure 75. Variation of line width with angle of rotation of crystals

rest of the absorption is actually taking place but quite undetectably. In a liquid, of course, the dipole-dipole interaction is not only altered at a high rate because of the rotational motion of the surrounding nuclei, but can also be changed very drastically by the rapid variation of the internuclear distance itself.

In this way *motional narrowing* can take place to almost any desired extent if the molecular reorientation and motion is rapid enough. The fact that an *averaging* takes place, rather than a summation of the incremental fields, due to all the different positions of the nuclei, is due to the fact that the time of one nuclear precession, which is about 10^{-7} sec, is much longer than the time of a molecular re-orientation, which takes place in about 10^{-11} sec. Thus the frequency of the radio-quantum actually absorbed, or emitted, must be determined by the average of many internuclear positions rather than by the extremes.

It should be pointed out, however, that interactions which do not vary with angle are not averaged out by molecular reorientation, and provided the internuclear distance does not change (i.e. the nuclei belong to the same molecule), these will still exist in the liquid state. This explains why the chemical shift and spin-spin coupling, described in the next sections, are still effective in liquids although the dipolar broadening has been eliminated.

10.3 THE CHEMICAL SHIFT

The great advantage of studying nuclear resonance in solutions is that very narrow line widths are obtained and hence high resolution is possible in the spectrum. Moreover, all the interactions which vary with angle have been averaged to zero, and hence the spectra themselves should be relatively straightforward to analyse since only the isotropic interactions will be present. It is the combination of these two features which has produced the great application of high resolution nuclear magnetic resonance to physical and organic chemistry.

In order to appreciate these various applications it is first necessary to consider two types of shift, or splitting, which occur in the spectra and which become observable with the high resolution attainable in solutions. The first of these is termed 'the chemical shift', and is the name given to the fact that the same nucleus will not always have its resonance absorption line at the same value of magnetic field for a given value of the applied radiofrequency. This fact, and the reasons for it, can probably be best illustrated by taking a specific example, such as ethyl alcohol.

This has the structural formula

$$
\begin{array}{ccc}
& \text{H} & \text{H} \\
& | & | \\
\text{H}-\text{C} & - & \text{C}-\text{OH} \\
& | & | \\
& \text{H} & \text{H}
\end{array}
$$

It will be noticed that the vast majority of carbon and oxygen atoms in these molecules have no nuclear spins or magnetic moments, since they are nearly all C^{12} or O^{16} nuclei. Hence the resonance is only to be expected from the hydrogen nuclei, i.e. the protons. On a low resolution nuclear resonance spectrometer a single absorption line at the magnetic field strength corresponding to the resonant frequency of the proton is therefore expected, each of the six protons in the molecule contributing to its intensity. If somewhat greater resolution is employed, however, and in practice this normally means using a magnet with a more uniform magnetic field, this single absorption line is found to split into three components, as illustrated in *Figure 76(a)*. Moreover, these three components are found to have intensity ratios of approximately 3 : 2 : 1. This fact itself suggests very strongly that they must be arising from the three different kinds of proton which are to be found in the alcohol molecule, i.e. the larger line corresponds to the resonance from the three protons from the CH_3 group, whereas the central line corresponds to the two from the CH_2 group, and the single line of smaller intensity arises from the one proton of the OH group.

The fact that these three different resonances are now observed for the same incident radiofrequency, indicates that these three different types of proton must be residing in slightly different values of magnetic field strength. Since the same external magnetic field is applied across all of the protons this in turn suggests that there must be different internal fields, arising within the molecule, which are responsible for these small splittings. These internal magnetic field contributions are produced by the different effects of the diamagnetic shielding, which is itself due to the outer electrons forming the chemical bonds of the three different chemical groups.

Thus the electron cloud distribution around the protons in the CH_3 group will be slightly different from that around the CH_2 groups, and the shielding that it produces from the applied magnetic field will in turn be slightly different. The fact that this shift, or splitting, is due to some mechanism such as this is confirmed by the dependence of the value of this splitting on the magnitude of the main applied d.c. magnetic field strength. Thus the higher the radiofrequency, and the larger the value of the main magnetic field, the larger will be the splitting between the three components lines. This is expected for any effect which depends on diamagnetism, since this itself is linearly dependent on the value of the applied magnetic field.

It now follows that such a shift, or splitting, can be used to characterize the particular group in which the proton resides. This concept of a chemical shift can be put on a quantitative basis by defining a parameter, δ, in terms of the following expression

$$\delta = \frac{H_{obs} - H_{ref}}{H_{ref}} \cdot 10^6 \qquad \ldots (10.2)$$

H_{obs} refers to the actual observed value of the magnetic field at which the particular resonance takes place, while H_{ref} refers to the magnetic field strength at which the signal from a standard resonance group occurs at the same applied radiofrequency. As long as the same substance is always taken as a reference, it is immaterial which group is actually employed to give the reference proton. Normally the protons in water are taken as the reference although sometimes the resonance position for the protons in benzene is taken as a sub-reference when organic solutions are being studied. The factor of 10^6 is added in the definition since the shifts which are obtained are extremely small, and a multiplying factor of this type then produces reasonable values for the actual shifts themselves.

A partial list of the different shifts that are obtained for protons in different chemical groups is given in *Table 10.1*. And it can be seen that these are of both positive and negative sign which indicates that the diamagnetic shielding for some groups is greater than that for the protons in water molecules, whereas for others it is less.

Table 10.1. Chemical Shifts for Protons in Different Chemical Groups

Group	δ	Group	δ
CH_3—C	+3·8	HC≡	+2·3
NH_2 (alkylamine)	+3·7	C—CH_2—X	+2·0
		CH_3—O—	+1·5
CH_2 (cyclic)	+3·5	H_2O	0
		CH_2=	−0·4
		OH	−0·5
CH_3—C=	+3·4	C=CH—C	−0·6
CH_3—	+3·3		−2·0
CH_3—$\overset{|}{C}$=	+3·2	NH_2 (amide)	−2·8
		X—CHO	−3·1
CH_3—N	+2·5	C—CHO	−4·7
		COOH	−6·5

It is evident that an accurate measurement of the actual magnetic field value at which the proton resonance occurs can be used as a means of identifying the chemical group in which the proton exists, and this can be an extremely powerful tool in any organic analysis. It should also be pointed out that exactly the same type of chemical shift also occurs for other nuclei than the protons, and qualitative and quantitative analysis of the different groups in which nitrogen or fluorine atoms exist can be made in the same way. One of the more recent developments in this field has been the detection and utilization of the chemical shift of C^{13} as it occurs in its natural abundance in organic compounds (see Section 12.5(a)). This analysis by chemical shift measurements is considerably assisted by the complementary analysis of another type of splitting, known as spin-spin coupling, which is now considered.

10.4 Spin-Spin Coupling

Soon after the chemical shift itself had been discovered and explained, it was noticed that, if higher resolution was employed, further splittings occurred in the spectrum which could not be attributed to different chemical environments. The particular case of ethyl alcohol can be taken as an example again, and the spectrum obtained under higher resolution is shown in *Figure 76(b)*. It can be seen that the CH_3 resonance line has now been split into three different components, with the centre component twice as intense as the two outer ones, while the CH_2 line has been split into four components with intensity ratios $1:3:3:1$. This further splitting can be explained in terms of the coupling between the spins of the protons in the two different groups of neighbouring sites.

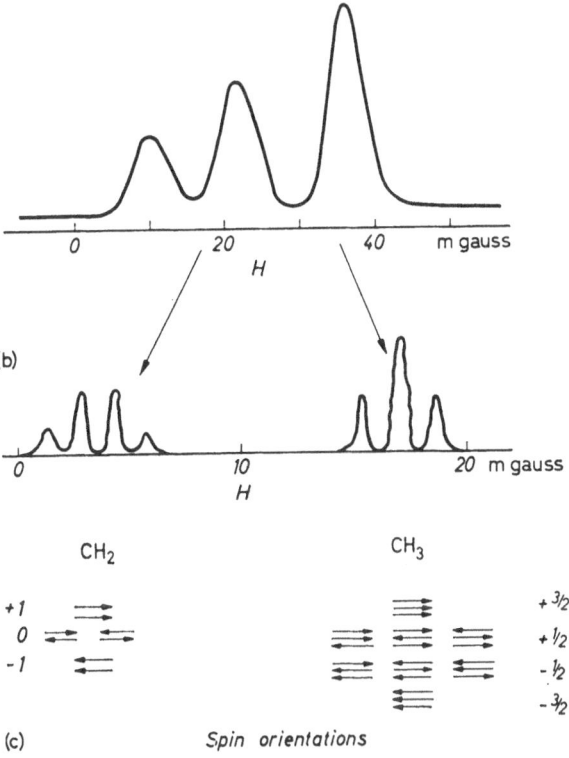

Figure 76. Spectra of ethyl alcohol : (a) under medium resolution, (b) under higher resolution, (c) spin orientations of CH_2 and CH_3 groups of protons, showing combined effects

Thus, all the three protons in the CH_3 group are not only affected by the external magnetic field and the diamagnetic shielding characteristic of the methyl group, but they are also quite close to the protons of the CH_2 group and these themselves act as small magnetic dipoles to produce a small additional field at the site of the protons of the CH_3 group.

There are various possible ways in which the two protons of the CH_2 group can be arranged. Their nuclear spins and magnetic

315

moments may both be pointing in the direction of the applied magnetic field, when they will produce the largest incremental effect in the direction of the field. It is possible, however, for both of these spins to be reversed as indicated in *Figure 76 (c)*, and in this case the same magnitude of incremental field would be produced at the methyl protons but in the opposite direction. It is also possible for the two spins and magnetic moments of the CH_2 protons to be pointing in opposite directions and thus cancel out, and produce no incremental effects at all. Moreover, it can be seen that there are two ways in which this null result can be achieved, whereas there is only one way in which either of the two other effects are produced. It therefore follows that if a statistical average is taken over all the ethyl alcohol molecules, there will be twice as many with their CH_2 protons cancelling than with their proton spins both lined up together, either in the direction of the applied magnetic field, or against it. This therefore explains why the additional splitting exists in the CH_3 proton resonance line, and also why the centre line, which corresponds to the unshifted resonance, is twice as intense as the two other lines.

This argument can now be applied in reverse to consider the effect of the CH_3 protons on the resonance produced by the protons of the CH_2 group. In this case there are four different ways in which the three proton spins and magnetic moment of the CH_3 group can be orientated. They can all be lined up in the direction of the field to produce the maximum shift in that direction, or alternatively all be lined up against the field. The two other possibilities correspond to two of the three protons cancelling, leaving one unpaired spin and magnetic moment. This single unpaired moment can. of course, be either with or against the direction of the field. Moreover as can be seen from *Figure 76(c)*, there are three different ways in which each of these two effects can be produced, explaining why the fourfold splitting of the CH_2 resonance line has an intensity distribution of $1 : 3 : 3 : 1$. There should also, of course, be a splitting of the OH proton resonance line due to the neighbouring CH_2 protons, which should again have an intensity ratio of $1 : 2 : 1$ although the splitting itself will not be so great as for the methyl protons owing to the further distance of the hydroxyl proton. It is shown in Section 10.6 that rapid exchange effects can sometimes average these splittings to zero.

It will be realized that the magnitude of this splitting is determined entirely by the values of the proton magnetic moments and the distances between the various chemical groups; in other words it is determined entirely by the molecular structure and not by the magnitude of any applied field or resonance frequency. It therefore follows that the spin-spin coupling produces splittings which are field-independent, and this fact can be used to differentiate them from chemical shifts in complex spectra. Thus if the high resolution

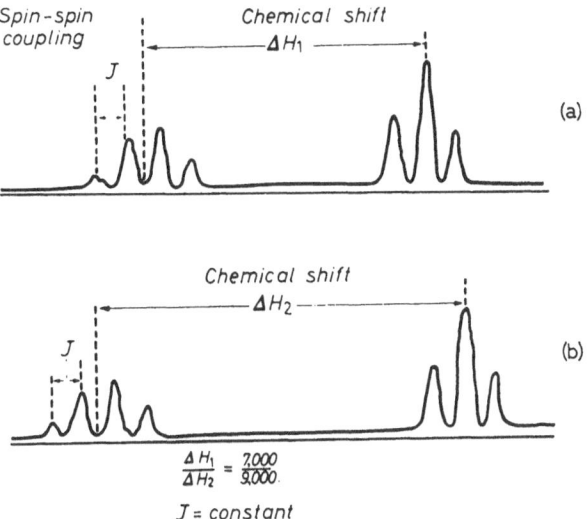

$$\frac{\Delta H_1}{\Delta H_2} = \frac{7,000}{9,000}$$

J = constant

Figure 77. Resonance from ethyl alcohol : (a) at 7,000 gauss, (b) at 9,000 gauss

spectrum of the ethyl alcohol is taken at two different radio-frequencies the spectra are as shown in Figures 77(a) and (b). It can be seen that the chemical shift increases directly with the magnitude of the applied magnetic field, whereas the splitting due to the spin-spin coupling remains the same at both frequencies. It is for this reason that it is sometimes advisable to study complex spectra at two widely different frequencies, so that the splitting due to spin-spin coupling can be quickly differentiated from those due to chemical shift. It should be pointed out here that in the liquid state the spin-spin splittings which are obtained are due to the isotropic part of the interaction, the angular variation having been averaged out as explained in Section 10.2.

This effect of spin-spin coupling is not only of inherent interest for its own sake, but it also provides a very powerful additional tool to assist in qualitative analysis of organic compounds. Thus it is now possible in principle not only to determine the nature of the chemical group in which the proton resides from the magnitude of its chemical shift, but also to tell which chemical groups are located close to it

317

from the splitting produced in its spectrum by the spin-spin coupling interaction. It is clear from the diagrams that the magnitudes of these splittings are, in fact, extremely small. For example the spin-spin coupling produces splittings of a few milligauss and hence it follows that magnetic fields must be employed which are homogeneous to at least this value.

10.5 APPLICATION TO ORGANIC ANALYSIS

The effects of chemical shift and spin-spin splitting have been used to a very great extent in recent years in a whole variety of structure determinations, and high resolution nuclear resonance can claim to be one of the most powerful analytical tools now available to the organic chemist.

The detailed way in which different NMR spectra can be analysed is beyond the scope of this book and those interested in following the subject further should consult a standard reference work such as *High Resolution Nuclear Magnetic Resonance* by POPLE, SCHNEIDER and BERNSTEIN[16]. The different types of spectra to be expected for different numbers of equally coupled nuclei, or unequally coupled nuclei are considered in great detail here, together with a general analysis of the theory of chemical shifts and spin-spin couplings.

A list of typical chemical shifts for protons in different chemical groups has been given in *Table 10.1* and for comparison a list of

Table 10.2. Spin-spin Coupling Constants between Proton Groups

Configuration	J c/s
H_2	280
CH_4	12·4 (obtained by deuteration)
C_2H_2	9·1 (obtained by deuteration)
CH_3CH_2X	7
	6
	3·5
	10

typical spin-spin coupling interactions between protons is given in *Table 10.2*. Since the spin-spin interaction is field independent there is no need to eliminate the effect of the field in the interaction constant in this case, and the values of this constant, J, are given as the splittings observed in the actual spectra.

It should be stressed, however, that this method of analysis is not confined to protons, but can be applied equally well to other nuclei possessing nuclear spins and magnetic moments. A large number of structural determinations have been made on molecules containing nitrogen and fluorine in this way, and a list of typical chemical shifts observed for the N^{14} nuclei is given in *Table 10.3*.

Table 10.3. Chemical Shifts of N^{14} relative to NO_3^-

Compound	δ p.p.m.	Compound	δ p.p.m.
NH_4^+	346	CN^-	112
N_2H_4	312	N_2	14
NH_3	290	NO_3^-	0
		$C_6H_5NO_2$	−2
SCN^-	151	$C_3H_7NO_2$	−26
CH_3CN	131	NO_2^-	−254

Measurements on C^{13} have also now become a practical possibility. This isotope only occurs in a natural abundance of 1 per cent, and most of the initial determinations were made on compounds with an artificially enriched composition of C^{13}. The advent of the double resonance technique, described in Section 12.5(a) has allowed an enhancement of such signals by a factor of about 500, however, and measurements on C^{13} in its natural abundance are

Table 10.4. Chemical Shifts and Spin-spin Coupling Constants of C^{13}

Compound	δ p.p.m.	J_{CH} c/s
CH_3I	150	151
CH_3Br	124	153
$CHBr_3$	115	208
CH_2Br_2	110	185
C_6H_{12}	100	140
CH_3OH	80	144
CH_2Cl_2	75	162
CH_3NO_2	58	137
$CHCl_3$	50	193
C_6H_6	0	159
$HCOOH$	−38	218
CH_3CHO	−72	174

now quite feasible. A list of some typical chemical shifts observed for C^{13}, using benzene as a reference, are given in *Table 10.4*, together with the spin-spin coupling constants for the $C^{13}H$ interaction in the same compound.

One or two specific examples of cases where nuclear resonance has played a crucial part in the determination of chemical structure may help to illustrate the methods of analysis that have been employed.

One very straightforward case that can be quoted is that of diketene. X-ray measurements by KATZ and LIPSCOMB[17] had shown that in the solid state diketene had the structural formula

$$
\begin{array}{c}
\text{H}_2\text{C}=\text{C}-\text{CH}_2 \\
| \quad\quad | \\
\text{O}-\text{C}=\text{O}
\end{array}
\qquad \ldots\text{(I)}
$$

From broadline nuclear resonance studies, FORD and RICHARDS[18] had confirmed this structural formula for the solid by analysing the dipole-dipole contributions to the line shape that would be expected from the two CH_2 proton groups. Vibrational infra-red spectra suggested, however, that in solution the diketene might exist as an equilibrium mixture between structure (I) above, and two other forms

$$
\begin{array}{c}
\text{H}_3\text{C}-\text{C}=\text{CH} \\
| \quad\quad | \\
\text{O}-\text{C}=\text{O}
\end{array}
\qquad \ldots\text{(II)}
$$

$$
\begin{array}{c}
\text{O}=\text{C}-\text{CH}_2 \\
| \quad\quad | \\
\text{H}_2\text{C}-\text{C}=\text{O}
\end{array}
\qquad \ldots\text{(III)}
$$

BADEN et al.[19] then carried out high resolution nuclear resonance studies on diketene to determine its form in the liquid state. It is evident, by simply considering the equivalent protons, that structure (I) should give two absorption lines of equal intensity and small separation, one CH_2 group being connected to two carbons while the other has a double bond to another single carbon. Structure (II), on the other hand, should give two well-separated absorption lines with an intensity ratio of 3 : 1, while structure (III) should give a single absorption line, all the protons being in equivalent chemical environments.

The actual spectra obtained are shown in *Figure 78*, under medium resolution which only shows the chemical shifts, and under high resolution where the spin-spin coupling is also observed. It is evident that two slightly separated absorption lines are obtained and, although they do not have quite the same width, or height, their integrated intensities are in fact equal. This suggests that all the diketene is still in form (I), even in the liquid state.

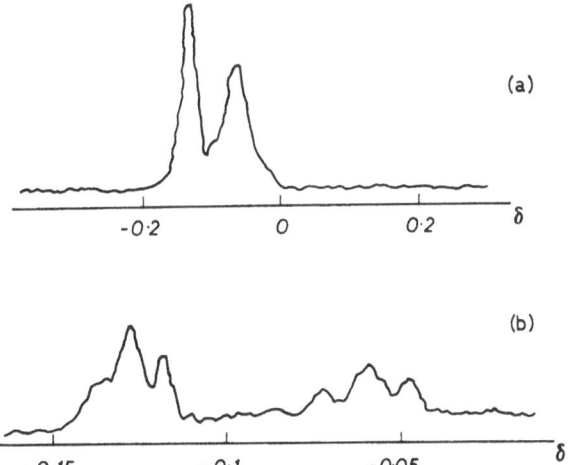

Figure 78. Spectra observed from diketene: (a) under medium resolution, (b) under high resolution (after Gutowsky, Williams and Yantwich[19])

The spin-spin splitting observed under higher resolution, as in *Figure 78(b)*, shows that each absorption is split into a triplet, as is to be expected if each has a CH_2 group as a neighbour. There is, no trace of the other forms of spin-spin splitting that would be expected from structures (II) and (III).

These high resolution nuclear resonance determinations thus give quite unequivocal evidence on the structure of the diketene in the liquid state, and rule out the suggestions of other structures that had come from the interpretation of the vibrational spectra.

An example of molecular structure determination from measurements on nuclei other than protons is afforded by the study of the P^{31} resonance in the monothiopyrophosphate ester, which can have two possible structural formulae — a symmetrical form

$$
\begin{array}{ccc}
\text{Et.O} & \text{O} \qquad \text{O} & \text{O.Et} \\
\diagdown \; \| & \quad \| \; \diagup & \\
\text{P} & \!\!-\text{S}-\!\! & \text{P} \\
\diagup & & \diagdown \\
\text{Et.O} & & \text{O.Et}
\end{array}
$$

or an unsymmetrical form

$$\begin{array}{ccc}
\text{Et.O} & \text{O} & \text{S} \quad \text{O.Et} \\
& \diagdown \, \| & \| \diagup \\
& \text{P--O--P} & \\
& \diagup & \diagdown \\
\text{Et.O} & & \text{O.Et}
\end{array}$$

Chemical evidence was not able to distinguish unambiguously between them, but it is clear that two separated P^{31} resonances should be obtained from the unsymmetrical structure, since the two phosphorus atoms are in slightly different chemical environments. On the other hand the symmetrical structure would only give a single P^{31} resonance from the two identical phosphorus atoms. Two P^{31} resonance lines were in fact observed experimentally[20], thus proving that the molecule had the unsymmetrical structure.

Another simple example of the use of phosphorus resonances in the elucidation of molecular structure is the measurement of the P—H spin-spin coupling parameters. The large values of these found in compounds of the general formula

$$\begin{array}{ccc}
\text{R} & & \text{O} \\
\diagdown & & \diagup\!\!\diagup \\
& \text{P} & \\
\diagup & & \diagdown \\
\text{R} & & \text{H}
\end{array}$$

shows that they must in fact have this structure, instead of that normally associated with trivalent phosphorus, i.e.

$$\begin{array}{c}
\text{R} \\
\diagdown \\
\quad\text{P--O--H} \\
\diagup \\
\text{R}
\end{array}$$

These particular examples have been given as typical nuclear magnetic resonance studies on simple molecules. In the majority of cases, however, high resolution nuclear resonance has made its greatest contribution in the elucidation of larger and more complex structures, and one example of such a case may be of interest.

The work of Corey, Burke and Remers[21], in which they employed nuclear magnetic resonance to determine whether a cyclo-heptatriene ring or a cyclopropane ring became fused to a cyclo-hexene ring, affords such an example. In order to check on this point a number of enol esters of eucarvone were prepared by acylation of sodioeucarvone, and their nuclear magnetic resonance spectra were studied. A typical spectrum obtained from the enol acetate is shown

322

in *Figure 79(a)*. It is seen that the ethylenic hydrogen absorption at *P* is well separated from the hydrogen absorption due to the $H_3C—C=X$ group at *Q* and the dimethyl hydrogen absorption at *R*. In particular there is no absorption in the place expected for the tertiary bridge hydrogens of a caradiene structure. In contrast to this the high resolution spectrum obtained from methyl caren

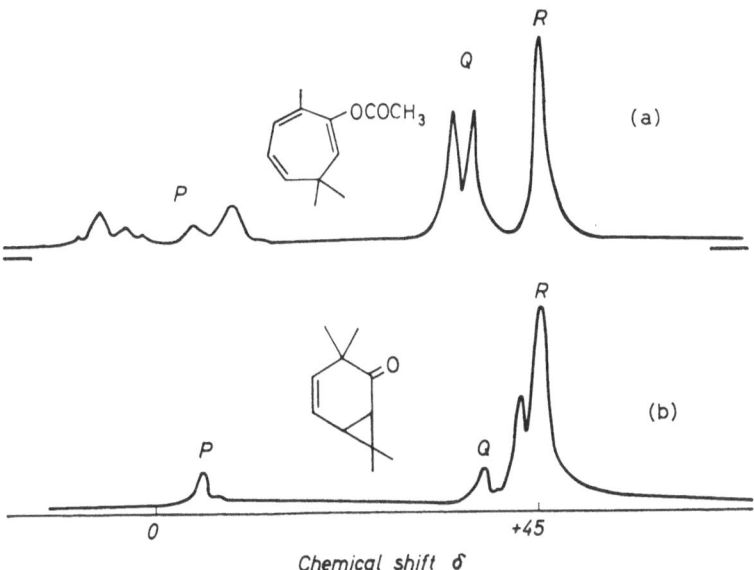

Figure 79. Spectra obtained from enol acetate derivatives: (a) enol acetate, (b) methyl caren
(after Corey, Burke and Remers[21])

formed by methylation of sodioeucarvone is given in *Figure 79(b)*. This shows the absorption due to the two tertiary bridge hydrogens quite clearly at *Q*, well resolved from the methyl hydrogen absorption at *R*, and of equal intensity with that of the ethylenic hydrogen at *P*.

This example has been given to show how nuclear resonance can often unambiguously distinguish between structures which are rather hard to differentiate on purely chemical grounds. Similar studies have also been made to check on the linkage of cyclopropane rings to sterculic acid[22] and to hypoglycin[23]. In the above examples most of the evidence has come from accurate measurements of the chemical shift, but sometimes the crucial factor is supplied by the spin-spin coupling interaction. Such an example is afforded by the work of VAN TAMELEN *et al.*[24], on photosantonic acid. In this

323

case the absorption lines due to the vinyl hydrogens are cleary split into three principal components, which identifies the grouping as

$$-CH_2-CH=C\overset{\diagup}{\diagdown}$$, and allows a definite assignment to be made for

the whole of the structure.

Many other examples are now to be found in the chemical literature of qualitative, quantitative and structural analysis by high resolution nuclear resonance, and the few given above may suffice to give some idea of the scope of the method.

10.6 EXCHANGE EFFECTS

It is shown in Section 10.2 that the rapid motion of solvent molecules averages out any anisotropic interaction and hence no splitting or line broadening occurs. This feature of averaging is a very common phenomenon in all resonance work. It may be stated in general that, if any internal field fluctuates rapidly compared with the Larmor frequency of the nucleus itself in the applied magnetic field, then the nucleus will only respond to the time average of the field set up by these fluctuating internal interactions; the resonance will not produce a broad line corresponding to all the different possible values of all the fluctuating parameters. This averaging process will, therefore, take place for all types of interactions which fulfil this condition, such as the interaction with conduction electrons in a metal, as well as for the rapid fluctuations produced by the motion of solvent molecules.

Another effect in which such an averaging occurs is when rapid exchange of protons or other nuclei takes place between different molecules. Thus a particular case may be taken of the resonance observed from a proton in molecule A, which would be observed at a magnetic field strength H_a, and that observed from a proton in molecule B, which would give a resonance at field strength H_b. As long as these two protons remain firmly attached to the two different molecules, both of these resonance lines will be observed and their intensities will, of course, correspond to the number of protons in the two molecules concerned. If, however, it is possible for the protons to exchange between these two molecules, and if this exchange takes place more rapidly than the frequency corresponding to the difference in energy between the two energy lines, then the field at which resonance is obtained will be at some value between H_a and H_b, the actual value depending on the proportion of protons of type A present to those of type B.

A simple but not entirely parallel analogy to this exchange effect can be given in the form of a camera taking photographs of rapidly changing events. If a football is being thrown between two players then, as long as the speed of the camera shutter is much faster than the rate at which the ball is being thrown between the two players, each photograph taken will show the ball either with one player or with the other. If, however, the two players are throwing the ball between themselves at a rate which is much faster than that of the camera shutter speed, then only a blurr will occur on the photo- graph, the position of the ball having been averaged in the process.

A good example of this averaging by exchange is afforded by the ethyl alcohol spectra already considered in some detail in Sections 10.3 and 10.4. It was mentioned there that the spin-spin interaction with the CH_2 group should split the OH line into a triplet, but, in fact, this splitting is usually not observed.

The reason for the absence of this splitting is that, unless the alcohol is very pure, the protons of its OH group rapidly exchange

(a)

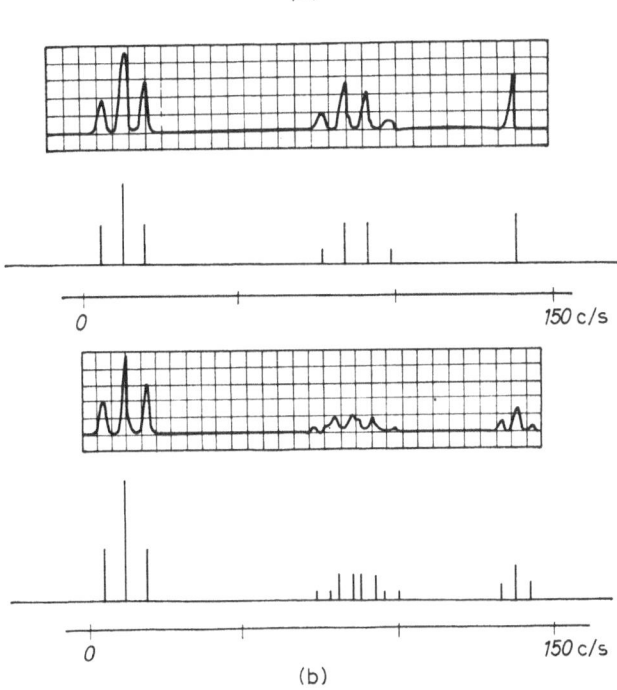

(b)

Figure 80. Arnold's initial results on ethyl alcohol: (a) impure, (b) very pure

325

with the protons of the hydroxyl groups of any water molecules or acids that may be present. If this exchange takes place at a frequency faster than that corresponding to the spin-spin splitting, this interaction is averaged out by the mechanism described above, and only one single absorption line is obtained.

Confirmation that this is the correct explanation of the absence of the triplet splitting in the hydroxyl line of ethyl alcohol was obtained by some elegant early studies of ARNOLD[25], who analysed the spectrum of ethyl alcohol in great detail. The comparison of the spectra he obtained from slightly impure and pure ethyl alcohol is shown in *Figure 80*. The triplet splitting of the hydroxyl peak, as shown in *Figure 80(b)*, is only obtained if the alcohol is extremely pure, and any excess of either OH^- or H^+ ions causes rapid exchange to take place, and averages the splitting to zero. This fact was confirmed quantitatively by adding very small concentrations of HCl to the pure alcohol with the results shown in *Figure 81*. It was estimated that for the smallest contamination shown in this figure ($3\cdot8 \times 10^{-6}$ HCl) the exchange times between the protons were about 1 sec, whereas at 5×10^{-5} contamination the exchange time had fallen to below $0\cdot01$ sec, and had thus averaged the splitting to zero.

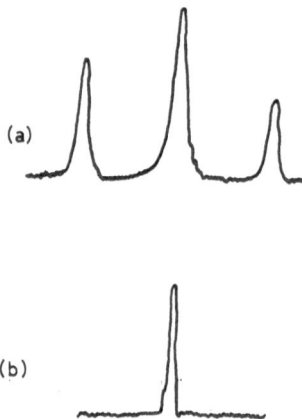

Figure 81. Spectra showing effect of dilute acid on OH exchange: (a) pure ethyl alcohol containing less than 4×10^{-6} N HCl, (b) ethyl alcohol with 5×10^{-5} N HCl added

The effect of this exchange between the protons of the hydroxyl groups can also be seen in another way, when the concentrations of water or other impurity become quite large. If proton resonance spectra are obtained from various mixtures of water and ethyl alcohol the position of the single hydroxyl line is found to shift.

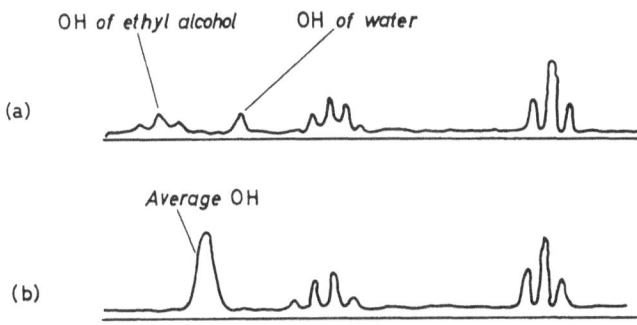

Figure 82. Spectra from ethyl alcohol and water mixtures: (a) 5 per cent water, (b) higher water concentration (after Weinberg and Zimmerman[26])

Figure 82(a) shows the spectrum that is observed from a mixture consisting mainly of ethyl alcohol but with about 5 per cent of water added[26]. It can be seen that the standard spectrum expected from ethyl alcohol is obtained, and also the single line due to the protons in the water molecule; there is a splitting of 32 c/s between these. If, however, the concentration of the water is increased it is found that the two resonances due to the hydroxyl protons merge into a single line, which lies between the position for the hydroxyl proton of the ethyl alcohol and that of the water molecule. The spectrum is then as shown in *Figure 82(b)*, and the resonance due to the OH proton can be clearly seen, its intensity having grown to correspond to the combined signals.

It is therefore evident that for small concentrations of water there are not enough hydroxyl protons of the water molecules present to allow the exchange, between these and those of the ethyl alcohol molecules, to take place at a rate sufficiently rapid to average out the 32 c/s splitting. It should be noted that this splitting is greater than the spin-spin splitting of the alcohol hydroxyl group itself, and hence higher exchange rates are required to average it to zero. On the other hand, once the concentration of the water approaches closer to that of the ethyl alcohol a more rapid exchange of the protons between the two different sites takes place and an averaging of the 32 c/s splitting is thus obtained. This change from the two spectra to an averaged single line occurs at a water concentration of 25 per cent. It can therefore be deduced that in this region of water concentration, the exchange of protons between the two molecules is taking place in a time of about 0·015 sec, since the mean lifetime in a given state is related to the splitting which has been averaged by the formula[27]

$$\tau = \frac{\sqrt{2}}{(\pi . \delta\nu)^{-1}} \qquad \qquad \dots (10.3)$$

where $\delta\nu$ is the splitting that is averaged out.

The above case is a very direct example of how rates of exchange can be measured from the disappearance of splittings in nuclear resonance spectra. It should be noted, however, that averaging by exchange is not the only type of averaging process that can occur. It is shown in Section 10.2 that rotation of nuclei in the solid state will produce an averaging of the dipolar broadening, and a similar kind of averaging will also take place in the liquid state if rotation sets in within a molecule.

A good example of this is the nuclear resonance spectra obtained from dimethyl formamide[28]. This molecule has the structural formula

The spectra that are obtained at different temperatures are shown in *Figure 83*.

The smaller line in the lower field is due to the single proton in

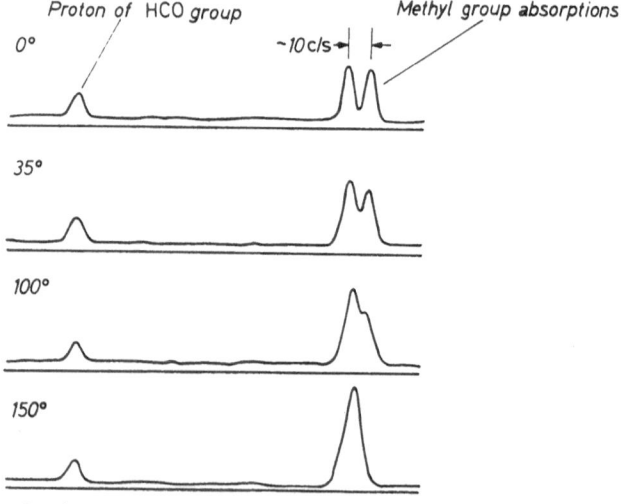

Figure 83. Spectra from dimethyl formamide at different temperatures (after Phillips[28])

the HCO group, and the two peaks, clearly resolved at the lower temperatures, are due to the two methyl groups. These are in slightly different environments, since the presence of the H and O atoms respectively will slightly alter the wave functions on the two sides of the molecule. It is seen, however, that at the higher temperatures the splitting of these two lines decreases and they merge into a single peak at 150° C. The reason for this is that the two groups of the molecule start rotating about their CN bond as this temperature is approached, and the differences in the environment of the two CH_3 groups are then averaged to zero.

It can be seen that in this case the averaging process and disappearance of spectral splitting has been used to determine the temperature for onset of molecular rotation, rather than the exchange of protons between molecules in solution. There are many different motional processes that can be studied in this way, including diffusion and hindered rotation in the solid state as well as proton exchange, nitrogen inversion[29] and similar interactions in the liquid state.

10.7 DOUBLE NUCLEAR RESONANCE

A technique was introduced by BLOCH[30] in 1954, which has been of very considerable assistance in analysing complex nuclear magnetic resonance spectra. This consists of a double irradiation experiment in which two different radiofrequencies are applied to the specimen at the same time. One of these frequencies is the same as the resonance frequency of one of the nuclei in the specimen, whilst the other is the frequency appropriate to the resonance of another nucleus in the molecule in the same magnetic field strength.

It is shown in the previous sections that the splitting due to the spin-spin coupling of neighbouring nuclei is only observed in the spectrum if the orientation of the nuclear moments remains fixed, or else changes at a rate which is small compared with the splitting, measured as a frequency. If, however, the nuclear moments are changing position rapidly compared with the frequency of the splitting, then the spin-spin interaction is averaged to zero and splitting disappears. The disappearance of such splittings can often be seen when compounds are heated and rotation or exchange occurs, but it can also be produced artificially. Thus it is possible to induce such a rapidly changing orientation by irradiating the sample with the frequency appropriate to the resonance of this second nucleus, and so cause a rapid change of the nuclei between the energy states in question.

The actual application of this technique of double irradiation can probably be best illustrated by two specific examples. One of the most striking changes in spectrum on double irradiation is obtained from diborane which has the structural formula

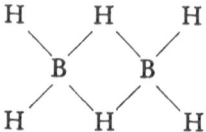

This structure itself was confirmed by the early nuclear resonance work of KELLY, RAY and OGG[31] who showed that the signal due to the bridge protons occurred at slightly higher fields than that due to the terminal protons. The spectrum itself is very complicated, however, since boron has two isotopes present in relatively high abundance, B^{11} with a nuclear spin of $\frac{3}{2}$ and B^{10} with a nuclear spin of 3. These couple to the protons and the four possible orientations of the first together with the seven possible orientations of the second combine to give a complicated set of overlapping and unresolved lines as in *Figure 84(a)*. This shows the normal proton resonance obtained from diborane at 30 Mc/s.

(a)

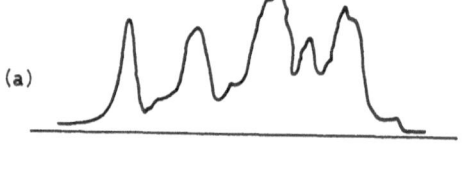

(b)

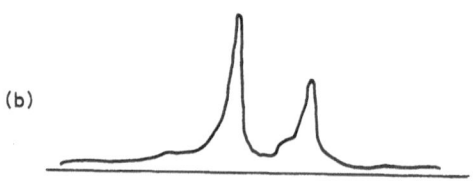

Figure 84. Spectra from diborane: (a) normal, (b) double irradiated at 9·626 Mc/s (after Shoolery[50])

The complexity due to the boron coupling can be removed by irradiating the sample at the same time with a frequency of 9·6257 Mc/s which is the resonance frequency of the B^{11} nuclei in the same magnetic field. When this is done the spectrum changes to that shown in *Figure 84(b)*, from which it can be seen that nearly all the boron splittings are removed leaving the two resonances due to the bridge and terminal protons. The latter give rise to the line in the lower field, as indicated by the intensity which is twice that of the signal due to the two bridge protons. The structure that still remains on these two lines is caused by the much more weakly coupled B^{10} nuclei which are not being resonance irradiated.

The drastic simplification of the spectrum in this case is a very good example of how such double irradiation techniques can radically assist in the analysis of complex spectra. It should also be noted that the measurement of the second irradiation frequency itself will give a very accurate value for the magnitude of the coupling between the B^{11} nuclei and the protons.

It should be stressed that the double irradiation technique is not limited to molecules containing different nuclei species, but can also be applied to similar nuclei in different chemical environments. It has been employed in this way for a large number of cases in which the proton resonance of one group is studied carefully at one frequency, while the compound is also irradiated with a slightly different frequency corresponding to the proton resonance frequency of another group within the same molecule.

A good example of this is afforded by some of the early work of ANDERSON[32] on 2,3-dibromopropene. This has the structural formula

$$\begin{array}{ccc} H & & H \\ | & & | \\ H-C-C=C \\ | & | & | \\ Br & Br & H \end{array}$$

and the normal proton resonance spectrum, obtained at 30 Mc/s is shown in *Figure 85(a)*. The sharp doublet on the right is due to the CH_2Br protons while the two other signals correspond to the two other protons attached to the carbon with the double bond. Because of the asymmetry in the molecule these two protons are not only in different diamagnetic screening fields, and thus have different chemical shifts, but also have a complicated structure due to the —CH_2Br coupling. These two lines are shown on an expanded scale in *Figure 85(b)* and it is seen that both lines have several components which are difficult to analyse unambiguously.

If, however, the sample is also irradiated with a frequency corresponding to the resonance of the CH_2Br proton most of this complicated pattern is eliminated, as shown in *Figure 85(c)*. The two doublet splittings which remain on both lines can now be

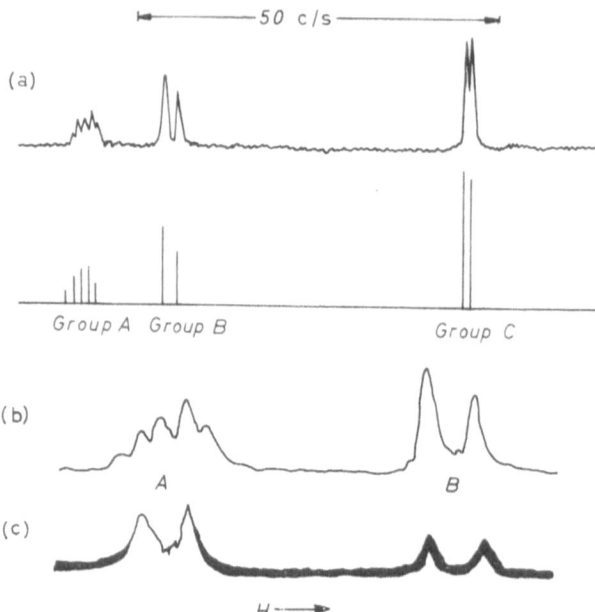

Figure 85. Spectra from dibromopropene: (a) normal, high resolution, (b) normal, expanded scale, (c) irradiated, expanded scale (after Anderson[32])

readily explained in terms of the coupling to the adjacent proton, while the double irradiation frequency has removed the coupling with the CH_2Br protons from this part of the spectrum.

It is evident that this method is only applicable if the chemical shift is large compared with the spin coupling constants, and a full theory of the effect has been given by BLOOM and SHOOLERY. [33] They showed that if a strong radiofrequency field is applied at the resonance frequency of the second nucleus, then the resonance of the first nucleus will occur at a field given by $H_0 + \Delta H$ where $\Delta H = 0$, or

$$(\Delta H)^2 = \frac{\pi^2 J^2(\gamma_1^2 - \gamma_2^2) + \gamma_1^2\gamma_2^2 H_2^2}{\gamma_1^2(\gamma_1^2 - \gamma_2^2)} \qquad \dots (10.4)$$

where γ_1 and γ_2 are the gyromagnetic ratios of the two nuclei. In other words, this theory predicts that there will be a central line corresponding to $\Delta H = 0$ and two other components. The intensity of these outer components will depend on the gyromagnetic ratios of the two nuclei, and if γ_1 is greater than γ_2 the two outer components will tend to spread out and become weaker as the strength of the applied radiofrequency field is increased, whereas if γ_1 is less than

γ_z they will tend to collapse and only the single central line will be observed. It therefore follows that, in general, a narrow single line is to be expected and this is, in fact, observed in practice.

It is probably fair to say that the main advantage of this technique is to remove splittings rather than to obtain additional information from them; but by so removing the splittings, and producing a more simple spectra, analysis is usually made quite unambiguous and can take place much more rapidly.

10.8 Nuclear Resonance in Metals

It is pointed out in Section 8.8(b) that nuclear resonance absorption lines obtained from metals are similar to those observed in liquids, in that they have a narrow width. This fact was taken as an indication that the spin lattice relaxation time is mainly determined by the conduction electrons, since these set up fluctuations in the field at the nucleus, and average out anisotropic contributions in a very similar way as occurs in the liquid state. In order to make some estimate of such correlation times for these fluctuations at the nucleus it is necessary to employ Fermi statistics, since the conduction electrons will obey this form of statistics rather than normal Maxwell–Boltzmann statistics. Hence only those electrons which are near the Fermi surface can contribute to the magnetic effects, since the others well below this energy will have their spins paired.

It is therefore possible to make an estimate of the time taken by a conduction electron when jumping from one atom to another in terms of the velocity at the Fermi surface and the length of the unit cell. The correlation time deduced from such an estimate is of the order of 10^{-16} sec and it is thus very much shorter than the inverse of the nuclear resonance frequency which is of the order of 10^{-7} sec. It follows that a statistical averaging will certainly be produced by the motion of the conduction electrons. In order to determine the net average effect of these electrons it is necessary to consider the density of the conduction band states, and the numbers of electrons which have their spins respectively parallel and antiparallel to the applied field. The Fermi–Dirac statistics are first applied to determine the ratio of these numbers and thus the net magnetization which is produced; from this Knight[34] was able to show that there will be a net shift of the resonance line given by the equation

$$H_N = H_0(1 + K) \qquad \ldots (10.5)$$

where K is the Knight shift and can itself be written as

333

$$K = \frac{8\pi}{3} \cdot \chi_p \cdot P_F \qquad \dots (10.6)$$

where χ_p is the susceptibility/atom. The quantity χ_p is, in fact, given in the free electron model by the Pauli expression

$$\chi_p = \frac{3\beta^2 p}{2E_F} \qquad \dots (10.7)$$

where E_F is the value of the Fermi energy level, and p is the number of valence electrons/atom. P_F is the coupling constant relating the wave function of the electron with Fermi energy, E_F, to the nucleus in question.

It follows that from a measurement of this shift in the resonance frequency, it is possible to obtain accurate quantitative estimates of the coupling between the conduction electrons and the nucleus, and this can often give very useful information on the energy band structure of the conduction electrons themselves. Since the relaxation mechanism in metals is also linked with the conduction electrons one would expect there to be a very definite relationship between the Knight shift of the frequency and the relaxation time, as deduced from the line width measurements themselves. Such relationships were deduced very early in these studies by KORRINGA[35] who showed that for a given nucleus a relationship should exist of the form

$$T_1 \left(\frac{\Delta H_0}{H_0} \right)^2 = \frac{\text{constant}}{T} \qquad \dots (10.8)$$

This general relationship has been found to be in fairly good agreement with experiment.

It should be noted here that there is also an effect on the shape of the resonance signal, observed from metallic specimens, which is due to the skin depth properties of the specimen rather than any atomic mechanism itself. If the specimen being investigated is very small compared with the skin depth, then the radiofrequency field penetrates the whole specimen with practically no attenuation, and hence the resonance conditions are more or less identical with those normally expected. If, however, the dimensions of the specimen are not small compared with the skin depth then the value of the field changes both in magnitude and phase as it penetrates the sample. Moreover, the change in magnetic susceptibility at resonance is itself accompanied by a change in skin depth, and as a result of these effects the total absorption observed is effectively a mixture of both absorption and dispersion curves. As a result the observed resonance line becomes shifted and distorted in shape. One other effect which will also influence the line width observed from metals is the self-diffusion of the metallic atoms through the lattice. In 1951 GUTOWSKI[36] observed that sodium shows a narrowing of its resonance line as the temperature is raised through 200° K and this can be explained by the self-diffusion of the sodium atoms. Further studies of these effects can give useful information on the onset of diffusion within a metallic lattice.

One other point of interest in these studies is that the resonance frequency observed from nuclei in a *liquid* metal is hardly shifted from that observed from the same nuclei in solid metals, and the small extra shift that does occur can be accounted for by the discontinuity in density at the melting point. This is striking confirmation of the fact that the electron energy states would appear to be very similar in the solid and liquid phases for metallic systems.

There is quite a significant change in the resonance frequency when external pressure is applied to the metal system, and a change of over 100 c/s has been observed in frequencies of the Na^{23} resonance as the pressure is increased up to 10,000 atm[37]. These results on the pressure dependence of K can be combined with Bridgemann's measurements on volume compression of the solid, to obtain a volume dependence of the Knight shift K. This can then be used to show how the nuclear conduction electron interaction P_F varies with volume, which is a parameter that can be compared very directly with theoretical prediction of such workers as BROOKS and HAM[38]. Good agreement between the theory and experiment is, in fact, obtained.

10.9 RESONANCE IN FERROMAGNETICS

The above section is concerned with metals which are not ferromagnetic, and where no extra fields exist within the specimen. In the case of ferromagnetic materials, however, very large internal fields also exist, which completely alter the resonance conditions. As a result of the very strong exchange interactions inside the ferromagnetic material, fields of the order of 500,000 gauss act on the nuclear moment, in addition to any external field that may be applied. It has been possible in recent years to obtain precise estimates of the value of these fields from the Mössbauer effect. Once known, the values can be used to calculate the frequencies that are required to observe the nuclear resonance signal in the internal field.

Thus HANNA[39] showed from Mössbauer effect measurements, that the splitting of the ground state of Fe^{57} must be due to the interaction of the magnetic nuclear moment with an internal magnetic field of about 330,000 gauss. Two groups of workers[40, 41] were then able to make a direct observation of the nuclear resonance of the Fe^{57} nucleus in the effective internal field, the actual resonance frequency being at about 45 Mc/s.

The theory of the origin of this field has been considered in detail by MARSHALL[42], who listed five different effects which would contribute to the total field, and showed that the one produced by core polarization of the 3s, 2s and 1s electrons is likely to be the most dominant for iron. This core polarization is in turn caused by the spin aligned d electrons.

These theories have also been checked by BENEDEK and ARMSTRONG[43] using pressure dependence measurements on the variation of the resonance frequency observed for ferromagnetic iron with increasing pressure; it would appear that the nuclear resonance frequency varies linearly with pressure up to very high values of applied pressure.

10.10 NUCLEAR PRECESSION IN LOW FIELD STRENGTHS

All the preceding sections of this chapter concern nuclear resonance signals obtained in the normal way—i.e. by absorption of the radio-frequency radiation from an applied electromagnetic field. It was pointed out in Section 8.7, however, that pulse and echo methods could be employed in nuclear resonance studies and had, in fact, been used very successfully by HAHN[44]. These echo techniques have been followed up in detail by other workers[45] and have proved particularly powerful in the measurement of relaxation times.

One other line of work in which similar type of pulse techniques are employed, is on nuclear precession in low magnetic field strengths. The initial work on this technique, and its basic principles, are given in Section 9.8(c). A large external polarizing field H_0 of about 100 gauss is applied until the nuclear magnetization is aligned along this direction. The polarizing field is then rapidly switched off, and the nuclear magnetization which remains for a time of the order of the spin-lattice relaxation time, then precesses around the small field which is to be measured. The frequency of this precession is determined directly by the strength of the field and can thus be used to measure its magnitude very precisely[46].

In the early experiments one large coil was used to apply the polarizing field, and a second was wound around the bottle containing the water, or other proton-containing liquid, at right angles to the first. This second coil picked up the signal from the precessing

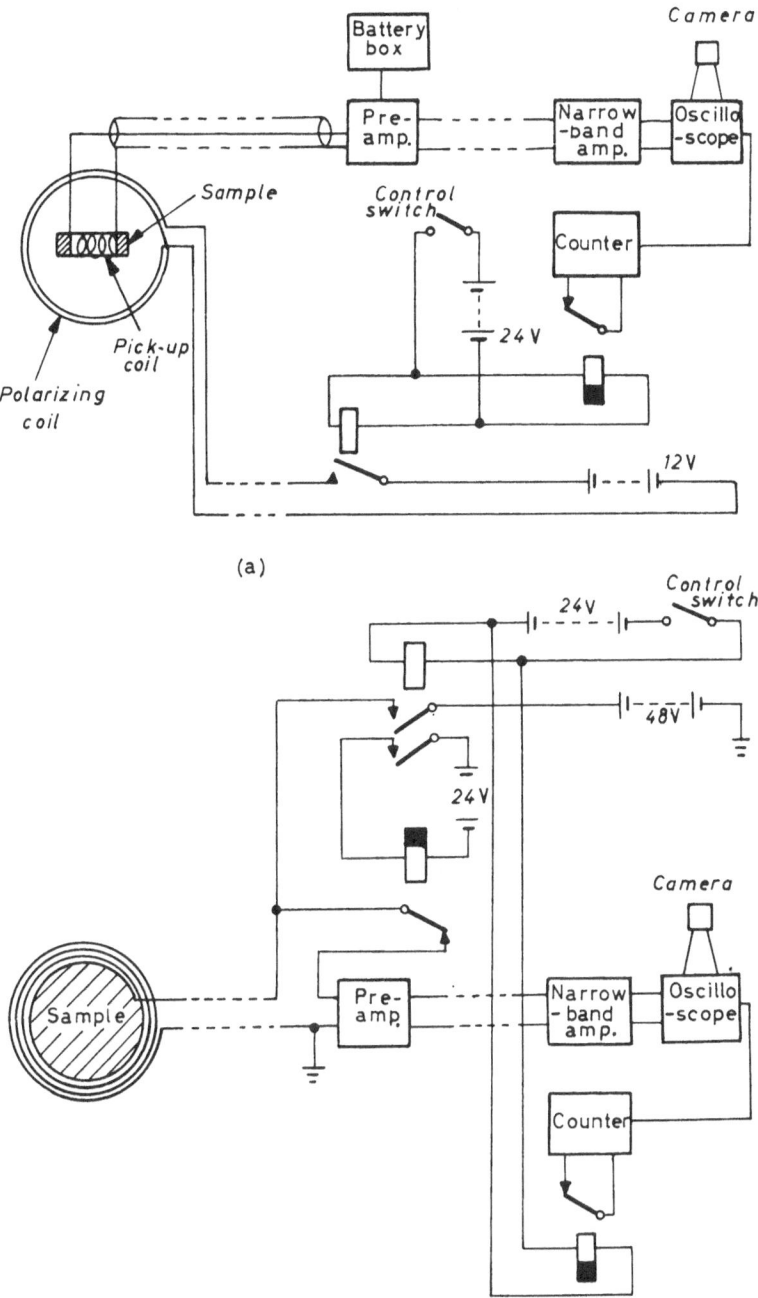

(a)

(b)

Figure 86. Nuclear resonance magnetometer for measuring the earth's field strength:
(a) double coil, (b) single coil

nuclear magnetization, as shown in the schematic diagram of *Figure 86(a)*. It was then found[47], however, that the same coil could be employed both to apply the polarizing field, and to pick up the signal from the precessing nuclei. Moreover small transistor counters could be constructed to determine the frequency of precession very precisely, and a simple self-contained instrument could thus be produced as shown in *Figure 86(b)*.

(a)

(b)

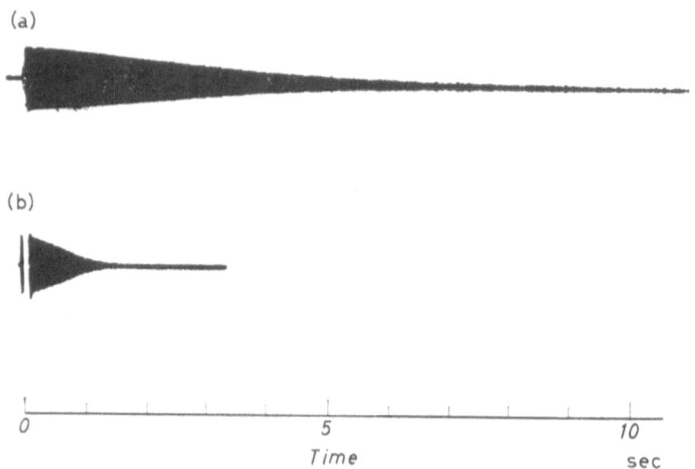

Time   sec

Figure 87. Nuclear precession signals obtained in the earth's magnetic field: (a) in very uniform field, (b) in inhomogeneous field of 40 microgauss/cm

The kind of resonance signals obtained from such a spectrometer are shown in *Figure 87*, (a) being for a specimen placed in the earth's field, with no disturbing objects nearby, while *Figure 87(b)* is for the same specimen with a razor blade placed a few centimetres distance. In the first case the nuclear precession signal lasts for a time given by the spin-lattice relaxation of the protons in the liquid, which can be of the order of 15 sec, as shown. In the second case, the field inhomogeneities cause the protons to lose phase coherence much more rapidly, with a corresponding rapid reduction in signal strength. It is evident that the magnetic field gradient, as well as its absolute magnitude, can be very readily measured by this technique.

The great application of these experiments has been to the accurate determination of the magnitude of the earth's magnetic field, and the variation of this over specific areas. Nuclear resonance magnetometers are now standard instruments in geological air surveys for oil prospecting, and have also been employed to locate buried ruins by archaeologists, and to locate sunken wrecks by salvage experts. In nearly all of these cases it is only the *variation* in the earth's magnetic field strength that is of importance, rather than its absolute magnitude. An extremely simple instrument can be made to detect and measure field variations, and takes the form of two specimen bottles with associated coils, mounted a metre apart on a wooden rod. The signals from these two different groups of protons are then mixed and their beat frequency detected. This can be fed to both a meter and a loudspeaker, and the onset of any significant changes in the earth's field strength can then be noticed immediately. Such an instrument can be built for a few pounds, and yet can measure changes in the earth's field strength of much less than a milligauss.

This method of initiating, and detecting, nuclear free precession signals has also been employed in the more academic studies of molecular structure, and nuclear coupling constant determinations. Beat frequencies can be detected from nuclei in different chemical environments[48], within the same molecule, because of the difference

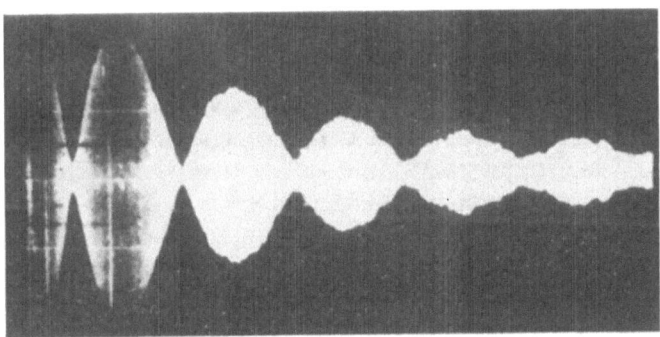

Figure 88. Beat signal from protons in fluorobenzene precessing in the earth's magnetic field

in the screening factors. *Figure 88* shows such a beat signal from protons in fluorobenzene precessing in the earth's magnetic field, the actual beat frequency being 5·8 c/s and the sweep shown being 1 sec. It is evident that this is an extremely simple and inexpensive method of measuring such chemical shifts. An example of recent work along these lines, is that of THOMPSON, BROWN and BLOOM[49] who observed signals from the protons in $HPO(OH)_2$ down to field strengths of 8×10^{-3} gauss. They were able to correlate all their measurements with a very refined theory of the interaction between two isotropically coupled nuclei of spin $I = \frac{1}{2}$.

One of the more interesting features of magnetic resonance studies in general has been the way in which practical applications, and advances in the more academic studies, have moved along side by side, each assisting the other. The particular developments in this field of nuclear precession afford a very good example of such an interaction in a case where the actual equipment is so extremely simple and inexpensive.

REFERENCES

[1] GOLAY, M. J. E. *Rev. sci. Instrum.* 29 (1958) 313

[2] LEANE, J. B., RICHARDS, R. E. and SCHAEFER, T. P. *J. sci. Instrum.* 36 (1959) 230

[3] BLOCH, F. *Phys. Rev.* 94 (1954) 496; ANDERSON, W. A. and ARNOLD, J. T. *ibid.* 94 (1954) 497

[4] ARNOLD, J. T. *ibid.* 102 (1956) 136

[5] PRIMAS, H. and GUNTHARD, H. H. *Rev. sci. Instrum.* 28 (1957) 510

[6] — —*Helv. phys. acta* 30 (1957) 315

[7] BAKER, B. B. and BURD, L. W. *Rev. sci. Instrum.* 28 (1957) 313

[8] BLOOM, A. L. and PACKARD, M. E. *Science* 122 (1959) 738

[9] PAKE, G. E. *J. chem. Phys.* 16 (1948) 327

[10] GUTOWSKY, H. S. and PAKE, G. E. *ibid.* 18 (1950) 162

[11] PAKE, G. E. *ibid.* 16 (1948) 327; *Solid St. Phys.* 2 (1958)

[12] ANDREW, E. R., BRADBURY, A. and EADES, R. G. *Nature* 182 (1958) 1659

[13] — —*ibid.* 185 (1959) 1802

[14] —and NEWING, R. A. *Proc. phys. Soc.* 72 (1958) 959

[15] —and JENKS, G. J. *ibid.* 80 (1962) 663

[16] POPLE, J. A., SCHNEIDER, W. G. and BERNSTEIN, H. J. *High Resolution Nuclear Magnetic Resonance*. New York: McGraw-Hill (1959)

[17] KATZ, L. and LIPSCOMB, W. N. *Acta cryst.* 5 (1952) 313

[18] FORD, P. T. and RICHARDS, R. E. *Disc. Faraday Soc.* 19 (1955) 230

[19] BADEN, A. R., GUTOWSKY, H. S., WILLIAMS, G. A. and YANTWICH, P. E. *J. Amer. chem. Soc.* 78 (1956) 2385

[20] JONES, R. A. Y., KATRITZKY, A. R. and MICHALSKI, *J. chem. Soc.* (1959) 321

[21] COREY, E. J., BURKE, H. J. and REMERS, W. A. *J. Amer. chem. Soc.* 77 (1955) 4941

[22] RINEHART, K. L., NILSSON, W. A. and WHALEY, H. A. *ibid.* 80 (1958) 503

[23] ROPP, R. S. *ibid.* 80 (1958) 1004

[24] VAN TAMELEN, E. E., LEVIN, S. H., BRENNER, G., WOLINSKY, J. and ALDRICH, P. *ibid.* 80 (1958) 501

[25] ARNOLD, J. T. *Phys. Rev.* 102 (1956) 136

[26] WEINBERG, I and ZIMMERMAN, J. R. *J. chem. Phys.* 23 (1955) 748

[27] GUTOWSKY, H. S. and HOLM, C. H. *ibid.* 25 (1957) 1228

[28] PHILLIPS, W. D. *ibid.* 23 (1955) 1363

[29] BOTTINI, A. T. and ROBERTS, J. D. *J. Amer. chem. Soc.* 78 (1956) 5126; 80 (1958) 5203

[30] BLOCH, F. *Phys. Rev.* 93 (1954) 944

[31] KELLY, J., RAY, J. and OGG, R. A. *ibid.* 94 (1954) 767

[32] ANDERSON, W. A. *ibid.* 102 (1956) 151

[33] BLOOM, A. L. and SHOOLERY, J. N. *ibid.* 97 (1955) 1261

[34] KNIGHT, W. D. *ibid.* 76 (1949) 1260

[35] KORRINGA, J. *Physica* 16 (1950) 601

[36] GUTOWSKY, H. S. *Phys. Rev.* 83 (1951) 1073

[37] BENEDEK, G. B. and KUSHIDA, T. *Physics Chem. Solids* 5 (1958) 241

[38] BROOKS, H. and HAM, F. *Phys. Rev.* 112 (1958) 344

[39] HANNA, S. S. *Phys. Rev. Lett.* 4 (1960) 177

[40] CROSSARD, A. C., PORTIS, A. M. and SANDLE, W. J. *Physics Chem. Solids* 17 (1960) 341

[41] ROBERT, C. and WINTER, J. M. *C. R. Acad. Sci., Paris* 250 (1960) 3831

[42] MARSHALL, W. *Phys. Rev.* 110 (1958) 1280

[43] BENEDEK, G. B. and ARMSTRONG, J. *J. appl. Phys. Supp.* 32 (1961) 106

[44] HAHN, E. L. *Phys. Rev.* 77 (1950) 297; 80 (1950) 580

[45] BENE, G. *Archs. Sci., Genève* 10 (1957) 200

[46] PACKARD, M. and VARIAN, R. *Phys. Rev.* 93 (1954) 941

[47] WATERS, G. S. and FRANCIS, P. D. *J. sci. Instrum.* 35 (1958) 88

[48] ELLIOTT, D. J. and SCHUMACHEN, R. T. *J. chem. Phys.* 26 (1957) 1350

[49] THOMPSON, D. D., BROWN, R. J. S. and BLOOM, M. *ibid.* 40 (1964) 3076

[50] SHOOLERY, J. N. *Disc. Faraday Soc.* 19 (1955) 215

11

RECENT ADVANCES IN ELECTRON RESONANCE

11.1 New Experimental Techniques

The basic principles underlying the design and construction of electron resonance spectrometers are given in Chapters 2 and 3. Actual examples of spectrometers are also analysed in some detail in Sections 2.7 and 4.10, including the provision of superheterodyne detection to obtain increased sensitivity.

In these earlier chapters it is made clear that the main limitation on the sensitivity is the excess noise generated by the detecting crystal, and most of the developments in spectrometer design have been aimed at reducing this effect. There have been suggestions in recent years that some of the newer devices, such as the maser or parameteric amplifier, might replace the crystal diode as a microwave detector, but this has not happened to any great extent and all the modern commercial spectrometers of high sensitivity still employ crystal detectors. The excess flicker noise in such crystals varies inversely with the frequency of detection, as is shown in *Figure 17*, and two methods have been successfully employed to overcome this effect. The first is the use of high frequency magnetic field modulation, usually at 100 kc/s, in place of the audiofrequencies employed in simple crystal-video detection. The second is superheterodyne detection, in which the frequency at which the crystal detects is the 'intermediate frequency' difference between the two klystrons, of about 30 Mc/s.

(*a*) *Spectrometers employing 100 kc/s field modulation.* High frequency magnetic field modulation has now been extensively developed and is used in most commercial spectrometers. A block diagram of a typical spectrometer system employing magnetic field modulation at 100 kc/s is shown in *Figure 89*. The 100 kc/s oscillator feeds an amplifier which in turn feeds the coil around the sample. A signal from this 100 kc/s oscillator is also taken as a reference to the phase-sensitive detector which enables a small effective bandwidth to be obtained in the detecting system. As in most modern spectrometers, a microwave bridge system is employed instead of the transmission system of the crystal-video spectrometer described earlier. The microwave bridge, discussed in detail in Section 3.7, works in exactly the same way as an ordinary bridge in lower frequency a.c. applications, and can be balanced by matching the cavity at the end of one arm with a load at the end of the opposite arm. When this balance has been effected all the incident power from the klystron is divided between these two arms and none is passed on to the crystal detector. When absorption occurs in the cavity, however, the bridge becomes unbalanced, and an out-of-balance signal is detected by the crystal.

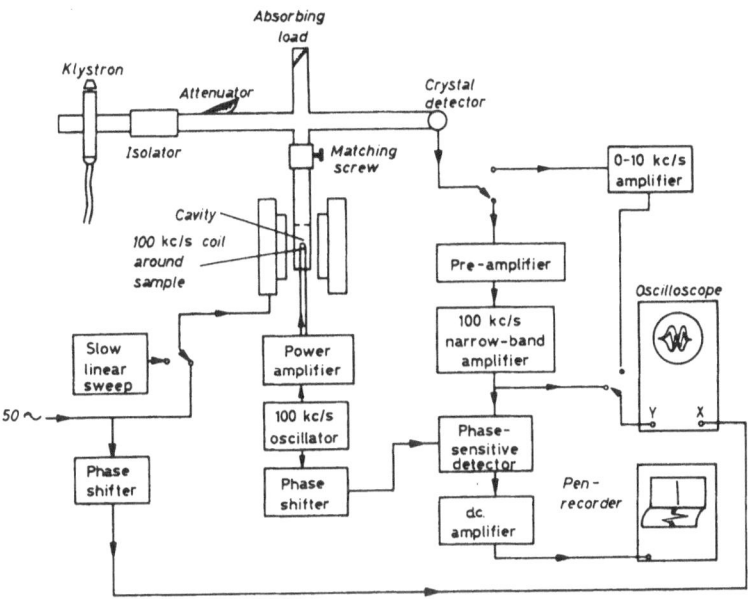

Figure 89. Block diagram of ESR spectrometer employing 100 kc/s magnetic field modulation

This system has two advantages over the simple transmission spectrometer. First there is only one waveguide leading down to the cavity resonator, and hence if work is being carried out at liquid helium temperatures there is much less heat conducted down to the cavity. Second, by varying the amount of unbalance in the bridge, the magnitude of the standing current through the crystal detector in the absence of resonance absorption can be suitably adjusted. This is quite an important point since there is always an optimum value of detecting current for a given crystal rectifier and much greater sensitivity is obtained if adjustment is made to secure this optimum.

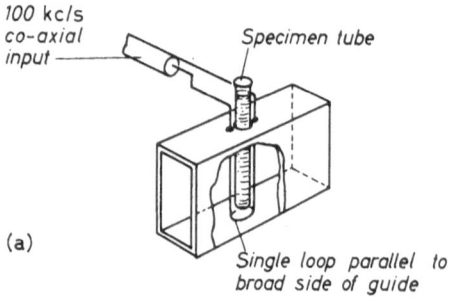

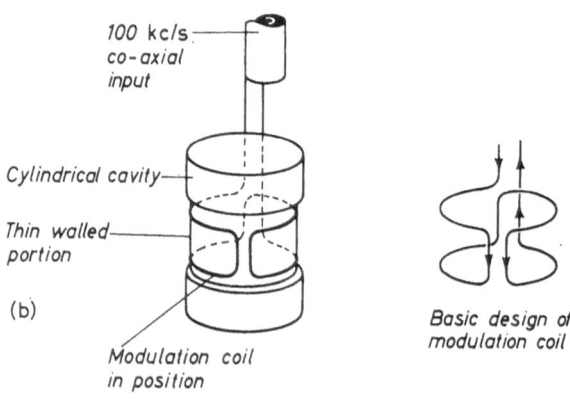

Figure 90. Coils for producing high frequency magnetic field modulation: (a) single loop, (b) Helmholtz pair

In order to apply the 100 kc/s magnetic field modulation to the sample, it is usual to replace the modulating coils around the magnet pole pieces by either a simple loop of wire wound around the sample inside the cavity resonator, or by a small pair of modulating coils, known as a 'Helmholtz pair', embedded in an insulating layer around the walls of the resonant cavity. Both of these possibilities are shown in *Figure 90*. The coils which are embedded in the cavity walls only produce an effective field at the site of the specimen if the walls of the cavity are sufficiently thin to let the 100 kc/s magnetic field pass through them without attenuation. This can be readily effected by coating the inner walls of a ceramic cavity with a silver lining, or casting an araldite mould around an electro-formed cavity wall, and in both cases arranging that the thickness of the wall is greater than the skin depth at the microwave frequency, but less than the skin depth at 100 kc/s.

One other general feature should also be mentioned here and this is that all these high frequency modulation systems use modulation sweeps which are smaller than the line width of the absorption to be

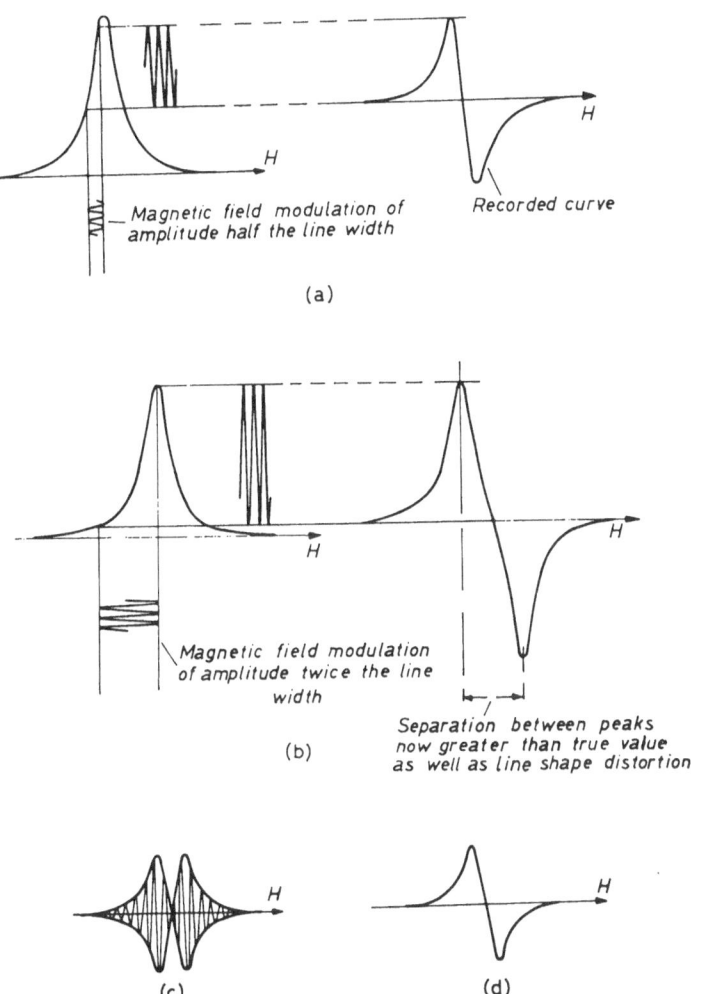

(a)

(b)

(c)   (d)

Figure 91. First derivative display produced by field modulation of magnitude: (a) under half the line width, (b) twice the line width, (c) oscilloscope presentation of first derivative before phase-sensitive detection, (d) pen-recorder trace after phase-sensitive detection

detected. This small magnitude sweep therefore responds to the gradient of the absorption line as indicated in *Figures 91(a)* and *(b)*, and as a result the magnitude of the detecting microwave system follows the first derivative of the absorption line instead of the actual magnitude of the absorption itself. If an ordinary 50 c/s sweep is also imposed on the magnetic field at the same time, and the output signal from a straightforward 100 kc/s narrow band amplifier is applied to the oscilloscope screen, then the pattern obtained is as shown in *Figure 91(c)*. Reference to *Figure 89* shows that this signal can easily be taken off the output of the narrow band amplifier and fed to the oscilloscope as shown. If the signal is instead fed into a phase-sensitive detector, which takes as its reference signal a voltage from the original oscillator via a suitable phase shifter, then the phase of the detected signal can also be retained. The output from this detector, suitably amplified by a d.c. amplifier, can be fed to a pen-recorder and the tracing obtained as shown in *Figure 91(d)* which is the true first derivative of the absorption line. This is the normal form of presentation nowadays in electron resonance spectra, the first derivative of the absorption being traced out in this fashion.

So far no attention has been paid to the stability of the whole system. It is evident, however, that this is a very important factor which will affect the ultimate sensitivity attainable, and hence attempts must be made to stabilize both the klystron frequency and the magnetic field value if the highest sensitivity is to be achieved. There are, in principle, two ways of stabilizing the klystron frequency. The first is to lock it to some external reference, either in the form of a quartz crystal feeding a frequency multiplying chain, or in the form of an external cavity of very high Q-value. The standard methods of automatic frequency locking and control can be employed as discussed in Section 2.9, in which case, a beat frequency is obtained between the frequency standard and the actual frequency emitted by the klystron itself. This beat frequency is fed to a phase-sensitive detector, which produces a correcting voltage of polarity and magnitude, to adjust the frequency of the klystron towards that of the fixed-frequency standard.

The second method is to stabilize the klystron by locking it to the resonant cavity in which the sample itself is placed. This has the disadvantage that the absolute frequency may then change, but the advantage that the klystron always remains in resonance with the actual frequency of the absorption cavity, and as a result no spurious changes in power due to a frequency shift between these is produced. Most workers in the electron resonance field have now come to the conclusion that it is far better to lock the klystron to the frequency of the cavity in which the specimen is placed rather than to an external frequency source. This locking can be accomplished by the relatively simple device of frequency modulating the klystron at, say, 465 kc/s and then using the resonance curve of the cavity to convert this frequency modulation into an amplitude modulation, in exactly the same way as the absorption line treated the magnetic field modulation shown in *Figure 91*. There will then be a zero 465 kc/s signal at the crystal detector if the klystron and cavity are exactly in resonance, whereas if the klystron drifts one way, a 465 kc/s signal of one polarity will be obtained, and if it drifts in the opposite direction the phase of the signal will alter. It is therefore only necessary to detect and amplify this 465 kc/s signal from the crystal detector and pass it to a phase-sensitive detector, which is also fed by the original 465 kc/s oscillator, and then use the output of this phase-sensitive detector as a correcting voltage to the reflector of the klystron.

Although the design of different spectrometers may vary in detail from the above description, this general account does cover most of the modern commercial spectrometers which employ 100 kc/s magnetic field modulation as their detecting mechanism.

There is one noticeable disadvantage of using high frequency modulation of this type for the detection and display of a resonance signal. This disadvantage arises from the modulation broadening that is produced by any modulation which is applied to the resonance condition as discussed in Section 4.2(f). This modulation broadening arises from sidebands which are formed in the same way as those produced in ordinary frequency modulation on a radio signal. The deviation of the sidebands from the frequency of the main signal is equal to the frequency at which the modulation takes place, and the amplitude of these sidebands depends on the magnitude of the modulation. It follows that this modulation broadening will produce a noticeable additional width to the absorption lines if the frequency of modulation is greater than, or of the same order of magnitude as, the width of the line without modulation. Thus if wide line spectra are being studied, high frequencies of modulation can be employed without any spurious effect on the line width; but if very narrow absorption lines are being studied the modulation broadening may well produce sidebands which will obscure the real lines of the spectra. If quantitative figures are applied to this relation, it can be shown that a frequency of modulation of 100 kc/s will produce a broadening of the order of 30 mgauss.

The hyperfine splittings of some free radicals in solution are of the order of 25 mgauss or less, and it follows that if these are to be

347

detected and resolved, modulation frequencies much lower than 100 kc/s must be employed. In practice, the 100 kc/s magnetic field modulation does not produce any noticeable additional broadening in most other cases; and only when very high resolution work has to be carried out, which usually implies free radical studies in solution, is it necessary to employ a spectrometer which does not produce this additional modulation broadening.

(b) *Spectrometers employing superheterodyne detection.* The other method of obtaining high sensitivity is by superheterodyne detection, which has been discussed in some detail in Section 4.10(b). The great advantage of this method is that no modulation broadening is produced, since the high frequency which carries the information and is detected by the crystal is produced by a direct beating of two microwave signals, and not by the modulation of either the magnetic field or the klystron signal frequency. This method can therefore be employed when very narrow line widths are expected in the spectrum, and provided the magnetic field is sufficiently homogeneous, there is no instrumental limitation on the resolution.

In theory the superheterodyne system should also be more sensitive than the 100 kc/s modulation technique, because the excess crystal noise is still falling in this frequency region. However, it has been found that above 100 kc/s most of the extra noise is contributed by the klystrons rather than the crystals.

As a result of this the 100 kc/s magnetic field modulation spectrometers and the superheterodyne spectrometers are of just about equal sensitivity, and the only real advantage of the superheterodyne system is that it does give much higher resolution. Superheterodyne spectrometers are however much more complex, and thus more expensive to purchase and more difficult to maintain, than the 100 kc/s systems. The use of superheterodyne spectrometers is only really justified when high resolution studies are to be made, such as in the investigation of free radicals in solution.

(c) *The use of circulators in spectrometer systems.* It is evident from the preceding sections that all forms of electron resonance spectrometers normally employ a microwave bridge system, and that the absorption cavity is matched against a balancing arm. This allows very sensitive adjustment of the power falling on the crystal detector, and the matching can usually be effected very simply by a slide-screw tuner in one of the bridge arms. It does suffer from the inherent drawback however, that half the power from the klystron signal is automatically lost in the matching arm, and is not available to produce resonance absorption in the cavity itself.

This situation has been altered by the advent of the microwave

348

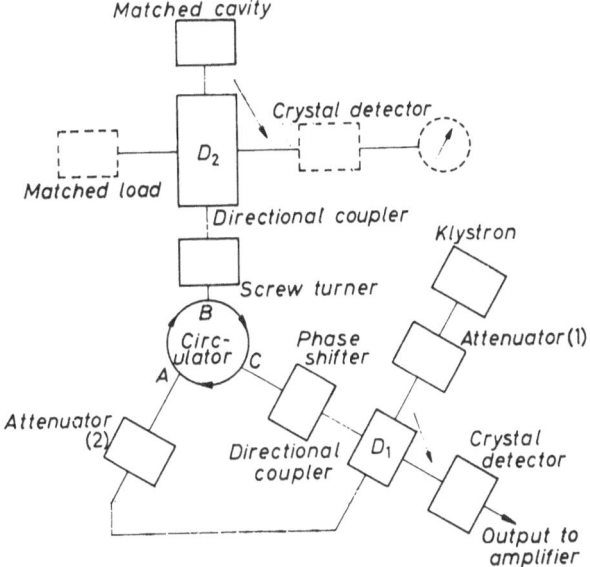

Matched cavity

Crystal detector

Matched load

D_2

Directional coupler

Klystron

Screw turner

B

Circ-ulator

C

Phase shifter

A

Attenuator(1)

Attenuator (2)

Directional coupler

D_1

Crystal detector

Output to amplifier

Figure 92. ESR spectrometer incorporating microwave circulator (after Faulkner[2])

circulator, which is a non-reciprocal device, based on the rotation of the plane of polarization of the microwaves by suitably magnetized ferrite inserts. The circulators can be of either the three-port or four-port type, and in each case the microwaves can only pass in one direction around the linking circle of waveguide, i.e. from input at A to output at B, from input at B to output at C, and so on. If such a circulator has the waveguide leading to a cavity resonator connected at one port (for example port B in *Figure 92*) then it is possible to differentiate between the microwave power flowing down the waveguide to the cavity, which must have come from input A, and the power flowing up the waveguide after absorption which must pass on to port C.

The great possibilities of circulators in spectrometer design were appreciated soon after their invention, and they were incorporated into maser systems very rapidly[1]. An example of their use in electron resonance spectrometer design is shown in *Figure 92*. This is a microwave circuit due to FAULKNER[2], and incorporates the additional feature of 'microwave bucking' to maintain the independent adjustment of microwave power level at the crystal detector.

Thus the klystron feeds its microwave power via attenuators (1) and (2) to port A of the circulator, and this power is fed out at port

B to the waveguide leading to the absorption cavity. After absorption has taken place the power re-enters the circulator at B, and passes out at C into the waveguide leading to the crystal detector. A directional coupler, D_1, also couples a certain fraction of power directly from the input line to the exit line. By varying attenuator (1) and the phase shifter it is thus possible to adjust the level of the microwave power actually reaching the detector crystal quite independently of the power fed to the cavity itself, which is controlled by attenuator (2). The system therefore has all the advantages of the normal magic-T bridge, but with the additional feature that all the available power can be fed to the specimen in the cavity instead of half of it being lost in the balancing arm. It follows that, in the absence of saturation, the theoretical sensitivity of this circuit is 6 dB better than that of the best magic-T circuit, and even in saturation limited conditions, it is still 3 dB better. The directional coupler D_2 and the rest of the circuit shown by dashed lines are only employed for monitoring the cavity tuning and match.

This same basic principle can also be applied to form a very neat 'homodyne' system of detection. A homodyne circuit is based on the same fundamental principle as a superheterodyne system, but instead of the second microwave frequency being obtained from a separate local oscillator klystron, it is actually derived from the same initial signal klystron. The shifted frequency is produced by feeding off some of the power from the klystron and passing it to a silicon crystal which is heavily modulated at the 'intermediate frequency'. Sidebands, carrying substantial microwave power, are then produced on either side of the signal frequency. One of these can be taken as the 'local oscillator' output to be mixed with the original frequency carrying the absorption signal, and a normal intermediate frequency signal is thus obtained. The great advantage of this system is that it only requires one klystron, and all the automatic frequency locking circuits between the two klystrons of a normal superheterodyne system are also eliminated. Conservation of microwave power in such a system is important however since this is effectively being used for two purposes at once, and it is in this connection that the incorporation of circulators in the microwave circuit can be so helpful.

Figure 93 shows the block diagram of such a homodyne circuit in which two circulators are used to separate the power flow and then mix the resultant signals as required. This circuit is also based on a design due to Faulkner[2]. The initial frequency is fed via the first circulator to the absorption cavity. The resonance in the cavity is modulated by the audiofrequency magnetic field modulation, and

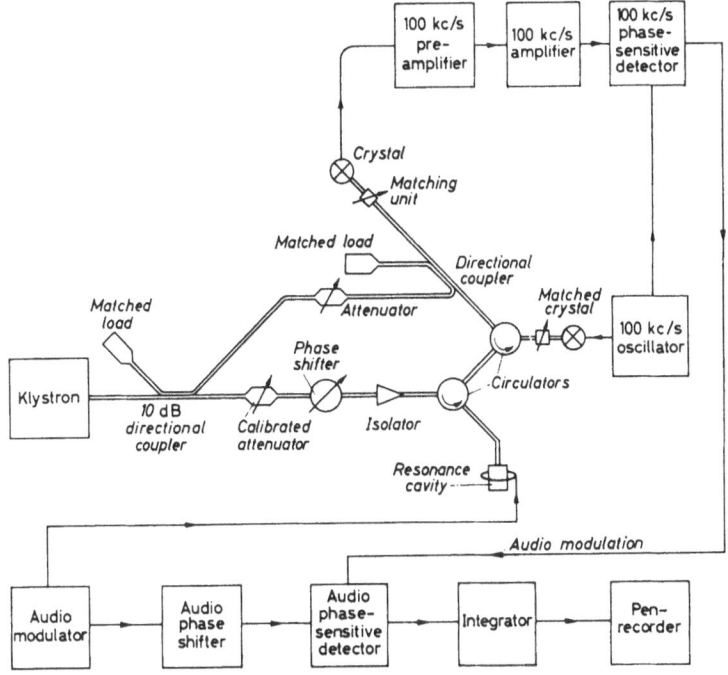

Figure 93. Homodyne ESR spectrometer employing only one klystron, but two circulators and directional couplers

this signal has its carrier frequency shifted by the crystal attached to the second circulator. It finally passes on to mix with some of the initial unchanged frequency, fed directly from the klystron via the two directional couplers. It is evident that this type of circuit includes all the advantages of superheterodyne detection, with the simplicity normally associated with the field modulation spectrometers, and also does not waste any microwave power in the balancing arms of magic-T bridges.

(*d*) *Helices as substitutes for cavity resonators.* One other development that has taken place in microwave circuit design over the last few years has been the employment of a helix, instead of a cavity resonator, as the absorption cell. The helix has two great advantages: (i) it is not so frequency-sensitive, and critical matching conditions are therefore avoided; and (ii) its open structure facilitates double irradiation techniques either with radiofrequencies or radiation in the visible region. It also has the advantage that the concentrated regions of microwave magnetic and electric fields are well separated,

351

and this can be very useful when aqueous or other lossy materials are being studied.

At first sight it might appear that the sensitivity of a system employing a helix must of necessity be much lower than that employing a cavity, because of the very high Q factor of the latter. A high concentration of microwave magnetic field strength occurs in a helix however, due to the effective slowing down of the wave velocity, and a complete analysis [3] of the field strengths concerned shows that the ratio of the power reaching the detector on and off resonance is nearly the same for the helix as for the cavity.

The idea of using a helix in ESR spectrometers was first suggested and tried by F. N. H. Robinson, and the theory was then developed and checked experimentally by WEBB [3], whose paper gives detailed expressions for the microwave field distributions within the helix and diagrams of spectrometers actually employing them. More recently, one firm has developed this system as the basis of a commercially available spectrometer, and a sensitivity of 10^{11} ΔH spins is quoted, which compares favourably with that of cavity spectrometers. Details of this spectrometer, marketed by the Alpha Scientific Laboratories, are to be found in Appendix II.

As a general rule, the helix does not give quite such good results as the cavity in straightforward electron resonance studies, unless very lossy solutions are to be studied. It does have a great advantage in all double resonance type experiments, however, where the same helix can be used as the nuclear resonance coil and as the microwave slow wave structure [3].

11.2 Rapid Recording Spectrometers and Signal Integration

It is shown in Section 4.10 that, once the excess noise in the spectrometer system is reduced to a minimum, the ultimate sensitivity attainable depends finally on the bandwidth Δf of the detecting system. Thus it can be shown [4] that if the microwave circuits are correctly matched, the optimum sensitivity of an electron resonance spectrometer expressed as the minimum susceptibility detectable χ''_{min} is given by

$$\chi''_{min} = \frac{1}{\pi . \eta . Q_0} \left[\frac{F . kT . \Delta f}{2P} \right]^{1/2} \qquad \dots (11.1)$$

where F is the noise figure of the microwave receiver, η is the 'filling factor' of the sample in the cavity, which has a factor Q_0, and P is the available power from the klystron. All of these factors are determined by the detailed spectrometer design, and insertion of typical values for an X-band spectrometer into the above equation then gives for the minimum number $N_{0_{min}}$ of detectable spins as

$$N_{0_{min}} \approx 3 \cdot 10^{10} \cdot (\Delta f)^{1/2} \qquad \dots (11.2)$$

It follows that the only way in which the ultimate sensitivity can be further increased is by reducing the bandwidth Δf of the detecting system. This explains why highly stable microwave sources and magnetic fields are required, since if either of these vary whilst the absorption is being traced out the signal will be lost. Since the time of tracing through the signal is determined by the time constants in the system, and must be approximately the inverse of the bandwidth of the detecting system, high sensitivity requires frequencies and fields which are highly stable over lengths of time of the order of seconds or minutes.

There is, in principle, no reason why the sensitivity should not be increased indefinitely by decreasing the bandwidth and therefore increasing the time taken to trace through the signal. This method of increasing the sensitivity can only be applied to completely static molecular systems however in which the unpaired electron concentration does not change. A large number of systems of interest possess rapidly changing concentrations of unpaired electrons, and quite often these are also the systems in which high sensitivity of detection is required—for example in measuring the kinetics of free radical or enzyme reactions, studying photochemical breakdown, or following the process of irradiation damage.

It would appear from the previous paragraph that high sensitivity can theoretically never be achieved in such systems where the unpaired electron concentrations are changing rapidly. Thus short time constants and large bandwidths are necessary to follow such rapid changes; but equation (11.2) shows that high sensitivity can only be attained by employing very long time constants in the recording system, with correspondingly small bandwidths. There is a direct conflict between these two requirements, and unless some method is devised for overcoming this conflict, rapid recording of signals will only be possible for those of very large strengths. Since most of the cases of interest, for example concerning irradiation damage or photochemical decomposition, involve relatively weak signals, it is essential that some other means are found to give reasonable sensitivity with rapid change of signal.

All the methods that have been devised to overcome this problem use the same essential principle. They record the particular event many times, and then integrate the recorded signal out of the noise background by using the coherence between its value and that of the synchronizing pulses. As a specific example the ultra-violet irradiation of a certain system, which then decomposes into free radicals, can be considered. The object of the experiment is to follow the free radical concentration in, say, the 100 msec following the initial ultra-violet flash. The electron resonance spectrometer has its frequency and field values set to the centre of the absorption line, and the detecting equipment is designed to plot the height of this resultant signal. In this way the line intensity can be monitored continuously. If it were recorded directly as a function of time on a photographic film or magnetic tape, signals would only be obtained above the noise level if they were of very large magnitudes. However the recording apparatus can be gated so that the signal is only recorded at an interval of 50–52 msec after the initiating flash, and then the whole process can be repeated a large number of times. The signal from the detecting equipment is fed back to the same position on the storage device each time, so that either the spot on a cathode ray tube or the element on a magnetic drum, corresponding to the 50 msec interval, is always the same. It is thus possible to integrate the signal level, corresponding to the electron resonance after 50 msec, out of the incoherent noise which is also present, but which does not sum in a linear fashion. By repeating this operation as a pulse sequence a large number of times, the narrowing of the bandwidth can be effectively achieved and hence sensitivity reclaimed.

It should be noted that this possibility of obtaining high sensivity even for a rapidly changing signal does not really contradict the fundamental requirement for a long time constant. It is only by many repetitions of the event that the signal can be recovered from the noise; hence a large overall time of detection is still required, even though it is now formed by a summation of very small elements instead of a single slow sweep.

Various devices were initially tried as storage and integrating systems, including magnetic tape and electronic storage tubes. This whole line of research was given great impetus however when KLEIN and BARTON[5] showed that a digital computer could be very effectively used as the storage device. The type of digital computer that is most readily available for such an application is a multi-channel pulse-height analyser, and a block diagram of his basic system is shown in *Figure 94*.

The voltage from the output stage of the electron resonance spectrometer, which represents the signal plus random noise, is converted into pulses in the 'voltage to frequency convertor', the pulse rate being proportional to the voltage. The channels in the

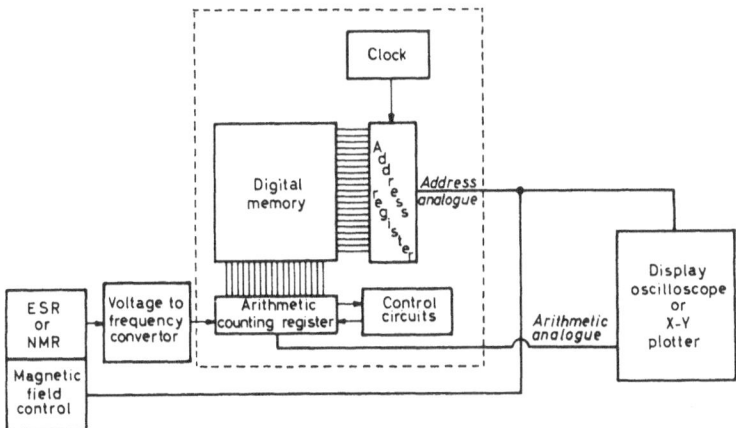

Figure 94. Enhancement of resonance signals by employing a computer of average transients (CAT) (after Klein and Barton[5])

pulse-height analyser itself are opened successively, in synchronism with the increase of the magnetic field if a static signal is to be enhanced. Hence each channel corresponds to a defined region of the magnetic field strength. The number of counts recorded in that channel is proportional to the output voltage from the spectrometer at the given field value or, more accurately, the time integral of that voltage. The control circuits are arranged to sweep the magnetic field successively through the spectrum and, during each successive sweep, the channels are gated in turn so that the same channel always receives the number of pulses representing the voltage at that particular magnetic field strength.

In such an arrangement as this the signal increases directly with the number of scans since it is coherent with the magnetic field sweep, while the noise only increases as the square-root of the number of scans since it is a statistical, incoherent phenomenon. It follows that the signal-to-noise ratio that is stored in the channels of the analyser increases very rapidly as the number of scans increases. Such pulse-height analysers have a maximum store of up to 10^6 pulses/channel, and so overloading or saturation effects are very unlikely. The enhanced signal can also be monitored at any stage, and hence the number of scans corresponding to an acceptable signal-to-noise ratio can easily be found.

None of the circuits employed have long time constants or narrow bandwidths, and hence the system can readily be applied to cases of rapid change in signal, the only requirement being the production of successive identical reactions which can be accurately synchronized. The method can be applied to such rapidly changing systems, in which case the magnetic field and microwave frequency are held constant at the resonance condition; the variable parameter corresponding to the different channels is then the time itself. The structure of the spectrum of a rapidly varying signal can of

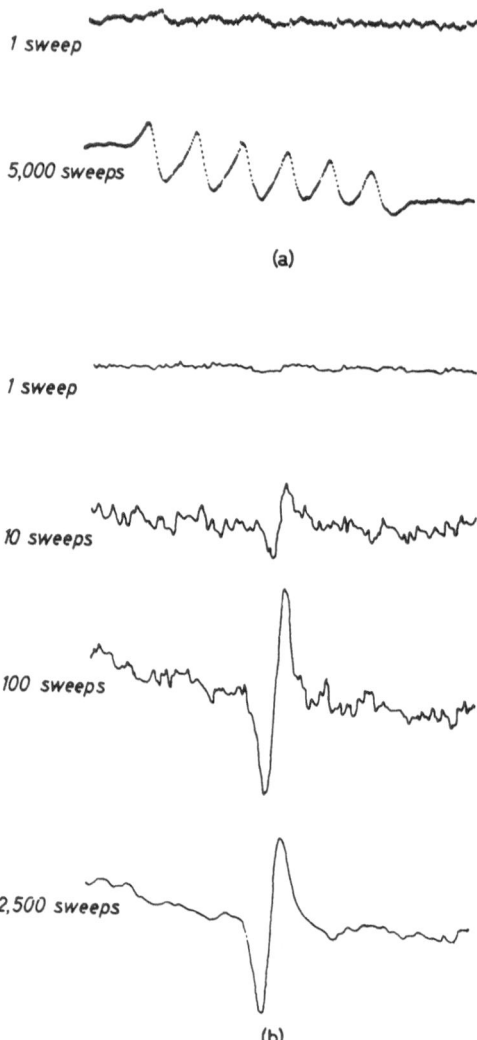

Figure 95. ESR spectra showing enhancement of resonance signals from: (a) 5×10^{13} Mn²⁺ ions in water, (b) hydrazyl

course be found by repeating the whole operation at different values of magnetic field, the field being held constant for each time display obtained from the analyser, and then changed after that particular enhancement has been completed.

In the early applications of this device to electron resonance spectra, weak static signals were used to show its effectiveness, and *Figure 95* shows two such examples, one taken from the original paper by Klein and Barton[5]. *Figure 95(a)* shows the spectrum obtained from a solution containing 5×10^{13} Mn^{2+} ions in water. The top trace is that observed with normal phase-sensitive detection employing a 0·01 sec time constant and with a sweep of 5 sec for the complete trace. The bottom trace is the spectrum obtained from exactly the same solution after 5,000 traverses have been fed into the pulse-height analyser and averaged in the way described. *Figure 95(b)* shows the spectrum of a dilute solution of hydrazyl, and the build-up of signal-to-noise ratio as the number of integrated sweeps is increased.

The striking potentialities of this technique, and the very great improvement in sensitivity that it gives, are immediately obvious from such examples. The same technique can also be applied to nuclear magnetic resonance, and to any other system in which large time constants can be effectively attained by a great number of successive sweeps. Because of the varied applications of such devices many commercial firms have now taken up their design and production. The control circuits and oscilloscope display units are usually incorporated with the pulse-height analyser itself in an instrument termed a 'Computer of Average Transients', and called a CAT for short!

Since the design and development of these instruments is progressing so rapidly at the moment (and their price is correspondingly falling), there is no point in summarizing any data on them in this book. The basic features which determine the performance of such an averaging system are, of course, (i) the number of separate channels available; (ii) the maximum number of pulses that can be accommodated in any one channel, and (iii) the rapidity with which the sequencing can be performed. This last point is not so important if enhancement is only required of static signals, but may be significant if rapidly changing spectra are to be studied.

11.3 FLOW SYSTEMS AND APPLICATIONS TO BIOCHEMISTRY

Apart from developments in the technique of electron resonance spectrometry which are of a general nature and applicable to any spectrometer, there have also been a number of more specific developments which have occurred in certain fields of research. Several of these have concerned cavity design, such as the provision of slits in the cavity wall for photochemical irradiation studies[6, 7]. Others have dealt with whole spectrometer systems which have been mounted on the end of high energy accelerators so that the effect of electron or heavy particle bombardment of specimens could be studied[8, 9].

The field in which there has probably been the most rapid growth of interest in the applications of electron resonance and which has produced some new and interesting techniques is that of biochemistry. The particular applications of electron resonance to biochemical and biophysical problems is summarized in Section 11.8, but a brief description of the new experimental techniques that have been introduced seems to be in place at this point.

There are two new techniques that have been developed specifically for biochemical investigations, and these are (i) the design of cavity resonators to study continuously flowing and reacting solutions; and (ii) the sudden freezing technique, which is also used in conjunction with transient reactions, but holds them static after a given time interval.

(a) *Continuous flow systems.* The continuous flow technique has been used in electron resonance investigations to study various biochemical reactions in which the intermediates of interest are formed a few seconds after the initial reactants are mixed. One field of investigation in which these techniques have found very great application is the study of enzyme reactions. The general idea behind this technique is shown in *Figure 96(a)*. The two solutions which are to be made to interact are stored in the reservoirs R_1 and R_2. The outlets from these reservoirs meet in a mixing chamber M, and the outlet from this leads through a reaction tube K. K can be made of a quartz tube suitable for insertion into a standard rectangular cavity of an X-band spectrometer. The distance x along K before the tube enters the microwave cavity can be altered and hence the time of reaction after initial mixing can also be adjusted. If unlimited amounts of the two reactants are available, they can be allowed to flow into the mixing chamber at suitable rates and the electron resonance spectrum of the resultant products can be followed as a function of time by gradually moving the quartz reaction tube further and further out of the cavity (i.e. increasing x).

In many of these studies however only relatively small amounts of the two reactants are available and this applies especially in work on enzymes. The simple technique described above must therefore

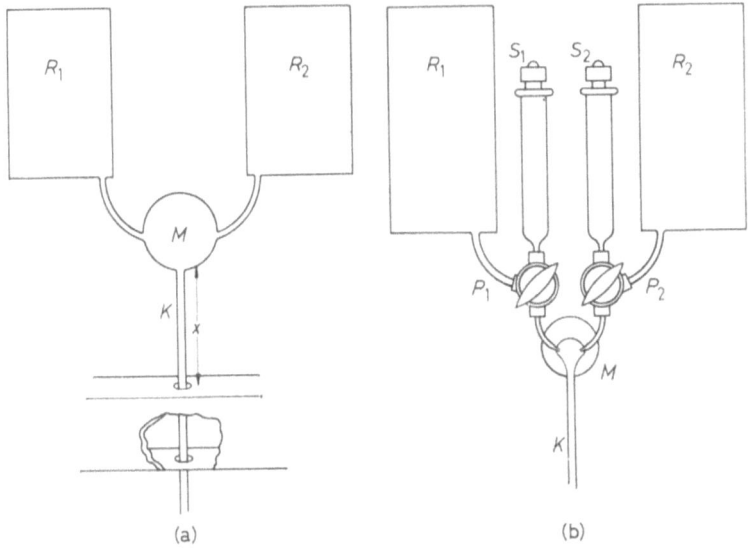

Figure 96. Continuous flow systems for the investigation of active radicals: (a) basic principle, (b) practical version

be modified so that continuous flow of the two reactants is not required. This may be done by inserting two syringes S_1 and S_2 between the reservoirs and the mixing chamber, as shown in *Figure 96 (b)*. Before taking resonance measurement the syringes are first filled from their respective reservoirs which are then turned off by the taps P_1 and P_2. The experiment is initiated by driving the syringes S_1 and S_2 down at a predetermined rate, to mix the two solutions in the mixing chamber, from which they are driven into the reaction tube. In practice the two reactants are introduced into the mixing chamber by a series of jets, and the actual nature of these jets and the design of the mixing chamber itself varies somewhat according to the proportions of the two reactants that are to be mixed. The general concept of this type of apparatus follows from the flow apparatus designed by HARTRIDGE and ROUGHTON[10] in 1923, and although the details of jet design in the mixing chamber have improved since then, the basic idea of a rapid thorough mixing of the reactants before passing on to the output tube has remained the same.

Most of the mixers which have been used to date in connection with electron resonance spectrometers, have incorporated simple jet systems, and a photograph of such a flow apparatus[11] positioned in

359

the magnet gap, is given in *Figure 97*. The 12-jet Lucite mixer and inlet tubes can be seen in front of the cavity, which is itself situated in the magnet gap. Several groups of workers have used flow equipment of this type in kinetic studies, and some of their results

Figure 97. Flow apparatus incorporating simple jet system

are discussed in Section 11.8. One of the main uses of this kind of study is in correlating the kinetics of appearance and disappearance of free radical intermediates with the kinetics of enzyme-substrate compounds.

(*b*) *Sudden freezing techniques*. The sudden freezing technique, mentioned as the second new development in these studies, is really an extension of the flow technique described above. It differs from the flow technique in that the kinetics of the reaction are measured at leisure, in a series of samples taken from the streaming and reacting solution after appropriate time lapses and suddenly deep frozen. The essential requirement is that the quenching process should be rapid compared with the actual reaction being studied. The most satisfactory way to achieve this appears to be the direction of a fine jet of the reacting solutions into another liquid held at a very low temperature, in which the two reactants themselves are not miscible. It is better to inject the mixture into a separate liquid rather than liquid nitrogen itself, since gasification of the nitrogen might otherwise easily retard the heat transfer. BRAY[12] has made an extensive study of the time required to freeze such dilute aqueous solutions and has come to the conclusion that, if the jet has a nozzle of about 0·2 mm diameter, a dilute aqueous solution at 20° C will only take about 10 msec to become deeply frozen when it is injected into hexane at −80° C.

The other general requirements for this technique can be summarized as follows. The flow of liquids should be accelerated very rapidly to a predetermined value and held there until sufficient of the reacting material has been produced; the flow should then be rapidly stopped. This procedure should give samples which are essentially homogeneous with respect to the time that elapses between mixing the solutions and freezing. In order to meet these requirements properly the syringes supplying the two reactants should be driven by a constant-speed hydraulic ram. A diagram of such equipment, designed and constructed by Bray[12] for use with an electron resonance spectrometer is given in *Figure 98*. The piston

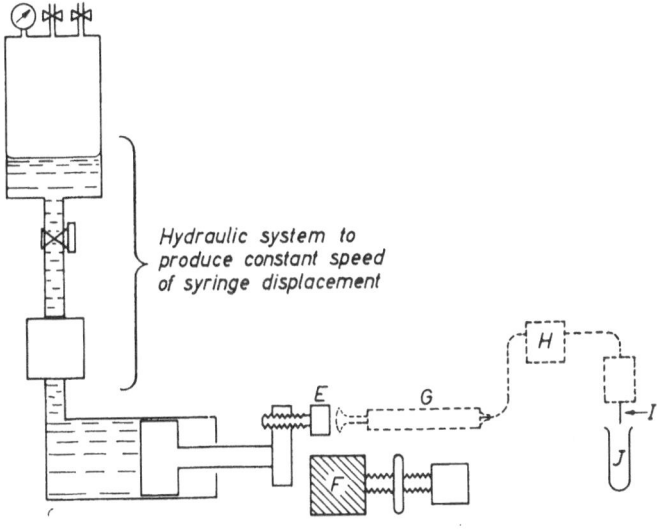

Figure 98. Sudden freezing technique

E holding the adjustable screws, bears on to the head of the syringe plunger *G* and in this way a continuous displacement of the reactant is driven into a mixing chamber *H* and from there through the jet *I*

into the container of colder liquid J. This type of apparatus has been used to produce homogeneous samples, deeply frozen at any desired stage, from reactions which occur in aqueous solution and have a half-life of the order of 100 msec.

The reacting liquid is actually injected into the cold hexane which is contained in a tube with a narrow tail of 3 mm diameter at the end. The frozen precipitate is then forced down into this bottom tail; when it is packed with material it can be inserted into the electron resonance spectrometer. The particular advantages of this technique over the continuous flow methods are (i) that it can work at smaller enzyme concentrations, and (ii) it gains in sensitivity from the lower temperatures that are employed.

Typical results obtained by using the experimental methods described above, are given in Section 11.8.

11.4 Free Radical Studies—Experimental

One or two brief references are made in Chapter 7 to the fact that electron resonance is being applied to the study of systems other than the transition group compounds, one example of this being the study of free radicals. Some free radicals can be grown as stable crystals and these form very suitable systems in which to make the initial measurements.

The particular case of diphenyl trinitropehnyl hydrazyl quoted at the beginning of Section 7.7 has served as a very useful system for investigation by electron resonance over a period of many years. In fact its usefulness in the study of free radicals compares with that of copper in the transition group atoms. Phenomena first discovered with electron resonance in copper salts but which also apply to the transition group compounds in general include: the existence of exchange forces[13], of hyperfine structure[14] and of the quadrupole interaction[15]; the concept of resonating crystal fields[16]; and the interactions between two coupled atoms[17]; see also Section 6.11(i). Similarly it is remarkable how frequently new effects in free radicals have been discovered, and followed up, by studying the spectra from hydrazyl under different conditions. Thus the presence of exchange narrowing[18] in free radicals, and of a hyperfine structure in solution[19], were first demonstrated in hydrazyl. Later work on solutions at higher dilutions showed that, under carefully controlled conditions in which the oxygen has been removed, the superhyperfine structure from the protons attached to the aromatic rings can also be observed[20]; these protons were employed in the first measurements of Overhauser effects in free radical systems[21].

Hydrazyl was one of the first stable free radicals to be studied as single crystals[18], and for which the hyperfine structure in the solid state was resolved[22] by dilution with a diamagnetic isomorphous compound, in this case hydrazine. The main reasons why hydrazyl has been used in so many experiments are probably the ease with which it can be prepared, and the fact that it is a very stable radical which can exist both as a crystalline solid and in a solution.

Following the early work on hydrazyl and a few other stable radicals[23], the experimental investigation by electron resonance was concentrated on one or two systems such as the semiquinones[24], and the aromatic negative and positive ions[25] that can be formed in suitable solutions. Solutions of these radicals, although not quite as stable as hydrazyl can be kept for some considerable time and were thus ideal for initial studies by electron resonance. The main importance of this work was probably not so much in the study of the free radicals themselves, but as a testing ground for the various theories on molecular orbitals and chemical structure. The precise electron resonance data allowed detailed experimental checking of the theoretical predictions.

This work on stable or semi-stable radicals in solution gave way to the more practical interests associated with transient radical species, and it became apparent that electron resonance could provide a powerful tool for following the kinetics of changing radical concentrations. These possibilities were followed up in particular by the biochemists, as well as by the physical chemists, and in some cases transient radical species could be correlated with changing valencies of metal atoms.

More recently attention has again concentrated on single crystals. Somewhat surprisingly, the study of single crystals was one of the later developments in free radical investigation, whereas crystals of transition group compounds were explored long before the solutions. The great advantage of studies on single crystals is that all the anisotropic interactions from the hyperfine structure also become available, and these can be of very great assistance in identifying the actual radical species that are formed. In particular, radicals formed by high energy irradiation can give quite ambiguous spectra if only solutions or polycrystalline spectra are available, whereas a detailed investigation of a single crystal, in which the radicals formed by the irradiation still maintain their orientation in the crystal structure, can give a very great deal of extra information. Single crystal studies were also important to the triplet state work in that they were essential for the successful observation of excited states. In attempting to provide a brief summary of the recent developments in electron resonance studies of free radicals it would probably be wise to group the different fields of investigation under four separate headings:

(a) The study of semiquinones and simple stable radicals
(b) Aromatic ions in solution
(c) Single crystal studies
(d) Free radical kinetics.

Although these four headings do not embrace all that has happened in this rapidly expanding field of research they do cover, together with the biochemical applications discussed in Section 11.8, the main developments and the investigations that have taken place during the last few years.

(a) *The study of semiquinones and simple stable radicals.* The initial measurements on hydrazyl, both as a crystalline solid and in solution, have been summarized in Section 7.7. It is pointed out that the hyperfine structure of five lines, with an intensity ratio of $1:2:3:2:1$, can be explained by assuming that there is an equal interaction of the unpaired electron with the two nitrogen atoms. Since most free radicals have a very small spin-orbit interaction, the main feature of interest in free radical spectra is the hyperfine structure rather than the g value variation, the latter usually being highly isotropic and close to the free spin value. The delocalization of the unpaired electron into a molecular orbital, as it occurs in a free radical structure, not only produces a small spin-orbit inter-action but also means that the unpaired electron may well interact with more than one nucleus, and hence hyperfine contributions are to be expected from several different nuclei. The situation is therefore somewhat different from the case of the transition group atoms where the unpaired electron is likely to interact with only one atom for most of the time. The hyperfine patterns which were expected from an interaction of the unpaired electron with several nuclei, were very well demonstrated by the work of VENKATARAMEN and FRAENKEL[24]

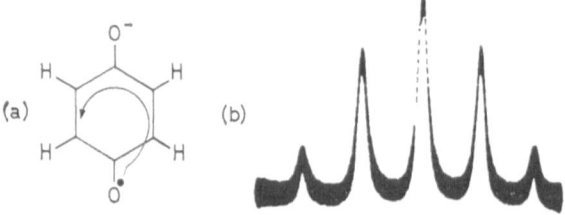

Figure 99. Electron resonance from p-benzosemiquinone: (a) structural formula, (b) hyperfine pattern observed

on the semiquinones, which followed as one of the early steps in electron resonance studies of free radical systems.

The structural formula of the simple p-benzosemiquinone which they initially studied is shown in *Figure 99(a)*. This semiquinone is formed by removing one of the hydrogen atoms from the OH group, and leaving an entirely symmetrical molecule. By reasons of symmetry alone it follows that the orbit in which the unpaired electron moves must be distributed equally over the four protons attached to the ring. The hyperfine structure is, therefore, entirely due to the interaction of the electron with the four protons around the aromatic ring. The pattern actually observed is shown in *Figure 99(b)*, and consists of five component lines, which are equally spaced but not of equal intensity, rising from the wings to the centre of the pattern. This kind of hyperfine pattern is typical of an interaction of the unpaired electron with a number of equally coupled identical nuclei. In order to explain the reason for this, the pictorial representation of hyperfine interaction as shown in *Figure 58(b)* is now extended to the case of several nuclei. To make this somewhat specific the three cases of an interaction with one proton, two protons and three protons are considered in turn, such as might occur in CH, CH_2 and CH_3 groups respectively.

If the unpaired electron is interacting with one proton then there are only two possibilities for the incremental fields which are produced by the nucleus at the site of the electron; these corresponding to the proton being aligned parallel or anti-parallel to the applied field. The two diverging energy levels of the simple electronic transition are therefore each split into two components as shown in *Figure 100(a)*, and a doublet hyperfine structure is obtained. This equal intensity doublet is, of course, only a simple example of the general rule that a single nucleus with spin I gives $(2I + 1)$ equally intense hyperfine components.

If the interaction with two equally coupled protons is now considered, the situation may be represented as in *Figure 100(b)*. In this case the position of the two original electronic levels is indicated on the left-hand side with the component $M_s = \pm \frac{1}{2}$. Both of these are split into two components by the interaction with the first proton as represented in *Figure 100(a)*, and then a further interaction occurs with the second proton so each of these split levels is again split. If the interaction between the two protons is equal, the splitting between the second set of levels is of equal magnitude to the original splitting and as a result the two central levels are brought together as shown. The energy level pattern obtained consists effectively of three different levels, but with the

365

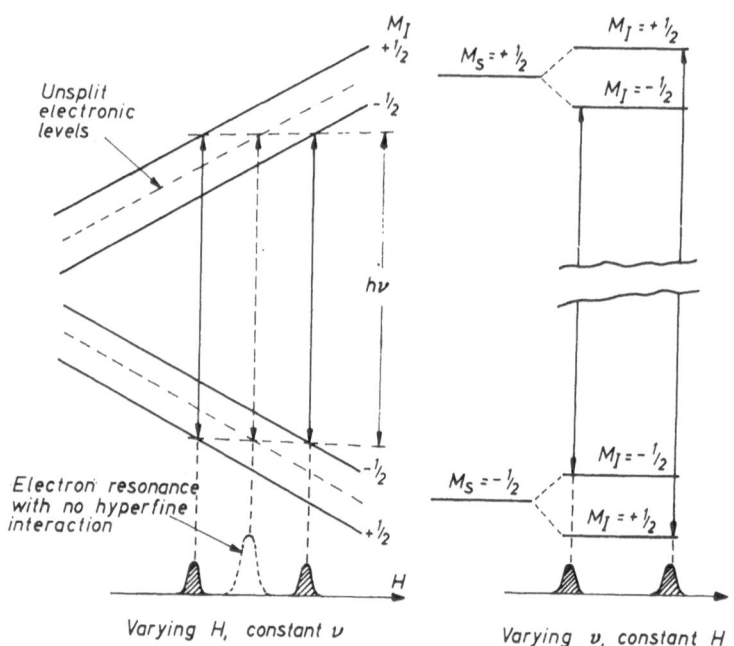

Varying H, constant ν

Varying ν, constant H

(a)

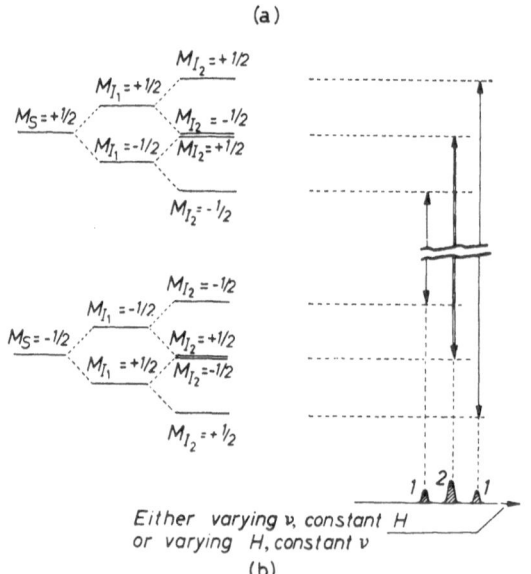

Either varying ν, constant H
or varying H, constant ν

(b)

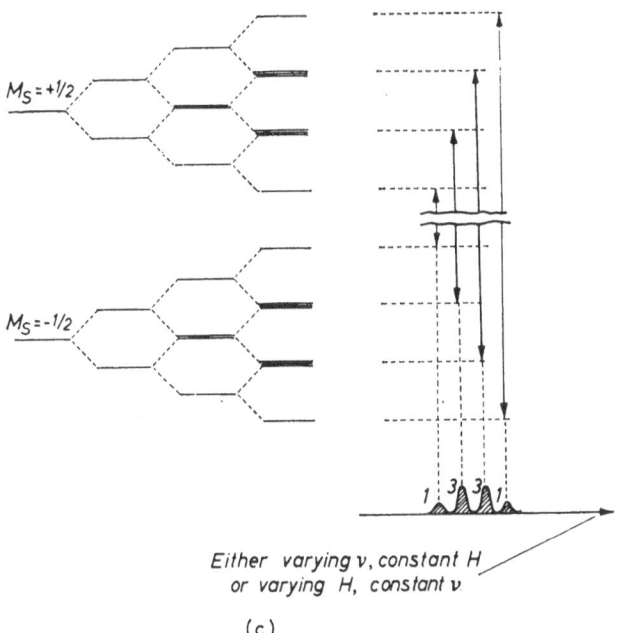

$M_S = +1/2$

$M_S = -1/2$

1 3 3 1

Either varying ν, constant H
or varying H, constant ν

(c)

Figure 100. Hyperfine patterns from interactions with equally coupled protons: (a) one proton,
(b) two protons, (c) three protons

centre level containing two components and thus being twice as populated as the two extreme levels. This form of energy level distribution is produced in both of the electronic energy groups, and when the microwave frequency is applied there are three permitted transitions resulting in a triplet hyperfine structure with the centre line twice as intense as the two outer lines.

The interaction with three equally coupled protons is shown in a similar manner in *Figure 100(c)*. The interaction, first with the single proton and next with the two protons, is represented by the two steps on the left. The interaction with the third proton further splits the three groups of levels formed by the interaction with the two protons, and this results in four separate sets of energy levels for each electronic transition. The two central levels now contain three components, and thus a quartet hyperfine structure is produced with the two central lines three times as intense as the two outer lines. It is obvious that this kind of representation can be extended indefinitely to n equally coupled protons, where n is any integer. It can quickly be shown that $(n + 1)$ equally spaced hyperfine lines are always obtained, and the intensity distribution of the lines follows that of a binomial curve, the expression for the intensities being given by

$$1 : n : n(n-1)/2 : \dots n!$$

This case of equal coupling of the unpaired electron in a molecular orbital to a number of protons is not only found in the semiquinones but also in a large number of other organic radicals including the aromatic ions discussed in the next section. It often happens, however, that as well as n protons being equally coupled to the electron, there is another set of, say, m protons which are also coupled to the electron equally, but with a different magnitude of interaction to that of the first group. If this second interaction is noticeably smaller than the first, then the $(n + 1)$ lines produced by the first group of protons are each split into $(m + 1)$ lines, with a binomial distribution in their magnitude as before. Examples of these different sets of equal splittings can be found in the next section.

The above analysis explains why the five lines were observed from the simple p-benzosemiquinone studied by Venkataramen and Fraenkel[26], and also accurately explains the observed intensity ratios of these lines. The directness and power of this method of hyperfine analysis was further demonstrated by the early work of WERTZ and VIVO[27] who showed that this five-line spectrum could be successively replaced by a four-line spectrum, a three-line spectrum, a two-line spectrum, and finally a single-line spectrum when the protons around the aromatic ring were successively replaced by one, two, three or four chlorine atoms. The hyperfine splitting expected from the chlorine atoms is much smaller than that from the protons, and therefore substitution of a chlorine atom effectively removes the hyperfine contribution from this particular site. The actual hyperfine patterns observed from the chlorinated derivatives are shown in *Figure 101*. The changes in spectrum and the predicted reduction in the number of lines caused by the decrease in protons are clearly visible. It may be noted that phase-sensitive detection with first derivative recording was used for this equipment, as is now normal in all free radical studies.

Soon, other chemical systems known to produce stable or semi-stable free radicals, including most of the classic examples such as triphenyl methyl[28], were also investigated by electron resonance. In a large number of these free radicals very well resolved hyperfine structure was obtained which could then be compared in detail with that predicted by the theoretical molecular orbital treatment of the radicals. Similar comparisons for the aromatic positive and negative ions have been particularly fruitful and this point is considered again in the next section.

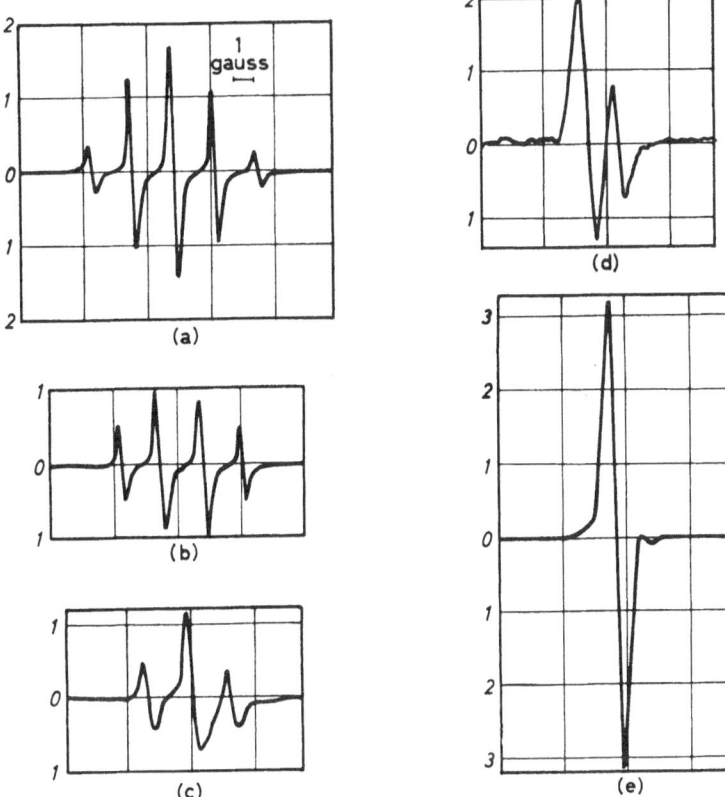

Figure 101. ESR spectra of chlorobenzosemiquinones: (a) unsubstituted benzosemiquinone, (b) monochlorosemiquinone, (c) 2,3-dichlorosemiquinone, (d) trichlorosemiquinone, (e) tetra-chlorosemiquinone. These are derivative tracings (after Wertz and Vivo[(27)])

The main importance of the study of the complex hyperfine structure associated with free radicals in solution, has been the precise data provided for checking molecular orbital theories and theoretical chemistry in general, but detailed measurements on the free radical systems have also been of interest to the theoretical physicist. Not only have concepts of relaxation phenomena and double resonance been tested, but the more refined treatment of the Zeeman effect has also been precisely checked. As a particular example, the measurements carried out by TOWNSEND, WEISSMAN and PAKE[(29)] on peroxyl amine disulphonate can be quoted. The interaction of the unpaired electron is entirely with the nitrogen atom in this radical, thus producing a three-line hyperfine splitting.

The splitting between these three lines is equal in large values of magnetic field, but if the splitting is followed back into small magnetic fields, the coupling between the quantum numbers changes and the splitting varies as predicted by the Breit-Rabi formula. [30]. Townsend, Weisman and Pake performed this experiment at radio-frequencies on solutions of the radical, and they traced the lines back into zero magnetic field.

(b) *Aromatic ions in solution.* It has been known for some time that negative ions of aromatic molecules can be produced in solution if they are dissolved in such solvents as tetrahydrofuran or dimethoxy-ethane and then reacted with metallic sodium. It was also discovered later[31, 32] that the positive ions of these aromatic molecules could be obtained if the compounds were dissolved in concentrated sulphuric acid. Both the negative and positive ions give rise to an electron resonance spectrum, and the only difference noticeable is a slight variation in the total splitting of the spectrum. In both cases complex hyperfine structures are obtained, and an analysis of the splitting predicts the actual electron spin densities present on the different carbon atoms of the aromatic structure. The initial comparison between experimental results and those predicted by the simple molecular orbital treatment of HUCKEL[33] gave encouraging

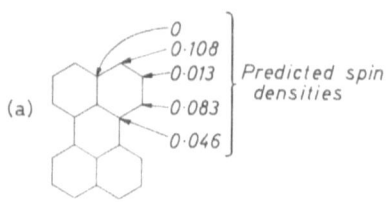

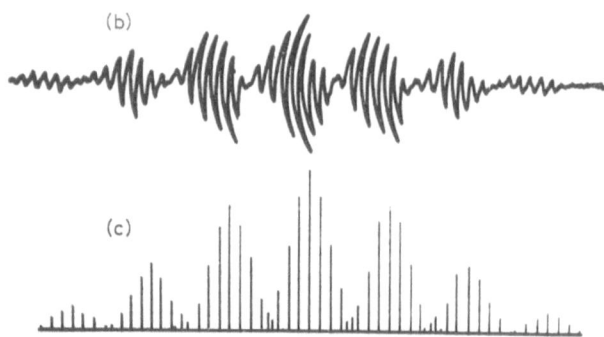

Figure 102. ESR spectrum of perylene: (a) structural formula, (b) observed pattern, (c) theoretically predicted lines

first-order agreement. An example of one of these aromatic ions is shown in *Figure 102*. In *Figure 102(a)* the structure of the perylene itself is given, together with the spin densities predicted by the Huckel theory; the observed hyperfine pattern is shown in *Figure 102(b)*, and in *Figure 102(c)* is the hyperfine pattern predicted from the spin densities.

The comparison between experimentally observed splittings and predicted theoretical values has now developed into a very sophisticated and precise procedure, in which second and third-order effects are inserted into the molecular orbital theory, and computer techniques are employed to evaluate the constants. In this way, a study of semi-stable radicals and their precise hyperfine patterns has served, not only to identify and characterize the radicals themselves, but also to check the accuracy of molecular orbital treatment. The resulting improvements in molecular orbital theory can then be applied in general to other compounds.

It is often found that the development of a subject depends on some crucial experimental technique or discovery, and this is certainly so in the study of the aromatic ions in solution. The work was revolutionized by HAUSSER[34], who discovered that the molecular oxygen normally dissolved in solvents causes considerable broadening of the electron resonance lines; if the solvents are carefully outgassed this extra source of broadening is removed, and much better resolution is then obtained. The realization of this fact, which should have been obvious many years before, enabled much better resolutions to be obtained in free radical studies in solution. Good examples of this are the recent measurements on hydrazyl in solution. The earlier work discussed in Chapter 7 showed that five lines could be obtained from hydrazyl in dilute solutions, these being due to the interaction of the unpaired electrons with the nitrogen atoms, but no trace was obtained of any interaction with the protons around the aromatic rings. The extra splitting on the spectrum due to these ring protons, was first observed by DEGUCHI[35], and the spectrum he observed is shown in *Figure 103*. A very large number of additional lines are

Figure 103. ESR spectrum obtained from hydrazyl in purified and degassed solution (after Deguchi[35])

superimposed upon the five main groups due to the interaction with the two central nitrogen atoms. This spectrum was only obtained after the solvent had been carefully purified, then degassed, and redistilled in the vacuum. Such treatment removes all the molecular oxygen and thus this source of broadening is eliminated. Such, degassing procedures are now standard when studying hyperfine spectra in solution and as a result much better resolution is obtained.

Another recent development in this field is the identification and analysis of the hyperfine splitting produced by C^{13}, as it occurs in its natural abundance of 1 per cent. The normal isotope of carbon C^{12} has no nuclear moment, and therefore no hyperfine structure is produced from the interaction of the unpaired electron with the vast majority of the carbon atoms. However C^{13} nuclei do have spins and magnetic moments, and hence will give rise to an observable hyperfine splitting provided the sensitivity and resolution of the spectrometer is sufficient. REITZ, DRAUNIEKS and WERTZ[36] were some of the first to observe this extra splitting due to C^{13} in its natural abundance, in their work on dihydroxysemiquinone. This has two protons on the central aromatic rings, the other four positions being taken by oxygen atoms. A triplet structure is therefore expected from the equal interactions with the two protons. This can be seen in the centre of *Figure 104*, the main lines due to the protons going right off the scale. In addition to the three main lines there are two pairs of weak triplets which can be seen on either side of the main pattern; these extra splittings can be attributed to the interaction of the unpaired electron with the C^{13} nuclei, their intensity being within about 5 per cent of the expected value.

The fact that such splittings can be detected from C^{13} in its

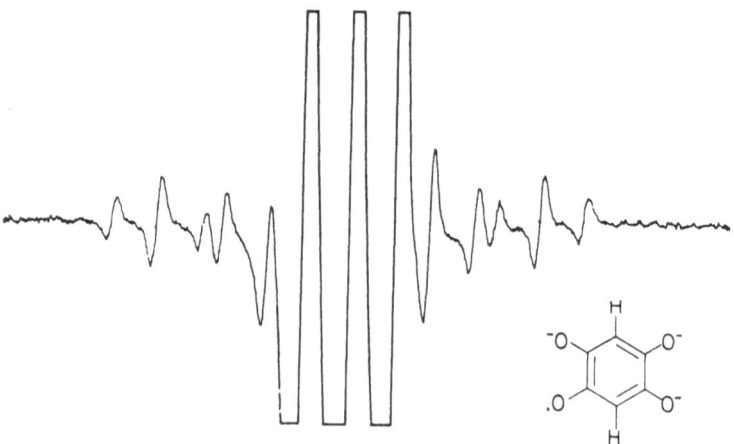

Figure 104. ESR spectrum from dihydroxysemiquinone, showing hyperfine splitting due to C^{13} in its natural abundance (after Reitz, Draunieks and Wertz[36])

natural abundance, offers the possibility of very precise additional checks on the calculations of the unpaired spin density over the whole molecule. It is shown in the next section that theoretical calculations had previously to assume two types of interaction between the carbon atom and the protons attached to them, i.e. either configurational interaction with protons in the same plane as the aromatic ring, or a form of hyperconjugation for the case of protons out of the plane. The calculation of spin densities on the carbon atoms from the splitting of the proton hyperfine pattern therefore involved some assumption as to the nature of this interaction. If, however, measurements are now available which give the spin density on the carbon atom directly, they will provide a far more accurate check on the actual distribution of the wave function over the central skeleton of the carbon structure.

From many aspects of theoretical physics and chemistry, therefore, the experimental investigation of the complex and well resolved hyperfine patterns obtained from semi-stable free radicals in solution, and from the aromatic ions in particular, has proved very fruitful. However, interest in the study of free radicals moved fairly rapidly to the investigation of systems of changing free radical concentration, in which transient species only existing for a relatively short time were formed. Before these kinetic studies are summarized, a brief review is given of the measurements that have been made on single crystals of free radical type compounds.

(c) *Single crystal studies.* The initial measurements on free radicals were made on polycrystalline samples of hydrazyl, but most subsequent studies dealt with free radicals in solution. The great advantage of studying radicals in solution was, of course, that the motional narrowing caused by the tumbling of the solvent molecules produced very narrow resonance lines, in which the anisotropic contributions to the hyperfine structure had been removed. The spectra were thus very simple to interpret and rapid progress in their analysis occurred. The success of this early work in which only the isotropic terms of the hyperfine interaction needed to be considered, is well demonstrated by the results on the semiquinone and the aromatic ions in solution. It soon became evident, however, that in some cases ambiguous interpretations of the observed hyperfine structure could be obtained, and that much more definite characterization of the free radicals would be possible if the anisotropic terms of the interaction could also be found.

The value of single crystal studies in obtaining extra information had already been demonstrated during work on the transition group compounds, and attention was therefore turned to this aspect of free radicals whenever possible. The first single crystals to be measured in detail were organic crystals which had been irradiated with γ-rays or similar high energy radiation [37]. The free radicals so formed were found to retain their orientation in the crystal lattice; they therefore remained aligned with respect to the different crystal and molecular axes, and all showed the same splitting and spectrum when the applied magnetic field made a particular angle with the crystal axis. These anisotropic contributions to the hyperfine splitting arose from the dipolar type interactions which are considered in Chapter 6 for the transition group atoms. The theory of such interactions for free radicals is basically the same as for the transition group atoms, the only major difference being that interaction takes place with several nuclei instead of predominantly with one. A number of different laboratories took up this work and applied it to many free radicals formed by γ-irradiation of solid crystals. ROWLANDS, WHIFFEN and their colleagues [38, 39] at the National Physical Laboratory made a systematic study, and were not only able to measure the anisotropic parameters associated with the hyperfine splittings of the various different nuclei included within the molecular orbital, but were also able to relate these to the nature of the free radicals themselves.

As an illustration of this type of analysis, the recent work of HOLMBERG and LIVINGSTON [40] on single crystals of hydrazyl is considered. The early resonance work on hydrazyl showed that if single crystals of the pure compound are grown, then because of the strong exchange interaction which eliminates the hyperfine splitting, only a single narrow resonance line is observed. However, if the distance between the individual molecules in the hydrazyl crystal is increased by growing mixed crystals with hydrazine, which is diamagnetic, the exchange interaction can be reduced to zero. The hyperfine interaction due to the two central nitrogen atoms can then be resolved even in the solid state. Since all the hydrazyl molecules will be aligned together in a crystal lattice, the hyperfine structure should be anisotropic and reflect the symmetry of the crystal lattice as a whole. Holmberg and Livingston [40] were in fact able to dilute the hydrazyl crystals in this way, and remove the exchange narrowing and make a detailed analysis of the anisotropic hyperfine splitting that was obtained.

In order to distinguish the interaction of the unpaired electron with the two nitrogen atoms they labelled one of the positions with an N^{15} nucleus. This has a spin of $\frac{1}{2}$ and hence produces a doublet structure as its hyperfine interaction, whereas the normal N^{14} nucleus has a spin of 1 and produces a triplet. The actual spectra

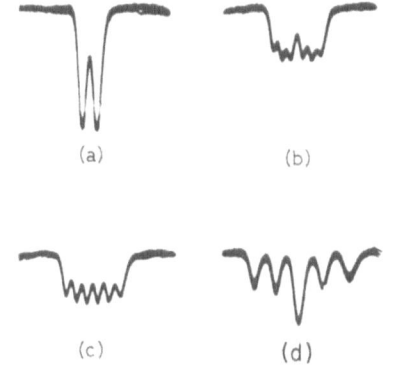

Figure 105. Spectra from single crystals of hydrazyl diluted with hydrazine and with one nitrogen position labelled with N^{15}. The angles between the field and crystallographic c axis were: (a) 70°, (b) 45°, (c) 40°, (d) 0° (after Holmberg, Livingston and Smith[40])

obtained from such an N^{15}-labelled single crystal suitably diluted with hydrazine, are shown in *Figure 105*. The crystal was rotated in the *ac* plane, and spectra for different angles between the field and the *c* axis are given. Spectrum (a) corresponds to the direction in which the hyperfine interaction with the normal N^{14} atom is zero, i.e. an angle of 54° 44′, as explained in Section 10.2. The only interaction left is that with the N^{15} nuclei and hence a doublet splitting is obtained. As the magnetic field is moved away from this orientation, however, the additional splitting due to the N^{14} nuclei becomes apparent and in spectrum (b) it can be seen that each of the two main lines in (a) has been split into three components. The ratio of this new splitting to the doublet splitting gives the ratio of the interaction with the N^{14} and N^{15} nuclei respectively. The six lines can also be clearly seen in spectrum (c) whereas in spectrum (d), which is for the magnetic field applied along the crystalline *c* axis there is a partial overlap between the two central lines.

These angular variations in the two hyperfine interactions can be summarized as in *Figure 106*. The differences of magnetic field from the centre of the pattern for each of the hyperfine components are plotted against the angle to the crystalline *c* axis. The variation of the splittings due to the N^{14} and N^{15} nuclei together with their respective angular variation can be clearly seen and they can then be related to the molecular orbital associated with the hydrazyl itself.

The analsysis of the more complex spectra produced by high energy irradiation of organic crystals is of course somewhat more

375

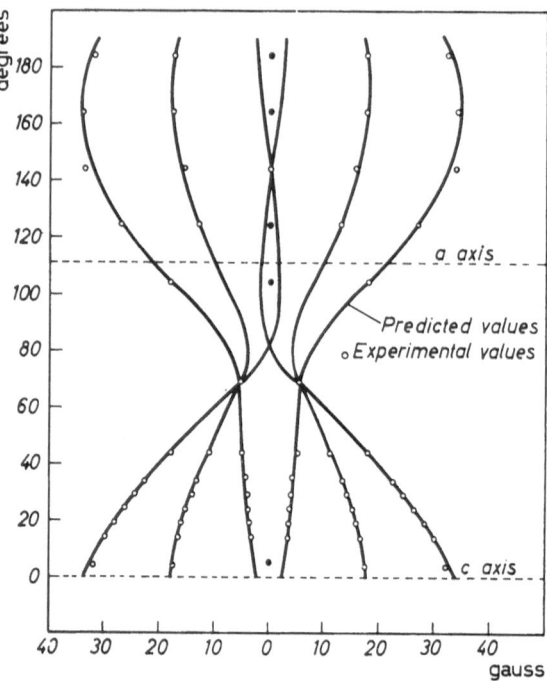

Figure 106. Angular variation of the combined hyperfine pattern due to N^{14} *and* N^{15} *nuclei (after Holmberg, Livingston and Smith[40])*

complicated than this example, but the general idea and principles remain the same. It is thus possible to write down a complete tensor representing the hyperfine interaction in all directions from such angular variations in the different crystallographic planes. From this tensor the detailed nature of the wave functions representing the spread of the unpaired electron over the radical or molecule can be deduced. The wave functions are in turn often critically dependent on the actual structure of the radical itself and hence give precise information on both the nature and structure of the free radicals.

(d) *Free radical kinetics.* The free radical systems which are of most practical importance are normally those associated with transient species taking part in some kind of chemical reaction. The initial electron resonance experiments which were designed to follow such changing radical concentrations, employed low-temperature techniques to slow up the whole reaction and facilitate the study of its kinetics at a reduced rate. One example of this which may be quoted is the early work on the production of secondary radicals in iso-propanol by hydroxyl attack[41], the hydroxyl radicals themselves being produced photolytically by ultra-violet irradiation. The two stages of the reaction can therefore be represented as follows[42]

$$H_2O_2 + h\nu \rightarrow 2 \cdot O\dot{H} \qquad \ldots\ldots(11.3)$$

and

$$(CH_3)_2 \cdot CH \cdot OH + O\dot{H} \rightarrow (CH_3)_2 \cdot \dot{C} \cdot OH + H_2O \qquad \ldots\ldots(11.4)$$

In the actual experiment the hydrogen peroxide was dissolved in low concentration in isopropanol and then this solution was deep frozen at liquid nitrogen temperatures and placed in the cavity resonator. The frozen solution was irradiated by ultra-violet light and the electron resonance spectrum observed at successive time intervals. The production of the hydroxyl radicals could be followed, and also the formation of the secondary isopropanol radicals as it occurred. The actual absorption lines obtained are shown in *Figure 107*, these being taken at 5 min intervals following the initiation of the ultra-violet irradiation. It should be noted that the gain of the amplifier was steadily reduced, as can be seen from the reduction in the noise level on the traces themselves, and hence the actual signal height was increasing far more rapidly than indicated by the traces. It is also evident that initially the electron resonance signal obtained was asymmetric in form and there was not much evidence of hyper-fine structure. As time progressed, however, the asymmetry of the signal disappeared and the seven-line derivative pattern expected from the six methyl protons of the isopropyl radical was obtained. The concentration of these secondary isopropanol radicals can be determined quite precisely from the intensity of the outer lines of the hyperfine pattern since these are not obscured by the initial asymmetric resonance due to the hydroxyl radicals.

The quantitative kinetics of this system may be analysed by allocating a rate constant k_1 to the first reaction, i.e. the ultra-violet photolysis, and a rate constant k_2 to the secondary reaction. The total number n_{OH} of hydroxyl radicals formed after time t, including those that have already reacted to form isopropanol radicals, is given by the expression[42].

$$n_{OH} = 2n_0[1 - \exp(-k_1 t)] \qquad \ldots\ldots(11.5)$$

where n_0 is the number of hydrogen peroxide molecules present initially.

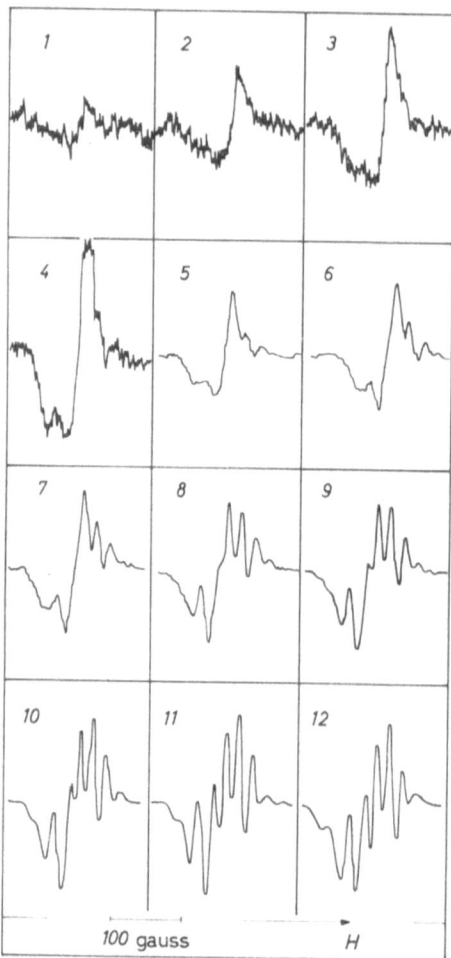

Figure 107. Growth and change of free radical signal during ultra-violet irradiation of isopropanol–hydrogen peroxide solution

The rate of formation of the secondary isopropanol radicals is proportional to the number n'_{OH} of hydroxyl radicals that are actually present in the solvent at a given time. This is equal to the total number n_{OH} of hydroxyl radicals that have actually been formed, less the number that have already reacted; but the number that have already reacted is, of course, equal to the number n_B of isopropanol radicals that have already been formed. The equation for the

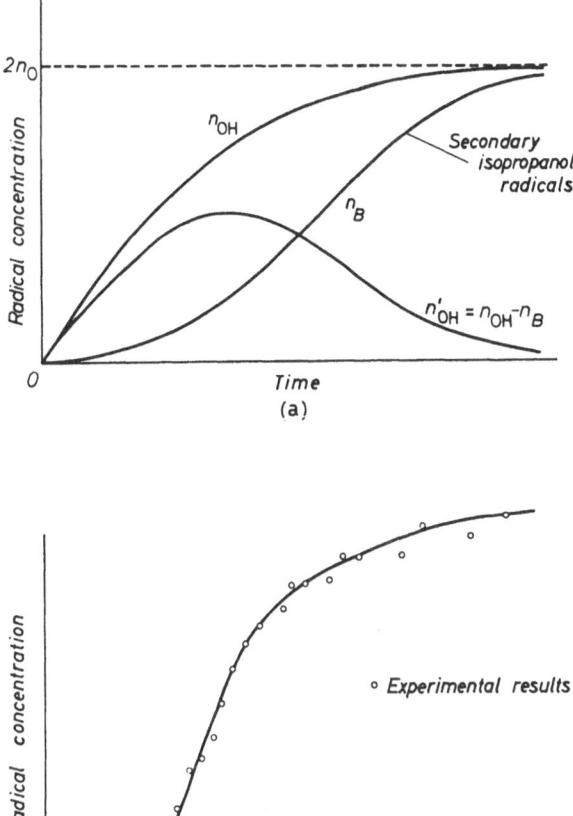

Figure 108. Growth of secondary isopropanol radicals: (a) *predicted*, (b) *observed*

formation of the secondary radicals is therefore given by

$$\frac{dn_B}{dt} = k_2(n_{OH} - n_B) \qquad \dots (11.6)$$

It is now possible to substitute for n_{OH} in equation (11.6) from equation (11.5), and to obtain an actual expression for the number of secondary radicals expected after any time t. This is given by

$$n_B = n_0 \left[1 - \frac{k_2 \exp(-k_1 t) - k_1 \exp(-k_2 t)}{k_2 - k_1} \right] \quad \ldots (11.7)$$

It is probably far easier to visualize these results graphically, and the variation of the primary and secondary radical concentrations are shown in *Figure 108(a)*. The initial rise, followed by a decay, for the hydroxyl radicals is clearly shown and it is seen that the concentrations of the secondary radicals grow rather slowly at first and then the curve rises steeply after which the growth is again slow. This latter curve may be compared directly with the experimentally observed growth of the secondary radicals as shown in *Figure 108(b)*, and it is quite clear that this experimental curve has the same general features as that predicted by the simple analysis carried out above. It is also evident that the actual values for the two rate constants k_1 and k_2, can readily be found by correlating the two curves. Determination of these constants would not be of any practical value in this particular case since the system is very artificial, but this early work showed that the normal methods of kinetic analysis could be applied to electron resonance results when changing free radical concentrations were being measured.

In more recent work[43], rapid recording spectrometers are being employed to follow radical concentrations which change in millisec time intervals. These same general principles of analysis are applied to systems which are of practical value and take place at room temperature, rather than in the frozen state; the experimental techniques employed are summarized in Section 11.2.

One example in which a dynamic concentration of transient radicals was built up at room temperature is afforded by FESSENDEN and SCHULER[44], who produced transient alkyl radicals in liquid hydrocarbon systems by bombarding them with 2·8 MeV electrons. The electron beam was brought down the axis of applied d.c. field in an evacuated pipe and then directed into the liquid contained in a normal rectangular cavity.

The kind of spectrum they obtained is shown in *Figure 109(a)* which is for liquid ethane irradiated at −180° C. The twelve numbered lines can be shown to be due to the ethyl radical, while the weak lines between (5) and (6) and between (7) and (8) are caused by the methyl radical. The electron resonance spectra not only enable the nature of the radicals formed in this way to be identified, but the kinetics of the reaction can also be followed from the dynamic concentration of radicals that is built up, and from their rates of disappearance. Thus the temperature dependence of the ethyl radical concentration in the liquid ethane is shown in *Figure*

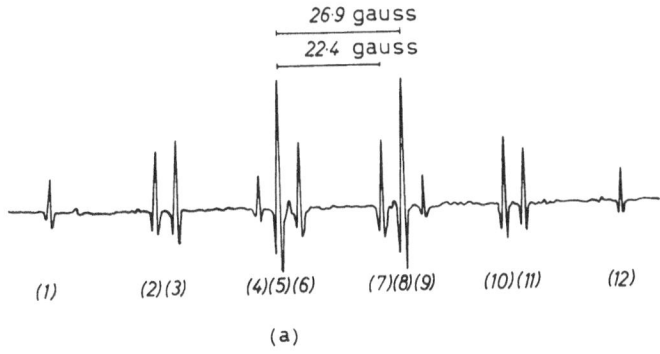

(a)

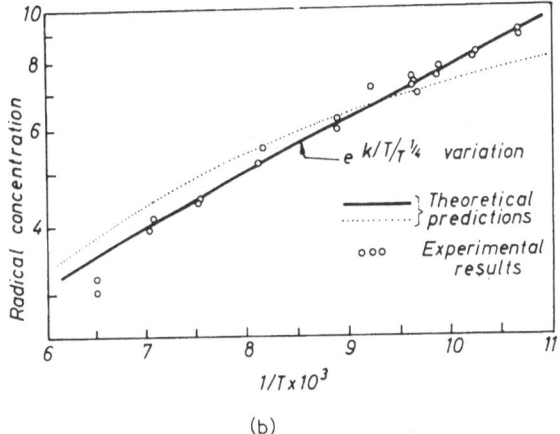

(b)

Figure 109. Radical production by electron bombardment : (a) ESR spectrum of liquid ethane, (b) variation of ethyl radical concentration with temperature of liquid ethane (after Fessenden and Schuler[44])

$109(b)$, as a plot of radical concentration against the inverse of the temperature T. A comparison is made between the experimental results and two theoretical predictions[44, 45], from which the first and second-order rate constants can be determined.

11.5 FREE RADICAL STUDIES—MOLECULAR ORBITAL THEORY

It is shown in earlier chapters that there are five basic parameters which characterize any electron resonance spectrum:

(1) The integrated area under the absorption line, which gives the actual number of unpaired electrons present in the sample.

(2) The width of the absorption line, which gives information on the interactions taking place between the electron and the lattice or molecule as a whole.

(3) The g value or 'spectroscopic splitting factor' as it is sometimes called, which is determined by the resonance field value and gives information on the interaction between the electron spin and angular momentum.

(4) The hyperfine interaction, due to the presence of nuclear magnetic moments.

(5) The electronic splitting which occurs, or may occur, in atoms which have more than one unpaired electron associated with them.

Although each of these parameters may in turn give information of considerable significance in the case of the transition group compounds, one or two of them can be effectively eliminated where free radicals studies are concerned. The g values are always very close to the free-spin value 2·0023, because the unpaired electron associated with a free radical is moving in a molecular orbital which is delocalized over the whole molecule, and thus strong spin-orbit interaction with any one atom does not take place. Although some significant variations have been noticed[46] in the departure of g values of free radicals from this free-spin value, on the whole, g value determination does not give very useful information in free radical studies. The electronic splitting can also be eliminated, since no free radicals possess more than one unpaired electron per molecule, with the particular exception of triplet states which are considered in detail in Section 11.7. Also, with the exception of studies on exchange narrowing, very little information has been deduced from studies on the line widths of free radical spectra. It is therefore clear that the main feature of an electron resonance spectrum in the study of free radicals is the existence and nature of the hyperfine pattern.

The basic theory for the hyperfine interaction in free radicals is exactly the same as that for transition group complexes given in Sections 6.7–6.8; it is pointed out in these sections that there are basically two types of interaction: one is the anisotropic splitting which arises from the classical dipole-dipole type of interaction between two magnetic moments; the other is the isotropic interaction, often referred to as the Fermi contact term, which arises from the finite probability of the unpaired electron being at the site of the nucleus itself. Although in general both of these interactions take place, the anisotropic interaction can be averaged to zero if the molecule being studied is in the liquid state and is re-orientating itself very rapidly in the applied magnetic field. If the rate of tumbling is noticeably faster than the hyperfine splitting, as measured in frequency units, then the anisotropic component of this hyperfine interaction is averaged to zero in a time associated with the microwave frequency, and as a result motional narrowing occurs. The detailed treatment of this phenomena follows the same steps as for the similar motional narrowing that occurs in nuclear resonance, which is treated in Section 10.2.

The existence of motional narrowing explains why free radicals in solution can give a very well resolved hyperfine pattern, consisting only of the isotropic interaction terms. It was these interactions which were studied and analysed in detail, when measurements were made on such systems as the aromatic ions in solution. A summary of this isotropic splitting therefore follows, and in particular a brief description is given of the two major mechanisms of interaction that occur in free radical systems, since these serve to bring out the interesting correlation that has been obtained between experiments and theory in this field of investigation.

(a) *Isotropic hyperfine interaction.* The isotropic component of the hyperfine interaction in the Spin Hamiltonian expression can be written as

$$\mathcal{H} = g_e \cdot g_N \cdot \beta \cdot \beta_N \cdot \left[\frac{8\pi}{3} \cdot \boldsymbol{S} \cdot \boldsymbol{I} \cdot \delta(r_e - r_N) \right] \quad \ldots (11.8)$$

In this expression the Dirac delta function $\delta(r_e - r_N)$ only has a non-zero value if the molecular orbital, occupied by the unpaired electron, does not vanish at the position of the nucleus in question. In the case of single atoms it is evident that p, d or f orbitals do not produce isotropic hyperfine structure since they have lobes of electron density pointing away from the nucleus in different directions, but a zero node at the nucleus itself. Only the s orbitals, which possess no asymmetrical lobes, have a finite density at the site of the nucleus itself. The same general comment also applies to the more delocalized molecular orbitals for free radicals. Thus the π orbitals, with one unit of angular momentum, have two lobes pointing in opposite directions corresponding to the atomic p orbitals, and also have a zero node at the central site of the nucleus. Only the σ molecular orbitals corresponding to the atomic s orbitals have any finite electron density at the nucleus, and hence produce any isotropic hyperfine splitting.

It follows from this that no isotropic hyperfine splitting is to be expected for any free radicals which do not possess σ type bonding in their molecular orbital. At first sight this would appear to automatically eliminate all the aromatic free radicals since, in order that the unpaired electron associated with these systems of plane ring structures can become delocalized and move over the whole molecule, it must be located in the π orbital. Therefore the density of the wave function of the unpaired electron must be confined to regions above or below the plane of the aromatic rings, and must have a zero value at the site of any nucleus in the aromatic structure. Similarly this applies to the protons held round the edge of aromatic hydrocarbons in the same plane as the skeleton of carbon atoms themselves. But, as shown in Section 11.4(b), aromatic hydrocarbons were found experimentally to have a hyperfine structure as noticeable and well-resolved as other compounds, such as triphenyl methyl.

Some additional interaction would therefore seem to be taking place, so that the wave function associated with the unpaired electron can take up a distribution which includes the site of the nuclei themselves. This type of interaction had in fact to be introduced in the theory of hyperfine splitting in transition group compounds, to account for the large isotropic splitting of about 100 gauss that was observed between successive hyperfine components of the Mn^{2+} spectra. It is pointed out in Section 6.10(b) that ABRAGAM and PRYCE[47] were able to explain this splitting in terms of a configurational interaction between the $(3s)^2(3d)^5$ ground state and the $(3s)^1(3d)^5(4s)^1$ excited state. As a result of this interaction the ground state wave function takes on a small amount of the distribution corresponding to the excited state. The wave function therefore includes a finite probability of unpaired electron density in the s orbitals at the nucleus, and the possibility of large isotropic hyperfine splitting arises.

This concept of configurational interaction can now be extended from the atomic orbitals of single paramagnetic ions to the molecular orbitals of the complex free radical. The mechanism produces the same result in principle, and thus some of the σ orbital character of an excited state becomes admixed with the pure π orbital character of the ground state to give a finite probability of the unpaired electron at the site of the nuclei. The detailed theory of this configurational interaction as it occurs in free radicals, has been undertaken by various authors including WEISSMAN[48], McCONNELL[49] and JARRETT[50].

The basic steps in this kind of theory can be briefly summarized. The ground state and the excited state molecular orbitals are first considered and may be represented as follows

Ground state	Excited state
[Filled orbitals]. $(\sigma_B)^2 . \pi$	[Filled orbitals]. $(\sigma_B)^1 . \pi . (\sigma_A)^1$

It should be noted that the σ_B orbital is made up of a carbon sp^2 hybrid orbital, and a $1s$ orbital of an attached proton, whereas the σ_A is exactly the same hybridization but with the opposite signs for the overlapping carbon and hydrogen orbitals, and is thus an anti-bonding instead of a bonding orbital. The configurational interaction may be considered as admixing some of the excited state orbital distribution into the ground state distribution, and the actual form of the resultant distribution may be calculated quantitatively by considering the actual wave functions of the states concerned. The wave functions ϕ corresponding to the distribution of the three electrons among the $(\sigma_B)^2 . \pi$ and the $(\sigma_B)^1 . \pi . (\sigma_B)^1$ orbitals may now be written down as follows

Unperturbed ground state
$$\phi_g = A \left\| \alpha . \sigma_B(1), \beta . \sigma_B(2), \alpha . \pi(3) \right\| \qquad \ldots . (11.9)$$

Excited state
$$\phi_1 = A \left\| \alpha . \sigma_B(1), \beta . \sigma_A(2), \alpha . \pi(3) \right\|$$
$$\phi_2 = A \left\| \beta . \sigma_B(1), \alpha . \sigma_A(2), \alpha . \pi(3) \right\| \qquad \ldots . (11.10)$$
$$\phi_3 = A \left\| \alpha . \sigma_B(1), a . \sigma_A(2), \beta . \pi(3) \right\|$$

In these expressions, A represents the anti-symmetrization and normalization operator, and α and β represent the two possible spin quantizations. The excited state is more accurately described by a linear combination of the three basic functions (11.10) which give two doublet states and one quartet state. Since the quartet state is of different spin muliplicity to the ground state they cannot admix. The resultant configurational admixture in the ground state, corresponding to the wave function ϕ_0, can thus be written as a sum of the ground state and contributions from the two doublet excited states:

$$\phi_0 = \phi_g + \lambda_x . \phi_x + \lambda_y . \phi_y \qquad \ldots . (11.11)$$

where λ_x and λ_y are the admixture coefficients.

The state given by ϕ_x corresponds to a combination of a singlet $(\sigma_B)(\sigma_A)$ with a π orbital; hence it has paired spins in the σ orbitals and no unpaired spin density at the site of the nucleus. The s hyperfine splitting is therefore produced entirely by the ϕ_y state, and the quantitative calculation of this splitting resolves into the determination of λ_y. If it is assumed that the π orbital is a linear combination of the carbon $2p_z$ orbitals, then an expression for λ_y can be written in terms of the electrostatic repulsion integral, usually denoted by G and corresponding to the repulsion term e^2/r_{12} between the bonding and anti-bonding orbitals. λ_y is therefore given by the expression

$$\lambda_y = -\tfrac{1}{2}\sqrt{6} \cdot \frac{G}{E_y - E_0} \cdot \rho_i \qquad \ldots\ldots (11.12)$$

where ρ_i is the density of the unpaired spin on the particular carbon atom. The hyperfine interaction energy ΔE can then be written in the form

$$\Delta E = \frac{32\pi}{3\sqrt{6}} \cdot \lambda_y \cdot \mu_N \cdot \mu_B \cdot \sigma_B(\mathbf{r}_N) \cdot \sigma_A(\mathbf{r}_N) \qquad \ldots\ldots (11.13)$$

where μ_N and μ_B are the magnetic moments of the nucleus and electron respectively. From this it is seen that the actual splitting is linearly dependent on the magnitude of λ_y. Since λ_y is linearly dependent on ρ_i, it follows that the hyperfine splitting ΔH, produced by a proton attached to a given carbon atom, should be linearly proportional to the unpaired electron density on that carbon atom. Thus there should be a very general relation for all aromatic hydrocarbons of the form

$$\Delta H = Q \cdot \rho_i \qquad \ldots\ldots (11.14)$$

The existence of an unpaired spin density at the proton also implies the existence of unpaired spin at the carbon nucleus, since both σ_B and σ_A are involved in the expression. Hence, if any C^{13} is present in the aromatic ring, it should produce a hyperfine splitting which can be directly related to that produced by the protons. Recent work[36] has shown that such hyperfine splitting from C^{13} can be observed, giving a very direct check on the theory of configurational interaction. The actual quantitative value Q can also be derived from the theory if various quantitative values are taken for the wave functions concerned. First-order calculations by Jarret[50] and McConnell[49], indicated that Q should be 28 gauss. This value was found experimentally to be very close to that observed for most aromatic free radicals, the majority of which have total overall proton hyperfine splittings varying from 23 to 29 gauss.

Some of the radicals however showed a hyperfine splitting considerably greater than the 28 gauss predicted, for example perylene gave an extreme splitting between the outermost hyperfine lines of 30 gauss. In order to correlate this with the theory outlined above, McCONNELL and DEARMAN[51] introduced the concept of negative spin density. This effect is produced by the influence of the unpaired electron on the orbitals of paired electrons close to it. It may be viewed as a form of spin repulsion, with the result that there is an effective partial unpairing of the previously balanced orbits. An extra odd electron density is thus produced, of opposite direction to that of the unpaired electron responsible for it; this new spin density can therefore be given a negative sign. Since the positive spin densities aligned in the same direction as the original unpaired electron also increase in magnitude, the algebraic sum of spin densities remains at unity. The hyperfine splitting, however, is independent of the sign of the spin density and is proportional to the sum of the moduli of the spin densities, which may now be greater than unity when summed over the whole molecule. Hence an overall hyperfine splitting of greater than 28 gauss can be expected.

These effective negative spin densities are not only apparent in the spectrum observed from the aromatic hydrocarbons but also from some of the aliphatic compounds such as allyl alcohol, in which the sp^2 lobes of a carbon hybrid orbital give a molecular structure very similar to that of the top half of an aromatic ring. This radical has the formula

$$
\begin{array}{ccc}
 & H & \\
 & | & \\
H & C & \\
\diagdown & \diagup \quad \diagdown & \\
C & & \dot{C} \\
| & & | \\
H & & H \\
\end{array}
$$

Simple molecular orbital or valence bond theories, when applied to the three central carbon atoms, predict an unpaired spin density of $\frac{1}{2}$ on each of the end carbon atoms with a zero on the centre carbon. If this were actually so in practice a four-line hyperfine structure would be expected from the three protons, which would then be coupled to the unpaired electron orbit. A more detailed calculation which takes into account all the possible π electrons and configurations and allows for the effect of negative spin densities, predicts, however, that the spin densities on the two outer carbon atoms should be about 0·6 while that on the central carbon atom should be about $-0·2$. Thus a significantly greater splitting is expected between the four main components of the hyperfine pattern, each of which should be split into two sub-components of separation about 5 gauss; this is in fact observed.

To summarize this work on the configurational interaction, the main success of the theoretical treatment has been in demonstrating the linear dependence of the hyperfine splitting on the unpaired spin density associated with the carbon atom to which the proton is attached. It has also shown that there should be a general constant of proportionality for this splitting in all aromatic-like proton bondings. The quantitative comparison of theory with experiment therefore resolves itself into the detailed calculation of the unpaired spin densities expected at each carbon atom, involving the possibility of negative spin densities as explained above. Such theoretical calculations were first made on the basis of the Huckel[33] theory of molecular orbits for the aromatic hydrocarbons, but more refined and sophisticated methods have since been tried using computers to analyse the actual details of the wave functions.

McLACHLAN[52] is one of the theoretical chemists who extended Huckel's original theory by introducing configurational interaction corrections and other effects. A large number of workers in this field have followed his methods to obtain very good agreement between the experimentally observed hyperfine splittings and the spin

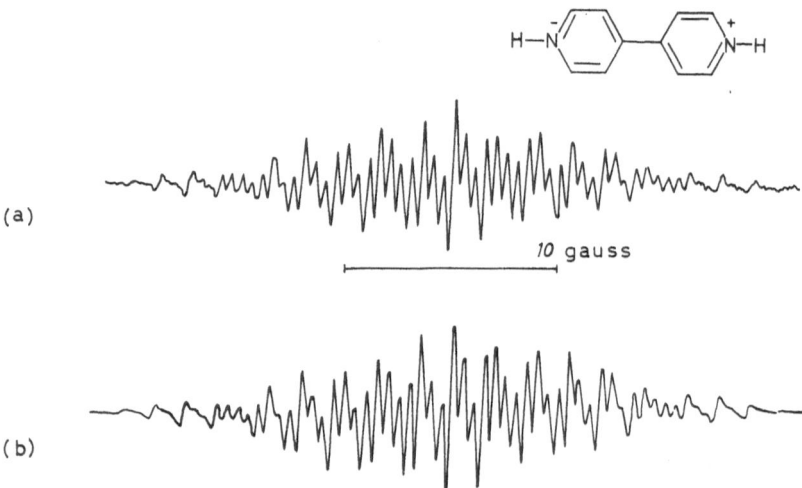

(a)

10 gauss

(b)

Figure 110. ESR spectrum of the bipyridyl radical cation in aqueous solution: (a) observed, (b) computed (after Johnston et al.[54])

densities predicted by the more sophisticated theories. An example of this is a full treatment of the spin density distribution in nitrile anion radicals by BOLTON and FRAENKEL[53], in which comparison was made with the nitrogen and the C^{13} hyperfine splitting constants, as well as those obtained for the protons.

The direct use of computer techniques in the analysis of such spectra is shown in Figure 110. Figure 110(a) is the spectrum obtained by JOHNSTON et al.[54] from the bipyridyl radical cation with the structural formula shown; the trace in Figure 110(b) was actually produced by a data plotter fed from an Illiac computer programmed to synthesize the spectrum from given splitting constants and line shapes.

It is evident that these theoretical calculations have now reached a high degree of sophistication and can be compared very precisely with the experimental results obtained from the high resolution spectra available.

(b) *Hyperconjugation.* The description of the mechanism of configurational interaction, given in the last section, explains how a finite hyperfine interaction can occur between the unpaired electron which is predominantly in the π orbital of an aromatic ring and edge protons which are situated around the edge of the aromatic ring. It does not explain how such an unpaired spin density can also be transferred to the protons attached to an aromatic ring but not in the plane of the ring itself.

An example of this is when methyl groups are attached to carbon atoms in the ring. The carbon atoms of the CH_3 groups then have four bonds and hence have sp^3 configurations, with lobes overlapping the three $1s$ orbitals of the protons and the one sp^2 lobe of a ring carbon. There are no free p orbitals to form a π bond system with the ring, and consequently no unpaired electron density is to be expected at the methyl groups, and no direct configurational interaction with the protons of the type considered in the last section is now possible. A similar situation also occurs in aliphatic compounds, where CH_3 groups may be directly coupled to another carbon atom which is itself part of a carbon chain. In both these cases an un-

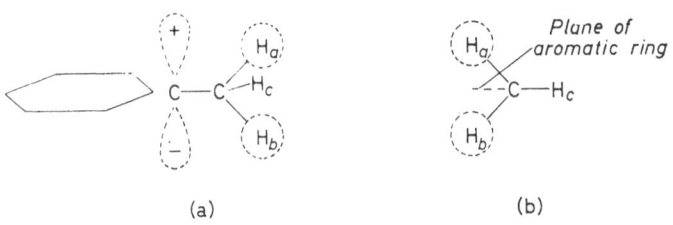

(a)   (b)

Figure 111. Hyperconjugation: (a) side view, (b) end view

paired electron may be considered as localized in the p orbit of the neighbouring carbon atom, and some mechanism of interaction whereby the spin density can be passed on to the proton attached to the neighbouring methyl group must now be postulated.

Quite a considerable interaction does exist, as demonstrated experimentally by the hyperfine splittings of some magnitude in radicals containing such methyl groups. The situation can be represented schematically as in *Figure 111*, where the unpaired electron is shown primarily located in the p orbit of the central carbon atom; the possible overlap of this with the s orbitals on the methyl protons are also indicated. Some interaction mechanism is therefore required which allows an admixture of the p orbit with the $1s$ orbits of the protons. If the $1s$ orbital of the lower proton is taken with a negative sign and combined with the positive orbital of the upper proton, then the resultant combined wave function has the same symmetry as the p_z orbital of the carbon atom. There is then a possibility of direct overlap between these two sets of orbitals, and since the p_z orbital possesses the unpaired electron, the unpaired electron density can be shared with the protons of the methyl group. So far this interaction has been confined to two protons of the methyl group, but if this group is rotating at high speed around the direction of the carbon/carbon axis, then all three protons pass through the overlap positions in turn. Hence they participate equally in the overlap and thus produce an equal hyperfine splitting in the observed spectrum.

This type of interaction, which depends on a spatial overlap of wave functions with the same symmetry, is termed 'hyperconjugation'. It depends essentially on correct spatial alignment and not on interaction involving excited levels, as in the case of configurational interaction. It is therefore a somewhat more direct and potent effect than the configurational interaction, and as a result very nearly the same hyperfine splitting is produced by each proton of the methyl group as for a single proton connected directly to a ring carbon atom. The basic equation definining this splitting can be written in the form

$$\varDelta H \approx 30 \, . \, \rho_i \text{ gauss} \qquad \qquad \dots (11.15)$$

where ρ_i is the unpaired spin density on the *neighbouring* carbon atom.

Although methyl groups have been used as an example of hyperconjugation, they are not the only configurations for which such couplings are possible. Any other groups in which suitable wave functions can be formed by a linear combination of proton orbitals can also take part in an interaction of this type. One common example is a CH_2 group where the carbon is part of an aromatic ring or is taking part in a similar π bond system.

In the discussions on configurational interaction and hyperconjugation, attention has been concentrated on the way in which the unpaired spin density, normally associated with a π bond or p orbital of the ring carbon, can be partially transferred to a $1s$ orbital of a proton. The basic hyperfine interaction which then takes place between this unpaired spin in the s orbital and the nucleus itself, is given by the Fermi contact interaction as discussed in Chapter 6. For all studies in solution, this isotropic splitting due to the Fermi contact term is by far the most important effect, since the anisotropic terms are either averaged to zero by the molecular motion or just appear as an extra broadening.

(c) *Anisotropic hyperfine interactions.* The anisotropic terms of the hyperfine interaction are given in the general Hamiltonian, and are also considered in great detail in Section 6.8 as applied to transition group atoms. The general angular variation produced is seen to contain a term $(3 \cos^2 \theta - 1)$, where θ is the angle between the direction of the applied magnetic field and the direction of quantization for the electron orbitals. To first approximation the anisotropic interaction can therefore be written as

$$\mathscr{H} = g_e \cdot g_N \cdot \beta \cdot \beta_N \cdot \sum_k (3 \cos^2 \theta_k - 1)/r_k^3 \text{ gauss} \quad \dots (11.16)$$

which can be simplified for the case of proton interaction to

$$\Delta H = B \cdot \sum_k (3 \cos^2 \theta_k - 1)/r_k^3 \text{ gauss} \quad \dots (11.17)$$

where r_k is the distance between the nucleus and the unpaired electron in Angstrom units, averaged over the wave function concerned.

If single crystals of the free radical compound are available, as has been possible in recent years for irradiated organic crystals[38, 39], then the anisotropic terms can often be measured directly from the observed spectrum, and hence the detailed nature of the wave functions deduced. If the free radical is studied in the solid state but not as single crystals the anisotropic hyperfine interaction tends to produce a broadening of the spectrum since its angular variation is summed over the random orientation of molecules present. As discussed earlier, however, if the molecules are either moving in a viscous medium, or undergoing hindered rotation or diffusion in a solid approaching the melting point, then it is possible for motional narrowing to occur.

It is evident that a whole range of intermediate conditions exist between the two extremes of (i) a dilute solution in which the free radical molecules are tumbling at a very fast rate, and (ii) single crystals in which all the molecules are accurately and permanently aligned together. The nature of the hyperfine structure seen in these intermediate stages, depends crucially on the comparison between the rate of the tumbling motion and the frequency of the hyperfine splitting which the anisotropic terms in the hyperfine structure produce. McConnell[55] carried out the quantitative treatment for the general case, giving an expression for the resultant line width or splitting $\Delta\omega$ as

$$\Delta\omega = \frac{1}{T_1} + \frac{1}{T_2'} \qquad \ldots (11.18)$$

where

$$\frac{1}{T_1} = \frac{\pi^2}{h^2}[9b_{mk}^2 + \tfrac{9}{2}b_{lk}^2]\frac{\tau_c}{(1 + 4 \cdot \pi^2 \cdot \nu^2 \cdot \tau_c^2)}$$

$$\left(\frac{1}{T_2'}\right)^2 = \frac{2\pi}{h^2} b_{lk}^2 \tan^{-1}\left(\frac{2 \cdot \tau_c}{T_2'}\right)$$

τ_c is the correlation time and can be calculated approximately from the relation

$$\tau_c = 4\pi \cdot \eta \cdot a^3/3kT \qquad \ldots (11.19)$$

where η is the viscosity of the medium and a is the effective radius of the free radical. The other constants b_{lk}, b_{mk} are derived from a time average over the wave functions and have orders of magnitude of $g \cdot \beta \cdot g_N \cdot \beta_N[1/r_k^3]_{AV}$. The variation of the resolution of electron resonance spectra with correlation time can be quite important when employed in this way in the study of glassy or polymer type systems.

11.6 Ligand Field Theory of Transition Group Complexes

One of the more significant advances that has taken place in the development of the theory of electron resonance in recent years is the application of ligand field theory to electron resonance studies. The analysis of the initial electron resonance measurements is considered at some length in Chapter 6, including a brief summary of how such workers as Abragam and Pryce[56] developed the crystal field theory initiated by Van Vleck[57] and others, and employed it to explain the g value variations obtained in the electron resonance measurements. The essential feature of this kind of treatment is to regard the transition group atom as located in an electric field, whose symmetry and magnitude are determined by the dipolar charges on the surrounding atoms or groups. The energy level splitting of the orbitals on the paramagnetic ion was determined by group theory methods as indicated in Chapter 6, and the results obtained for the first transition group ions are summarized in *Figure 60*. The concept of the Spin Hamiltonian was also introduced then making it possible to derive a good general theory for the angular variation of the observed g values, and to account for a large amount of the data available on hyperfine splittings.

A year or so after the initial measurements on hyperfine structure were obtained from diluted paramagnetic salts, further super-hyperfine structure was observed in various compounds[58, 59]. One of the first substances studied was iridium chloride, in which a pronounced hyperfine pattern was obtained from the surrounding chlorine atoms[58]. Thus it was soon clear that the additional splittings arose from an interaction of the unpaired electron with the nuclei of the ligand atoms, rather than from an interaction purely with the nucleus of the paramagnetic atom. The unpaired electron could therefore not be considered as spending the whole of its time on the paramagnetic atom, since its wave function must be embracing some of the surrounding ligand atoms. The straightforward crystal field theory could not account quantitatively for the additional interactions, but the alternative approach of molecular orbital theory, in which the molecular orbitals embracing different atoms are specifically considered, appeared to have great potentialities.

The general idea of the molecular orbital method of theoretical analysis can be summarized as, 'an attempt to find orbitals which fulfil the same function for molecules as the well-defined s, p, d, orbitals do for atoms'. The electronic structure of the molecule can then be described by assigning electrons, two at a time, to these molecular orbitals. In any particular problem the determination of the molecular orbitals themselves is accomplished by attempting to combine the orbitals on the central paramagnetic atom with those on its ligand, following the general rule that the symmetry of the orbitals on the atoms must match. It is often found that only certain combinations of ligand orbitals have the same symmetry as, for instance, the d orbitals on the central metal atom. Combinations of the appropriate central metal atom and ligand orbitals can then be made to give a resultant molecular orbital picture of the complex itself. As a specific example, the energy level diagram for copper phthalocyanine[60] is shown in *Figure 112(b)*.

Copper phthalocyanine is a simple molecule, in the sense that the copper atom is surrounded by a square of four nitrogen atoms, which themselves reside in a plane composed of other organic groups[61]. Each molecule therefore consists of symmetrical planar structure with the copper atom at the centre as shown in *Figure 112(a)*. The determination of suitable molecular orbitals for this particular complex therefore resolves itself into a comparison of the d orbitals on the copper with the π and σ orbitals on the surrounding nitrogens. This construction of the molecular orbitals and their corresponding

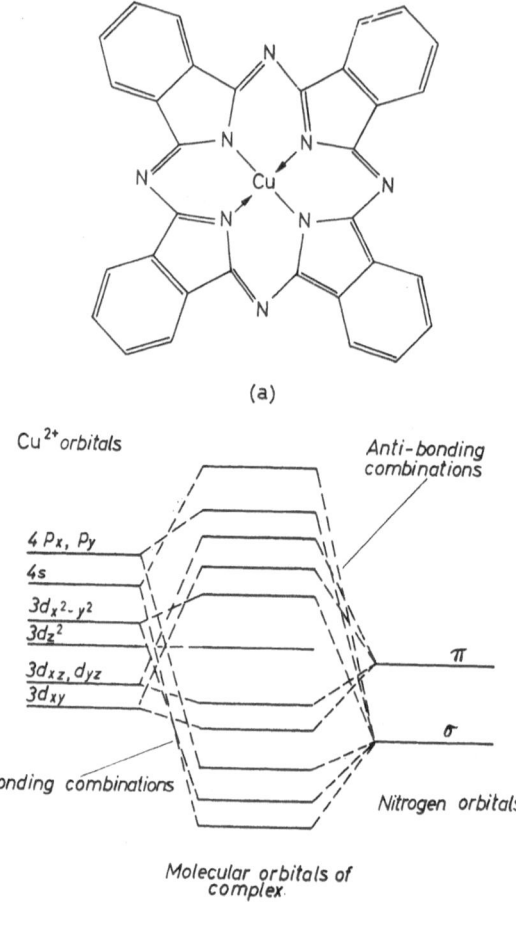

(a)

(b)

Figure 112. Copper phthalocyanine: (a) structural formula, (b) postulated energy level scheme

energy level system, from the orbitals and energy levels of the copper
and ligand atoms are shown in *Figure 112(b)*. It may be seen from
this diagram that when two orbitals combine to form a molecular
orbital, new orbitals are produced, one of which is more stable and
of lower energy than either of the original atomic orbitals, while the
other is higher. These two molecular orbitals are termed the
'bonding' and 'anti-bonding' orbitals respectively.

It gradually became obvious in applying these theories that neither the crystal field theory by itself[62], nor the simple molecular orbital theory, could fully account for all the experimental results obtained by electron resonance studies of transition group compounds. The crucial point in either theory is the energy level system which it predicts; the correct assignment of these energy levels is more important than a precise determination of whether the energy level splittings arise from an internal crystalline field effect, or from linear combinations of atomic orbitals. A dual approach has therefore developed embodying the concepts of both theories.

The acceptance of this dual approach to the problem of theoretical analysis has given rise to the term 'ligand field theory'. This concentrates on the energy level splittings produced for the complex itself, and considers the total observed splitting as arising from a combination of effects, including the electrostatic effect of the internal field and those due to σ and π bonding; these latter can be taken as a direct contribution from the molecular orbitals. As a practical example of this dual approach, the expected energy level systems for the d orbitals of the first transition group atoms can be considered when the ions are placed in different crystalline electric fields.

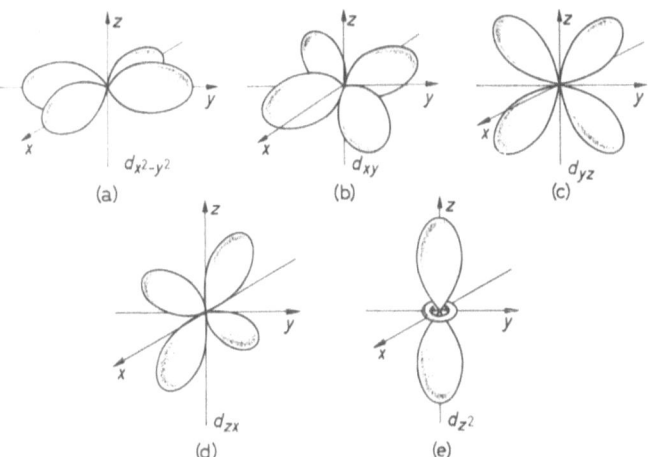

Figure 113. Orbitals of the d electrons

First, the form of the d orbitals is considered for a completely free paramagnetic atom, with no external fields to influence it. The physical distribution of these orbitals in space is shown in *Figure 113*, from which it can be seen that five different spatial distributions are in fact possible. *Figure 113(a)* corresponds to the four electron density lobes pointing along the x and y axes respectively, and is labelled the $d_{x^2-y^2}$ orbital. The next is obtained by rotating this through $45°$ so that the electron density lobes are still in the xy plane but bisect the x and y axes; this is termed the d_{xy} orbital. Two more similar orbitals, d_{yz} and d_{zx}, exist where the electron density lobes bisect the axes in the yz and zx planes respectively. The only remaining possible configuration, shown in *Figure 113(e)*, occurs when the electron density is mainly concentrated along the z axis, with a small annulus of probability density around this axis in the xy plane. The first four of these five orbitals are very similar, and consist of four lobes at right angles to each other. Only this last orbital, designated the d_{z^2} orbital, is significantly different in shape from the others. The difference arises because the z axis is chosen as the axis of quantization.

In the absence of any perturbing electric or magnetic field these orbitals all have the same energy, since there is no direction to differentiate between them. It should be noticed in this connection that if all five orbital distributions are added together, a spherically symmetric distribution of charge is produced similar to an s orbital distribution; this is of course to be expected for all closed shells. The particular energy level splittings caused by the presence of surrounding ligand atoms are now considered.

The simplest case is one central paramagnetic atom surrounded by six ligand atoms, with the two members of each pair at equal distances from the central paramagnetic atom and in opposite directions along the x, y and z axes; this is indicated diagrammatically in *Figure 114*. It is clear that the $d_{x^2-y^2}$ and the d_{z^2} orbitals are now in proximity to the electron concentration on the ligand atoms, and in quite a different configuration from the other three. In the case of the $d_{x^2-y^2}$ orbital, the electron spin density of the d orbital is concentrated so that it points along the x and y axes. There is thus considerable repulsion between the $d_{x^2-y^2}$ orbital and any containing an electron spin density which resides on a point along the x or y axis. The d_{xy} orbital does not suffer this form of repulsion since its lobes are pointing out along the bisector of the xy axes, and consequently it does not interact strongly with the orbitals of the surrounding ligand atoms located on the axes. Similarly this applies to the d_{yz} and d_{zx} orbitals. As a first-order effect, the atoms surrounding the ion complex would therefore be expected to split the energy level system, moving the $d_{x^2-y^2}$ level noticeably above the d_{xy}, d_{zx} and d_{yz} levels.

The d_{z^2} orbital is affected in the same way as the $d_{x^2-y^2}$ orbital, since its two lobes are also pointing towards positions of high electron

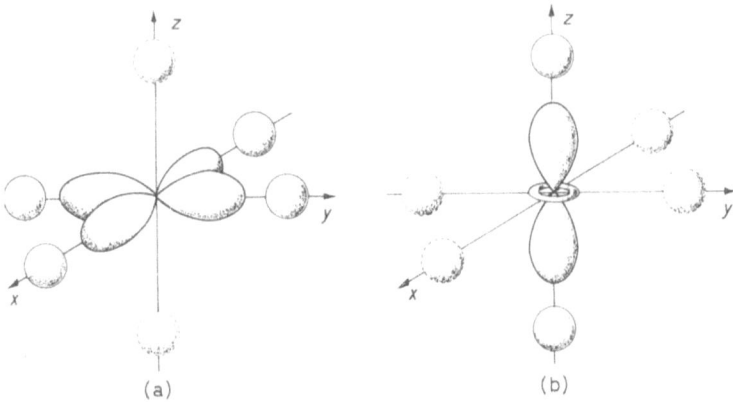

Figure 114. Orbitals of an octahedral complex: (a) interaction of $d_{x^2-y^2}$ with four ligand atoms, (b) interaction of d_{z^2} with other two ligand atoms

density, located on the ligand atoms along the z axis. Thus the first effect of six surrounding ligand atoms is to split the energy level pattern of the d orbitals into two groups. The d_{xy}, d_{yz} and d_{zx} are together in the lower energy group; the $d_{x^2-y^2}$ and d_{z^2} are both raised in energy by the repulsion associated with the electrons on the ligand atoms. These two groupings are often referred to as the t_{2g} and the e_g groups of orbitals from the nomenclature of group theory, which can also be applied to determine these sets of levels in a very general fashion.

In the derivation of this splitting of the energy level pattern into two groups, the distribution of electron density on each of the six surrounding ligand atoms is assumed to be spherically symmetrical. In practice, however, it is more likely that some of the ligand atoms may have p orbitals associated with them, especially if they are taking part in a π bond system. It is therefore interesting to see how this second form of perturbation may also affect the energy level pattern. By referring to *Figure 115(a)* and (b) respectively, it is clear that the p orbitals on the four ligand atoms in the horizontal plane possess electron densities which closely approach those of the d_{yz} and d_{zx} orbitals, whereas in the plane of the d_{xy} orbital they all have a null, as indicated in *Figure 115(c)*. This general consideration suggests that in the presence of π bonding around the ligand atoms themselves, the d_{xy} orbital would be more stable and have lower energy then either the d_{yz} or d_{zx} orbitals. Thus the lower group of levels can be further subdivided, with the orbital at the bottom and the remaining two together above.

397

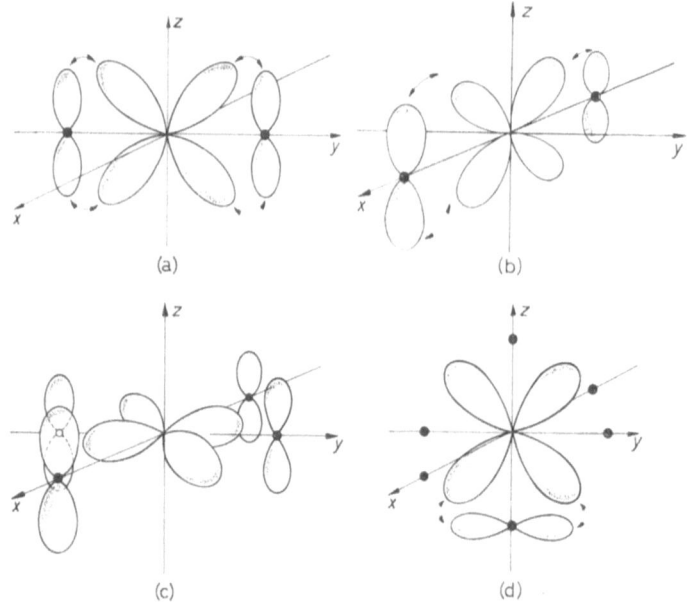

Figure 115. Effect of p orbitals and π bonding of atoms: (a) interaction of d_{yz} orbital with p orbital along y axis, (b) interaction of d_{zx} orbital with p orbital along x axis, (c) d_{xy} orbital not approaching p orbital of any atom, (d) further interaction of d_{yz} but not d_{zx} orbital with p orbital of sixth ligand atom

A detailed consideration of the p or π orbital on the atoms at the fifth or sixth coordination point shows that the subgroup t_{2g} can be further split. If one of the atoms at, say, the fifth coordination point is taking part in a π bond system, its p orbital is at right angles to the plane of the π bond system, and hence approaches either the d_{zx} or d_{yz} orbital more closely than the other. If the plane of the π bond system in which this atom is taking part is as shown in *Figure 115(d)*, it is clear that the d_{yz} orbital is repelled to a higher energy than the d_{zx} orbital.

The degeneracy of all three levels in the t_{2g} group is now removed, and a detailed quantitative molecular orbital treatment will enable the actual splitting of these levels to be related to the structural symmetry of the immediate surroundings of the transition group metal atom.

The splitting of the d orbitals into the two t_{2g} and e_g groups immediately explains quite a large amount of the magnetic behaviour of the transition group complexes, and is another way of classifying ionic and covalently bound derivatives. Thus the total number of unpaired electrons in a given transition group atom now depends on the energy required for pairing electrons in a single orbital, compared with the energy required to raise electrons from the lower t_{2g} groups to the e_g group of orbitals. For example, if there were three unpaired electrons around the transition group atom they would each be able to reside in a d orbital in a different t_{2g} group, and the upper levels of the e_g group would remain empty. The complex would therefore have a resultant S of $\frac{3}{2}$. In the case of four unpaired electrons, it is not immediately clear whether the additional electron would go into the lower t_{2g} group, and therefore pair with an electron already in one of the orbitals, or whether it would go into the higher e_g group of orbitals, with its spin still aligned in the same direction as the three initial electrons. The interaction which tends to align all the electron spins in the same direction, arises from the exchange force between the electrons and can be quite strong. On the other hand, if the splitting between the two groups of orbitals is very great, this may be more than the energy required to pair the spins in the lower orbital. In such a case, the fourth electron would obviously go into one of the t_{2g} orbitals, pairing with one electron there, and giving a resultant $S = 1$. The whole problem of whether the four electrons in the transition group atom produce a net unpaired electron component of two or four with S equal to 1 or $\frac{3}{2}$, therefore depends crucially on the splitting Δ between the two groups of levels, compared with the strength of the exchange interaction between the electrons themselves.

This description of the spatial distribution of the d orbital wave functions, together with the interactions which they experience and the resultant energy level scheme produced, has been followed through in some detail since it is basic to an understanding of the theoretical treatment of most electron resonance measurements on the transition group ions.

In the earlier theoretical treatment little attention was paid to the spatial distribution of the orbitals themselves. Consideration was concentrated instead on the splitting produced between the energy levels by the internal crystal field, and theoretical expressions of g values and hyperfine interaction constants were derived in terms of the ratio of this quantity to the strength of the spin-orbit interaction. Thus in the particularly simple case of $(3d)^3$ or $(3d)^8$ configurations, with the orbital singlet of the ^3F state lying lowest, the g value expected is isotropic and of magnitude

$$g = 2 \cdot 0023 - \frac{8\lambda}{\Delta} \qquad \qquad \dots (11.20)$$

where λ is the spin-orbit coupling parameter, and Δ is the splitting between the orbital levels produced by the internal crystalline field. This equation also holds for the g value parallel to the crystal field axis of a $(3d)^9$ complex in a distorted octahedron, such as $[Cu(H_2O)_6]^{2+}$. The g values calculated for other complexes by this simple crystal field approach, are summarized in Section 6.11, including for example the detailed expression for the g values of the $(3d)^1$ configuration of Ti^{3+}.

When allowance is made for the possibility that the unpaired electron may move out from its orbital around the paramagnetic atom to share in the ligand orbitals, these first-order calculations have to be modified. Thus, not only may hyperfine structure be observed from the ligand nuclei, but the actual g values obtained will be influenced by the percentage of covalent binding present. In order to calculate the magnitude of these corrections, the distribution of the electrons between the t_{2g} and e_g orbitals must first be determined. The modification that is produced on the energy of these orbitals by the bonding is then calculated, and new expressions for the g value can be obtained in terms of the admixture coefficients, which measure the percentage of the original d orbital in the final molecular orbital.

For the simple d^3 or d^8 case considered above, the molecular orbital $\phi_{x^2-y^2}$ is made up from $\alpha \cdot d_{x^2-y^2}$ due to the metal atom and $\dfrac{(1-\alpha^2)^{1/2}}{2} \Sigma \psi(\sigma)$ from the ligands; the expression for the g value becomes modified to

$$g = 2 \cdot 0023 - \alpha^2 \cdot \frac{8\lambda}{\Delta} \qquad \ldots (11.21)$$

Other changes are also produced by the delocalization of the unpaired electron to the ligands. The four general effects that can always be expected when a significant amount of covalent bonding occurs in a complex can be summarized as:

(i) A reduction in the orbital contribution to the g value as indicated above.

(ii) A change in the magnitude of the hyperfine structure from the central ion.

(iii) An increase in the spin-lattice relaxation time.

(iv) The appearance of superhyperfine splitting from the magnetic moments of the ligand nuclei.

These effects were explained in a general way by STEVENS[63], who extended VAN VLECK's[64] theory and demonstrated that in suitable circumstances the mixing of the ligand wave functions has the effect of lowering all the matrix elements of the angular momentum L, by a small factor which can be related to the amount of the admixture. These reductions lead to a more complete quenching of the orbital momentum and thus produce the first three effects listed above. The magnitude of the superhyperfine structure can be calculated by assigning the electrons to the molecular orbital levels of the complex; the magnetic electrons are then found to spread out into the anti-bonding orbitals and produce the observed transferred hyperfine interaction.

It is evident that the basic steps that must be undertaken when applying ligand field theory to any particular complex are therefore:

(1) To tabulate the energy level diagrams of the central metal ion and of the ligand atoms separately on either side of a diagram, as is done for copper phthalocyanine in *Figure 112(b)* and in more detail in *Figure 116*.

(2) To construct the energy level diagram of the complex by admixing the d wave functions and orbitals of the central ion with those of the ligand atoms having the appropriate symmetry. This produces a set of bonding orbitals with the main component from the ligand atoms, and a set of anti-bonding orbitals with the main component from the d wave functions, as shown in the centre of *Figures 112(b)* and *116*.

(3) To feed the total number of electrons, belonging to the $3d$ shell of the ion and to the outer orbits of the ligands, into the molecular orbital pattern, from the bottom up in pairs. The bonding orbitals then become filled with paired electrons and do not contribute to the magnetic effects, while the unpaired electron, or electrons, reside in the anti-bonding orbitals.

(4) To concentrate attention on the molecular orbital containing the unpaired electron, and on the orbitals immediately above it. Detailed expressions can then be derived for the g values and hyperfine interaction parameters in terms of the splittings of these orbital levels, the spin-orbit coupling constant of the paramagnetic ion, and the admixture coefficients.

A large number of workers[65-68] have carried out detailed calculations on many complexes in this way, and it is not possible to summarize them all here. One example, showing how these steps are followed through may however be helpful. The case of copper phthalocyanine is considered, since its energy level diagram has already been discussed in some detail. Reference back to *Figure 112(b)* shows that steps (1) and (2) outlined above are completed

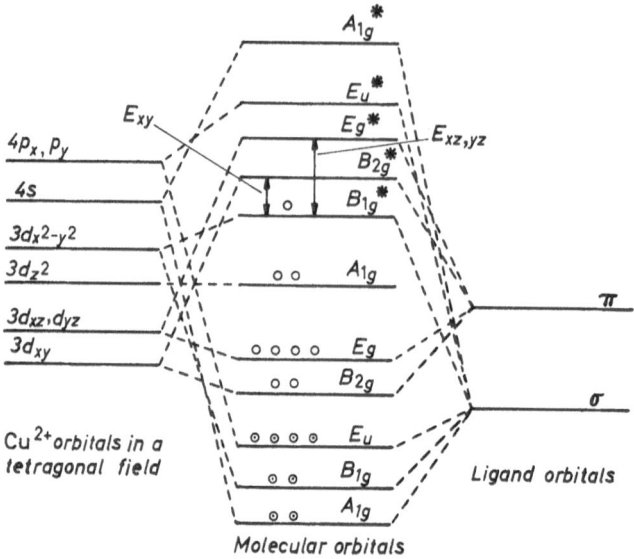

Figure 116. Postulated energy levels of copper phthalocyanine

there, but the diagram is repeated in *Figure 116* with details of the splittings involved and the electron distribution in the orbitals. The d_{xy} orbital of the copper admixes with the π orbitals of the ligand nitrogens to give a bonding B_{2g} orbital, and an anti-bonding B_{2g}^* orbital. Similarly, the $d_{x^2-y^2}$ copper orbital admixes with the σ orbitals on the nitrogens to give a bonding B_{1g} and an anti-bonding B_{1g}^* orbital. The molecule is planar and hence the d_{z^2} orbital does not admix at all, its lobes pointing towards the copper atoms in the molecules above and below.

The molecular orbital pattern of the complex having been built up in this way, step (3) can now be taken and the electrons fed into the energy level system. This is shown in *Figure 116*, each of the bonding orbitals having a pair of electrons (the E_g has two pairs since it derives from both the d_{xz} and the d_{yz} copper orbitals). It is evident that the single unpaired electron then resides in the anti-bonding B_{1g}^* orbital as indicated.

Step (4) deals with the B_{1g}^* orbital and anti-bonding B_{2g}^* and E_g^* orbitals above it; expressions must be obtained for the g values and hyperfine interaction parameters of an electron in this B_{1g}^* orbital, with admixture from these two orbitals above. The wave functions ψ for the three separate molecular orbitals must first be written down, in terms of the admixture coefficients relating the copper

d orbitals to the ligand orbitals. These can be expressed by

$$\psi_{B_{1g}^*} = \mathcal{N} . d_{x^2-y^2} - \frac{\lambda_s}{2}(-s_1+s_2-s_3+s_4) - \frac{\lambda_p}{2}(-p_1+p_2-p_3+p_4)$$

$$\psi_{B_{2g}^*} = \beta_1^2 d_{xy} - \frac{(1-\beta_1^2)^{1/2}}{2}(p_{y_1}-p_{y_3}+p_{x_2}-p_{x_4}) \qquad \dots(11.22)$$

$$\psi_{E_g} = \beta_2^2 d_{xz} - \frac{(1-\beta_2^2)}{2}(p_{z_1}-p_{z_3})$$

In these expressions s and p represent the nitrogen $2s$ and $2p$ orbitals, and λ_s, λ_p measure the amount of ligand orbital admixed; $\mathcal{N}$ measures the general amount of σ bonding, while β_1 and β_2 are the admixture coefficients which determine the amount of π-bonding that is also taking place.

The general theoretical treatment of Stevens[63] and TINKHAM[69], can be followed once expressions for the g values and hyperfine parameters are obtained. i.e.

$$g_{\parallel} = 2 \cdot 0 + 8 . \lambda . \mathcal{N}^2 . \beta_1^2/E_{xy} \qquad \dots(11.23)$$

$$g_{\perp} = 2 \cdot 0 + 2 . \lambda . \mathcal{N}^2 . \beta_2^2/E_{xz} \qquad \dots(11.24)$$

$$A = P[\tfrac{4}{7}\mathcal{N}^2 - (g_{\parallel}-2) - \tfrac{3}{7}(g_{\perp}-2) + K] \qquad \dots(11.25)$$

$$B = P[\tfrac{2}{7}\mathcal{N}^2 + \tfrac{11}{14}(g_{\parallel}-2) - K] \qquad \dots(11.26)$$

λ is the spin-orbit coupling constant for the copper; E_{xy}, E_{zx} are the splittings between the B_{1g}^* and the B_{2g}^*, and the E_g^* orbital levels respectively. K is a parameter determining the core-polarization of the copper atom, and P has its normal value of $2g . \beta . g_N\beta_N <r^{-3}>_{AV}$.

These expressions have the same general form as those deduced without any allowance for the effect of electron transfer on to the ligands, but are modified by the admixture coefficients $\mathcal{N}$, β_1, and β_2. Substitution of the experimentally determined g-values and hyperfine parameters for the copper ion[70], enable these admixture coefficients to be evaluated as

$$\mathcal{N}^2 = 0 \cdot 79, \quad \beta_1^2 = 0 \cdot 65 \quad \beta_2^2 = 0 \cdot 63$$

indicating that there is an appreciable amount of both σ and π bonding present.

The above hyperfine parameters were in fact for the copper nuclei, but a superhyperfine structure is also expected from the ligand nitrogens. Each N^{14} nucleus has a spin $I = 1$ and the four equally coupled nitrogens should therefore give a superhyperfine pattern of nine lines on each of the copper hyperfine components.

These are obtained quite distinctly, as shown in *Figure 117*. This spectrum was traced out by a Q-band spectrometer working at 8 mm wavelength, in order to obtain as large a separation as possible between the two sets of hyperfine patterns, one from the Cu^{63} and one from the Cu^{65} nuclei. The four groups of lines from one of these isotopes is clearly visible, with the nitrogen superhyperfine structure on each group, and the subsidiary splitting due to the two copper isotopes. (It is interesting to compare this spectrum of the two copper isotopes with the much earlier one in *Figure 61(b)*.)

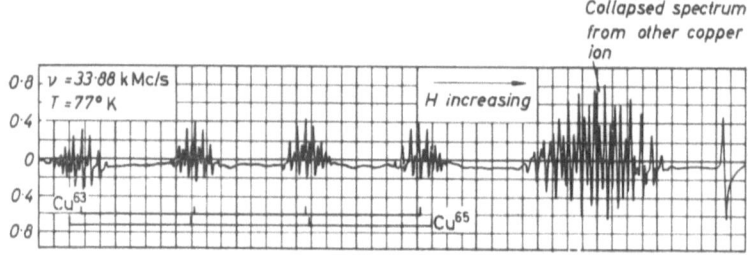

Figure 117. Superhyperfine structure of copper phthalocyanine spectrum

The magnitude of the superhyperfine splitting can be employed to determine the values of λ_s and λ_p in the expression for the wave function for the B_{1g}^* orbital. These parameters measure the amount of ligand orbital admixed, and hence the larger their magnitude, the bigger will be the superhyperfine splitting. The value of these parameters can be expressed as

$$A = \frac{16\pi}{3}\cdot\gamma\cdot\beta\cdot\beta_N\mid S(0)\mid_{2s}^2\cdot\frac{\mathcal{N}^2\cdot\lambda_s^2}{4} + \frac{8}{5}\gamma\cdot\beta\cdot\beta_N\cdot\left\langle\frac{1}{r^3}\right\rangle_{2p}\cdot\frac{\mathcal{N}^2\cdot\lambda_p^2}{4}$$
$$\ldots\,(11.27)$$

with a similar equation for B.

Substitution of typical values for the nitrogen wave functions gives a ratio of λ_s/λ_p of 30/56 which agrees well with the expected sp^2 hybridization of the ligand bonds. These superhyperfine splittings have also been remeasured by ENDOR techniques[70], and afford a good example of small energy differences in which such double resonance techniques are especially appropriate.

The particular case of copper phthalocyanine has been considered in some detail to illustrate both the different steps in a typical calculation, and the way in which the experimentally determined g values and hyperfine constants are correlated with the theory.

Readers who would like to follow the whole development of

ligand field theory through in more detail are referred to *Ligand Field Theory* by ORGEL [71] as an introductory text, and to *Theory of Transition Metal Ions* by GRIFFITH [72] as a more advanced work on the transition group ions in particular.

11.7 TRIPLET STATE STUDIES

Electron resonance studies on molecules subjected to different types of irradiation have been very extensive. Investigations on photochemical decomposition, and radical formation by high energy electron beams, are mentioned in Section 11.4(*d*), and some examples are also given in the next section of irradiation damage on molecules of biological importance.

One particular field of study in these radiation experiments

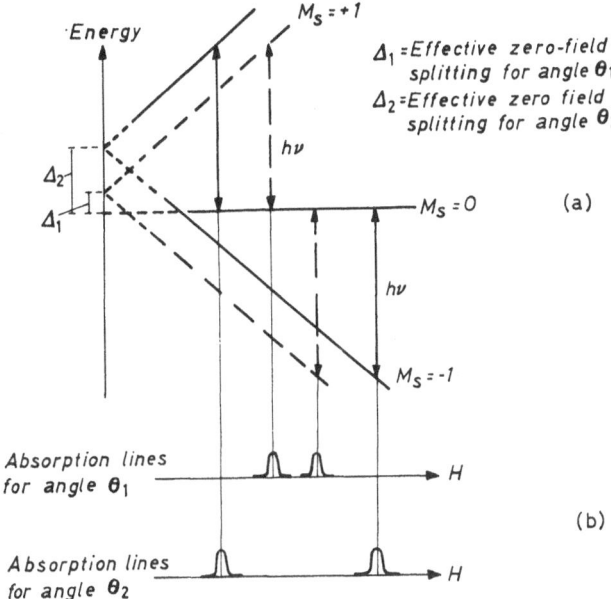

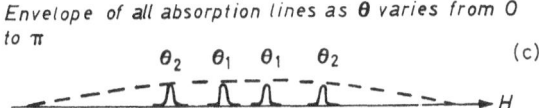

Figure 118. Energy level splitting of triplet state: (a) divergence of three energy levels for two different crystal settings, (b) the resulting doublets, (c) smeared out spectrum

deserves especial mention, however, and that is the investigation of excited triplet states by electron resonance. The term 'triplet state' is given to the condition that exists in a molecule when one electron from a normal electron-pair orbital, is excited to a higher level with its spin aligned in the same direction as that of the electron remaining in the original orbital. Such a molecule has two unpaired electrons and an effective $S = 1$. These unpaired electrons can be detected by magnetic susceptibility measurements[73], and it was therefore expected that they would also be easily detected by electron resonance. A large number of different laboratories set up apparatus to perform such experiments, but for many years all attempts met with complete failure, even when at least 10^{15} unpaired electrons were known to be present in the specimen.

The reason for this failure becomes apparent when the energy level system of the triplet state is considered in detail. There is considerable interaction between the two unpaired electrons, mainly of a dipolar type between their spins. This interaction splits the three levels of the $S = 1$ system even in zero magnetic field, as shown in *Figure 118*. Moreover, if the interaction is dipole-dipole in nature it has an angular variation of $(3 \cos^2 \theta - 1)$. It follows that the zero-field splitting between the $M_s = 0$ and $M_s = \pm 1$ levels varies with angle, sometimes having a value of Δ_1 and sometimes of Δ_2, as indicated in *Figure 118(a)*. Thus not only are two separate resonance absorption lines produced for each angle but the separation between these varies with angle as shown in *Figure 118(b)*. Hence if glassy type or polycrystalline samples are studied, with random orientation of their molecules, a complete range in the values of Δ is obtained, and the over-all absorption is smeared right out along the field axis, as in *Figure 118(c)*. It was therefore not surprising that all the initial efforts to detect triplet states by electron resonance ended in failure, since most of these had been made on gels or low-temperature glasses. The triplet state was only likely to be successfully detected if careful measurements were made on fluorescent molecules carefully aligned in a crystal lattice—diluted if possible to reduce broadening effects.

The first successful detection of the triplet state by electron resonance was achieved by HUTCHISON and MAGNUM[74] in 1958, who employed the exact kind of conditions postulated in the last paragraph. They used single crystals of durene containing a small percentage of naphthalene, and these were irradiated with light from a high-pressure mercury arc lamp. A schematic diagram of their apparatus is shown in *Figure 119*, and the actual microwave cavities employed at the two different wavelengths of 3 cm and

1 cm, are shown in *Figure 120*. The light from the lamp was focused by the quartz lens on to a front-silvered mirror, which reflected the radiation up through a quartz window into the cavity itself. It then fell on the crystal, which was mounted centrally in a cylindrical

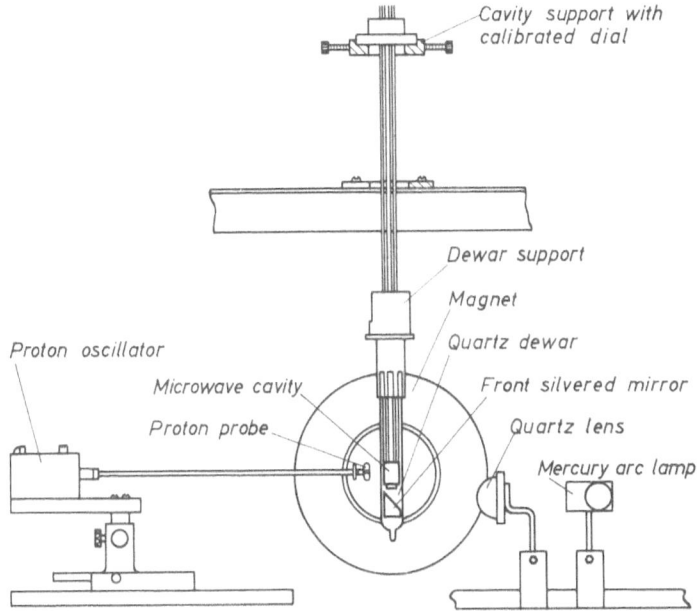

Figure 119. ESR Spectrometer used to observe triplet state excitation in naphthalene (after Hutchison and Magnum[74])

H_{011} cavity for the 1 cm experiments, and at the far end of a rectangular H_{101} cavity for the 3 cm wavelength measurements. The crystal was mounted on plastic wedges, cut at different angles, so that the angular variation of the spectra could be studied in different crystallographic planes.

Measurements at both wavelengths showed that each naphthalene molecule produced a doublet spectrum on irradiation which could be represented by the Spin Hamiltonian [75]

$$\mathcal{H} = g \cdot \beta \cdot \mathbf{H.S} + D \cdot S_z^2 + E(S_x^2 - S_y^2) \qquad \ldots . (11.28)$$

with the parameters $S = 1$, $g = 2\cdot0023$, $D = 0\cdot1003 \pm 0\cdot0006$ cm^{-1} and $E = 0\cdot0137 \pm 0\cdot0002$ cm^{-1}.

Typical plots of the angular variation between these two electronic transitions are shown in *Figure 121*, two different pairs of lines

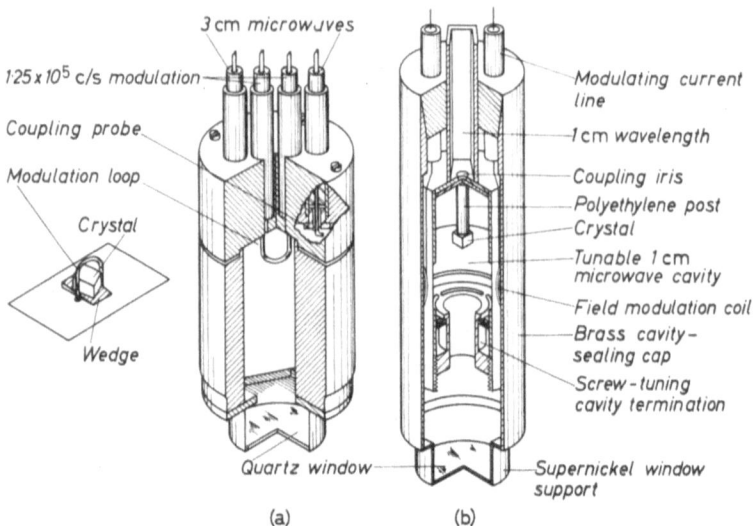

Figure 120. Microwave cavities for triplet state studies: (a) at 3 cm wavelength, (b) at 1 cm wavelength

being obtained because the durene crystal has two molecules per
unit cell. Hutchison and Magnum[75] were able to carry out a complete analysis of the spectra they obtained, and showed that the
zero-field splitting of the triplet has its origin in the dipole-dipole
interaction between the two unpaired electrons in the triplet state.
A small hyperfine structure was also observed on the spectrum and
was attributed to a C—H fragment, which confirmed the spectrum
as arising from the naphthalene.

This series of successful measurements came as a fitting reward

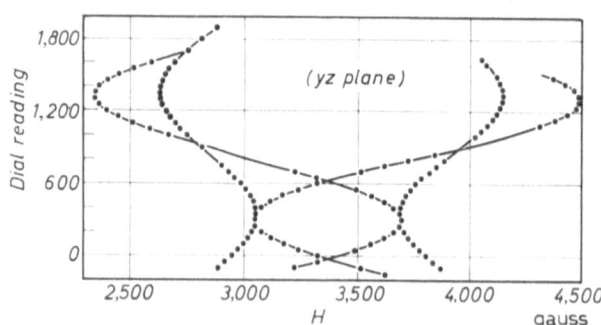

Figure 121. Angular variation of electronic splitting of triplet state in naphthalene

for a long and painstaking search for the triplet state, and it is interesting to note that the experiments were only successful when diluted single crystals were carefully irradiated in this way.

The next major step forward in the triplet state studies came when VAN DER WAALS and DE GROOT [76] reported that it was possible to observe the $\Delta M_s = \pm 2$ transitions for some cases. These would be forbidden on a simple first-order theory, but second-order effects can produce quite a significant transition probability. The great advantage of studying these transitions is that they take place between the highest and lowest of the three electronic levels, as indicated in *Figure 118*. Their resonance field value is therefore not altered to any great extent by the angular variation of the zero-field dipole-dipole splitting, and they should consequently be detectable from solid solutions or other media with a random orientation of the molecules. Also, the field value for resonance is about half that of the $\Delta M_s = \pm 1$ transitions.

Van der Waals and de Groot [76] obtained their first signals from naphthalene in a rigid glass solution in glycerol and irradiated by ultra-violet at 77° K. Following this, they were able to detect similar resonances from quite a number of polynuclear hydrocarbons under the same conditions [77].

These triplet state studies have been continued by many workers, and one of the more interesting recent developments has been the detection of ground state triplets.

11.8 BIOPHYSICAL AND BIOCHEMICAL APPLICATIONS

The biophysical and biochemical applications of electron resonance are probably advancing more rapidly now than any other branch of this field of studies. The potentialities of the electron resonance technique have become known to many biochemists who have employed it in a great variety of studies, not only to locate and identify biochemical reactions involving free radicals or unpaired electrons, but also to correlate these with valency changes of metal atoms which often occur at the same time. In an attempt to summarize these many new applications, the free radical studies are considered first, grouped together under four headings; and a final section is devoted to work on transition group complexes. However, many other biochemical investigations which may not fit under any one of these headings are constantly being reported.

(*a*) *Studies on metabolic activity.* Studies on the relation between free radical concentration and metabolic activity represent an attempt by biologists to survey the general processes taking place within the living cell. The measurements need to be considered with the utmost caution, since there are often several explanations for observed signals, and unambiguous interpretation may not be possible.

The work was initiated by the classic paper of COMMONER, TOWNSEND and PAKE [78], in which the free radical concentrations present in different metabolic tissues were measured by the electron resonance technique. The samples were removed from the different animals under consideration, and the tissue freeze dried to prevent too much damping being caused by the original water content. Evidence was found for a higher concentration of free radicals in the tissues associated with the more active metabolic processes, but it was very difficult to reach any quantitative conclusions because of the possible production of free radicals by the freeze drying process itself.

Attention was therefore turned to the design of cavity resonators into which aqueous solutions could be inserted, and in which free radical concentrations could be measured in conditions much closer to those of the living specimen. Rectangular cells are normally employed in such cavities to keep the aqueous solution out of the electric field as much as possible, while filling the maximum volume of effective microwave magnetic field. Despite these precautions, the sensitivity is somewhat reduced by the damping due to the high dielectric constant and by the loss associated with the liquid water. COMMONER and HOLLOCHER [79] are continuing these studies, and are trying to establish the kind of conditions necessary for a meaningful interpretation of these types of measurements. Since the systems are so complex it will obviously be some time before definite identification of the radicals can be made, but a general knowledge of their concentration and kinetics should be extremely useful in interpreting the biochemical aspects of different types of metabolic process.

(b) *Studies on photosynthesis.* Several precise and methodical series of investigations by electron resonance on photosynthesis have been carried out, including those by two very active groups of research workers: COMMONER, HEISE and TOWNSEND [80] on the one hand and SOGO, PON and CALVIN [81] on the other. In the initial experiments [80] green leaves were lypholized before insertion into the cavity resonator; only static measurements of concentration could be made, no kinetic studies on the actual photosynthetic system being possible.

In later experiments, however, it proved possible to study aqueous solutions of the chloroplasts, *in situ* in the cavity, and to follow the kinetics of the free radical concentration as the conditions of illumination were varied. The solutions were placed in small specimen tubes of $\frac{1}{2}$ mm × 1 mm cross-section, and these were illuminated through suitable slits in the cavity wall. Only very small free radical concentrations could be detected in the absence of any illumination, but this concentration increased by a factor of six as soon as the high intensity light source was switched on. The free radical concentration was plotted in detail against the illumination, and was shown to undergo a saturation effect similar to that of the photosynthetic process itself. The resonance signals had a width of about 10 gauss, and were thus very like those of the free radicals observed in most metabolically active tissues. The sudden signal growth at the commencement of illumination was also traced out by rapid recording techniques, and the concentration was found to rise exponentially to a steady value with a time constant of about 12 sec.

SOGO, JOST and CALVIN[82] studied these kinetics in more detail, and in particular investigated the variation of the time constants with temperature. They were able to show that quite often the times of growth and decay were the same at liquid nitrogen temperatures as at room temperature, and thus were able to eliminate mechanisms depending on chemical reactions for the formation of the unpaired electrons. As a result of this work, they suggested that the more likely explanation for the presence of the electrons was a form of semi-conductor trapping technique, in which the electrons were trapped below a metastable state and energy from this state became available in the photosynthetic process. Quantitative estimates on the number of unpaired electrons produced by the incident light quanta were also possible from these kinetic studies and a figure of about 0·03 was obtained[83].

These accurate kinetic studies, and the quantitative figures available from them, are of considerable significance in trying to derive a detailed interpretation of the photosynthetic mechanism. The exact role of the photo-induced unpaired electrons in this mechanism is still to be finally established, but it seems significant that the kinetics of their formation and decay are closely related to those of the main photosynthetic process itself.

(c) *Enzyme studies.* One of the main problems associated with these studies is the nature of the actual linkage that occurs between the enzyme and its substrate while the catalytic action of the enzyme takes place. The enzymes themselves can be considered as forms of polypeptide chains held together by a variety of types of bonding such as the interaction of polar groups, or the presence of hydrogen bonds. In an enzymic reaction, the enzyme acts in a catalytic

411

fashion on a specified compound, called the substrate, to produce certain reaction products, and then reforms itself at the end of the interaction. During this process the enzyme must form some kind of complex with the substrate, and the actual mechanism whereby this complex is formed and held together is currently a matter of considerable interest. It is known that in some cases S—S bonds produce this union of the enzyme with its substrate, but it is also thought that in some cases transition metal atoms are involved. If this is so, there will be a change in the valency state of the metal ion during the enzymic reaction: this change in valency, and the free radical concentration which may be produced at the same time can both be followed by electron resonance. It is clear that electron resonance affords a very powerful technique in studies like these, where the simultaneous reactions of the free radical and metal atom of the enzyme can be followed and plotted together.

One example of such a system is the enzyme peroxidase. An iron atom which can readily undergo a change of valency is contained in a haem plane of configuration in the central portion of the enzyme; and the work of YAMAZAKI, MASON and PIETTE[84] has also shown quite conclusively that free radicals are formed from substrates during this enzymic oxidation–reduction process. Xanthine oxidase is another enzyme of which the free radical concentration and the valency of the associated metal ions have been measured in detail by electron resonance; this was studied by BRAY and his colleagues[85, 86]. A brief account of these particular investigations serves to illustrate the kind of techniques employed and the information that can be obtained from them.

The reduction of xanthine oxidase by substrates and other reducing agents is accompanied by complex changes in the visible absorption spectrum. Part of this change occurs rapidly, whereas another part occurs more slowly and appears to be associated with the reduction of an 'inactive' component. In 1959, the early work[85, 86] of electron resonance on these compounds showed that resonance signals were obtained when the enzyme was treated with either xanthine itself or with sodium dithionite, and it was also shown that the growth of the resonance signals on reduction of the enzyme contained rapid and slow components. Moreover, the rates of growth could be correlated with the nature and concentration of the substrate, and with the composition of xanthine oxidase sample.

More recent studies by BRAY, PALMER and BEINERT[87] have shown that four basic kinds of experiments can be carried out on these systems. The first is termed 'a single turnover' experiment in which the enzyme is treated with one equivalent or less of substrate in the presence of an excess of oxygen. Under these conditions no steady state is reached, since each enzyme active centre reacts either once only, or else not at all; hence the enzyme remains reduced for an absolute minimum of time, so that the possibility of secondary changes, such as the reduction of an 'inactive' enzyme by an 'active' one, is effectively eliminated. Thus this single turnover technique enables the kinetics to be interpreted in a fairly straightforward manner with no secondary complications. An example of the results obtained for catalytic action by xanthine oxidase in such an experiment is shown in *Figure 122*. The time variation for the appearance

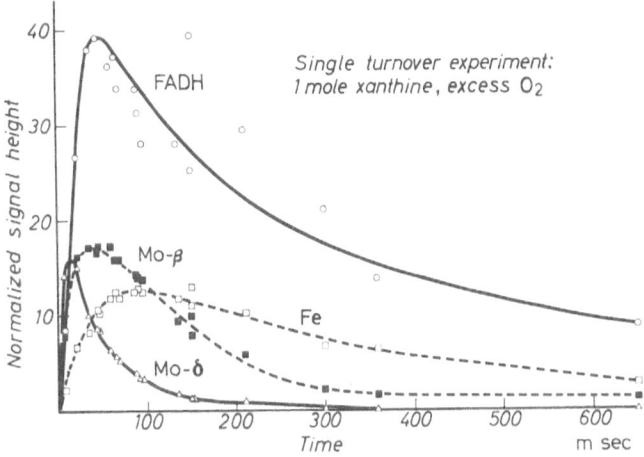

Figure 122. Time variation of concentration for different free radical and transition group components of the xanthine oxidase reaction (after Bray, Palmer and Beinert[87])

and disappearance of each of the four signals is clearly quite different: the molybdenum-δ signal appearing and disappearing very rapidly with a maximum of about 15 msec, whereas the molybdenum-β, the FADH, and the iron signals take much longer and have maximum times of 40, 45 and 100 msec respectively. These results can be combined with those from other studies of the reaction, including the observations of prolonged steady states, following of the anaerobic reduction of the enzyme and investigations on its reoxidation. From all these results in which the kinetics of radical and ion valency states are plotted by ESR, a detailed picture of the mechanism of enzyme reaction can be built up[87].

The potentialities of electron resonance studies in following not only the appearance and kinetics of the free radicals associated with enzyme activities, but also the valency changes associated with the metal ions, is obvious from such a study as this. Considerable advances are anticipated in this particular field of study in the relatively near future.

Another recent investigation of metal ions in enzymic activity was by EHRENBERG and YONETANI[88] on cytochrome oxidase. They were able to obtain a signal which was readily identified as coming from a copper atom, since the four-line hyperfine structure could be resolved in its wings. In some cases, it even proved possible to pick up the superhyperfine structure from the nitrogen surrounding the copper atom. The interpretation of electron resonance signals from biochemical specimens is always considerably facilitated if additional precision can be obtained by the presence of hyperfine or superhyperfine lines. When such precise identification is not possible, it is wise to treat the interpretation of the electron resonance results with some caution at the moment, since it will probably be some time before completely unambiguous deductions can be drawn from them. As more definite conclusions are established from the kinetic studies however, it should be possible to obtain increasingly more definite and unambiguous results from future investigations.

Enzymes are not the only compounds of biological importance in which transition metal atoms are found. Such atoms can also play a crucial part in other biochemical processes, for instance the combination and transport of oxygen around the blood stream by haemoglobin and its derivatives. Electron resonance has been used to study such compounds, and to obtain information of structural interest concerning the parts of the molecule surrounding the central iron atom. This work is summarized briefly in Section 11.8(e), since this information has been obtained from a detailed study of the transition ion complex.

(d) *Studies of irradiation damage in biological systems.* GORDY, ARD and SHIELDS[89] initiated the irradiation studies of biological material in 1955, and a large number of other workers have since followed this line of research. The great difference between these studies and those on irradiation of inorganic material, is the very large molecular weight and complex structure associated with biological material. As a result, the hyperfine structure obtained on the electron resonance spectrum is likely to be very complex and difficult to analyse. Nevertheless it is sometimes possible to identify some characteristic groupings from the hyperfine pattern and to associate them with a definite bond-type or radical.

414

One particular example of this is the molecules containing sulphur, where a hyperfine pattern with a line noticeably shifted from the free-spin g value is obtained. This shift of g value can be attributed to an unpaired electron localized at an S—S bond, since the spin-orbit coupling is quite significant for the sulphur atoms, as is also the case for coupling to oxygen atoms. The shifted g value was noticed first by COHN[90] in his early work on cysteine, for which a g value of 2·025 was obtained. More recently, HENRIKSON[91] carried out a systematic study of high energy irradiation on sulphur containing compounds such as the thiols and disulphides. In all these cases a complex structure was obtained, with a g factor shifted appreciably from the free-spin value, and occurring at the same value of 2·025 as had been observed earlier for cysteine. Basically similar patterns were in fact obtained for cysteine, cysteamine hydrochloride, and dithiopentaerythrite; more complex patterns were obtained in several other thiols and disulphides, indicating that in these other radical species had also been formed. A systematic quantitative survey[92] was carried out on the decay rates of these radiation-induced free radical centres. The results indicated that, although the primary energy absorption producing irradiation damage must be evenly distributed through the compound, the odd electrons which are formed initially in proteins and peptides

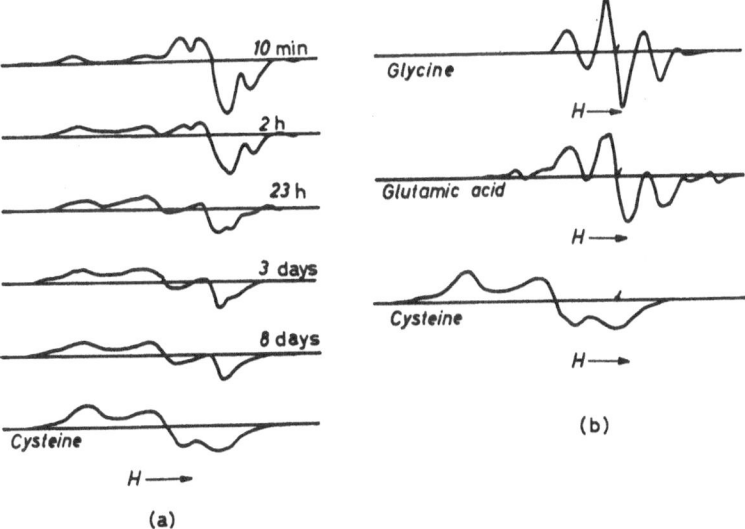

Figure 123. Radicals produced by radiation damage: (a) time variation of spectra from irradiated glutathione, (b) spectra from similarly irradiated basic amino acids

415

migrate to specific groupings such as glycyl-glycine groups, or to the sulphur of the cysteine.

In order to check on these ideas HENRIKSEN and PIHL[92] carried out more detailed studies on the migration processes taking place in glutathione, and *Figure 123(a)* shows how the spectra they observed after irradiation changes with time. As a comparison with this set of spectra, measurements were also made on the basic amino acids and results are given in *Figure 123(b)*. These two sets of traces suggest that as late as 10 min after the end of exposure to radiation, the radical centres in the glutathione are distributed among three component amino acid residues, in more or less random fashion. As time progresses however the electrons must move from the glutanic acid and the glycine residues towards the cysteine, and finally the resultant spectrum takes on the form very similar to that obtained from the cysteine alone.

These initial measurements do indicate that specific interpretations can sometimes be obtained from a study of the more complex biological materials, and they may have considerable significance in the theories of the mechanism of radiation protection[93]. Compounds containing S—S bonds may be able to provide an 'electron sink' for electrons produced during irradiation damage, and thus prevent them attacking other groups around the molecule.

(*e*) *Transition group complexes of biological importance.* Although most of the electron resonance studies in biochemistry have so far been associated with free radicals, increasing attention is now being focused on the role that transition group atoms play in such reactions. Some examples of this are given in Section 11.8 (*c*) in the discussions on enzyme studies. The kinetics of the reactions and changes of the metal ion valency, have mainly been studied, but in some cases very significant information has been obtained on the detailed internal structure and bond-nature of the molecule in question. One example of this will show the potentialities of electron resonance in this work.

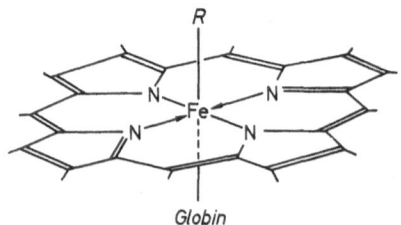

Globin

Figure 124. Central structure of haemoglobin molecule

The haemoglobin molecule is a compound of very considerable biochemical importance, and is also one for which electron resonance has been able to give some very important information on its structure and binding. The structural formula of the haemoglobin molecule is represented schematically in *Figure 124*, from which it is seen that a single iron atom is located at the centre of a large plane, and is immediately surrounded by four nitrogen atoms. This plane is called the 'haem' or 'porphyrin' plane, and the rest of the protein in the molecule is attached below, via a bond to a nitrogen atom of a histidine ring, as indicated. The sixth coordination point of the iron atom is occupied by an oxygen molecule when the haemoglobin has just passed through the lungs and become oxygenated, or alternatively by a carbon dioxide molecule after the oxygen has been used in the body. Other atoms or groups can also be attached to this sixth coordination point, and when the haemoglobin is crystallized from a suitable buffer solution, a water molecule is normally present. The derivative is then termed the 'acid-met' haemoglobin.

There are some groups such as CN which, once attached to the iron atom are impossible to remove, as is probably readily appreciated. The electron resonance investigations on haemoglobin were initiated to find out as much detail as possible about the bonding between the iron atom and the different groups on this sixth coordination point. This bonding must involve a very precise balancing of the energies concerned, since the actual number of unpaired spins changes significantly when the oxygen is associated with the iron atom. Since this bond is formed and dissociated by the simple process of breathing in and out, the energies cannot be of a very large magnitude. During the early investigations however it became apparent that because of the very large g value variation obtained in certain derivatives, the electron resonance could also give very important information on the actual structure of the haemoglobin molecule itself.

The iron atom at the centre of the porphyrin plane can be of either a ferrous or a ferric valency; the binding can also be classified approximately as 'ionic' or 'covalent' according as the three d orbitals are used by the electrons of the ligand atoms or are left empty for the $3d$ electrons of the iron atom itself. The different situations which can thus arise in these four cases are summarized in *Table 11.1*, which also indicates the resultant overall spin value for the electrons in the $3d$ shell. There are no unpaired electrons left in the covalently bound ferrous derivatives, which are hence diamagnetic. Both the ionically and covalently bound ferric derivatives possess unpaired electrons however, and in the ionically bound case the five unpaired electrons couple together with their spins parallel, to produce a resultant $S = \frac{5}{2}$. This case is therefore identical to the iso-electronic Mn^{2+} compounds, and three degenerate doublets are to be expected in the solid state crystal,

Table 11.1. Spin State of Iron Atoms in Haemoglobin

		Electron configuration			Total spin
		3d	4s	4p	
Ionic bonds	Fe^{2+}	(⇅)(↑)(↑)(↑)(↑)	○	○○○	2
	Fe^{3+}	(↑)(↑)(↑)(↑)(↑)	○	○○○	$\frac{5}{2}$
Octahedral covalent $d^2 sp^3$ complexes	Fe^{2+}	(⇅)(⇅)(⇅)○○ ○ ○○○			0
	Fe^{3+}	(⇅)(⇅)(↑)○○ ○ ○○○			$\frac{1}{2}$

corresponding to the possible orientations of the $S = \frac{5}{2}$ resolved along the axis of crystalline electric field with $M_s = \pm\frac{1}{2}$, $\pm\frac{3}{2}$ or $\pm\frac{5}{2}$. This situation is indicated on the left-hand side of *Figure 62* (p. 165) where the splitting between these three doublets in zero magnetic field is shown. This figure also demonstrates how these levels may be considered as moving apart on the application of the external magnetic field, until they are arranged as indicated on the right-hand side, with the $\pm\frac{5}{2}$ level at the top and the others arranged in order beneath. Under these conditions five electronic transitions between the six electronic levels are obtained, as discussed in Section 6 and as shown in *Figure 61(c)*.

This behaviour of the energy levels and resultant production of five electronic transitions is only obtained if the size of the microwave quantum is significantly greater than the splitting of the three doublets in zero magnetic field. If this zero-field splitting is very much larger than the microwave quantum being employed for the resonance, then the separate energy levels do not reach the position indicated on the right-hand side of *Figure 62* but are instead as in *Figure 125(a)*. Here it can be seen that although the application of the external magnetic field removes the degeneracy of all three doublets, the only transition that is observed by application of a microwave frequency is between the two resolved doublets of the $S = \pm \frac{1}{2}$ level. Transitions between the two components of the other doublets are not allowed since they correspond to a quantum jump of greater than one. Unless very high values of magnetic field are employed, none of the other levels are brought close enough in energy to a level having a quantum number differing by one, and hence no other microwave transitions are induced. Only one single electronic resonance line is therefore observed, corresponding to the doublet of $S = \pm \frac{1}{2}$.

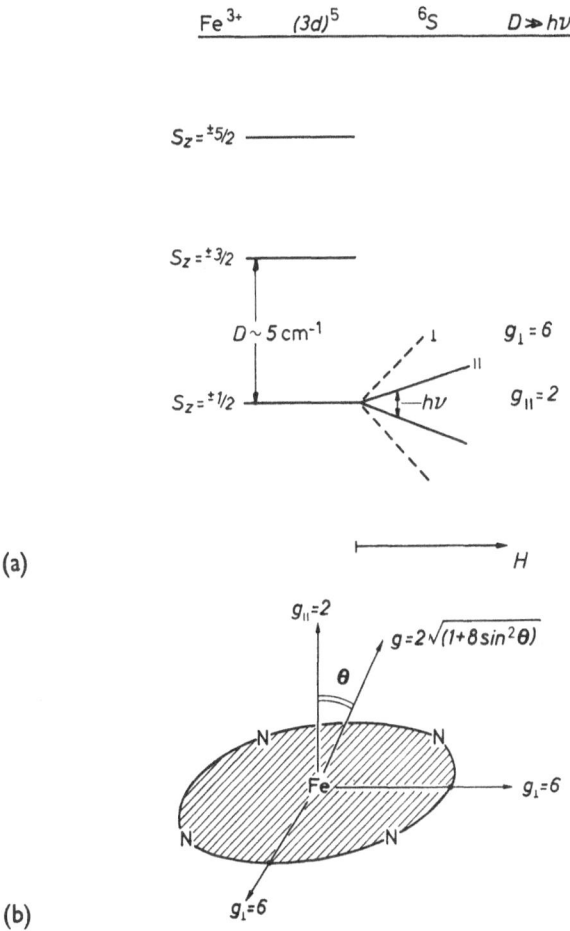

(a)

(b)

Figure 125. Ionic ferric haemoglobin: (a) energy level configuration, (b) resulting g value variation with respect to axis through the haem plane

This however no longer has an isotropic g value equal to 2, but a g value which varies from 2·0 along the axis of the crystalline electric field to 6·0 for any direction perpendicular to this, i.e. for any direction lying within the porphyrin plane itself. This g value variation from 2 to 6 is produced by the admixture of the higher levels, and the detailed theory is given by GRIFFITH[94]. The qualitative reason for such a variation can probably be best visualized as follows.

The fact that the ground state is $M_s = \pm\frac{1}{2}$ means that the total spin vector of $S = \frac{5}{2}$ must be precessing close to the porphyrin plane, with only a small component resolved along the axis of the crystalline field normal to this plane. When the magnetic field is applied along the axis of this crystalline field, it is only interacting with the small resolved component of the spins, and hence a g value of 2·0 is obtained. If the magnetic field is applied in any direction in the porphyrin plane however, the direction corresponds to that in which the resolved component of the total $S = \frac{5}{2}$ has its greatest value, and consequently the larger magnitude $g = 6$ is produced. It follows from this that a wide g variation from 2 to 6 is to be expected in all ferric derivatives which have large zero-filled splittings between their three doublet components.

The actual spatial variation of this g value variation for the haemoglobin molecule is shown in *Figure 125(b)*, and it is now evident that this variation provides a method for accurately locating the orientation of the porphyrin planes. When these electron resonance investigations were carried out, the X-ray studies of KENDREW[95] and PERUTZ[96] had not been completed; it was possible to locate accurately the orientation of the porphyrin planes by electron resonance measurements, and then hand this information on to the X-ray crystallographers to assist in their complete analysis of the rest of the molecule.

In principle, all that is required to locate the orientation of the porphyrin planes is to mount a crystal in the cavity resonator and rotate it in all possible directions until a g value equal to 2·0 is obtained. The magnetic field is then being applied along the direction normal to the porphyrin plane, and hence the orientation of the plane is located. In practice it is very difficult to move a crystal in three directions at once in a cavity resonator, and the crystals are therefore mounted in different crystallographic planes in turn, and the g value variation for the different planes is plotted out.

The g value variation for the three crystallographic planes of the myoglobin molecule is shown in *Figure 126*. The particular derivative

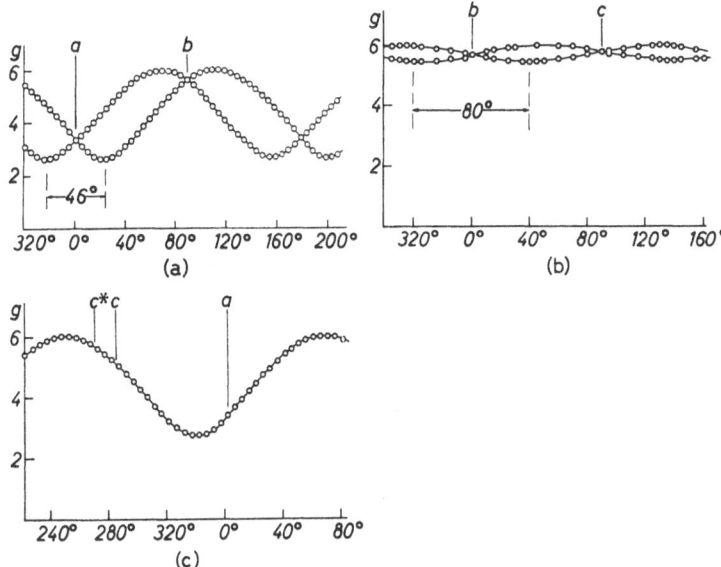

Figure 126. Variation of g value in myoglobin crystal: (a) in ab plane, (b) in bc plane, (c) in ac plane

being studied in this case is the acid-met derivative of a type *A* crystal, which is of a monoclinic form with two molecules per unit cell. Each myoglobin molecule only contains one iron atom and porphyrin plane, and hence only two resonance lines are expected, one corresponding to each molecule of a unit cell. The *g* value variations for the two different molecules can be clearly seen in *Figure 126(a)*. Moreover the cross-over points in this figure locate the directions of the crystallographic *a* and *b* axes precisely, and hence the alignment of the crystal can be determined from the electron resonance measurements themselves, without the need to accurately align the crystal in the cavity by optical means.

It can be seen from these results in the *ab* plane that the minimum *g* value is 2·64 and occurs at 23° to the *a* axis. This particular value corresponds to $\theta = 18°$ when substituted into the general relation $g = 2(1 + 8 \sin^2 \theta)^{1/2}$ quoted in *Figure 125*. It follows that the angle between the haem normal and the *c* axis is 72°, and hence this one measurement of the position and magnitude of the minimum *g* value in the *ab* plane is sufficient to locate the direction of the normal to the porphyrin plane.

Measurements are always carried out in other crystallographic planes in order to cross-check the calculations. *Figure 126(b)*

421

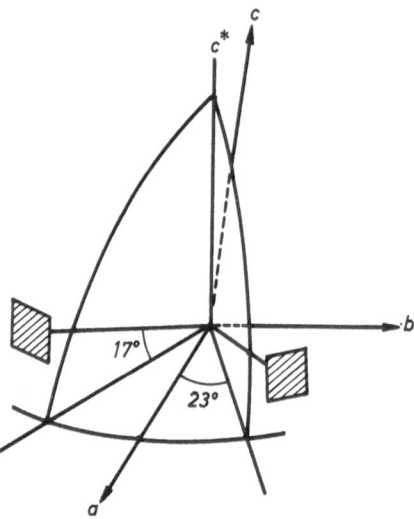

Figure 127. Haem plane orientations in myoglobin type A crystals

summarizes the *g* value variation in the crystallographic *bc* plane, and shows that this plane is in fact quite close to the two porphyrin planes themselves, since the *g* value remains close to 6·0 in all directions. A quantitative analysis can be made from the *g* value minima in this plane as before, and also from the *g* values obtained along the direction of the axes. Thus it is possible to cross-check the orientations and determine the axes perpendicular to the porphyrin planes with considerable accuracy. The results obtained from the type *A* myoglobin crystals are shown in *Figure 127*, in the form of a three-dimensional perspective plot, with the planes indicated by

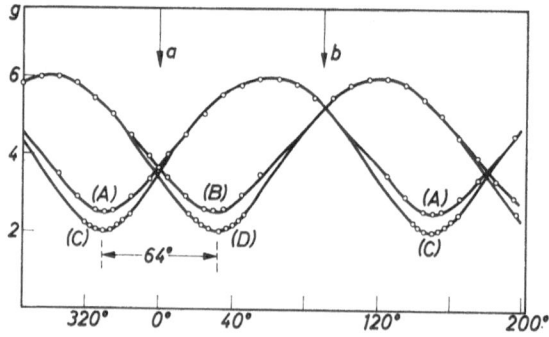

Figure 128. Variation of g value in haemoglobin crystal

shaded squares. As mentioned above, the determination by electron resonance of the orientation of these porphyrin planes, assisted in the final X-ray crystallographic analysis of the myoglobin structure.

Crystals of haemoglobin were then studied in the same way, each haemoglobin molecule containing four iron atoms and porphyrin planes. It had been assumed before the electron resonance measurements were carried out, that the four porphyrin planes in the haemoglobin molecule were probably parallel to one another, but the initial measurements on single crystals of the haemoglobin indicated immediately that this was not so. The g value variations observed in the (100) plane of the acid-met haemoglobin are shown in *Figure 128*, from which it is evident that there are four separate g value variations, corresponding to the four haem planes per molecule (the different molecules per unit cell are all orientated in the same way in this particular type of crystal). Two of these g value variations reach their minimum of 2·0 in the (100) plane, and hence the directions of the normals corresponding to these two porphyrin planes are immediately located. The other two planes can be found by a quantitative analysis as outlined for the myoglobin above; the results are summarized in three-dimensional perspective in *Figure 129(a)*, and as a stereographic plot in *Figure 129(b)*. The four different planes A, B, C, D are again indicated by shaded squares.

These determinations of the precise orientation of porphyrin planes within the myoglobin and haemoglobin molecules serve as

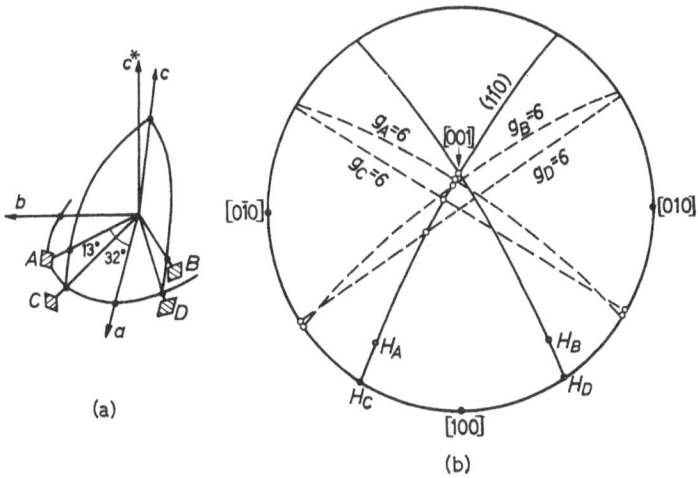

(a)

(b)

Figure 129. Haem plane orientations in haemoglobin molecule : (a) three-dimensional perspective, (b) stereographic plot

423

very good examples of the possibilities of electron resonance in the investigation of compounds of biochemical interest. It will be appreciated that in these particular studies the g value variations have been used simply as a probe of the symmetry surrounding the iron atom. It has not been necessary to invoke any detailed theory on molecular orbitals themselves and the whole argument has been based on one of pure symmetry.

Additional information can often be obtained if further measurements are made on different derivatives, and if molecular orbital theory is applied to analyse second-order effects. A good instance is the measurements made on the myoglobin and haeomoglobin covalent derivatives, such as the azide complex in which three nitrogen atoms are linked to the sixth coordination point of the iron atom. Thus this derivative has a resultant spin of $S = \frac{1}{2}$, as shown in *Table 11.1*. In terms of the more precise molecular orbital picture discussed in Section 11.6, this is equivalent to a large splitting produced by the t_{2g} and e_g levels so that the five electrons fill the d_{xy}, the d_{yz} and d_{xz} orbitals and none enter the e_g levels. The single unpaired electron should therefore produce one electronic transition per iron atom from the lowest Kramer's doublet, and the g value should spread across the free-spin value. From a detailed analysis of these g values it is possible to deduce which of the t_{2g} levels is lying lowest, and to obtain some idea of what type of σ or π bonding is to be associated with the iron atom. Quite apart from the information on the chemical bonding, further details on the structural symmetry of the surroundings of the iron atom can also be found. This is possible because the variation in the g value no longer has axial symmetry as in the ionic derivatives, but varies with direction in the porphyrin plane itself as shown in *Figure 130(a)*. The maximum value of 2·80 is in the direction of the normal to the porphyrin plane, and the two other principal axes of the g value variation lie in the porphyrin plane.

A theoretical analysis of these g values shows that they are only consistent with an energy level diagram in which the d_{xy} orbital must lie lowest in the t_{2g} group, with the d_{zx} next and the d_{yz} at the top of the group. The particular order of these energy levels may be explained by considering the interaction of the magnetic d electron on the iron atom with the p electron density on the surrounding nitrogen atoms. The p orbitals of the nitrogens in the porphyrin plane form part of the π bond system of the conjugated rings and are therefore all aligned at right-angles to the porphyrin plane. It follows that the d_{xy} orbital which lies in the porphyrin plane has its electron density pointing between the lobes of the p orbitals, and

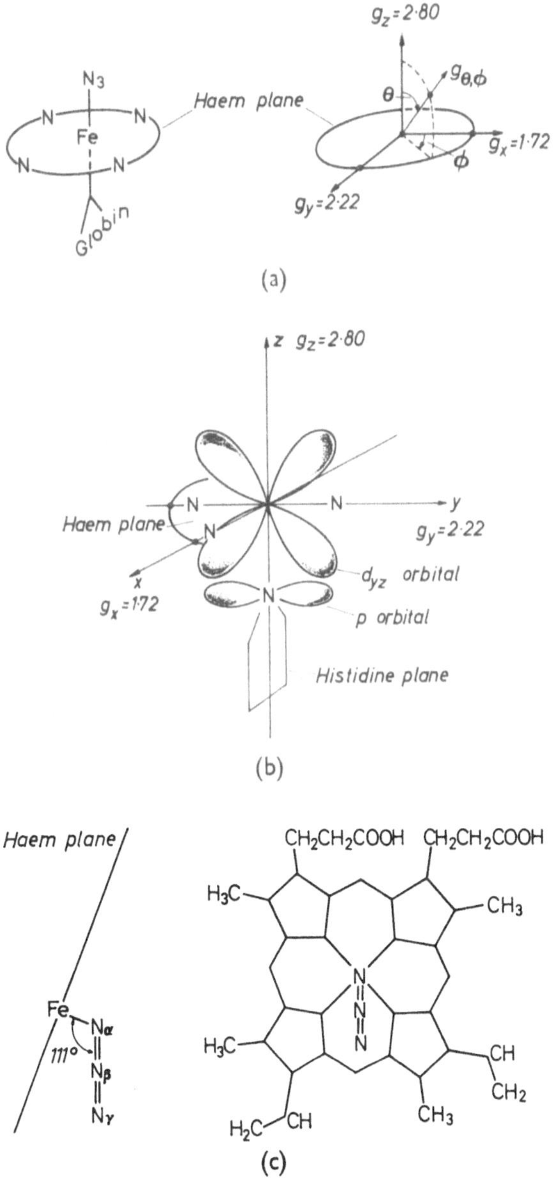

(a)

(b)

(c)

Figure 130. Variation of g value in azide derivative of haemoglobin: (a) within haem plane,
(b) electron orbitals showing interaction of p orbital in the histidine nitrogen with d_{yz} orbital,
(c) location of azide group in relation to haem plane (after Stryer, Kendrew and Watson[98])

thus is repelled only a little by them. Hence electrons can be fed into this orbital without requiring very much extra energy, and the d_{xy} orbital consequently lies lowest. Both the d_{yz} and the d_{zx} orbitals have a density distribution which approaches the p orbitals of the nitrogen in the porphyrin plane. Thus there is a noticeable repulsion between the electrons in both of these orbitals and those in the π bond system of the porphyrin plane, causing the d_{yz} and the d_{zx} orbits to have higher energies than the d_{xy}.

On the other hand there, is as yet no apparent reason why the d_{zx} and d_{yz} orbitals should be separated in energy. Such an energy separation must be produced by something other than the proximity of the p orbitals to the porphyrin plane. The only other obvious interaction is with the p orbit of the nitrogen belonging to the histidine ring below[97]. If the p orbit of this nitrogen interacts with the d_{yz} orbit, it must follow that the histidine plane, which is at right-angles to the p orbit of its nitrogen, is along the direction of the x axis. The fact that the energy level of the d_{yz} orbit is noticeably higher than that of the d_{zx} indicates that such an interaction between a magnetic electron and the p orbit of the nitrogen on the histidine ring must indeed be taking place. Thus the direction of the histidine plane must be parallel to the x axis of the g value variation. This is indicated in *Figure 130(b)* where the orientations of the different orbitals can be clearly seen.

The main step in this additional analysis has been to use the asymmetry of the g value variation in the porphyrin plane as a probe of the asymmetry in the structure of the molecule below the plane. It has proved possible to locate the orientation of the histidine ring by its effect on the electron distribution above it. A certain amount of molecular orbital theory was required to determine which of the t_{2g} orbitals corresponded to the higher energy level, but once this determination had been carried out the reasoning was again one of spatial symmetry considerations.

It is, of course, possible that the asymmetry in the g value variation around the haem plane is produced by the particular configuration of the azide group above the plane, rather than by the orientation of the histidine plane below it. In the above analysis it was assumed that the three nitrogens of the azide group were positioned along a line vertical to the haem plane—i.e. along the haem normal itself —and would thus not cause any asymmetry in the g value around the plane. Recent work[98] has shown, however, that they are orientated along a line which makes an angle of 111° with the haem normal, as in *Figure 130(c)*. Moreover, the projection of this $N{=}N{=}N$ orientation on to the haem plane, as shown at the right of this figure, corresponds very closely with that of the principal g value; it is therefore very likely that the electron distribution associated with this $N=N=N$ group also alters the interaction energy of the d_{yz} and d_{zx} orbitals.

There are quite a number of compounds of biochemical interest which contain ferric atoms in a central position, and it is likely that the same kind of analysis as was applied to haemoglobin and its derivatives will soon be applied to these other molecules. It is obvious that a maximum amount of information will only be obtained if single crystal specimens are available for investigation, and at the moment this is one of the limitations on some of the studies on organic crystals. By extending the techniques of crystal growing which had been employed by the X-ray crystallographers it was possible to grow crystals of the haemoglobin derivatives large enough for electron resonance investigations. These crystal growing techniques are now being developed in several laboratories and it should not be long before a large number of other derivatives and molecules are available for electron resonance in the same way.

REFERENCES

[1] ARAMS, F. R. and KRAGER, G. *Proc. Inst. Radio Engrs.* 46 (1958) 912

[2] FAULKNER, E. A. *J. sci. Instrum.* 39 (1962) 135; *Lab. Pract.* 13 (1964) 1065

[3] WEBB, R. H. *Rev. sci. Instrum.* 33 (1962) 732

[4] WILMSHURST, T. H., GAMBLING, W. A. and INGRAM, D. J. E. *J. Electron. Control* 4 (1962) 339

[5] KLEIN, M. P. and BARTON, G. W. *Rev. sci. Instrum.* 34 (1963) 754

[6] PIETTE, L. H. *NMR and EPR Spectroscopy.* Oxford: Pergamon (1960) p. 243

[7] HUTCHISON, C. A. and MAGNUM, B. J. *J. chem. Phys.* 34 (1961) 908

[8] FESSENDEN, R. W. and SCHULER, R. H. *ibid.* 39 (1963) 2147

[9] ——*Disc. Faraday Soc.* 36 (1963) 147

[10] HARTRIDGE, H. and ROUGHTON, F. J. W. *Proc. Roy. Soc.* A104 (1923) 376

[11] YAMAZAKI, I., MASON, H. S. and PIETTE, L. *J. biol. Chem.* 235 (1960) 2444

[12] BRAY, R. C. *Biochem. J.* 81 (1961) 187

[13] BAGGULEY, D. M. S. and GRIFFITHS, J. H. E. *Nature* 162 (1948) 538

[14] PENROSE, R. P. *ibid.* 163 (1949) 992

[15] INGRAM, D. J. E. *Proc. phys. Soc.* A62 (1949) 664

[16] BLEANEY, B. and INGRAM, D. J. E. *ibid.* 63 (1950) 408; ABRAGAM, A. and PRYCE, M. H. L. *ibid.* 63 (1950) 409

[17] BLEANEY, B. and BOWERS, K. D. *ibid.* 65 (1962) 667

[18] TOWNES, C. H. and TURKEVITCH, J. *Phys. Rev.* 77 (1950) 148; HOLDEN, A. N. KITTEL, C., MERRITT, F. R. and YAGER, W. A. *ibid.* 77 (1950) 147

[19] HUTCHISON, C. A., PASTOR, R. C. and KOWALSKY, A. G. *J. chem. Phys.* 20 (1952) 534

[20] DEGUCHI, Y. *ibid.* 32 (1960) 1584

[21] BELJERS, H. G., VAN DER KINT, L. and VON WIERINGEN, J. S. *Phys. Rev.* 95 (1954) 1683

[22] HOLMBERG, R. W., LIVINGSTON, R. and SMITH. W. T. *J. chem. Phys.* 33 (1960) 541

[23] JARRETT, H. S. *ibid.* 21 (1953) 761

[24] VENKATARAMEN, B. and FRAENKEL, G. K. *J. Amer. chem. Soc.* 77 (1955) 2707

[25] WEISSMAN, S. I., TOWNSEND, J., PAUL, D. E. and PAKE, G. E. *J. chem. Phys.* 21 (1953) 2227

[26] VENKATARAMEN, B. and FRAENKEL, G. K. *ibid.* 23 (1955) 588

[27] WERTZ, J. E. and VIVO, J. S. *ibid.* 23 (1955) 2441

[28] WEISSMAN, S. I. and SOWDEN, J. C. *J. Amer. chem. Soc.* 75 (1953) 503; JARRETT, H. S. and SLOAN, G. J. *J. chem. Phys.* 22 (1954) 1783

[29] TOWNSEND, J., WEISSMAN, S. I. and PAKE, G. E. *Phys. Rev.* 89 (1953) 606

[30] BREIT, G. and RABI, J. I. *ibid.* 38 (1931) 2082

[31] WEISSMAN, S. I., DE BOER, E. and CONRADI, J. J. *J. chem. Phys.* 26 (1957) 963

[32] YOKOZAWA, Y. and MIYASHITA, I. *ibid.* 25 (1956) 796

[33] HUCKEL, E. $Z.$ *Phys.* 70 (1931) 204

[34] HAUSSER, K. H. *Proc. IXth Colloq. Ampère* (1960) 239

[35] DEGUCHI, Y. *J. chem. Phys.* 32 (1960) 1584

[36] REITZ, D. C., DRAUNIEKS, G. and WERTZ, J. E. *ibid.* 33 (1960) 1880

[37] UEBERSFELD, J. and ERB, E. *C. R. Acad. Sci., Paris* 242 (1956) 478

[38] CHANTRY, G. W., HORSFIELD, A., MORTON, J. R., ROWLANDS, J. R. and WHIFFEN, D. H.: *Molec. Phys.* 5 (1962) 233

[39] ROWLANDS, J. R. and WHIFFEN, D. H. *Nature* 193 (1962) 62

[40] HOLMBERG, R. W., LIVINGSTON, R. and SMITH. W. T. *J. chem. Phys.* 33 (1960) 541

[41] GIBSON, J. F., INGRAM, D. J. E., SYMONS, M. C. R. and TOWSEND, M. G. *Trans. Faraday Soc.* 53 (1957) 914

[42] INGRAM, D. J. E., FUJIMOTO, M. and SAXENA, M. C. *Archs Sci., Genève* 12 (1959) 185

[43] FESSENDEN, R. W. *J. phys. Chem.* 68 (1964) 1508

[44] FESSENDEN, R. W. and SCHULER, R. H. *J. chem. Phys.* 39 (1963) 2147

[45] ——*Disc. Faraday Soc.* 36 (1963) 147

[46] STONE, A. J. *Molec. Phys.* 7 (1964) 311

[47] ABRAGAM, A. and PRYCE, M. H. L. *Proc. Roy. Soc.* A205 (1951) 135

[48] WEISSMAN, S. I. *J. chem. Phys.* 25 (1956) 890

[49] McConnell, H. M. *ibid.* 24 (1956) 764

[50] Jarrett, H. S. *ibid.* 25 (1956) 1289

[51] McConnell, H. M. and Dearman, H. H. *ibid.* 28 (1958) 51

[52] McLachlan, A. D. *Molec. Phys.* 2 (1959) 271

[53] Bolton, J. R. and Fraenkel, G. K. *J. chem. Phys.* 40 (1964) 3307

[54] Johnston, C. S., Visco, R. E., Gutowsky, H. S. and Hartley, A. M. *ibid.* 36 (1962) 1580

[55] McConnell, H. M. *ibid.* 25 (1956) 709

[56] Abragam, A. and Pryce, M. H. L. *Proc. Roy. Soc.* A206 (1951) 173

[57] Van Vleck, J. H. *Theory of Electric and Magnetic Susceptibilities.* Oxford: Clarendon Press (1935)

[58] Owen, J. and Stevens, K. W. H. *Nature* 171 (1953) 836

[59] Tinkham, M. *Proc. Roy. Soc.* A236 (1956) 535, 549

[60] Deal, R. M., Ingram, D. J. E. and Srinivasan, R. *Proc. XIIth Colloq. Ampère* (1963) 239

[61] Robertson, J. M. *J. chem. Soc.* (1935) 615; (1936) 1195

[62] Schlapp, R. and Penney, W. G. *Phys. Rev.* 42 (1932) 666

[63] Stevens, K. W. H. *Proc. Roy. Soc.* A219 (1953) 542

[64] Van Vleck, J. H. *J. chem. Phys.* 3 (1935) 807

[65] Griffiths, J. H. E., Owen, J. and Wood, I. M. *Proc. Roy. Soc.* A219 (1953) 526

[66] Owen, J. *Disc. Faraday Soc.* 19 (1955) 53; 26 (1958) 53

[67] Baker, J. M., Bleaney, B. and Hayer, W. *Proc. Roy. Soc.* A247 (1958) 141

[68] Harrison, S. E. and Assour, J. M. *J. chem. Phys.* 40 (1964) 365

[69] Tinkham, M. *Proc. Roy. Soc.* A236 (1956) 535, 549

[70] Deal, R. M., Ingram, D. J. E. and Srinivasan, R. *Electronic Magnetic Resonance and Solid Dielectrics.* Amsterdam: North Holland Publishing Co. (1963) p. 246

[71] Orgel, L. E. *Ligand Field Theory.* London: Methuen (1960)

[72] Griffith, J. S. *Theory of Transition Metal Ions.* London: Cambridge University Press (1961)

[73] Evans, D. F. *Nature* 176 (1955) 777

[74] Hutchison, C. A. and Magnum, B. J. *J. chem. Phys.* 29 (1958) 952

[75] ——*ibid.* 34 (1961) 908

[76] Van der Waals, J. H. and de Groot, M. S. *Molec. Phys.* 2 (1959) 333

[77] de Groot, M. S. and Van der Waals, J. H. *ibid.* 6 (1963) 545

[78] Commoner, B., Townsend, J. and Pake, G. E. *Nature* 174 (1954) 689; *Science* 126 (1957) 57

[79] —and Hollocher, T. C. *Proc. nat. Acad. Sci.* 46 (1960) 405

[80] — , Heise, J. J. and Townsend, J. *ibid.* 42 (1956) 710

[81] Sogo, P. B., Pon, N. G. and Calvin, M. *ibid.* 43 (1957) 387

[82] — , JOST, M. and CALVIN, M. *Radiat. Res. Suppl.* 1 (1959) 511

[83] — , CANTER, L. A. and CALVIN, M. *Free Radicals in Biological Systems*. New York: Academic Press (1961) p. 311

[84] YAMAZAKI, I., MASON, H. S. and PIETTE, L. *J. biol. Chem.* 235 (1960) 2444

[85] BRAY, R. C., MALMSTRAM, B. G. and VANGUARD, T. *Biochem. J.* 73 (1959) 193

[86] — *ibid.* 81 (1961) 196

[87] — , PALMER, G. and BEINERT, H. *J. biol. Chem.* 239 (1964) 2667

[88] EHRENBERG, A. and YONETANI, T. *Acta chem. scand.* 15 (1961) 1071

[89] GORDY, W., ARD, W. B. and SHIELDS, H. *Proc. nat. Acad. Sci.* 41 (1955) 983; 41 (1955) 996

[90] COHN, M. *Biochemistry* 2 (1963) 21

[91] HENRIKSEN, T. *Free Radicals in Biological Systems*. New York: Academic Press (1961) p. 279

[92] — and PIHL, A. *Nature* 185 (1960) 307

[93] ZIMMER, K. G. and MULLER, A. *Current Topics in Radiation Research* (edited by Ebert, M. and Howard, A.). Amsterdam: North Holland Publishing Co. (1965) Vol. 1, p. 1

[94] GRIFFITH, J. S. *Proc. Roy. Soc.* A235 (1956) 23

[95] KENDREW, J. C. *Science* 139 (1965) 1259

[96] PERUTZ, M. F. *Nobel Prize Lecture, Stockholm* (1963); *Proc. Roy. Soc.* A265 (1961) 15

[97] GRIFFITH, J. S. *Nature* 180 (1957) 30

[98] STRYER, L., KENDREW, J. C. and WATSON, H. C. *J. molec. Biol.* 8 (1964) 96

12

DOUBLE RESONANCE — MASERS AND LASERS

12.1 COHERENT AND INCOHERENT RADIATION

IN this chapter consideration is given to the various different methods and techniques of double resonance, and their applications in *masers* and *lasers* and other similar devices. In all of this work the normal energy level populations of the given atomic or molecular systems are disturbed, and the essential properties of the devices come from the modified or inverted energy level populations. The reason why the actual relative populations of the energy levels in an atomic system are so important is connected with the different ways in which radiation can interact with matter. In particular, stimulated emission can only be made greater than normal absorption if there is a larger number of atoms in the excited state than in the ground state.

Stimulated emission is the means by which amplification of incoming signals can be effected; it is also the basic mechanism required for atomic or molecular oscillators with an output of coherent radiation. The conditions necessary for producing an increase or excess of stimulated emission from any given atomic system are therefore crucial to most double resonance or maser/laser type systems. Before the phenomenon of stimulated emission itself is considered, however, it might be wise to discuss the actual nature of radiation that is produced, and in particular the difference between coherent and incoherent radiation.

The concept of coherent radiation is very natural in the radio or microwave region, where it is normal to think of an aerial system emitting a continuous wave train, which maintains time coherence with all successive emissions. It is also well established that phase coherence can be readily obtained in a spatial distribution by connecting a synchronizing link between different oscillators or different aerial systems. The synchronization is such that either all the oscillators produce their maximum voltage swings at the same time or, if the aerials are spaced by any finite amount, then the time lag in the voltage swing corresponds to the path distance between the aerials. It is by using such multi-aerial systems, all driven by

431

the same master oscillator, that large aerial arrays can be constructed with very precise directional properties. In all of the designs it is assumed without question that the wave trains emitted, both by one aerial over a considerable length of time and by all of the aerials together, are phase coherent and without any sudden or random changes in phase. The basic reason for this is that the electrons themselves are moving up and down in phase in the aerial conductors; it is only because such electron movement is phase-synchronized that the resulting electromagnetic radiation is itself coherent, and can be modulated with precision.

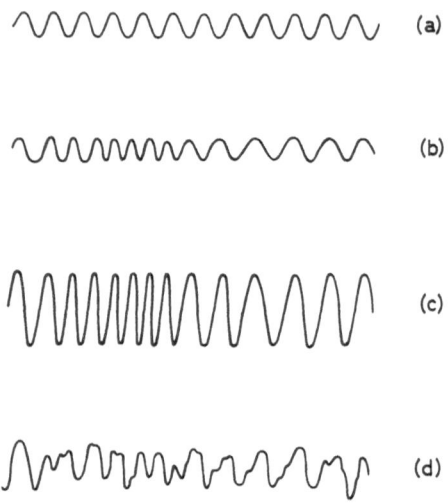

Figure 131. Coherent and incoherent mixing: (a) unmodulated carrier wave, (b) frequency modulated carrier, (c) four modulated carriers added with phase coherence, (d) four modulated carriers added without phase coherence

The importance of phase coherence is clearly seen when a question of modulating the wave train is considered. A simple example is shown in *Figure 131*. In *Figure 131(a)* an unmodulated carrier frequency is shown and in *Figure 131(b)* this same carrier frequency is shown with frequency modulation superimposed on it. If four such carrier signals of the same phase are now taken then the resultant signal is that shown in *Figure 131(c)*, whereas if the four modulated carriers are initially of random phase then the resultant signal has some such form as that in *Figure 131(d)*. It is clear from this simple example that all the precise information available in the frequency modulation is completely lost unless there is phase coherence in the carrier waves. In fact the only type of modulation possible with an incoherent wave train is a gross form of amplitude modulation, which is limited to very small information content.

The extra information that can be transmitted on a coherent beam becomes greater and greater the higher the frequency of the carrier beam. For example, in order to transmit all the information required to build up a television picture, a bandwidth covering 4 Mc/s per sec is necessary, and since the carrier frequency must be an order of magnitude larger than the modulation frequency it follows that the television programmes are normally broadcast on frequencies of 40 Mc/s or above. Thus, in a frequency range between 40 and 50 Mc/s it would be possible to include three different television programmes. If coherent radiation is available, then all the information represented by these different television bands can be transmitted at the same time and still be selected at the receiving end. If on the other hand a microwave frequency is used for transmission, then a coherent beam of microwave frequencies extending from 4,000 to 5,000 Mc/s is able to carry 100 times the amount of information transmitted at the lower frequency, and thus in principle 300 simultaneous television programmes. Passing on to the particular case of the lasers operating in the visible region, it follows that if coherent visible radiation with a wavelength stretching from 4,000 Å to 7,000 Å were available, then there would be a bandwidth of 3×10^7 Mc/s per sec for transmission of the information; this corresponds to over a million television programmes being transmitted simultaneously along the same beam. This simple example gives some idea of the immense potential applications of coherent visible radiation in the field of communications.

Until very recently however all the available sources of radiation in the visible region produced incoherent radiation, and were therefore not suitable for any type of precise modulation or detailed transmission of information. The emission or absorption of visible radiation normally corresponds to a transition in atomic energy levels involving the outer electronic structure. The electron of an atom is first excited to the higher level by some means such as impact with fast moving electrons, or thermal vibrations, and after staying in the excited state for a certain length of time, the electron returns to the ground state, emitting the visible quantum as it does so. This emission of radiation produced when the atom returns to the ground state in the absence of any external stimulus, is known as 'spontaneous emission'. It may be considered as very similar to radioactive emission in that the actual time or phase of the emission for any one atom cannot be predicted, although a definite half-life can be given for the average over the atoms in the upper level.

However, the wave trains corresponding to these emitted quanta of radiation are entirely uncorrelated with similar emissions from other atoms in the excited level, and hence there is a completely random distribution of phase between all the emissions produced in this way. In other words, the spontaneous emission obtained from atoms in the normal excited states is quite useless so far as any precise modulation or transmission of detailed information is concerned.

In comparing this situation with that for the radiowaves emitted from an aerial system, it is obvious that the feature lacking in the visible radiation is the correlating, or coordinating, link between the different oscillators. If such a link could be arranged, so that each atom in the excited level was made to radiate its wave train in synchronism with the phase of others which had undergone the same transition just before, then a coherent constructively interfering wave train would be produced.

12.2 THE INTERACTION OF RADIATION WITH MATTER

In order to see how coherent radiation can be produced in different regions of the spectrum, it is now necessary to consider in more detail the nature of the three basic processes concerned with the interaction of radiation with matter. These are represented in *Figure 132*. *Figure 132(a)* corresponds to the spontaneous emission just discussed, and represents the return of the excited atom to the ground state in the absence of any external stimuli, resulting in the production of a series of quanta of random phase. The inverse of this process of spontaneous emission is the process of absorption, represented in *Figure 132(b)*. The atom, or electron, which is initially in the ground state absorbs the incoming radiation and in so doing jumps to the excited level. Once in the excited level, however, there is another process by which it can return to the ground state if radiation of the correct frequency is also present. This is the process known as stimulated emission, as shown in *Figure 132(c)*. In this case an incoming quantum of resonance radiation stimulates an atom already in the excited state, to return to the ground state and emit a second quantum in the process. The emission is thus basically of the same frequency as the spontaneous emission, but the presence of the stimulating radiation not only causes the stimulated emission to take place more quickly than the spontaneous emission, but also ensures that the phase of the emitted quantum is the same as that of the stimulating radiation. As a result, two phase coherent quanta are produced which can travel on through the medium and in the same way stimulate more atoms to drop from the excited state to the ground state. Hence a phase coherent wave front consisting of emissions from many atoms is built up.

There are effectively, therefore, two possible interactions for an incoming quantum of the correct frequency. If the atom is in the ground state, the quantum is absorbed and the atom raised to the excited level, whereas if the atom is already in the excited level, the incoming radiation can cause stimulated emission and as a result

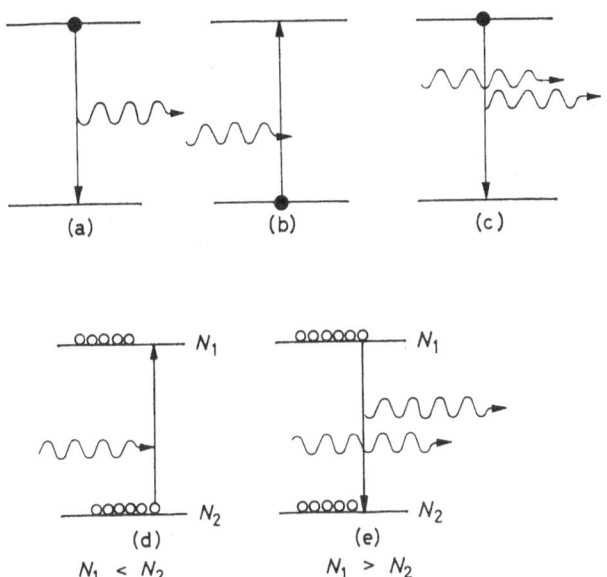

Figure 132. Different forms of emission and absorption: (a) spontaneous emission, (b) absorption, (c) stimulated emission, (d) normal population distribution resulting in absorption, (e) population inversion resulting in stimulated emission

two phase coherent quanta are produced. Einstein showed in his early treatment of radiation theory that the coefficients of absorption and of stimulated emission were in fact equal. Hence the actual probability of the incoming quantum being absorbed, or causing stimulated emission, depends entirely on the actual numbers of atoms in the ground and excited levels respectively.

This situation can be represented diagrammatically as in *Figures 132(d)* and *(e)*. *Figure 132(d)* represents the state of affairs normally occurring under thermal equilibrium: the number of atoms N_1 in the excited level is less than the number N_2 in the ground level, and hence the incoming quantum of radiation is absorbed and its energy used to raise one atom from the ground state to the excited level.

435

However, if the energy level population can be inverted as in *Figure 132(e)*, and N_1 made greater than N_2, then the incoming quantum of radiation has a higher probability of producing stimulated emission than of being absorbed, and a resulting amplification of the incoming radiation is obtained. Moreover, these two quanta can now go on and interact with other excited atoms to produce further quanta all in phase with each other, and thus a beam of coherent radiation is produced. It follows that the whole principle of producing coherent radiation in either the microwave or the visible region is bound up with the possibility of producing an inverted population between the energy levels. The history of the maser and laser development can be followed in the various methods devised for obtaining such inversion of the energy level populations.

The development of the ideas of double resonance are now considered from a more formal and historical viewpoint, starting with the general theory of saturation of a given spectroscopic transition.

12.3 Saturation and Relaxation Effects

The general principle of saturation effects is discussed in Section 4.2(*e*) in connection with microwave gaseous spectroscopy. It is shown there that if the power density becomes too high, the normal population distribution between the upper and lower levels of the microwave transition is affected and, as a result, the absorption line tends to broaden and to be reduced in amplitude. This type of interaction and broadening is, of course, not limited to the gaseous spectroscopy but applies in general to any microwave transition. Its possible occurrence in electron resonance is briefly discussed in Section 4.8(*e*), although it is indicated that in the early measurements on the electron resonance of transition group ions there were no very noticeable saturation effects. This is because interactions are then via the magnetic dipole, and not the electric dipole as in gaseous spectroscopy, and hence the transition probabilities are very much smaller. However, it became clear in later work that saturation effects could become quite significant in some of the free radical studies and also in transition group complexes at very low temperatures. In both of these cases the spin-lattice relaxation time can become very long, preventing a rapid return of the excited atoms to the ground state. Since the theory given for the gaseous absorption is not directly applicable to the case of the magnetic resonance transitions, the complete theory for the case of electron resonance is now briefly summarized.

In any spectroscopic transition two competing factors must be considered. On the one hand the number of atoms in the excited state is increased by the presence of the incoming microwave radiation; on the other hand, this increase in population tends to be reduced by the spin-lattice interaction, which attempts to re-establish conditions of thermal equilibrium. Saturation occurs when the spin-lattice interaction is not strong enough to return the excited atoms to the ground state sufficiently fast to maintain the normal distribution of population between the two levels, and hence the transition probabilities between the two levels begin to change. This situation can be treated quantitatively by considering two energy levels A and B, where A is the excited level, with the number of atoms in the two levels given by $\mathcal{N}_1$ and $\mathcal{N}_2$ respectively. In thermal equilibrium and in the absence of any incoming radiation, the ratio between the energy level populations is given by the normal Maxwell–Boltzmann distribution, i.e.

$$\frac{\mathcal{N}_1}{\mathcal{N}_2} = \exp\left(-h\nu/kT_L\right) \qquad \ldots\,(12.1)$$

where T_L is the temperature of the lattice. When the microwave resonance radiation is applied, transitions are induced from the ground state to the excited state and are also stimulated to take place from the excited state to the ground state. The net effect of both of these transitions can be represented by the term $(dn/dt)_{\text{r.f.}}$, and the total rate of change of population can be written

$$\frac{dn}{dt} = \left(\frac{dn}{dt}\right)_{\text{r.f.}} + \left(\frac{dn}{dt}\right)_{\text{s.l.}} \qquad \ldots\,(12.2)$$

where $n = \mathcal{N}_2 - \mathcal{N}_1$, and $(dn/dt)_{\text{s.l.}}$ represents the action of the spin-lattice interaction in returning the excited atoms to ground state. The value of $(dn/dt)_{\text{r.f.}}$ can be obtained directly from classical radiation theory and is given as

$$\left(\frac{dn}{dt}\right)_{\text{r.f.}} = -\tfrac{1}{4}\pi \cdot \gamma^2 \cdot H_1^2 \cdot g(\omega - \omega_0) \cdot n \qquad \ldots\,(12.3)$$

where γ is the gyromagnetic ratio of the electron, H_1 is the strength of the microwave magnetic field, and $g(\omega - \omega_0)$ is the shape function of the absorption line. This is normalized so that

$$\int_0^\infty g(\omega - \omega_0) \cdot d\omega = 1 \qquad \ldots\,(12.4)$$

The expression for the rate of transitions produced by the spin-lattice interaction can be represented by the equation overleaf

$$\left(\frac{dn}{dt}\right)_{\text{s.l.}} = \frac{n_0 - n}{T_1} \qquad \ldots (12.5)$$

where T_1 is the spin-lattice relaxation time, and n_0 is the value of n at thermal equilibrium and is given from equation (12.1) as

$$n_0 = \mathcal{N}_{2_0} - \mathcal{N}_{1_0} \approx \frac{\hbar \cdot \omega_0}{k T_L} \cdot \mathcal{N}_{1_0} \qquad \ldots (12.6)$$

When equilibrium conditions have set in $(dn/dt)_{\text{total}}$ must be zero, and hence

$$-\tfrac{1}{4}\pi \cdot \gamma^2 \cdot H_1^2 \cdot g(\omega - \omega_0) \cdot n + \frac{n_0 - n}{T_1} = 0 \qquad \ldots (12.7)$$

Therefore

$$n = n_0 [1 + \tfrac{1}{4}\pi \cdot \gamma^2 \cdot H_1^2 \cdot g(\omega - \omega_0) \cdot T_1]^{-1} \qquad \ldots (12.8)$$

The rate P_a at which energy is absorbed from the magnetic field may now be calculated. Thus

$$P_a = - h\nu \left(\frac{dn}{dt}\right)_{\text{r.f.}} \qquad \ldots (12.9)$$

and substituting (12.3) and then (12.8) into this gives

$$P_a = \hbar \cdot \omega \cdot \tfrac{1}{4} \cdot \pi \cdot \gamma^2 \cdot H_1^2 \cdot g(\omega - \omega_0) \cdot n_0 \cdot [1 + \tfrac{1}{4}\pi \cdot \gamma^2 \cdot H_1^2 \cdot g(\omega - \omega_0) T_1]^{-1}$$

$$= \tfrac{1}{2}\omega \cdot \omega_0 \cdot \left\{ \frac{\gamma^2 \cdot \hbar \cdot n_0}{2\omega_0} \right\} \cdot H_1^2 \cdot \left[\frac{\pi \cdot g(\omega - \omega_0)}{1 + \tfrac{1}{4}\pi \cdot \gamma^2 \cdot H_1^2 \cdot T_1 \cdot g(\omega - \omega_0)} \right]$$

$$\ldots (12.10)$$

This rate of energy absorption is also given by the expression for the complex susceptibility χ'', thus

$$P_a = \tfrac{1}{2}\omega \cdot \chi'' \cdot H_1^2 \qquad \ldots (12.11)$$

Therefore equating (12.10) and (12.11), and substituting χ_0, the d.c. susceptibility, for $\gamma^2 \cdot \hbar \cdot n_0 / 2\omega_0$ gives

$$\chi'' = \omega_0 \cdot \chi_0 \left[\frac{\pi \cdot g(\omega - \omega_0)}{1 + \tfrac{1}{4} \cdot \pi \cdot \gamma^2 \cdot H_1^2 \cdot T_1 \cdot g(\omega - \omega_0)} \right] \ldots (12.12)$$

At the resonance frequency, $g(\omega - \omega_0)$ has its maximum value, and from the definition of the spin-spin relaxation time T_2, this maximum value is equal to T_2/π. Hence

$$\chi''_{\omega_0} = \chi_0 \cdot \omega_0 \cdot T_2 \left[\frac{1}{1 + \tfrac{1}{4} \cdot \gamma^2 \cdot H_1^2 \cdot T_1 \cdot T_2} \right] \qquad \ldots (12.13)$$

In the absence of saturation, χ''_{ω_0} is given by

$$\chi''_{\omega_0} = \chi_0 \cdot \omega_0 \cdot T_2 \qquad \dots (12.14)$$

and the term 'saturation factor Z' is therefore given to the remainder of the expression in (12.13), i.e.

$$Z = \frac{n_s}{n_0} = \frac{1}{1 + \frac{1}{4} \cdot \gamma^2 \cdot H_1^2 \cdot T_1 \cdot T_2} \qquad \dots (12.15)$$

where n_s is the population difference between the two levels under saturation conditions.

This ratio indicates the way in which the absorption of power varies, since the amount of absorption decreases as n_s is reduced. Hence the absorption is reduced by large values of the spin-lattice relaxation time T_1, and by increases in the level of the microwave power H_1. This decrease in the expected absorption obviously occurs first in the centre of the lines where the greatest power is absorbed, and only affects the wings of the lines as the value of the microwave radiation rises still further. The effect of saturation is therefore not only to reduce the expected power absorption, but also to alter the line shape, flattening it in the centre before it does so in the wings and thus increasing the apparent width. This can be seen quantitatively in equation (12.12), where $g(\omega - \omega_0)$ is multiplied by H_1 in the denominator but not in the numerator, indicating that the line changes shape as H_1 increases. This theory applies to all cases of 'homogeneous broadening', i.e. where the line is initially broadened by interactions within the spin system, or from an external interaction which is fluctuating rapidly compared with the time taken for a spin transition. It therefore includes dipole spin-spin interaction, spin-lattice interaction, and motional or exchange narrowing.

The theory does not apply to external interactions varying with a period which is long compared with the time of a spin transition. Such interactions may be from unresolved hyperfine structure, or from inhomogeneities in the magnetic field. In these cases, any broadening which occurs due to saturation takes place for the individual lines or components separately; hence the overall line shape, which is the envelope of several real absorption lines, does not change. Each of the component individual lines has its peak height reduced by the same saturation factor, the net result being that such an 'inhomogeneously broadened' line does not change shape or width on saturation, but the expected power absorption falls in the same proportion across the whole of the line shape.

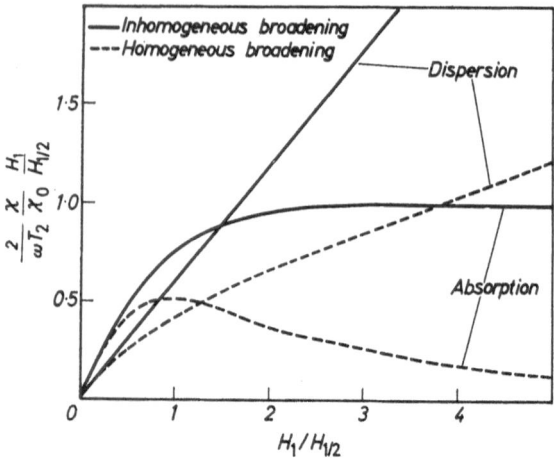

Figure 133. Saturation effects for homogeneous and inhomogeneous broadening (after Portis[1])

This difference in behaviour between homogeneously and inhomogeneously broadened absorption lines under saturation conditions can be used to differentiate between them, and to elucidate the mechanism responsible for line broadening. This is illustrated in *Figure 133* where the variation of both the absorption and dispersion of the two kinds of lines is plotted against the value of the applied microwave magnetic field strength. The abscissa is actually $H_1/H_{\frac{1}{2}}$ where $H_{\frac{1}{2}}$ is the value of the microwave field strength which makes the saturation parameter equal to $\frac{1}{2}$; the ordinate is $\left[\dfrac{2}{\omega T_2} \cdot \dfrac{\chi}{\chi_0} \cdot \dfrac{H_1}{H_{\frac{1}{2}}}\right]$. The complete theoretical and experimental treatment of this problem was given early in the development of the theory of saturation by PORTIS[1].

12.4 SPIN-LATTICE INTERACTION AND POPULATION DISTRIBUTION

It is evident from the calculations of the preceding section, that the onset of saturation can be used not only to determine the different types of line broadening, but also to calculate quantitatively the value of the spin-lattice relaxation time from the decrease in the absorbed power. The ratio of the actual power absorbed to that expected from the known microwave magnetic field strength, is equal to the saturation factor Z. This ratio can be determined experimentally by plotting the magnitude of the absorbed power against H_1^2 and noting the deviation of this plot from a straight line. The value of T_1 can be calculated from the value of $\mathcal{Z}$ determined in this way, and from the known values of H_1 and T_2.

The actual value of the microwave field H_1 of the specimen can be calculated if the Q of the cavity and the magnitude of the power entering the cavity are known. Thus for a rectangular cavity carrying an H_{01} mode, the microwave magnetic field at its centre can be expressed as

$$H_1^2 = \left[Q \cdot P_g \cdot \sqrt{1 - \left(\frac{\lambda}{2a}\right)^2} \, \right] \Big/ (30 \cdot \pi \cdot a \cdot b) \quad \dots (12.16)$$

where P_g is the power in watts in the waveguide, a is the cavity resonance factor, and the waveguide dimensions a and b are in centimetres.

Although these saturation studies have been mainly used in electron resonance spectroscopy to determine spin-lattice relaxation times and similar parameters, they have also become very important in wider applications, because of the actual change in the energy level populations which saturation produces. It has already been seen that the essential effect of the saturating field is to disturb the normal thermal equilibrium and thus reduce the difference in population between the ground level and the excited state. The ratio of the difference in population which actually exists between these two states and that which would exist under thermal equilibrium conditions is equal to the saturation parameter $\mathcal{Z}$, as seen in equation (12.15).

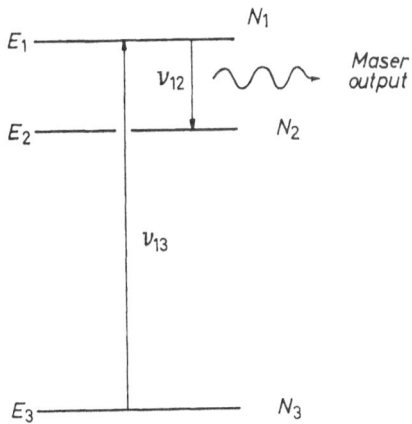

Figure 134. Populations in three-level maser system

441

The reason why this change in energy level population can be so important, is probably best illustrated by a brief reference to the three-level maser system as illustrated in *Figure 134*, although the idea introduced here is also fundamental to most systems of double resonance. The three energy levels E_1, E_2, and E_3 in *Figure 134* are drawn with their normal populations N_1, N_2 and N_3, on the right-hand side of the figure, and under thermal equilibrium $N_1 < N_2 < N_3$. Consider now the situation which develops when a large magnitude of microwave power is fed in at the resonance frequency between the first and third level, i.e. ν_{13}. If the value of H_1 for this applied microwave power is sufficient to cause appreciable saturation between these two levels, then the population distribution of levels E_1 and E_3 is altered and N_1 begins to approach N_3 in magnitude. If the saturation is sufficiently strong, it is quite possible for the number of atoms in level E_1 to become greater than that in level E_2. In this case the normal difference in population between the energy levels E_1 and E_2 is not only noticeably modified but actually reversed in sign. It is then possible for the stimulated emission from these two levels to become greater than the normal absorption of the resonance radiation, and hence amplification or oscillation at the frequency ν_{12} becomes possible.

The more formal mathematical treatment of this three-level maser system can be summarized as follows, indicating the importance of the spin-lattice relaxation time between the different sets of levels, or the corresponding probabilities p_{12}, p_{13}, etc. In thermal equilibrium the actual populations of the three levels can be written

$$N_1^0 = \frac{N}{F} \cdot e^{-E_1/kT}$$

$$N_2^0 = \frac{N}{F} \cdot e^{-E_2/kT} \qquad \qquad \dots\dots (12.17)$$

$$N_3^0 = \frac{N}{F} \cdot e^{-E_3/kT}$$

where N_1^0, N_2^0, N_3^0 indicate the number in levels E_1, E_2, E_3 under conditions of thermal equilibrium, N is the total number of atoms in the system, and F is the partition function given by

$$F = \sum_i e^{-E_i/kT} \qquad \qquad \dots\dots (12.18)$$

The changes in these populations are now considered under the conditions appropriate to maser action, i.e. in the presence of an intense radiation field of frequency ν_{13}, and a very weak one of frequency ν_{23}. If the pumping radiation is very intense, complete saturation of this transition can be assumed and then

$$\mathcal{N}_1 = \mathcal{N}_3 = \tfrac{1}{2}(\mathcal{N}_1^0 + \mathcal{N}_3^0) \qquad \ldots (12.19)$$

The spin-lattice relaxation mechanisms between the different energy levels can be characterized by transition probabilities p_{12} such that

$$p_{12} = p_{21} \cdot \exp\left(-h\nu_{12}/kT\right) \qquad \ldots (12.20)$$

This equation arises from the principle of detailed balancing, which states that for a system in thermal equilibrium, the rate of transitions between two energy states in one direction must be equal to the rate in the opposite direction. Since the numbers of atoms in the two levels are different it follows that the transition probabilities must also be different to compensate.

The rate of change of the populations in the three levels may then be written as

$$\dot{\mathcal{N}}_1 = (p_{31}\mathcal{N}_3 + p_{13}\mathcal{N}_1) + (p_{21}\mathcal{N}_2 - p_{12}\mathcal{N}_1) + A_{31}(\mathcal{N}_3 - \mathcal{N}_1) + A_{21}(\mathcal{N}_2 - \mathcal{N}_1)$$
$$\dot{\mathcal{N}}_2 = (p_{12}\mathcal{N}_1 - p_{21}\mathcal{N}_2) + (p_{32}\mathcal{N}_3 - p_{23}\mathcal{N}_2) + A_{21}(\mathcal{N}_1 - \mathcal{N}_2)$$
$$\dot{\mathcal{N}}_3 = -\dot{\mathcal{N}}_2 - \dot{\mathcal{N}}_1 \qquad \ldots (12.21)$$

where the A's are the actual radiation induced transition probabilities and $A_{13} = A_{31}$, etc.

If the normal approximation for the exponential is made and the above equations are solved for steady state conditions, it can be shown that the actual emission of radiation between the first and second levels is given by

$$\frac{N \cdot h^2 \cdot \nu_{12}}{3kT} \cdot \frac{(p_{32}\nu_{23} - p_{21}\nu_{12}) \cdot A_{21}}{p_{32} + p_{21} + A_{21}} \qquad \ldots (12.22)$$

It is thus possible to calculate the actual conditions that must be fulfilled to make this expression positive and hence to obtain amplification or stimulated emission from the system.

The use of a high power saturating pumping frequency to produce stimulated emission between two other levels is one of the essential principles behind a large number of maser devices, and is considered in more detail in later sections of this chapter.

12.5 Overhauser and Double Resonance Effects

If any form of coupling exists between electrons and nuclei in a system, then it is more than likely that an applied radiation field causing some saturation between the ground level and one of the excited states, and hence changing the population distribution between atoms, will also cause some change in the population of the other excited states coupled to the one directly affected. The first suggestion that this might be so, and that it might be possible to observe the change in population distribution for the second set of levels, was put forward by OVERHAUSER[2]. It is pointed out in Section 8.10, that the particular interaction with which Overhauser was concerned was due to the hyperfine coupling between the electron and nuclear spins, and his initial theoretical considerations were confined to conduction electrons in metals. He showed that the relaxation processes which act to restore equilibrium after the absorption of energy by the electron resonance, also induce nuclear transitions via the hyperfine coupling, and as a result alter the normal Boltzmann distribution between the nuclear hyperfine levels. The result of this interaction is that the two energy levels between which nuclear resonance can be observed then have a large difference in population, instead of the normal small difference. The transition probability for nuclear resonance is thus considerably increased, and enhanced nuclear resonance signals should be obtained. This effect was quickly confirmed experimentally, the first measurements being made on the alkali metals[3] and later on other compounds including free radicals such as hydrazyl[4].

The possibility of altering the resonance signal observed at one given frequency by irradiating the specimen with a different frequency, has given rise to a large number of different types of double resonance techniques. In this connection it is a little difficult to give a very precise definition of the 'Overhauser effect'. The most general definition can probably be phrased as follows: 'if, in a system containing a group of energy levels, saturation is produced between two of them, it should be possible to observe a change in population between two other levels of this group.' The existence of this effect necessarily implies some cross-coupling or cross-relaxation between the energy levels within the group, and the actual mechanism of this interaction can of course vary. The general theory behind the effect can now be briefly summarized.

A scalar coupling M is assumed to exist between the electron spin S and the nuclear spin I, and this coupling is also assumed to be *smaller* than the Zeeman splitting of both S and I (thus the case of paramagnetic ions is ruled out for the moment). M is assumed to have a fluctuating component, implying that there must be a third system with a continuous spectrum, such as the lattice. Energy can be either given to, or taken from, this reservoir which is required to ensure conservation of energy when the spin 'flip-flop' process takes place. This flip-flop process is the transition occurring when the electron spin and nuclear spin interact in such a way that the electron spin changes to a higher level while the nuclear spin changes down to a lower energy level, or vice versa. It can be represented symbolically by the expression

$$S_+ I_- \rightleftharpoons S_- I_+ \qquad \dots (12.23)$$

The energy normally involved in the electron spin change is evidently not the same as the energy involved in the nuclear spin change; hence the flip-flop process must involve an interaction with a third system, which can either supply or take up the energy difference. In the case of the conduction electrons in a metal, this reservoir is supplied by the kinetic energy of the electrons. It should be noted that the existence of this reservoir is one of the essential requirements for the Overhauser effect.

The main cause of relaxation of the nuclei is assumed to be the coupling M discussed above. If the nucleus is to change its spin orientation, and thus alter the energy level populations, it must meet an electron which will also undergo the corresponding spin flip; the combined process can therefore be written in the form of an equation

$$\frac{dn}{dt} = - n_- . \mathcal{N}_+ . W_{+-} + n_+ . \mathcal{N}_- . W_{-+} \qquad \dots (12.24)$$

and in the steady state condition this leads to an expression for the actual population of the two nuclear energy levels of the form

$$\frac{n_+}{n_-} = \frac{\mathcal{N}_+}{\mathcal{N}_-} . \left[\frac{W_{+- \rightarrow -+}}{W_{-+ \rightarrow +-}} \right] \qquad \dots (12.25)$$

In these expressions, n_+ and n_- are the numbers in the two nuclear levels, whilst $\mathcal{N}_+$ and $\mathcal{N}_-$ are the numbers in the two electron levels. W indicates the transition probability between the levels indicated.

The probabilities $W_{+- \rightarrow -+}$ and $W_{-+ \rightarrow +-}$ are *not* in general equal, since the system is in equilibrium with the lattice; the ratio of the two can be written as $\exp [-\beta_L \hbar (\omega_S - \omega_I)]$, where β_L represents $1/kT_L$, and T_L is the lattice temperature. It can be seen that $\hbar(\omega_S - \omega_I)$ is the energy difference in the flip-flop process, ω_S and ω_I being the angular resonance frequencies of the electron and nuclear spins respectively.

In the same notation the ratio of the population of the two electron spin states can be written as shown overleaf

$$\mathcal{N}_+/\mathcal{N}_- = \exp\left(-\beta_S \cdot \hbar\omega_S\right)$$

and the ratio for the nuclear spin populations can be written

$$n_+/n_- = \exp\left(-\beta_I \cdot \hbar\omega_I\right)$$

In general, however

$$\frac{n_+}{n_-} = \frac{\mathcal{N}_+}{\mathcal{N}_-} \cdot \exp\left[-\beta_L \cdot \hbar(\omega_S - \omega_I)\right]$$

Therefore

$$\exp\left(-\beta_I \cdot \hbar\omega_I\right) = \exp\left(-\beta_S \cdot \hbar\omega_S\right) \cdot \exp\left[-\beta_L \cdot \hbar(\omega_S - \omega_I)\right]$$

Therefore

$$\beta_I = \frac{\omega_S}{\omega_I}\beta_S + \frac{1}{\omega_I} \cdot \beta_L(\omega_S - \omega_I)$$

Or

$$\beta_I = \beta_L - \frac{\omega_S}{\omega_I}(\beta_L - \beta_S) \qquad \ldots\ldots(12.26)$$

This equation may be considered as a basic statement of the Over-hauser effect. If $\beta_L = \beta_S$ then of course thermal equilibrium conditions are obtained, but if the temperature of the electron spins and the nuclear spins are not equal to T_L then Overhauser effects may be observed and although saturation of electron resonance is often the easiest way to produce a difference between β_L and β_S it should be noted that it is not the only way.

The above treatment has assumed a *scalar* interaction, whereas in a large number of cases the most important interaction may be of a tensor form, for instance a dipolar interaction. Such an interaction introduces new terms into the considerations above, since it allows for a 'flip-flip' process as well as the flip-flop process. This flip-flip process can be represented by the transition probabilities $W_{+-\to--}$, and the net result of this is to add an expression for the dipolar interaction of the form

$$\beta_I = \beta_L + \tfrac{1}{2}\frac{\omega_S}{\omega_L}(\beta_L - \beta_S) \qquad \ldots\ldots(12.27)$$

Thus, the general effect can be summarized by a dynamic equation written in the form

$$\frac{dI}{dt} = \frac{1}{T_1}\{(I - I_0) + m.f.(S - S_0)\} \qquad \ldots\ldots(12.28)$$

where $m = -1$ for a scalar interaction and $= +\frac{1}{2}$ for a dipolar inter-action; I_0 and S_0 are the average components of the nuclear and electron spin quantum numbers respectively in the direction of quantization; f is the leakage coefficient. This theoretical treatment has not considered the problem of experimental observation of the various parameters, but it should be noticed that measurements will lead to the values held by S and I under the saturation conditions. Thus S can be determined by direct measurement of the electron polarization, by means of the Knight shift for instance, and I can be determined by a direct measurement of the enhancement of a nuclear resonance signal.

So far, the coupling between electron spins and nuclear spins only has been considered, but it is also possible for such coupling to exist between one type of nucleus and another. The theoretical steps are then exactly the same as those outlined above, resulting in a dynamic equation of the form

$$\frac{dI'}{dt} = \frac{1}{T_1}[(I' - I_0') + mf' \cdot (S - S_0) + \alpha(I - I_0)] \quad \ldots (12.29)$$

where I' represents the energy level population of one nucleus, say C^{13}, and I represents the energy level population of the other nucleus such as H. If the electron spins are saturated then it follows that

$$I' = mf'S_0 - \alpha I \quad \ldots (12.30)$$

There are several general points following from this theoretical treatment which are sometimes not appreciated. For instance, by employing the Overhauser effect it is normally considered possible to obtain enhancement of the nuclear resonance signal by a factor equal to the ratio of the electron gyromagnetic ratio to that of the nucleus. In principle, it is in fact possible to obtain a greater enhancement than this if the field seen by the electron is greater than that experienced by the nucleus. Thus, for the resonance of the unpaired electron in a free radical in the earth's field, the electron experiences a field of about 20 gauss or so due to the nucleus, while the nucleus itself only experiences the earth's field. As mentioned previously, the basic equation only predicts the necessity of altering β_L compared with β_S, but this does not necessarily imply the use of saturating electron resonance frequencies. Thus it is possible to alter β_L instead of β_S, and this has been achieved by the use of electric fields in such compounds as indium antimonide.

(a) *Practical applications.* It is evident from the theoretical treat-ment outlined above, that measurements on Overhauser effects enable such parameters as the relaxation times to be determined quite accurately, and are in fact a very direct method of studying cross-relaxation between the energy levels in a given system. The effect also has a very practical application in the enhancement of nuclear resonance signals, and is now being used for this purpose by physical chemists to improve the sensitivity of ordinary nuclear resonance spectroscopy. In this work it is normal to dissolve organic free radicals in the liquid whose nuclear resonance is to be studied, and then to saturate the electron resonance signal of the free radicals. The coupling between the unpaired electrons of the free radical and the nuclei of the liquid then changes the population distribution of the nuclei in their energy levels, as the electrons relax back to equilibrium conditions. A dipolar type of interaction should produce an inverted and enhanced nuclear resonance signal; the enhancement ratio predicted for different nuclei are

$$
\begin{array}{ll}
\text{H} & 330 \\
\text{P} & 810 \\
\text{C}^{13} & 1{,}300
\end{array}
$$

RICHARDS and WHITE[5] at the Physical Chemistry Laboratory, Oxford, were two of the first to utilize the Overhauser effect in the practical enhancement of nuclear resonance spectroscopy and *Figure 135* shows the enhanced proton resonance signal from dimethyl formamide containing a small amount of free radical semiquinone. The four small humps on the left are the successive traces of the signal with no microwave pumping, while the four large inverted signals are produced when the microwave power is applied.

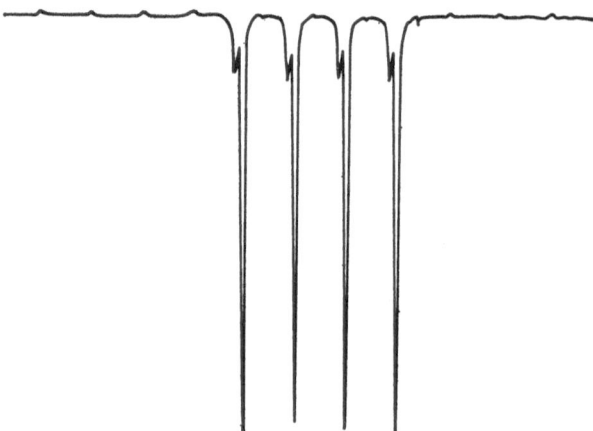

Figure 135. Enhancement of NMR signal (after Richards and White[5])

Theoretically all that need be done in this technique is to apply the microwave radiation at the correct frequency, and there is no need to detect or display it. However, it is often better to build a complete electron resonance spectrometer in order to check that the electron resonance is being obtained and saturated. Experiments have been performed on the protons of benzene and toluene and both of these give 'extrapolated' enhancements of the correct order. The actual enhancement obtained in a given situation depends on both radical concentration and the incident microwave power; in order to compare the experimental values with theory these enhancements should be extrapolated to infinitely high radical concentration and infinitely large microwave power. In practice, the normal enhancement obtained with protons at 3,300 gauss (i.e. working at X-band electron resonance) is of the order of 150. Under the same conditions an actual enhancement of about 500 is expected for C^{13}.

12.6 ENDOR

A year or two after the initial demonstration of the feasibility of double resonance measurements by CARVER and SLICHTEN's verification of the Overhauser effect[3], a new technique was introduced by FEHER[6] which became known as the ENDOR method, standing for Electron Nuclear DOuble Resonance. As a very brief qualitative description of the difference between the Overhauser and the ENDOR techniques it may be said that the Overhauser effect

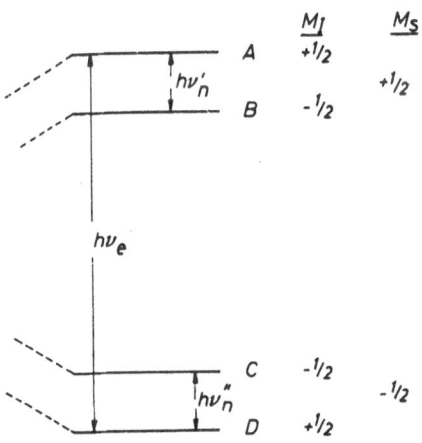

Figure 136. Principle of ENDOR technique

enables the high sensitivity of electron resonance to be made available in the actual observation of the nuclear resonance signal, while the ENDOR technique allows the high resolution of the nuclear resonance technique to be attained in the actual observation of electron resonance spectra.

The main application of the ENDOR technique is in making measurements on electron resonance lines in which unresolved hyperfine structure is suspected. The nuclear resonance frequency is then used to pick out these unresolved hyperfine splittings. The principle of the method is probably best illustrated by a simple example in which a hyperfine interaction of a nucleus with spin $\frac{1}{2}$, such as a proton, is considered with each electronic energy state as illustrated in *Figure 136*. Each of the electronic levels is then split into two and the energies of the four resulting levels can be written as

Level	Energy
A	$\frac{1}{2}g\beta H + \frac{1}{4}A - \frac{1}{2}g_N\beta_N H$
B	$\frac{1}{2}g\beta H - \frac{1}{4}A + \frac{1}{2}g_N\beta_N H$
C	$-\frac{1}{2}g\beta H + \frac{1}{4}A + \frac{1}{2}g_N\beta_N H$
D	$-\frac{1}{2}g\beta H - \frac{1}{4}A - \frac{1}{2}g_N\beta_N H$

$$\dots \dots (12.31)$$

The second term in these expressions for the energy represents the normal hyperfine interaction, whereas the third term arises from the direct interaction of the applied magnetic field with the magnetic moment of the proton. It can be seen from these detailed expressions that the energy difference between levels A and B is not quite the same as that between C and D.

In the ENDOR technique a high power electron resonance frequency is first appied to saturate one of the electron resonance transitions, such as that between A and D in *Figure 134*. The result of this saturation is to increase the energy level population of level A and hence this gives level A a higher population than level B. Whilst this saturation is taking place, a radiofrequency is also applied to the sample, with a frequency $h\nu_{\text{r.f.}}$ equal to a splitting between A and B; this then stimulates transitions from A to B and the populations of the two levels return to their normal equilibrium values. Consequently the saturation of the electron transition is removed, and a strong electron resonance line is suddenly obtained in place of the weakened saturation condition. The net result is that if the detecting system is kept set on the electron resonance signal, a sudden increase in this is obtained when the nuclear resonance signal sweeps through the condition

$$h\nu_{\text{r.f.}} = \frac{1}{2}A - g_N\beta_N H \qquad \dots \dots (12.32)$$

A similar situation also arises when the radiofrequency sweeps through the resonance value corresponding to the nuclear transition between levels C and D. The saturation of the electron resonance will have reduced the number of atoms in the level D, but when the nuclear resonance transition is induced by the radiofrequency field the population of levels C and D will be more or less equalized; thus the electron resonance signal being observed will suddenly become desaturated and a large signal will be obtained. It follows from this that if the radiofrequency signal is slowly swept through a range of values centred on $\nu = A/2h$, a large increase in the electron resonance signal will be obtained when the frequency satisfies either of the conditions

$$h\nu_{\text{r.f.}} = \tfrac{1}{2}A \pm g_N \beta_N H \qquad \qquad \ldots . (12.32a)$$

From these two values of the radiofrequency resonance signal, both A and g_N can be deduced very accurately.

The ENDOR measurements were initially applied by Feher[6] to study resonance from doped atoms in silicon, but were also rapidly extended to other systems[7], such as those formed by F-centres in irradiated KCl. Normal electron resonance techniques had not been able to resolve any hyperfine structure from such crystals, but by employing the double resonance technique Feher was able to resolve not only the hyperfine splitting from the two chlorine isotopes but also from the small quantity of K^{41}. The spectrum he obtained is shown in *Figure 137* and is typical of a double resonance tracing of this type. The ordinate represents the actual absorption produced by the electron resonance transition, and both the microwave frequency and the value of the applied magnetic field are held constant throughout the experiment at this resonance condition. The abscissa corresponds to the changing value of the applied radiofrequency field, and absorption lines are obtained whenever this frequency corresponds to an actual hyperfine splitting present in the overall energy level pattern. It should be noticed in this connection, that the position of the lines correspond to *splittings* and hence no symmetry is to be expected in the pattern as it is traced out. Since the factors which determine whether or not the desaturation is to occur are those which determine the width of the nuclear resonance line, it can be seen that the effective resolution has been increased by a factor of about 1,000; this spectrum is an excellent example of the very great increase in resolution that can be obtained by the ENDOR method.

This ENDOR technique has now been applied to a variety of different atomic systems. Most of the initial studies were concentrated

451

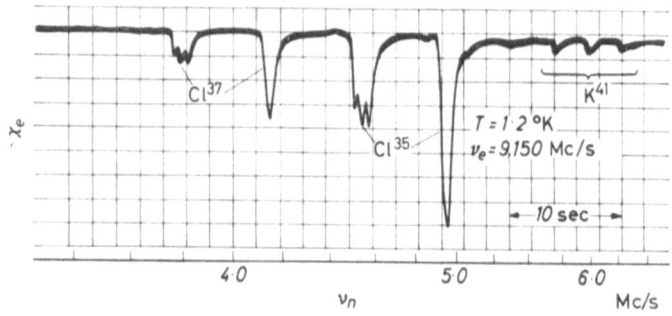

Figure 137. ENDOR spectrum of F-centres in irradiated KCl (after Feher[7])

on either doper atoms or defects in crystal lattices, or on transition group ions in single crystals. Thus the work of Feher on phosphorus atoms in silicon and on F-centres in KCl was followed by measurements on the nuclear hyperfine coupling constants of Sb^{121} and Sb^{123} in silicon[8], and by measurements on the different types of F-centres produced in LiF by bombardment with neutrons or 50 kV X-rays[9]. The great advantage of employing ENDOR techniques in the study of these systems is the very much higher accuracy obtained in the determination of hyperfine coupling constants. This, of course, also applies to the measurements made on the hyperfine splittings in the transition group complexes, such as those on neodymium in diluted lanthanum chloride[10].

The next development in the ENDOR field was the successful application of the technique to free radical studies. This was first accomplished for free radicals orientated in a crystalline lattice, as exemplified by the work of COLE, HELLER and LAMBE[11], on the free radical $COOH.CH.CH_2.COOH$, formed when succinic acid is irradiated by X-rays. Earlier straightforward ESR measurements by HELLER and McCONNEL[12] and by POOLEY and WHIFFEN[13] had shown that there was appreciable but unequal coupling to the two protons of the CH_2 group, together with an anisotropic coupling to the CH proton; the ENDOR measurements enabled precise determinations of these coupling constants to be obtained.

More recently COOK and WHIFFEN[14] introduced a form of 'double ENDOR' technique by which the signs, as well as the magnitude, of the coupling constants can be determined. This has been applied to the two CH_2 protons of the radical mentioned above, the couplings to which were found to have the same sign as each other, but opposite to that of the CH proton. In principle it is possible to determine the absolute sign of a coupling constant if it can be discovered whether the larger ENDOR frequency transition occurs in the 'electron spin parallel' or 'electron spin anti-parallel' sub-set of energy levels. This is not easy to ascertain in a simple ENDOR experiment. If however two simultaneous ENDOR transitions are induced, then they reduce each others efficiency if they are in the sub-set of states with the same electron spin, whereas they enhance each other if they are in different sub-sets.

This double ENDOR technique can be simply applied by injecting a second unmodulated radiofrequency into the same ENDOR loop in the cavity and sweeping this through the expected frequency range. Note is taken of the ENDOR signals which are enhanced and of those which are reduced by the presence of the second transition, and thus the sub-sets of energy levels can be rapidly sorted out.

Another recent development is the observation of ENDOR for free radicals in solution, such as sodium in liquid ammonia[15], and also for strongly coupled protons in Coppingers radical[16]. These studies promise to open up a wide and interesting field of investigation.

It would probably be wise to end this section with a word of caution, however. It is often not easy to detect ENDOR signals, even in systems where they are to be expected and the coupling constants are already approximately known. The conditions for observation are often rather critically dependent on the relaxation times involved, and a careful search may have to be made at different temperatures and with applied radiation fields of different magnitudes.

12.7 Inversion of Energy Level Populations

In the Overhauser and ENDOR double resonance experiments the population of the upper energy level is artificially increased, but not actually inverted with respect to the ground state. However, it was seen, in Section 12.2 that if stimulated emission is to become greater than absorption, such an inversion of the energy level population must be produced, and this is therefore the necessary requirement for the practical production of any devices for amplification, or oscillation, such as the 'maser' or the 'laser'.

In this connection, the great practical advantage of the *maser* is its low noise figure, since there is no extraneous noise arising from macroscopic effects such as bad electrical contacts or hot electron streams. Thus, any amplifying device which can use atomic or molecular energy levels as its basic mechanism of amplification should be inherently much less noisy than other devices employing semi-conductor contacts or modulation and demodulation of electron beams. This inherently low noise factor which is the great feature of microwave maser devices, resulted in their being used for such applications as transatlantic television relay amplifiers.

In contrast to this, the main practical application of the *laser* has been in the production of coherent radiation in the optical region, the great advantages of which is explained in Section 12.1. In both cases, however, the basic requirement is the possibility of producing an inverted energy level population, and the ways in which this has been brought about are now discussed.

The first method by which the continuous inversion of energy level populations was actually achieved was that used by GORDON ZEIGER and TOWNES[17] in the ammonia maser. The energy level population was obtained by an actual spatial separation of the excited and ground state molecules, those in the ground state being dispersed away from the entrance to the cavity resonator, whereas those in the excited state were concentrated into a small beam and fired into a cavity tuned to the appropriate frequency. The possibility of carrying out such a physical separation arises from the Stark effect of the ammonia molecule, and the way in which the two energy levels change on the application of an external electric field is shown in *Figure 138(a)*. The energy value of the excited molecules (X) rises as the electric field strength increases, whereas that of the lower level (Y) falls. Thus, if an inhomogeneous electric field is produced over the region in which the ammonia molecules are moving as represented in *Figure 138(b)*, those in the excited state will tend to move towards the region of lower electric field strength, whereas those in the lower state will tend to move towards regions at higher electric field strength. In the quadrupolar field distribution illustrated in *Figure 138(b)*, the low values of the electric field are in the centre of the region and the high values of electric field are to be found between the wires themselves. Hence if the beam of ammonia molecules is fired down between the four wires producing the quadrupolar field along an axis at right angles to the plane of the figure, the molecules in the excited state will be focused towards the axis, but those in the ground state will be dispersed away from the centre towards the positions between the wires.

This is the basis of the separator used by Gordon, Zeiger and Townes in the design and construction of the first ammonia maser, and it is also discussed briefly in Section 5.11(*d*). *Figure 56* there indicates the arrangement of the rest of the equipment; in this way a high concentration of excited ammonia molecules can be obtained in the cavity resonator, causing much greater stimulated emission than absorption in this region.

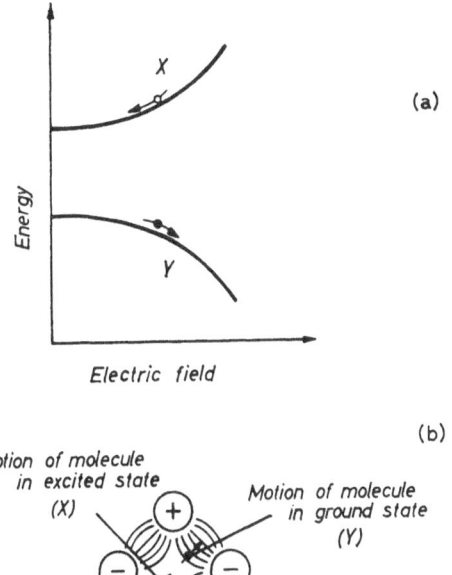

Figure 138. Population inversion in ammonia maser: (a) energy level variations in applied electric field, (b) separation of excited and ground state molecules by quadrupolar field

There are three practical applications of this particular kind of device: (i) as a very high resolution spectrometer, since the Doppler broadening normally associated with the molecules has been removed (see Section 5.11(*d*)); (ii) as an amplifier; (iii) as an oscillator. The particular feature which determines whether it operates as an amplifier or an oscillator is the amount of inversion produced between the energy level population, as compared with the losses which occur in the cavity walls themselves. If the energy level population inversion is not very great, then the ordinary losses in the microwave system will damp out the resulting stimulated emission which will thus not be sufficient to produce permanent oscillations. In such a case, however, an input signal at the resonance frequency will be amplified, and hence the device may be used as a very narrow band amplifier. If the energy level population inversion is sufficiently great, sufficient stimulated emission will be produced within the cavity to overcome the losses, and a molecular oscillator will result. This is initiated by a noise phonon of the correct frequency.

The main importance of the ammonia maser is as an oscillator rather than an amplifier, since it only operates at the frequency of the ammonia resonance line and can thus not be tuned over any reasonable frequency range. As an oscillator however it provides a very precise frequency standard, the frequency of the oscillations being determined entirely by the atomic and molecular constants of the ammonia molecule itself. A very considerable amount of research has been concentrated on the development of this ammonia maser as a frequency standard, and the whole equipment has been miniaturized so that it can be transported in aeroplanes or satellites.

This kind of physical separation between the atoms in the excited and ground states is not possible in the solid state, of course, and in the solid state masers and lasers other methods of inversion have to be employed. The first attempts at this inversion were made in the form of pulse systems, often employing rapidly swept magnetic fields. These methods inverted the population between two energy levels for a certain length of time, and relied on the stimulated emission taking place within the time interval for which the inversion was effective. Such two-level maser systems have now been replaced to a large extent by the three-level systems mentioned in Section 12.2, but the principles of these two methods are first briefly described.

12.8 Two-level Masers

All two-level masers are essentially the same as the ammonia maser, in that they are concerned with a system having only two energy levels between which the interactions of interest can take place. Therefore, by some means or other, the normal population distribution between these levels must be inverted. The actual spatial separation applied in the ammonia maser obviously cannot be used in the solid state, where the atoms or molecules are all locked in a crystal lattice. It follows that inversion of the population of a two-level system in the solid state must be in the form of pulse operation, and some means must be devised whereby the majority of the atoms concerned can be excited to the higher level and immediately after this maser action can take place. This action will die away, however, as the energy level population returns to its normal thermal equilibrium value. Three time intervals for the operation of such a device can be defined: (i) the 'exciting time' interval, in which the majority of the atoms are thrown up into the excited state and thus become available for stimulated emission processes; (ii) the actual 'operating time' when the maser action can take place as the excited atoms return to the ground state; (iii) the 'recovery time' in which the atoms are returning to their normal equilibrium value, before being re-excited at the beginning of the next pulse.

This normally means that the actual operational time of the two-level maser is only a small fraction of the total time available, since the recovery time is usually much longer than the useful operational time during which linear amplification is possible. Various means have been suggested for reducing the length of this recovery time, such as irradiating the crystal with light to restore the equilibrium more rapidly once the useful period of amplification has ceased. It nevertheless follows from these basic considerations, that the two-level maser system is automatically limited to pulse operation, and in this sense has very great drawbacks compared with the three-level systems discussed later.

In compensation for the limitation of pulse operation, however, the two-level maser scheme has some advantages not possessed by continuously operating masers, such as the possibility of producing much higher frequency radiation than that employed in the excitation process. Thus it is possible to invert the energy level population between two levels when the atom is in a moderate magnetic field strength, and then suddenly apply a very large magnetic field strength so that the energy difference between the two levels is very considerably increased. The stimulated emission which then occurs will be of very high frequency, in the millimetre or sub-millimetre region. In fact, the main practical application of the two-level maser system seems to lie in the field of very high microwave frequency generation rather than of normal amplification or use as a coherent oscillator.

Two methods have been suggested and successfully applied to the production of energy level population inversion in the two-level maser system and each of these is now briefly described. The first is known as the '180° pulse method' and the second as the 'adiabatic fast passage method'. Both of these cases are concerned with paramagnetic material in which the electron spins are relatively loosely coupled to the remainder of the crystal lattice; it is necessary to have this relatively weak spin-lattice interaction if the inversion of population is to be retained for any significant time. Consequently most of the substances used for two-level maser operation are formed either by electrons trapped in F-centres or other defects, or by electrons in semi-conductors, since in both of these cases the coupling to the lattice can be quite small.

(a) *The 180° pulse method of inversion.* If a paramagnetic system with weak spin-lattice interaction is considered in the presence of an externally applied field, then the normal thermal distribution of

457

electrons between the two energy levels gives rise to a net magnetic moment $N\beta^2 H/kT$ in the direction of the applied field. The essential principle of the method of '180° reversal' is simply to reverse the direction of this magnetization M, in a time which is short compared with the spin-lattice relaxation time. There is then a significant period in which more electrons are lined up against the field than with the field, and hence the normal energy level population distribution is inverted. This method of inverting the direction of a magnetization was developed earlier for nuclear magnetic resonance, and gave rise to the spin-echo method employed so successfully by Hahn and discussed in some detail in Section 8.7(b).

The easiest way to view these echo and inversion techniques is to use the classical description, in which the magnetization vector M can be considered as slowly spiralling down to a 180° orientation, and then back again to its original orientation, as the resonance process takes place. This form of motion continues as long as the applied magnetic field and microwave frequency satisfy the resonance condition, and the angular velocity of the precession, as viewed in a rotating frame of resonance, is given by

$$\omega_1 = g\beta H_1/\hbar \qquad \ldots(12.33)$$

where H_1 is the magnitude of the microwave field strength of the sample. In order to invert the magnetization it is necessary to switch off the microwave driving field at a time when it has rotated to π radians; in other words, to apply the microwave field for a time just long enough for half a rotation to occur. Although a rotation of 3π would also be possible in principle, a loss of signal would be caused by the spins interacting amongst themselves or with the thermal vibrations during this longer time interval. The time t_π for production of 180° reversal can therefore be written in the form

$$t_\pi = \frac{h}{2 g . \beta . H_1} \qquad \ldots(12.34)$$

It is evident from the above equation that the successful accomplishment of inversion depends on the correct adjustment of two parameters, i.e. H_1 the actual magnitude of the microwave field strength, and t_π, the time for which the field is applied. Substitution of quantitative values into this equation shows that the product $H_1 . t_\pi$ is of the order of 10^{-7}; in practice it is normal to operate with H_1 of the order of 1 gauss and t_π of the order of 10^{-7} sec. The value of t must be quite small, since the inversion must occur in a time much shorter than the normal relaxation time of the electron spins, and it should also take place in a time which is short compared with the spin-spin relaxation time. In this connection it is also important that the magnitude of H_1 should be greater than the random internal field experienced by the individual electron spins due to their mutual interaction, since otherwise the magnetization vector M will not maintain any fixed magnitude during the inversion. The magnitude of this internal field is given directly by the line width of the observed electron resonances, and if crystals can be obtained with resonance lines of 10–100 mgauss a driving field of 1 gauss is sufficient.

The other practical parameter which comes into these considerations is the Q of the resonance cavity. The value of this determines the actual power input P required to produce the driving field of 1 gauss at the microwave frequency. The relation between the two can be written in the form

$$P = \frac{VH_1^2 \cdot \nu}{4Q} \qquad \ldots(12.35)$$

where V is the cavity volume. This is not the only consideration that applies, however, since if the Q is too high, the cavity may become shock excited by the pulse and will 'ring'. Hence a low value of Q is really required for the excitation process, although once amplification is taking place as high a Q as possible is normally desirable. These conflicting requirements have led to the development of some cavities whose Q values can be switched rapidly from a high to low value, by such mechanisms as a spark discharge or an avalanche breakdown within the semi-conductor material itself. It follows from these general considerations that the excitation must take place in the low Q cavity, and hence a relatively high power microwave driving source is required in the form of either a large klystron or magnetron.

Two-level solid state masers designed on the principles outlined above have in fact been constructed and operated, employing either semi-conductors[18] or defects in crystals[19] as the active medium.

(b) *The adiabatic fast passage method of inversion.* The 180° pulse inversion method must be carried out with the applied magnetic field held at a constant value, the only parameter that is adjusted, once the level of the microwave field has been decided, is the time for which this is applied. Thus no change in the main d.c. magnetic field component occurs in the whole operation, but just a steady pulsing of the microwave field for determined time intervals.

In the method of obtaining population inversion known as the adiabatic fast passage method, the inversion is obtained by sweeping the main d.c. magnetic field through the resonance condition. Thus the inversion is produced by slowly changing the direction of the effective magnetic field seen by the electrons through 180°, while the magnetization vector precesses about the effective field. The term 'adiabatic' implies that any change of the energy of the system should be slow relative to the motions which already exist, and this can be expressed quantitatively by the expression

$$\frac{dH}{dt} > \omega_p H_1 \qquad \qquad \dots (12.36)$$

where ω_p is the precessional velocity of the magnetization vector about the effective field, and is approximately equal to γH_1. The condition for adiabatic change can therefore be written

$$\frac{dH}{dt} \gg \gamma H_1^2 \qquad \qquad \dots (12.37)$$

Where γ is the gyromagnetic ratio of the electron in question. The same considerations on the actual magnitude of H_1 as in the preceding section, can be applied, i.e. that its value should be considerably larger than that due to the internal fields, and hence about ten times the line width of the resonance concerned. If a value of about 1 gauss is taken for H_1, then equation (12.37) suggests that the change in the static field should take place at a rate much slower than 10^7 gauss/sec which, of course, can be easily achieved. The condition on the other extreme is that the field must be changed in a time relatively small compared with the spin-lattice relaxation time, as otherwise the thermal motion will destroy the inversion as soon as it occurs. It follows that the method can only be applied to crystals with relatively long spin-lattice relaxation times; this is not too serious a limitation since quite a number of crystals can be obtained which at liquid helium temperatures have values of more than 1 msec for this constant.

In this method the magnetization vector precesses around the effective field during the whole process of turning the field through 180°. Thus it is quite distinct from the pulse reversal method in which the precession is only allowed to take place for half a cycle. In this adiabatic fast passage method, the precession occurs for quite a noticeable time, in fact throughout the period when the external field is being swept from above resonance to below, or vice versa this sweep taking the order of 1 msec, as discussed above. After the external field has been swept through resonance in this manner, the net magnetization is aligned at 180° to its original direction, and therefore the energy level population is inverted compared with the initial distribution, this inversion lasting until the spin-lattice relaxation destroys it. Thus in the intervening period the system is available for maser action by stimulated emission.

This method has been used by a variety of workers, including COMBRISSON, HONIG and TOWNES[20] who made the earliest experiments in 1956 with silicon-doped materials. In 1958, FEHER et al.[21] were able to obtain oscillation from such a system by working at 1·2 °K with a cavity of Q 20,000, and using silicon doped with phosphorus atoms. CHESTER, WAGNER and CASTLE[22] used MgO and quartz crystals in which defects such as F-centres had been produced; as in the previous method, it is essential to deal with electrons which are only loosely coupled to the surrounding nuclei and lattice.

12.9 THREE-LEVEL MASER SYSTEMS

The general principle of the three-level maser systems is outlined at the end of Section 12.4. *Figure 134* shows how a pumping power which saturates the transition between two levels E_1 and E_3 can be employed to produce a higher population in energy level E_1 than exists in energy level E_2, permitting stimulated emission to take place at the intermediate frequency ν_{12} between these two levels. The possibility of using three electronic energy levels in a solid para-magnetic compound in this way, was first put forward theoretically by BLOEMBERGEN[23] in 1956, and the first successful operation of such a maser by SCOVIL, FEHER and SEIDEL[24] followed later in that same year. They used a rare earth crystal of lanthanum ethyl sulphate containing about ½ per cent gadolinium. This gives eight electronic levels, only three of which were used for the actual maser operation.

From the experimental point of view, there are three basic conditions that must be met in the operation of such a three-level device: (i) a suitable energy level system must be chosen, usually one of the transition group compounds; (ii) it is normally necessary to cool the specimen to liquid helium temperatures in order to produce a long enough spin-lattice relaxation time and thus allow saturation to take place between the pumped levels; (iii) it is necessary to have a microwave structure for the cavity which can be resonant at both the pumping frequency and at the frequency of amplification. The way in which this was achieved in the first maser is shown in *Figure 139*.

The lanthanum ethyl sulphate crystal is the long cylinder in the centre of the structure, and the structure itself is composed of a normal rectangular cavity which is resonant at the higher frequency

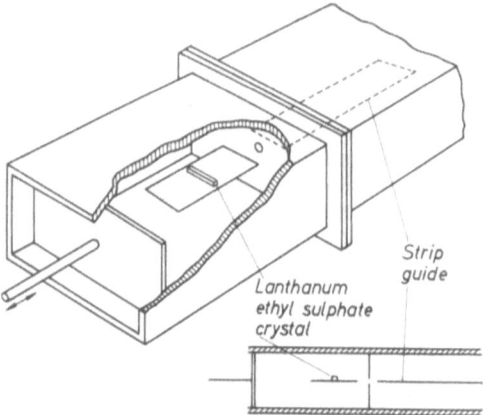

Figure 139. First three-level maser (after Scovil, Feher and Seidel[24])

pumping power. Into this cavity a strip conductor is inserted, so that the combined system also has a resonance mode at the lower frequency, corresponding to the signal to be amplified. This signal frequency is fed down a strip wave guide as shown, making it possible to feed in both the pumping power and the signal frequency at the same time, and to produce resonance modes for both in the same cavity. The initial experiments of Scovil and his colleagues[24] were only really designed to demonstrate experimentally that the ideas of Bloembergen[23] could be achieved in practice; but they also showed that the values of the spin-lattice relaxation time could be crucial in the design of such devices.

Since this initial success a large number of different transition group atoms have been investigated for possible maser systems, and one that has proved extremely popular has been the trivalent chromium atom, as found in the aluminium oxide lattice which constitutes artificial rubies. These chromium atoms have four levels, and three of these can be selected to make a very suitable maser system with relaxation times of the correct order. Most of the masers in everyday use at the moment, such as those used on the Atlantic television relay link[25], are composed of ruby type material.

The great advantage of these masers in their microwave applications is the inherently low noise figure which they possess. There are no semi-conductor contacts involving the large excess figure noise of silicon detectors, nor are there any hot electron beams as in normal valve devices. The noise arises solely from the Johnson noise of the effective resistance at a temperature of operation, and since this is

462

usually 4°K the inherent noise figure of the amplifier can be made extremely small. One of the initial drawbacks of maser operation was the necessity of operating at 4°K, since the long spin-lattice relaxation times required could only be obtained at these low temperatures.

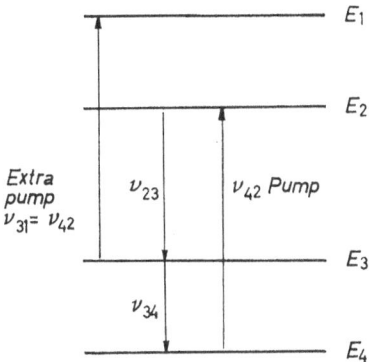

Figure 140. Four-level pumping

Later work at the Royal Radar Establishment, Malvern[26] showed, however, that it was possible to carry out three-level solid state maser action at liquid oxygen temperatures by using a double pumping system. This is illustrated in *Figure 140*, where the four energy levels for the trivalent chromium atoms, as in ruby, are indicated. If the angle θ which the applied magnetic field makes with the crystal axis is chosen carefully it is possible to produce an energy level system in which the difference in energy between E_1 and E_3 is the same as that between E_2 and E_4. The bottom three levels can then be used for normal maser operation, the pump frequency being ν_{42} and the signal frequency ν_{23}. At the same time as this is taking place, the pumping frequency is also exciting electrons from level E_3 to level E_1, and thus reducing the energy level population of E_3. This additional pumping effect increases very considerably the inversion of the energy level population between levels E_2 and E_3, and hence even at liquid oxygen temperatures, a sufficiently large dynamic difference in population can be built up to make maser action possible.

Although in some applications such operation at the higher temperatures of liquid oxygen may be of considerable advantage, if signals are being received from space, however, it is preferable to work at the lower liquid helium temperatures which produce much less inherent noise. It does appear that in future most practical masers

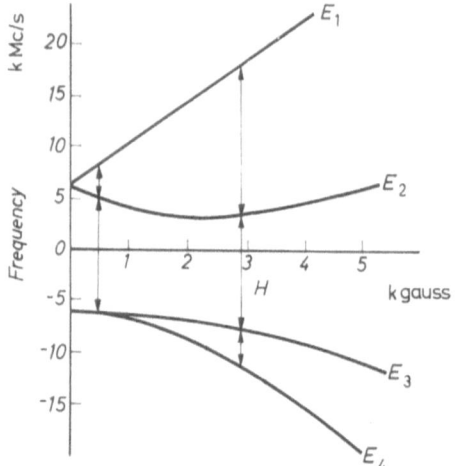

Figure 141. Ruby maser energy levels

of this type will be operated at liquid helium temperatures, especially since liquid helium itself is now becoming more readily available.

As a definite example of a three-level maser now in continuous operation, the one at Goonhilly Downs Receiving Station for

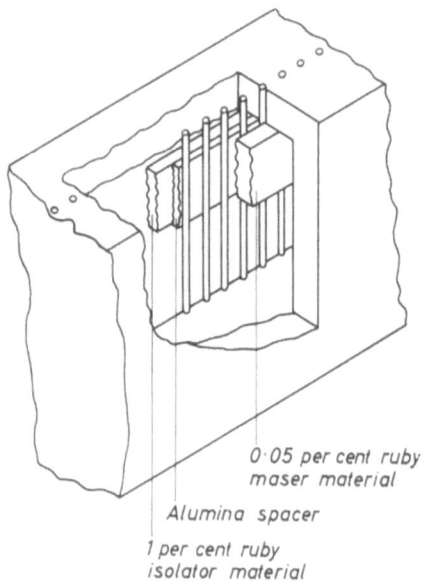

Figure 142. Comb structure in travelling wave maser

satellite communication can be quoted[27, 28]. This has ruby as an active element, and the actual energy levels and their variation with magnetic field, together with possible frequencies of pumping and amplification, are shown in *Figure 141*. The microwave design of these devices has also been developed since the first experiments, and instead of a single crystal of ruby placed at the centre of the resonance structure, the more modern masers employ a travelling wave type of structure shown in *Figure 142*. The comb type metal system is used to slow the microwaves by a factor of about 100 and to give a bandwidth of about 370 Mc/s[28]. The ruby itself is mounted along one side of the comb structure and the incoming signal is continuously amplified as the wave traverses this active material. This particular maser was found to have performance figures very close to those calculated from the theoretical analysis of its known bandwidth, and also to have a noise temperature of the expected[27] order of 10° K.

12.10 THE CONCEPT OF THE LASER

There is no reason why all the ideas and arguments of the last few sections concerning the possibility of inverting energy level populations and producing stimulated emission in the microwave region, should not be transferred in principle to any other frequency region of the spectrum. Thus it should be possible to obtain stimulated emission instead of absorption for any desired frequency, provided : (i) a system of three energy levels or more is available, such that the energy gap between two of these is appropriate to the desired frequency; (ii) it is possible by pumping or other such means to invert the normal thermal population distribution between these two levels. It was appreciated by several scientists quite soon after the successful work on the masers, that this same basic idea could also be applied to the generation of coherent radiation in the visible region. Hence the term 'laser' was born which stands for Light Amplification by Stimulated Emission of Radiation.

It was also appreciated very rapidly that such coherent radiation in the visible region might have immense practical applications, since the information capacity of such a beam of radiation would be extremely high. A simple quantitative example of the kind of information carrying capacity of such a coherent light beam is given in Section (12.5), showing that, in principle, about a million television channels would be transmitted down it simultaneously. The potentialities of such systems for communication become immediately obvious, and this is one of the reasons why such a large amount of research work has been devoted to the study of lasers. They also have other practical applications, arising from the fact that the wave fronts produced are of a coherent character, and hence the radiation can be focused very precisely. Thus the power density can be increased to values which are very much greater than those normally encountered with incoherent radiation. This introduces the possibility of studying the effect of extremely powerful electric and magnetic fields on the properties of solids, and their interatomic and intermolecular bonding forces.

If a laser system is to be developed and operated in the visible region, then the same general principles apply as for the masers in the microwave region; there are really three basic requirements which can be summarized as follows:

(1) A working medium must be available with a suitable energy level system, which must contain two levels with an energy difference corresponding to the frequency of desired operation, and must also contain other levels between which saturation pumping or other means of excitation can be effected.

(2) Some means of producing the inverted energy level population must be devised so that there are more atoms in the upper level of the optical transition than in the ground state, and thus there is a higher probability of stimulated emission rather than absorption.

(3) A sufficiently large number of atoms in the excited level must be available to the incoming radiation if any noticeable beam of coherent emitted radiation is to be produced. This require-ment is met in the microwave masers by the use of a cavity resonator which effectively reflects the microwaves to and fro thousands of times and thus enables them to interact with many atoms in the upper state. Some such system must also be employed in any laser if a coherent beam of any measurable intensity is to be obtained.

A suggestion as to how this last requirement could be met in practice was made in 1958 by Schawlow and Townes[29]. Their paper represents the first step in the practical development of the laser as an experimental possibility. They suggested that sufficient path length could be obtained by using a reflecting system very similar to that employed in a normal Fabry–Perot etalon. Two mirrors are aligned to be very accurately parallel to one another, so that a ray of light travelling in a plane normal to the mirrors is reflected to and fro between them a large number of times. Thus if the gain due to excess stimulated emission in the activated medium is greater than the losses which occur on reflection at the mirrors, there is a net amplification of radiation during each passage; hence a large amplitude is built up as a result of the many reflections. It also follows that such a system will produce a plane wave front with very little angular divergence, since any radiation which starts off at a slight angle from the normal to the mirrors will leave the activated medium after relatively few reflections and will be lost to the general beam of coherent radiation. This system as suggested by Schawlow and Townes[29] should fulfil all the necessary requirements for a laser outlined above, and provided a suitably activated medium can be inserted between the mirrors, coherent monochromatic radiation with highly directional properties should be produced. The radiation would be initially triggered off by a noise phonon of the right frequency.

The optical systems employed in nearly all the successful lasers have been of this type. The technical requirements for such Fabry–Perot reflecting plates are very severe, since they must not only be aligned accurately parallel but must also be flat to a high degree of precision. These points have been considered in detail by HEAVENS[30], who showed that for a reflectance of 0·90 at the mirrors surface the variation in the flatness of the surface must be less than $\lambda/30$. The parallelism and accuracy of the reflecting windows can be tested by setting up a system of fringes corresponding to those normally observed with the Fabry–Perot etalon.

12.11 THE RUBY LASER

The general principles of laser systems can be applied to solid, liquid or gaseous systems. The first lasers actually built operated with a solid state active medium, namely ruby which had proved so successful in the microwave maser. The energy levels of the

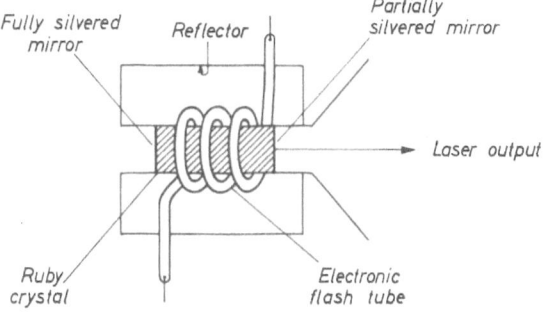

Figure 143. First operational laser (after Maiman[31])

doper chromium atoms were again used to provide the necessary quantum transitions, although the transitions were of course much larger and took place between the $^4A_2$ ground state, and the $^4F_2$ and ^2E excited states.

The first such laser to operate successfully did so in July 1960 and was designed and constructed by MAIMAN[31] of the Hughes Aircraft Company. He employed a ruby crystal as the activated medium and the parallel reflecting surfaces were obtained by accurately polishing the two ends of this crystal so that they were parallel to better than six seconds of arc. If a solid crystal is used as the active medium, the demands of homogeneity for the optical properties of the crystal itself are also very great. This is in fact one of their main limitations. Maiman's laser is shown schematically in *Figure 143*, from which it can be seen that the back end of the crystal is completely silvered whereas the front end is only partially silvered. The laser is pumped by light from the electronic flash tube, and it is through the semi-silvered end that the pulse of stimulated coherent radiation is emitted.

The energy level of chromium atoms, which were present in about 0·05 per cent concentration, is shown in *Figure 144*. The crucial level in this system is level C which is a metastable state[32], and the atoms excited into this level have a lifetime of about 10^{-2} sec, if they are not stimulated to emit beforehand. The transition from level C to the ground state gives a wavelength of 6,943 Å, and it is this radiation which constitutes the red fluorescence seen in ruby

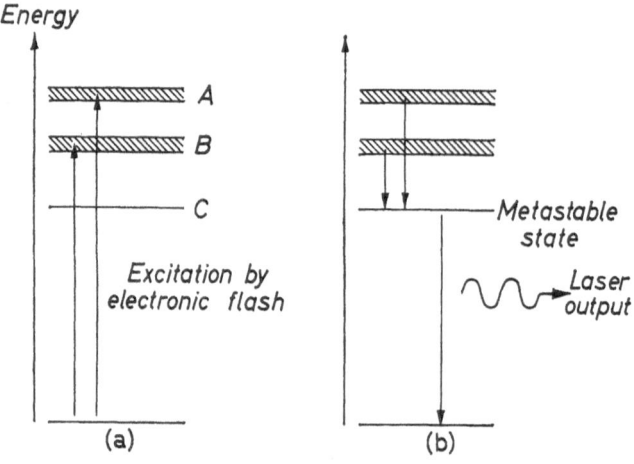

Figure 144. Ruby laser energy levels: (a) excitation from ground state, (b) falling back to metastable state, from which laser emission occurs

crystals. In order to produce laser action in this system, sufficient chromium atoms must be raised to the metastable level for the stimulated emission which they produce to be greater than the losses due to all types of absorption.

The chromium atoms can be raised to the metastable level by pumping them to the higher levels A or B and then letting them decay to level C, which they will do relatively quickly. All that is required is therefore a source of radiation with a frequency range which will cover the transition from the ground state to level A, and also of sufficient power to excite a large number of chromium atoms to the upper levels. The situation is then as indicated in *Figure 144(b)*. The excited atoms quickly fall to the metastable level C and are there available for laser action when stimulated by incoming radiation of the correct frequency, or by a noise phonon if oscillation rather than amplification is required. Maiman's particular contribution was the discovery that all the required pumping power could be supplied by an electronic flash tube wound around the ruby crystal as shown in *Figure 143*.

Owing to the large amount of internal heating which occurs with such a flash tube the system could only be used to produce pulses of coherent radiation instead of acting as a continuous source. The power level of the electronic flash tube was found to be quite critical, since below a certain level only the incoherent red phosphorescence of the ruby was obtained, spread over its normal wide range of frequencies. Once the level of the electronic flash was increased above the critical value, however, an intense beam of red light was emitted within a period of 10^{-4} sec after the electronic flash; this indicated that sufficient chromium atoms had been pumped up to the excited state to produce stimulated emission greater than the losses occurring at the ends of the optical system[33]. This practical demonstration that coherent optical radiation could be produced via a process of stimulated emission confirmed the basic ideas and theories concerning laser action. Quite a number of other lasers using similar single crystals have now been built and operated, but all of these suffer from two inherent drawbacks: (i) since the activated medium is in the solid state there is a definite limit to its homogeneity and therefore to the directional coherence of the emitted radiation; (ii) the thermal stresses set up in the solid by the absorption of radiation from the electronic flash tube prevent continuous operation and hence the use of efficient modulation techniques. Both of these limitations can be overcome if a gaseous medium is used to supply the activated atoms, and the successful operation of such a gaseous laser was the next advance in laser development.

469

12.12 Gaseous Lasers

The first successful laser system using gas molecules was that proposed by Javan of the Bell Laboratories in 1959, and successfully operated by him in 1961[34]. This produced a continuous coherent output and only required 50 watts of input power.

The optical system is again based on the Fabry–Perot reflecting plates discussed above; these are mounted at either end of a glass tube containing the gas and can be adjusted for parallelism by metal bellows which connect them to the main tube. These features can be seen in *Figure 145*. The energy level system employed is that of the neon atom as shown on the right-hand side of *Figure 146*. The crucial level in this case is the 2S level at the top of the system, and laser action is produced by stimulated emission between this level and the next, the 2P level. The large number of excited atoms in the 2S level is obtained by an indirect method employing helium gas. Helium has a long-lived metastable state 2^3S, shown on the left of *Figure 146*, and atoms in the ground state of the helium atom can be excited to this metastable state by processes such as electron impact in a gas discharge. Since the excited helium atoms are then in the energy state corresponding to the 2S level of the neon atoms, it is very simple for them to transfer their energy by a direct collision with the neon atoms and excite a neon atom directly to the top 2S level. In a suitable mixture of helium and neon gas it is therefore

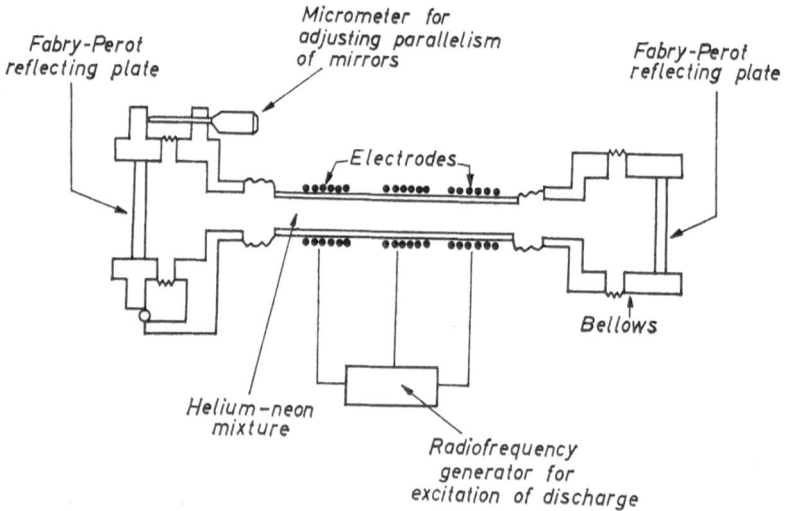

Figure 145. First gaseous laser (after Javan, Bennett and Herriott[34])

470

possible to build up a dynamic equilibrium in which there are a large number of atoms in the 2S level of neon, although this is not itself a metastable state.

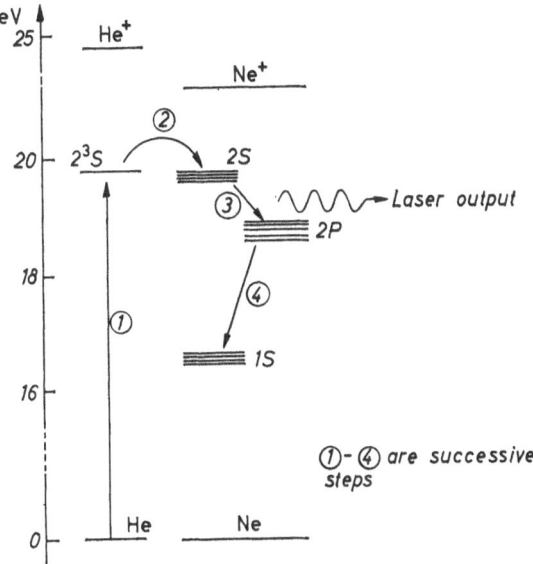

Figure 146. Energy levels of helium–neon system

Two features distinguish this energy level system from that used in the ruby laser. First, the excess population in the higher level associated with the laser transition is achieved not because of the particular properties of this level itself, but because it can be fed from a level of equal energy in another atom with metastable properties. Secondly, the pumping energy is supplied not directly by a photon of higher energy as in the electronic flash tube, but by collision with an accelerated electron in a gas discharge. It is possible in principle, of course, to use any method of excitation.

The chamber containing the helium–neon mixture consists of a quartz tube 80 cm long and with inside diameter 1·5 cm. The quartz tube is terminated at each end with larger metal chambers holding the Fabry–Perot reflecting plates, which can be adjusted by the flexible bellows as shown in *Figure 145*. The separation of these plates is 1 metre, and the discharge is excited by external electrodes fed from a 28 Mc/s generator. These first laser oscillations from a gas have now been followed by very powerful CO_2 lasers.

471

All the lasers so far described have been operating as oscillators rather than light amplifiers. The initiation of the stimulated emission was by a noise phonon of the correct frequency, and the number of atoms in the excited state was sufficiently great for the stimulated emission that resulted to overcome the absorption losses in the system. For a large number of applications, however, an *amplifier* rather than an oscillator is required, and the first successful operation of a laser amplifier therefore merits description. This gas laser used caesium vapour as its active medium, and was also of importance in demonstrating that coherent light beams could be modulated and detected quite readily.

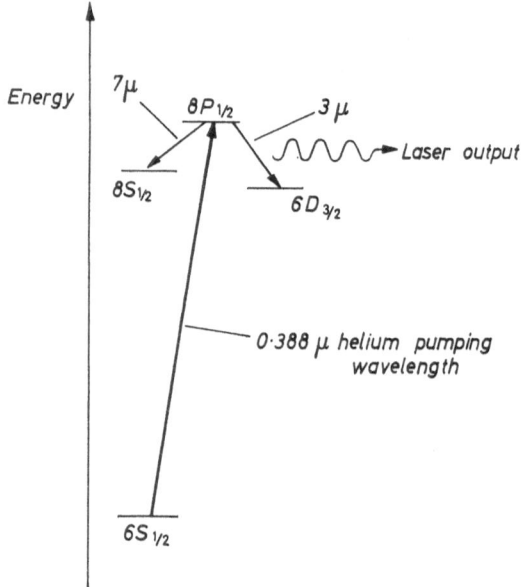

Figure 147. Energy levels of caesium system

This first amplifier was designed and constructed by JACOBS, GOULD and RABINOWITZ [35], and the caesium vapour was excited by selective optical pumping following a method suggested by Schawlow and Townes [29]. In this method of excitation the caesium vapour is pumped with the helium wavelength of 3,888 Å. The helium discharge lamp emits an intense radiation at this wavelength, and it also overlaps three of the caesium energy levels as shown in *Figure 147*. It can be seen from this figure that the incident radiation from a helium discharge lamp excites caesium atoms from the $6S_{1/2}$ to the upper $8P_{1/2}$ energy level. The population of the $8P_{1/2}$ level can therefore be made larger than that of either the $8S_{1/2}$ or the $6D_{3/2}$ levels. Consequently it should be possible to obtain laser action at either of the wavelengths corresponding to these two transitions, which are 7 μ and 3 μ respectively. The experimental measurements were actually carried out at the wavelength of 3·2 μ since better techniques of detection employing lead sulphide cells were available here.

In order to prove that coherent light beams were being amplified, it was necessary to have a source of coherent radiation to feed the amplifier. This source was also formed from activated caesium vapour pumped by the same helium line, in other words a laser which oscillated rather than amplified. The output from this source was modulated at a frequency of 105 c/s and fed to the laser amplifier itself, which consisted of a caesium cell 90 cm long illuminated by a long helium discharge lamp mounted behind it. The output from the amplifying cell was fed to the lead sulphide detector, and the modulation on the light beam was detected in a phase-sensitive amplifier with a final time constant of $\frac{1}{3}$ sec. Signal-to-noise ratios of 100:1 were obtained, corresponding to a power of 10^{-9} watts of coherent radiation. The amplifying action of the caesium cells was confirmed by switching off the various activating elements in turn, and it was shown that a gain of 6 per cent was obtainable with the cell of 90 cm length.

A theoretical treatment shows that the amplification coefficient $k(\nu)$ is related to the populations of the two energy levels concerned by the equation

$$k(\nu) = \frac{(h\nu)}{c}[B_{21}\mathcal{N}_2 - B_{12}\mathcal{N}_1] \cdot S(\nu) \qquad \ldots (12.38)$$

where $S(\nu)$ is the normalized line shape function and B is the coefficient of induced emission. For a Doppler profile on the line this expression reduces to

$$k(\nu_0) = \left\{\frac{\log_e 2}{16c^2\pi^3}\right\}^{1/2} A_{21} \cdot \frac{\lambda_0^4}{\varDelta\lambda_d} \cdot \left\{\mathcal{N}_2 - \frac{g_2}{g_1} \cdot \mathcal{N}_1\right\} \qquad \ldots (12.39)$$

where A is the spontaneous emission coefficient, g_1 and g_2 are the statistical weights, and $\varDelta\lambda_d$ is the Doppler width.

The crucial parameter in the above expression is of course the difference in the population of the two levels, $\mathcal{N}_2 - \mathcal{N}_1$. A calculation of the energy level populations when the system is pumped by intense radiation from the mercury lamp is summarized in *Table 12.1*, from which it is seen that it should be possible to produce

Table 12.1. Relative Population of Caesium Levels when Pumped with 3,888 Å Helium Light

Level	$8P_{1/2}$	$8P_{3/2}$	$8S_{1/2}$	$6D_{3/2}$	$5D_{3/2}$
Population	100	30	20	3	200

30 times as great a population density in the $8P_{1/2}$ level compared with the $6D_{3/2}$. Although the experimentally observed gain was not as good as that predicted by the simple theory above, the fact that amplification of light could be produced in this way did complete the second essential step in the design and construction studies of laser techniques.

12.13 SEMI-CONDUCTOR LASERS

The next significant advance in the development of lasers came with the discovery that energy level populations could be inverted in ordinary semi-conductor type material, enabling the production of a continuous output of coherent radiation in the visible region from a solid semi-conductor. The frequency of the emitted radiation corresponds to the energy jump associated with the forbidden gap of the semi-conductor material, and the radiation is generated by the recombination of electrons and holes in the semi-conductor.

There are, of course, other means whereby such electrons and holes can lose their energy and not give up radiation in the process, including direct interaction with a lattice vibration or with other electrons, holes or lattice defects. In certain cases, however, it is found that the probability of the electron hole recombination producing radiation is much higher than the probability of the other interactions taking place, because the maximum energy of the valence band and the minimum of the conduction band both occur for zero value of the momentum vector, and high transition probabilities for this dipole interaction are therefore produced.

The first semi-conductor for which this type of laser action was observed was gallium arsenide[36, 37], and the features of the laser are outlined in *Figure 148*. It consists of a single crystal of gallium arsenide, with its opposite faces accurately polished to provide the normal Fabry–Perot type of reflection system and allow the radiation to be reflected to and fro many times. The coherent radiation emerges at right-angles to the polished face as shown, and observation of the interference fringes produced by such radiation indicates that there is a higher degree of coherence over the whole of the emitting region.

The emission of stimulated radiation from such gallium arsenide junctions was first observed by HALL *et al.*[36] working at the General Electric Company in New York, and by NATHAN *et al.*[37] working at the I.B.M. Laboratories, New York. In both cases the proof that stimulated emission was, in fact, taking place was provided by the sudden narrowing of the line width of the emitted radiation once the injection current rose above a threshold value.

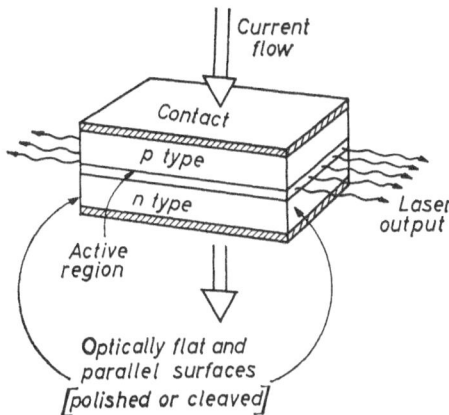

Figure 148. Basic features of semi-conductor laser

In analysing the characteristics of their device, Hall *et al.*[36] pointed out that there were several requirements which must be fulfilled if such stimulated emission is to be observed:

(1) The electron and hole populations within the active region from which the emission is to take place, must be large enough for their Fermi levels to be separated by an energy greater than that of the radiation.
(2) Losses due to absorption and other processes such as those mentioned above, must be small relative to the gain produced by the stimulated emission.
(3) The active region must be contained within a cavity having a resonance which falls in the wavelength range expected, as is provided by the polished surfaces at the side of the crystal.

The second of these conditions is the most difficult to fulfil, since the actual transition region forming the junction between the *n* and *p* type material, and in which the stimulated radiation propagates, is extremely thin. The stimulated radiation is thus likely to extend into the passive regions on either side; hence losses may predominate.

The actual population inversion is produced by the injection of carriers from the degenerate n and p type regions into the transition region itself. It is probable that the electrons pass from the conduction band of the n type material through the junction barrier to a virtual level in the p type material spectrum, before making transitions to levels at the top of the valency band. The coherent radiation is emitted during this last step. It follows that if this type of system is to operate successfully as a laser, an excess population must be produced in the upper conduction band level over that in the lower valency band level. In order to obtain this a large injection current must be applied.

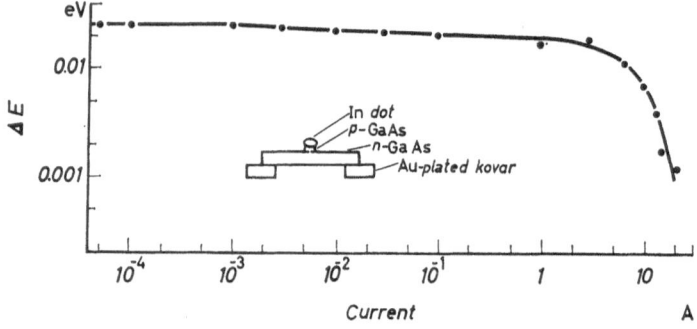

Figure 149. Threshold current in semi-conductor laser (after Nathan et al.[37])

The existence of such a threshold current is shown in *Figure 149*, which is taken from the first paper of Nathan *et al.* [37] reporting the observation of this stimulated emission. It can be seen from this figure that there was a very sudden decrease in the observed line width of the emitted radiation once the injection current rose above a value of 10 A. This 10 A current corresponded to a value of about 10^4 or 10^5 A/cm² for the current densities through the transition region; HALL [38] also showed that a sudden narrowing in the spectral distribution of the emitted radiation occurred when the current density in his specimens exceeded about 10^4 A/cm².

The original diodes were approximately cubic with a 0·4 mm edge, the junction lying in horizontal plane at the centre. The current was passed through the junction by means of direct contact attached to the top and bottom faces, and the front and back faces were polished parallel to each other and perpendicular to the plane of the junction itself[39]. The original experiments were carried out at liquid nitrogen temperatures and pulse systems were employed. In the first experiments[36] normal gallium arsenide junctions were used and these produced coherent radiation in the infra-red region at 8,400 Å wavelengths. Shortly after this, however, HOLONYAT and BEVACQUA[40] showed that it was possible to obtain coherent visible light from semi-conductor diodes grown not from pure gallium arsenide but from gallium and a mixture of arsenic and phosphorus with the general formula $Ga(As_{1-x}P_x)$. These diodes have voltage–current characteristics similar to those of the GaAs $p–n$ junctions, but the band gap varies with the arsenic and phosphorus ratio. The particular diodes used, produced coherent stimulated emission at a wavelength of 7,100 Å and the critical current density for which stimulated emission took place was about 18,000 A/cm².

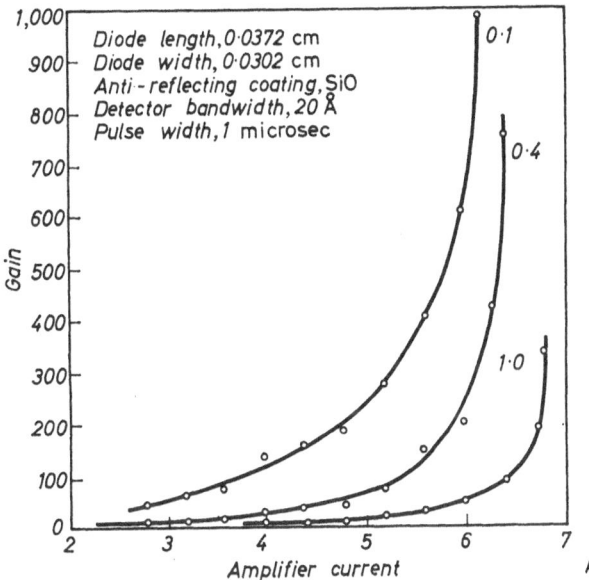

Figure 150. The gain of a semi-conductor laser amplifier (after Crowe and Craig[41])

The great advantage of these lasers over those described earlier, is the *direct* conversion of electrical energy into visible radiation which makes it an extremely simple matter to modulate the output radiation. The whole subject of semi-conducting lasers has developed extremely rapidly since the first observations, and they have been developed not only as emitters of coherent radiation but also as amplifiers. One example of the early measurements on semi-conductor lasers used as amplifiers is shown in *Figure 150*. The curves[41] were obtained from an ordinary gallium arsenide laser

477

diode and indicate the variation of gain with current for three different values of input intensity. Amplification factors greater than 1,000 were obtained when small input signals from a laser oscillator were employed, and the bandwidth of the amplified light was found to be about 15 Å and to remain constant for amplifier currents up to 10 A. These curves were in fact obtained by focusing the light from the gallium arsenide oscillator through a microscope objective on to the junction of the amplifier diode which was itself mounted below a small dewar, both diodes being operated at liquid nitrogen temperatures. The output of the amplifier diode was then focused through a second microscope on to the entrance slit of a monochromater, and the light amplification hence determined.

Since the whole field of lasers is developing so rapidly, it would serve no useful purpose to summarize any more of the devices that have been produced recently. It is evident, however, that the laser has now become a practical solid state device which can be used both as an easily modulated source of coherent oscillation and also as an amplifying device in its own right.

12.14 OTHER DOUBLE RESONANCE SYSTEMS

In most of the double resonance systems discussed so far, the two frequencies concerned have been in approximately the same region of the spectrum. Thus in the three-level maser both the pumping frequency and the amplifying frequency are in the microwave region; in the ruby laser the excitation of atoms to the higher state is produced by visible or ultra-violet radiation from a flash tube, while the coherent emission also takes place in the visible region. It is true that the Overhauser and ENDOR effects are concerned with interactions between radio and microwave frequencies, but these are in the same general wavelength region.

There is no reason, however, why double resonance effects should not be obtained between two frequencies of very different magnitudes if the absorption of one noticeably alters the population distribution between the two energy levels corresponding to the other. A good example is the double resonance experiments employing optical and radiofrequency radiations carried out by Kastler[43]. By combining the better resolution of the radiofrequency region with the higher sensitivity of the optical region, these experiments have enabled very precise measurements to be made on hyperfine splitting constants, and similar parameters.

The first double resonance experiment of this type was carried out by BROSSEL and BITTER[42], following the theoretical predictions of BROSSEL and KASTLER[43]. They studied mercury vapour in an external magnetic field, and optical radiation was applied to excite the mercury atoms from the $6\,{}^1S_0$ ground state to the $6\,{}^3P_1$ excited state. Plane polarized light of the resonance wavelength 2,537 Å was used, with its electric vector parallel to the direction of the applied magnetic field. Such a radiation system corresponds to π excitation with the corresponding selection rule $\Delta M_J = 0$, as indicated in *Figure 151*.

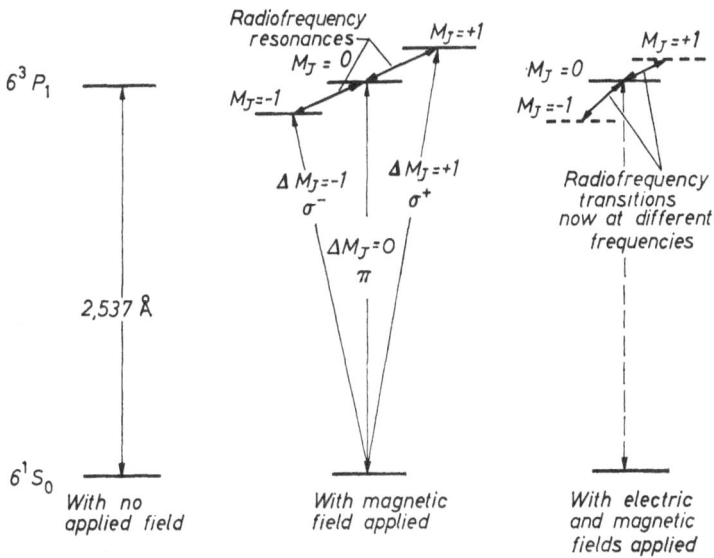

Figure 151. Double resonance in mercury vapour, showing possible transitions between the $^1S_0$ ground state and the three components of the $^3P_1$ excited state

This figure represents the different energy changes involved in the double resonance experiment. The three top levels belong to the excited $6\,{}^3P_1$ state, the splitting between them being produced by the Zeeman effect of the applied magnetic field. The $6\,{}^1S_0$ state is not split by a magnetic field, since its J equals zero, and the three optical transitions that can occur in the presence of a magnetic field are therefore as shown in the figure. The two transitions with $\Delta M_J = \pm 1$ correspond to σ polarization, while the $\Delta M_J = 0$ transition corresponds to the π polarization, as explained above. The radiofrequency transitions take place between the $\Delta M_J = \pm 1$ and $M_J = 0$ levels of the excited state. It is evident from the figure that such radiofrequency transitions decrease the population of the

479

$M_J = 0$ level and increase that of the $M_J = \pm 1$ levels, bearing in mind that the optical irradiation only excites to the $M_J = 0$ level. Thus the presence of a radiofrequency resonance can be detected in practice by an increase in the σ and a decrease in the π components of emitted light.

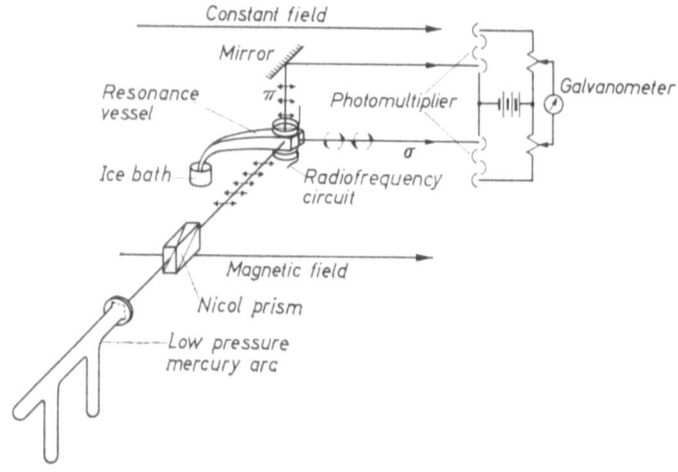

Figure 152. Optical radiofrequency double resonance experiment (after Brossel and Bitter[42])

The actual experimental arrangement set up by Brossel and Bitter[42] is shown diagrammatically in *Figure 152*. The exciting 2,537 Å radiation passed from the low pressure mercury arc through the nicol prism, so that the electric vector was plane polarized horizontally parallel to the d.c. magnetic field, and fell on the absorption cell along an axis normal to this field. The radiofrequency fields were applied by a pair of Helmholtz coils mounted above and below the absorption cell, and in this initial experiment the radiofrequency ν was kept constant while the applied magnetic field was changed slowly until the resonance condition $h\nu = g_J \beta H$ was attained.

As explained above, the radiofrequency resonance itself could be detected by the sudden increase in the σ component of the emitted 2,537 Å radiation. This change was detected in practice by mounting photomultipliers to receive light in the direction of, and perpendicular to, the field respectively, with their currents mixed in opposition as indicated in the diagram. The advantage of this arrangement was that fluctuations in the strength of the light source were eliminated and the signal strength was effectively doubled.

A comprehensive series of measurements were made on the shape

480

of the resonance line and on its dependence on such factors as the radiofrequency field strength. The really important contribution of this experiment, however, was in demonstrating the feasibility of such double resonance experiments, in which the high resolution of radiofrequency techniques could be combined with the greater sensitivity of the optical detectors. The experimental value of the g factor for the mercury $^3P_1$ state was in fact found to be

$$g_J = 1 \cdot 4838 \pm 0 \cdot 0004.$$

Further studies by BLAMONT and BROSSEL[44]) with the same equipment, but with a pair of electrodes inserted into the mercury vapour to produce an electric field of 50 kV/cm, were then carried out to investigate the Stark effect of the $^3P_1$ state. The electric field shifts both of the $M_J = \pm 1$ levels in the same direction and hence alters their energy splitting from the central $M_J = 0$ state, as shown by the dashed energy levels in *Figure 151*. A doubling of the radiofrequency resonance is thus obtained, and the Stark effect interaction can be calculated from the separation of the peaks of this doublet. It is also possible to obtain accurate estimates for the lifetime of the excited states by a detailed analysis of the line shapes[45].

A large number of experiments on other excited states of mercury[46] and on excited states of other atoms[47] followed these initial investigations, and Kastler and Brossel in Paris were mainly responsible for opening up this fascinating field of double resonance studies. A further development came when BOGLE, DODD and McLEAN[48] were able to determine the hyperfine structure coupling constants for the odd isotopes of mercury, by employing microwave and waveguide techniques in place of the radiofrequency techniques used by Brossel and Bitter[42]. When these higher frequencies are used F ceases to be a good quantum number and the separation of the Zeeman levels becomes unequal, since the curved region of the energy level divergence is reached, as indicated in the general case of *Figure 69*. It is then possible to deduce a value for the hyperfine structure coupling constant from the frequency differences between the various absorption lines observed. A series of double resonance measurements were also made in zero external magnetic field to determine hyperfine structure constants and nuclear electric quadrupole constants, the $P_{3/2}$ levels of the alkali metals being particularly suitable for these experiments[47, 49, 50].

In all of the cases so far considered the redistribution of population between the energy levels was brought about by irradiating the sample with visible light, this usually being plane polarized. It is possible to use other means of population inversion, and controlled electron impact has been successfully employed (in a way similar to its use in the gaseous lasers). This was the method used by LAMB and RETHERFORD[51, 52] in studying the $2\,{}^2S_{1/2}$ to $2\,{}^2P_{1/2}$ interval in atomic hydrogen, a classic experiment which was the first successful radiofrequency experiment of an excited atomic state. Electron impact was used to populate the excited levels and the emission of optical radiation from these levels was studied; changes in the polarization of the emitted light by applied radiofrequency fields was also investigated, in a manner exactly similar to the simpler double resonance experiments already described. It is not possible to summarize all the developments in this field of research, but a comprehensive survey of *Radiofrequency Spectroscopy of Excited Atoms* has been given by SERIES[53], and full details may be found there.

Attention has been concentrated in this section on the use of the double resonance techniques in accurately determining atomic *g* values and hyperfine structure coupling constants. The techniques also have some very practical applications, in which optical radiation can be used to invert the energy level population for ground state radiofrequency and microwave transitions. These particular applications have come to be known as 'Optical pumping' techniques[54], and they have provided very useful sources for frequency standards.

12.15 NON-LINEAR EFFECTS

This chapter is concluded with a brief mention of the various non-linear effects now being discovered, which are proving to have considerable practical importance in laser and other double resonance studies. The first observation of the creation of light of a second harmonic frequency was by FRANKEN *et al.*[55] in 1961, who found that 3,472 Å radiation could be obtained from a laser beam of 6,943 Å wavelength if this was passed through suitable crystals. From the quantum point of view this phenomenon can be considered as a process of annihilation of two light quanta and the creation of one of twice the energy. It was also soon shown that production of third harmonic radiation could be effected in the same way.

The production of higher order Raman effects by laser beams was then observed, in which coherent radiation was obtained at both the Stokes and anti-Stokes lines of several orders. It should be pointed out that the frequencies of higher order Stokes and anti-Stokes lines are given by $(\nu_p \pm n \,.\, \nu_R)$ where ν_p is the frequency of the laser pump and ν_R is the frequency of an actual atomic transition. It follows that, by employing suitable mixing techniques and combining the different order Raman and Stokes radiation from a high power laser beam, it should be possible in principle to obtain coherent radiation at any frequency from the far infra-red to the ultra-violet region.

One other device of considerable potential practical interest is the 'quantum counter' first proposed by BLOEMBERGEN[56]. This device requires a suitable ion embedded in a crystal lattice such that four energy levels are available with $E_4 > E_3 > E_2 > E_1$, and the temperature of operation is such that $h\nu_{12} \gg kT$ (i.e. all the atoms are normally in the ground state E_1). Continuous pump power is fed into the specimen at ν_{24}, but since E_2 is normally empty no great absorption of this power occurs. If, however, a quantum of radiation ν_{12} is also fed into the system, the ion will be raised to level E_2 and then immediately to the highest level E_4 by the pumping frequency. From E_4 it will decay to E_3, emitting a quantum of radiation at frequency ν_{34} in the process. The net result of this whole operation is therefore to absorb a quantum of frequency ν_{12}, and emit one at frequency ν_{34}.

The great practical application of such a device is in the detection of infra-red radiation. The frequency ν_{12} can be adjusted to the infra-red region, and ν_{34} can then be altered to a frequency range in which normal photomultipliers are highly sensitive, thus effectively extending their range above $1\,\mu$. The successful practical development of such a system requires a good quantum efficiency, as well as a suitable energy level structure. Consequently the work to date has shown that infra-red radiation can be detected in this way, and even converted directly into optical radiation, but at the moment the quantum efficiency is very low, (i.e. a large number of input quanta at the infra-red frequency are required for each output quantum at the optical frequency).

The first experimental demonstration that a quantum counter could be made to work was by PORTER[57], who obtained such action in Pr^{3+} doped into $LaCl_3$ operating at liquid helium temperatures. BROWN and SHAND[58] then followed this and showed that it was possible to obtain the effect for Pr^{3+} in various fluoride lattices when operating at both nitrogen and room temperatures. They further showed[59] that a quantum counter could be made from Er^{3+} doped at 1 per cent into various fluoride lattices, and that although both pump and signal frequencies were in the near infra-red, the fluorescence from the combined scheme produced radiation visible to the naked eye. In this way direct conversion of infra-red radiation to visible radiation was achieved by the quantum counter, although its efficiency still needs to be considerably improved.

REFERENCES

[1] PORTIS, A. M. *Phys. Rev.* 91 (1953) 1071

[2] OVERHAUSER, A. W. *ibid.* 92 (1953) 411

[3] CARVER, T. R. and SLICHTEN, C. P. *ibid.* 92 (1953) 212

[4] BELJERS, H. G., VAN DER KINT, L. and VAN WIERINGEN, J. S. *ibid.* 95 (1954) 1683

[5] RICHARDS, R. E. and WHITE, J. W. *Proc. Roy. Soc.* A269 (1962) 287, 307; A279 (1964) 474, 481

[6] FEHER, G. *Phys. Rev.* 103 (1956) 500, 834

[7] — *ibid.* 105 (1957) 1122

[8] EISINGER, J. and FEHER, G. *ibid.* 109 (1958) 1172

[9] LORD, N. W. *ibid.* 105 (1957) 756; *Phys. Rev. Lett.* 1 (1958) 170

[10] HALFORD, D., HUTCHISON, C. A. and LLEWELLYN, P. M. *Phys. Rev.* 110 (1958) 284

[11] COLE, T., HELLER, C. and LAMBE, J. *J. chem. Phys.* 34 (1961) 1447

[12] HELLER, C. and McCONNELL, H. M. *ibid.* 32 (1960) 1535

[13] POOLEY, D. and WHIFFEN, D. H. *Molec. Phys.* 4 (1961) 81

[14] COOK, R. J. and WHIFFEN, D. H. *Proc. phys. Soc.* 84 (1964) 845; *J. chem. Phys.* 43 (1965) 2908

[15] CEDEQUIST, A. Thesis, Washington University (1963)

[16] HYDE, J. S. and MAKI, A. H. *J. chem. Phys.* 40 (1964) 3117

[17] GORDON, J. P., ZEIGER, H. J. and TOWNES, C. H. *Phys. Rev.* 95 (1954) 282

[18] COMBRISSON, J. and TOWNES, C. H. *Onde élect.* 36 (1956) 989

[19] BOLEF, D. and CHESTER, P. F. *Trans. Inst.-Radio Engrs.* 6 (1958) 47

[20] COMBRISSON, J., HONIG, A. and TOWNES, C. H. *C. R. Acad. Sci., Paris* 242 (1956) 2451

[21] FEHER, G., GORDON, J. P., BUEHLER, E., GERE, E. A. and THORMOD, C. D. *Phys. Rev.* 109 (1958) 221

[22] CHESTER, P. F., WAGNER, P. E. and CASTLE, J. G. *ibid.* 110 (1958) 281

[23] BLOEMBERGEN, N. *ibid.* 104 (1956) 324

[24] SCOVIL, H. E. D., FEHER, G. and SEIDEL, H. *ibid.* 105 (1957) 762

[25] WALLING, J. C. and SMITH, F. W. *Br. Commun. Electron.* 9 (1962) 596

[26] DITCHFIELD, C. R. and FORRESTER, P. A. *Phys. Rev. Lett.* 1 (1958) 448

[27] WALLING, J. C. and SMITH, F. W. *Philips tech. Rev.* 25 (1965) 289

[28] — *Low Noise Electronics, Report on Oslo Conference.* Pergamon Press (1961) p. 225

[29] SCHAWLOW, A. L. and TOWNES, C. H. *Phys. Rev.* 112 (1958) 1940

[30] HEAVENS, O. S. *Hilger J.* 6 (1961) 63

[31] MAIMAN, T. H. *Nature* 187 (1960) 483

[32] — *Phys. Rev.* 123 (1961) 1145

[33] — , HOSKINS, R. H., D'HAENEUS, I. J., ASAWA, C. K. and EUTUHOV, V. *ibid.* 123 (1961) 1151

[34] JAVAN, A., BENNETT, W. J. and HERRIOTT, D. R. *Phys. Rev. Lett.* 6 (1961) 106

[35] JACOBS, S., GOULD, G. and RABINOWITZ, P. *ibid.* 7 (1961) 415

[36] HALL, R. N., FENNER, G. E., KINGSLEY, J. D., SOLTYS, T. J. and CARLSON, R. O *Phys. Rev. Lett.* 9 (1962) 366

[37] NATHAN, M. I., DUMKE, W. P., BURNS, G., DILL, F. H. and LASHER, G. *Appl. Phys. Lett.* 1 (1962) 62

[38] HALL, R. N. *Solid-St. Electron.* 6 (1963) 405

[39] NATHAN, M. I. *ibid.* 6 (1963) 425

[40] HOLONYAK, N. and BEVACQUA, S. F. *Appl. Phys. Lett.* 1 (1962) 82

[41] CROWE, J. W. and CRAIG, R. M. *ibid.* 4 (1964) 57

[42] BROSSEL, J. and BITTER, F. *Phys. Rev.* 86 (1952) 308

[43] — and KASTLER, A. *C. R. Acad. Sci., Paris* 229 (1949) 1213

[44] BLAMONT, J. E. and BROSSEL, J. *ibid.* 243 (1956) 2038

[45] GUIOCHON, M. A., BLAMONT, J. E. and BROSSEL, J. *ibid.* 243 (1956) 1859

[46] BROSSEL, J. and JULIENNE, C. *ibid.* 242 (1956) 2127

[47] RITTER, G. J. and SERIES, G. W. *Proc. Roy. Soc.* A238 (1957) 473

[48] BOGLE, G. S., DODD, J. N. and McLEAN, W. L. *Proc. phys. Soc.* B70 (1957) 796

[49] KRUGER, H. and SCHEFFLER, K. *J. Phys. Radium* 19 (1958) 854

[50] MEYER-BERKHOUT, V. *Z. Phys.* 141 (1955) 185

[51] LAMB, W. E. and RETHERFORD, R. C. *Phys. Rev.* 79 (1950) 549; 81 (1951) 222; 86 (1952) 1014

[52] — *Rep. Progr. Phys.* 14 (1951) 23

[53] SERIES, G. W. *ibid.* 22 (1959) 280

[54] GUESIC, J. E. and SCOVIL, H. E. D. *ibid.* 27 (1964) 241; SERIES, G. W. *ibid.* 22 (1959) 280

[55] FRANKEN, P., HILL, A. E., PETERS, C. W. and WEINREICH, G. *Phys. Rev. Lett.* 7 (1961) 118

[56] BLOEMBERGEN, N. *ibid.* 2 (1959) 84

[57] PORTER, J. F. *ibid.* 7 (1961) 414

[58] BROWN, M. R. and SHAND, W. A. *ibid.* 11 (1963) 366

[59] — — *ibid.* 12 (1964) 367

APPENDIX I

RESONANCE EQUATIONS AND DEFINITIONS

A. GASEOUS SPECTROSCOPY

I. Rotational spectra

(i) Diatomic Molecules

$$\nu = 2B(J+1) - \frac{16B^3}{\omega^2}(J+1)^3 \qquad \dots (1)$$

(ii) Linear Molecules

$$\nu = 2B(J+1) - 4D(J+1)^3 \qquad \dots (2)$$

(iii) Symmetric-top Molecules

$$\nu = 2B(J+1) - 4D_J(J+1)^3 - 2D_{JK}(J+1)K^2 \qquad \dots (3)$$

II. Inversion spectra

Semi-empirical for Ammonia

$$\nu = \nu_0 \cdot \exp[a \cdot J(J+1) + b \cdot K^2 + c \cdot J^2(J+1)^2$$
$$+ d \cdot J(J+1)K^2 + f \cdot K^4] \qquad \dots (4)$$

III. Hyperfine splitting

(i) Linear Molecules

$$\Delta E = \left(\frac{\partial^2 V}{\partial z^2}\right) \cdot eQ \cdot \left[\frac{\frac{3}{8}G(G+1) - \frac{1}{2}I(I+1)J(J+1)}{I(2I-1)(2J-1)(2J+3)}\right] \qquad \dots (5)$$

(ii) Symmetric-top Molecules

$$\Delta E = \left(\frac{\partial^2 V}{\partial z^2}\right) \cdot eQ \times$$

$$\times \left[\frac{3K^2}{J(J+1)} - 1\right] \cdot \left[\frac{\frac{3}{8}G(G+1) - \frac{1}{2}I(I+1)J(J+1)}{I(2I-1)(2J-1)(2J+3)}\right] \qquad \dots (6)$$

In above two equations:

$$G = F(F+1) - I(I+1) - J(J+1)$$

IV. Stark splitting

(i) Linear Molecules

$$\Delta E = \frac{\mu^2 \cdot \epsilon^2}{2h \cdot B} \cdot \left[\frac{J(J+1) - 3M_J^2}{J(J+1)(2J-1)(2J+3)}\right] \qquad \dots (7)$$

(ii) Symmetric-top Molecules

$$h \cdot \Delta\nu = 2\mu \cdot \epsilon \cdot \left[\frac{K \cdot M_J}{J(J+1)(J+2)}\right] \qquad \dots (8)$$

(iii) Ammonia Molecule

$$\Delta\nu = 0.5065 \cdot \frac{\mu^2 \cdot \epsilon^2}{\nu_0} \cdot \left[\frac{K \cdot M_J}{J(J+1)}\right]^2 \qquad \dots (9)$$

(ν in megacycles, ϵ in volts per centimetre)

V. Definitions

ν = Frequency of observed absorption line

$\Delta\nu$ = Shift of frequency from normal value

ΔE = Change in energy produced by extra inter-action

$\mathcal{J}$ = Total rotational energy quantum number

K = Quantum number defining component of angular momentum along the molecular figure axis

M = Quantum number defining component of angular momentum along applied magnetic or electric field

I = Value of nuclear spin

F = Quantum number equal to vector addition of I and $\mathcal{J}$

$B = \dfrac{h}{8\pi^2 I_b}$ where I_b is the moment of inertia of the molecule about an axis perpendicular to that joining the nuclei

D = Distortion constant, produced by centrifugal stretching

ω = Fundamental vibrational frequency of the molecule

$a, b, .. f$ = Empirical constants

Q = Nuclear electric quadrupole moment

$= \dfrac{1}{e} \cdot \int (3z^2 - r^2) \cdot dq$ over nucleus

$\left(\dfrac{\partial^2 V}{\partial z^2}\right)$ = Gradient of electric field at nucleus

$= \int \left(\dfrac{3z^2 - r^2}{r^5}\right) \cdot dq$ outside nucleus

μ = Molecular dipole moment, in Debye units

ϵ = Applied electric field

B. ELECTRON SPIN RESONANCE

I. With effective spin of $\frac{1}{2}$

(i) Electronic transition

$$h \cdot \nu = g \cdot \beta \cdot H \quad \text{or} \quad H = \frac{h \cdot \nu}{g\beta} = \frac{21 \cdot 4184}{g \cdot \lambda} \qquad \ldots (10)$$

(H in kilogauss, λ in centimetres)

(ii) With nuclear interactions
Hamiltonian:

$$\mathcal{H} = \beta \cdot [g_\| H_z S_z + g_\perp (H_x S_x + H_y S_y)] + A \cdot S_z I_z$$
$$+ B(I_x S_x + I_y S_y) + Q' \cdot [I_z^2 - \tfrac{1}{3} I(I+1)] \quad \dots (11)$$

(a) Along the Oz axis:

$$H = H_0 - A \cdot M_I - \frac{B^2}{2H_0} \cdot [I(I+1) - M_I^2] \qquad \dots (12)$$

(b) At an angle θ to the axis:

$$H = H_0 - K \cdot M_I - \frac{B^2}{4H_0} \cdot \left[\frac{A^2 + K^2}{K^2}\right] \cdot [I(I+1) - M_I^2]$$

$$- \frac{1}{2H_0} \cdot \left(\frac{A^2 - B^2}{K}\right)^2 \cdot \left(\frac{g_\| g_\perp}{g^2}\right)^2 \cdot M_I^2 \cdot \sin^2 \theta \cdot \cos^2 \theta$$

$$- \frac{2Q'^2}{K} \cdot \left(\frac{A \cdot B \cdot g_\| \cdot g_\perp}{K^2 g^2}\right)^2 \cdot [4I(I+1) - 8M_I^2 - 1]$$
$$\times M_I \cdot \sin^2 \theta \cos^2 \theta$$

$$+ \frac{Q'^2}{2K} \cdot \left(\frac{B \cdot g_\perp}{K \cdot g}\right)^4 \cdot [2I(I+1) - 2M_I^2 - 1] M_I \cdot \sin^4 \theta$$
$$\dots (13)$$

where $\quad H_0 = \dfrac{h\nu}{g\beta} \quad K^2 g^2 = A^2 g_\|^2 \cos^2 \theta + B^2 g_\perp^2 \sin^2 \theta$

$$g^2 = g_\|^2 \cdot \cos^2 \theta + g_\perp^2 \cdot \sin^2 \theta \quad \dots (14)$$

II. With effective spin greater than $\tfrac{1}{2}$
Hamiltonian:

$$\mathcal{H} = \beta \cdot [g_\| \cdot H_z S_z + g_\perp \cdot (H_x S_x + H_y S_y)] + D \cdot [S_z^2 - \tfrac{1}{3} \cdot S(S+1)]$$
$$+ A \cdot S_z I_z + B(I_x S_x + I_y S_y) + Q'[I_z^2 - \tfrac{1}{3} \cdot I(I+1)] \quad \dots (15)$$

(i) Electronic transitions

(a) Along the Oz axis:

$$H = H_0 - 2D(M_s - \tfrac{1}{2}) \qquad \dots (16)$$

(b) At an angle θ to the axis:

$$H = H_0 - D(M_s - \tfrac{1}{2}) \cdot \left\{3 \cdot \frac{g_\|^2}{g^2} \cdot \cos^2 \theta - 1\right\}$$

$$+ \frac{D^2}{2H_0} \cdot \left(\frac{g_\| g_\perp}{g^2}\right)^2 \cdot [4S(S+1) - 24M_s(M_s - 1) - 9]$$
$$\times \sin^2 \theta \cos^2 \theta$$

$$- \frac{D^2}{8H_0} \cdot \left(\frac{g_\perp}{g}\right)^4 \cdot [2S(S+1) - 6M_s(M_s - 1) - 3] \cdot \sin^4 \theta$$
$$\dots (17)$$

(ii) With nuclear interactions

(a) Along the axis:

$$H = H_0 - 2D\left(M_s - \tfrac{1}{2}\right) - A \cdot M_I - \frac{B^2}{2H_0} \cdot [I(I+1) - M_I^2]$$

$$- \frac{B^2}{2H_0} \cdot [M_I(2M_s - 1)] \qquad \ldots (18)$$

(b) At an angle θ to the axis:

The following terms are added to the right-hand side of equation (17):

$$-K \cdot M_I - \frac{B^2}{4H_0} \cdot \left(\frac{A^2 + K^2}{K^2}\right) \cdot [I(I+1) - M_I^2]$$

$$- \frac{B^2}{2H_0} \cdot \frac{A}{K} \cdot [M_I(2M_s - 1)]$$

$$- \frac{1}{2H_0}\left(\frac{A^2 - B^2}{K}\right)^2 \cdot \left(\frac{g_\parallel g_\perp}{g^2}\right)^2 \cdot M_I^2 \cdot \sin^2\theta \cdot \cos^2\theta$$

$$- \frac{Q'^2}{2K \cdot M_S(M_S - 1)} \cdot \left(\frac{A \cdot B \cdot g_\parallel g_\perp}{K^2 g^2}\right)^2$$

$$\times [4I(I+1) - 8M_I^2 - 1] \cdot M_I \cdot \sin^2\theta \cdot \cos^2\theta$$

$$+ \frac{Q'^2}{8K \cdot M_S(M_S - 1)} \cdot \left(\frac{B \cdot g_\perp}{K \cdot g}\right)^4$$

$$\times [2I(I+1) - 2M_I^2 - 1] \cdot M_I \cdot \sin^4\theta \qquad \ldots (19)$$

where H_0, g and K are defined, as before, in equation (14).

III. Cyclotron resonance

(i) For free ions in a gas:

(a) Electrons:

$$\omega = \frac{eH}{mc} \quad \text{or} \quad \nu = \frac{e}{2\pi mc} \cdot H \qquad \ldots (20)$$

(b) Positive ions:

$$\omega = \frac{eH}{A_P Mc} \quad \text{or} \quad \nu = \frac{e}{2\pi A_P Mc} \cdot H \qquad \ldots (21)$$

(ii) For electrons in a solid:

$$\omega = \frac{eH}{m^* c} \quad \text{or} \quad \nu = \frac{e}{2\pi m^* c} \cdot H \qquad \ldots (22)$$

where m^* is the effective mass, often $\ll m$.

IV. Definitions

g = Spectroscopic splitting factor
For a free electron this is equal to the spin gyro-magnetic ratio of 2·0023

$g_{||}, g_{\perp}$ = Values of spectroscopic splitting factor, parallel and perpendicular to the Oz axis

β = Bohr magneton $\left(\dfrac{eh}{4\pi mc}\right)$

λ = Wavelength of radiation

S = Effective spin quantum number

I = Value of nuclear spin

M_S = Quantum number defining resolved component of electronic spin momentum in direction of applied field

M_I = Quantum number defining resolved component of nuclear spin momentum in direction of applied field

A, B = Coefficients measuring the separation between successive hyperfine components, parallel and perpendicular to the Oz axis. Measured in units of cm^{-1}

D = Coefficient measuring the electronic splitting. Separation of successive electronic transitions along Oz axis equals $2D$. cm^{-1}

Q' = Quadrupole interaction coefficient, equals

$$\frac{3e}{4I(2I-1)} \cdot \left(\frac{\partial^2 V}{\partial z^2}\right) \cdot Q$$

where Q is the nuclear electric quadrupole moment

ω = Angular frequency for cyclotron resonance

m = Mass of the electron

M = Mass of the proton

A_P = Atomic mass number of positive ion

Note: $g . \beta = 2 \times$ Electron magnetic moment, for free electron

$\therefore$ Intrinsic magnetic moment of electron

$$= 1{\cdot}001145 \text{ Bohr magnetons.}$$

C. NUCLEAR RESONANCE

I. Nuclear paramagnetic resonance

(i) Resonance condition:

$$h . \nu = g_I . \beta_N . H \quad \text{or} \quad \omega_0 = \gamma_I . H \quad \quad \dots (23)$$

(ii) Power absorbed at resonance:

$$P(\omega_0) = \frac{M_0 . \omega_0^2 . H_1^2/(H_0 T_2)}{1/T_2^2 + (\omega_0 H_1/H_0)^2 (T_1/T_2)} \quad \dots (24)$$

$$\text{and} \quad M_0 = \frac{I(I+1) g_I^2 \beta_N^2}{3kT} . N_0 H_0 \quad \dots (25)$$

II. Pure quadrupole nuclear resonance

(i) Energy levels with axial symmetry:

$$E_Q = \frac{eQ}{4} . \left(\frac{\partial^2 V}{\partial z^2}\right) . \left[\frac{3M_I^2 - I(I+1)}{I(2I-1)}\right] \quad \dots (26)$$

(ii) Resonant frequencies:

$$h\nu = e . Q . \left(\frac{\partial^2 V}{\partial z^2}\right) . \left(\frac{3}{4I(2I-1)}\right) . [2|M_I| - 1]$$
$$\dots (27)$$

where $|M_I|$ is the larger value in the
$\Delta M_I = \pm 1$ transition

III. Definitions

g_I = Nuclear g factor
γ_I = Nuclear gyromagnetic ratio

Note. There is some confusion in the literature on the use of the above two terms. The term 'nuclear g factor' is best reserved for the parameter obtained from the resonance measurements, and, as defined in equation (23), it is a pure number, with value varying from about 0·3 to 6·0.

Thus: $\quad g_I = \dfrac{\text{Magnetic moment in Nuclear Magnetons}}{\text{The spin of the nucleus } (I)}$

The term 'gyromagnetic ratio' is defined as the ratio of the nuclear magnetic moment to the angular momentum. It thus measures the same essential quantity as g_I, but is usually expressed in different units. As defined in equation (23), it is measured in radians, seconds^{-1}, gauss^{-1}, and this definition is equivalent to:

$$\gamma_I = \frac{\text{Nuclear magnetic moment in erg . gauss}^{-1}}{I . h/2\pi}$$

β_N = Nuclear magneton $\left(\dfrac{eh}{4\pi Mc}\right)$

ω_0 = Angular frequency for resonance, or Larmor precessional frequency

M_0 = Nuclear magnetization

H_1 = Magnitude of radiofrequency magnetic field

H_0 = Strength of d.c. field at resonance

T_1 = Spin-lattice relaxation time

T_2 = Spin-spin relaxation time

N_0 = Number of nuclei per unit volume

$Q, I, M_I, \left(\dfrac{\partial^2 V}{\partial z^2}\right)$ defined as in Sections A and B

D. NUMERICAL VALUES

$$h = 6.6252 \pm 0.0005 \times 10^{-27} \text{ erg . sec}$$
$$c = 299792.9 \pm 0.8 \text{ km . sec}^{-1}$$
$$\beta = 0.92732 \pm 0.00006 \times 10^{-20} \text{ erg . gauss}^{-1}$$
$$\beta_N = 0.50504 \pm 0.00004 \times 10^{-23} \text{ erg . gauss}^{-1}$$
$$k = 1.38042 \pm 0.00010 \times 10^{-16} \text{ erg . deg}^{-1}$$
$$N = 6.02472 \pm 0.00036 \times 10^{23} \text{ per gramme-mole}$$
$$g_{\text{electron}} = 2.002292 \pm 0.000025 \text{ [1]}$$
$$g_{\text{proton}} = 5.58554 \pm 0.00012 \text{ [2]}$$
$$\gamma_{\text{proton}} = 2.67527 \pm 0.00008 \times 10^4 \text{ radian . sec}^{-1} \text{ gauss}^{-1} \text{ [3]}$$

$\dfrac{1 \text{ eV. of}}{\text{energy}} \equiv 11605.9 \pm 0.6° \text{ K}$

$\equiv 2.41812 \pm 0.00009 \times 10^{14} \text{ c/s}$

$\equiv 8065.98 \pm 0.30 \text{ cm}^{-1}$

REFERENCES

[1] KOENIG, S. H., PRODELL, A. G. and KUSCH, P. *Phys. Rev.* 88 (1952) 191

[2] HIPPLE, J. A., SOMMER, H. and THOMAS, H. A. *ibid.* 76 (1949) 1877; 80 (1950) 487

[3] THOMAS, H. A., DRISCOLL, L. and HIPPLE, J. A. *ibid.* 75 (1949) 902; 78 (1950) 787; THOMAS, H. A. *ibid.* 80 (1950) 901

All other values taken from:

DUMOND, J. W. and COHEN, E. R. *Rev. mod. Phys.* 25 (1953) 691

APPENDIX II

DETAILS of commercially available electron spin resonance and nuclear magnetic resonance spectrometers are given on the following pages, and are listed alphabetically by the name of the manufacturer. All prices are for 1965.

ELECTRON SPIN RESONANCE SPECTROMETERS

(1) ALPHA SCIENTIFIC LABORATORIES
940 Dwight Way
Berkeley, California

Alpha Laboratories offer a variety of ESR spectrometers, ranging from the ALX-10 which is an advanced research instrument to the 340 series which consists of relatively cheap spectrometers operating in the 300 Mc/s range and very suitable for demonstration or teaching purposes.

Model ALX-10

This is available for X or Q-band operation, and can be supplied with low temperature probe and dewars and a linear field control (employing a Hall feedback loop). A Computer for Average Transients (CAT) can also be incorporated and a digitally converted numerical display is available for direct computer entry. Special resonant systems in the form of helices or dielectric cavities are available, and allow large sample sizes to be accepted. Sensitivity is 10^{11} ΔH spins. Prices for X-band spectrometers are from \$21,000 depending on magnet power supply and accessory choice.

340 Series

These spectrometers employ an oscillator, detector and amplifier, and a probe unit which operates in the frequency range 320–350 Mc/s. A helix sample probe is used and no special tuning adjustments are required once the resonance is established. 100 kc/s field modulation and phase-sensitive detection is employed, with additional 60 c/s field modulation of variable amplitude from 0 to 150 gauss. The main magnetic field can be automatically or manually driven, and the spectrometers include chart recorders with oscilloscopes as optional extras. Sensitivity is 10^{14} spins/gauss line width, with a resolution of better than 1 part in 10^5.

Prices, including magnet system, are from \$5,750. Basic electron resonance heads, but without 100 kc/s phase-sensitive detection or magnet systems, are available at \$1,365.

(2) Decca Radar Limited   Hersham Division, Lyon Road
Walton-on-Thames, Surrey

Comprehensive *X*-band spectrometer systems, built from self-powered standard units, allow ready extension from one type of system to another. All units are available separately for inclusion in any apparatus. Three basic spectrometers are available:

Type X1. 100 kc/s field modulation and phase-sensitive detection with simple (homodyne) microwave detection.

Type X2. As X1 with the addition of superheterodyne microwave detection (switched changeover).

Type X3. As X2 with the addition of 33 c/s field modulation and phase-sensitive detection with dual modulation facility.

All spectrometers have a choice of Y–*t* or X–Y recorder.

Three magnet systems are supplied with these spectrometers.

High senstitivy of the spectrometers is coupled with simplicity of operation and accurate calibration of spectra. The klystron signal source is phase-locked to a harmonic of a quartz crystal oscillator with or without correction for cavity drift, and may also be locked entirely to the cavity. The magnet systems have a magnetic flux integrator which is triggered by a proton resonance magnetometer and gives out a signal directly proportional to magnetic field change. Spectra plotted on an X–Y recorder may be calibrated directly in terms of magnetic field, measurements of *g* values may be made within a few parts in 10^5, and line widths and separations may be read directly from the recorder chart.

Accessories include a flat sample cell, a flow mixing adaptor, a liquid nitrogen finger dewar, and a variable temperature cavity insert covering the temperature range $-175°$ C to $+300°$ C.

The following prices are approximate and include all transport, installation and commissioning. All magnetic prices also include turntable, trolley and rails.

Spectrometers		*Magnet systems*			*Pen recorders with trolley and control box*	
X1	£6,800	*M2*	£4,600	Y–*t*		£370
X2	£8,900	*M4X*	£8,300	X–Y		£870
X3	£9,800	*M5X*	£10,690			

(3) JAPAN ELECTRON OPTICS LABORATORY COMPANY LIMITED
U.K. agents: Delviljem Ltd.
4 Shakespeare Road, London, N.3

JES-3BS-X, JES-3BS-K and JES-3BS-Q

All these instruments are designed for multipurpose use by addition of the appropriate microwave bridge assembly, and are capable of carrying out measurements at X, K and Q-bands respectively. Main general features are: sensitivity 1×10^{11} spins/ gauss (X-band, 100 kc/s modulation), 1×10^{10} spin/gauss (Q-band, 100 kc/s modulation); field modulation 80 c/s and 100 kc/s; low impedance magnet with gap of 60 mm. The magnetic field is varied by a field dial, and has a range of 500–13,000 gauss for a 60 mm gap. Digital indication of magnetic field intensity is given.

Attachments. JES-VT-2 temperature controller; JES-UCT-2X variable temperature adaptor; liquid helium for X, K and Q-bands; JES-SH-30X superheterodyne adaptor; universal cavity for ultra-violet radiation with sample rotation.

JES-P10-S Radical Detector

This detector is compactly designed exclusively for measurement at the X-band by 100 kc/s field modulation. Its main features are: sensitivity 1×10^{11} spins/gauss; low impedance magnet; range of variable magnetic field 1,000–5,000 gauss. A linear field sweep system is used, and digital indication of magnetic field intensity.

Attachments. As for JES-3BS-X with the exception of liquid helium facilities.

(4) MICROSPIN   Hilger and Watts Ltd., 98 St. Pancras Way, Camden Road, London, N.W.1

ESR spectrometers are available in three basic models:

ESR 1. X-band Spectrometer
ESR 2. Q-band spectrometer
ESR 3. X and Q-band spectrometer

Any one of the three models can easily be converted to either of the other two, as the majority of the instrumentation is common to all of them.

All three models are high sensitivity spectrometers employing 100 kc/s field modulation with recorder display, and also low frequency modulation for crystal-video display.

The standard magnets used with the instrument are 4 in., 7 in., 8 in. and 11 in. pole diameter made by Newport Instruments Ltd. An automatic field measuring and linear field display system can also be supplied.

Automatic frequency control of the microwave frequency is employed, stabilization against the ESR sample cavity or reference wavemeter being dictated by application conditions.

The normal output presentation of the first derivative of absorption or dispersion mode on the recorder can be directly converted to the absorption or dispersion mode by using the Integrator and Differentiator Unit accessory. Additional facilities available allow the relative spin concentration to be directly displayed, and the double differential of the ESR signal to be obtained to assist in the interpretation of complex spectra.

A superheterodyne attachment is available at X-band. For low temperature work a nitrogen gas cryostat system is used to give a controllable temperature range. Liquid helium cryostat also available.

A range of cavities is available for the following applications:
(i) General purpose work with samples in the solid, liquid and gas states.
(ii) Operation with the sample at low temperatures.
(iii) Aqueous solution samples.
(iv) Irradiation studies.
(v) Anisotropic sample studies.

Approximate prices are   *ESR 1*   £8,500
                          ESR 2   £9,500
                          ESR 3   £10,500

(5) SPECTROSPIN   Badenerstrasse 701, Zurich.
(Formerly Trub Tauber, now in association
with Bruker-Physik of Karlsruhe)

ESR spectrometers are available for X and Q-band operation,
with frequency modulation channels at 713 c/s, 4,040 c/s and 100
kc/s. All spectrometers may be furnished with optional extras of
$x-y_1-y_2$ recorders for simultaneous recording of the spectra and its
integral, and digital time averaging computers are also available.
Sensitivity is 1.5×10^{11} spins/gauss half-width for X-band system.

Price of X-band spectrometer B-ER402, *without* magnet system
is about £6,500.

Price of Q-band spectrometer B-ER404, *without* magnet system is
about £7,500.

Various magnet systems are available:

Electromagnet B-E 22 C 3

Standard model, weight approx. 800 kg; fixed yoke 45° mount; pole diameter 22·5 cm.	£1,500
Power Supply B-MN 120	£1,000
TOTAL	£2,500

Electromagnet B-E 25 C 8

Heavy model, weight approx. 2,000 kg for high field intensities; variable air gap (0–12 cm); yoke rotating ± 180°; pole diameter 24/29 cm.	£2,400
Power Supply B-MN 100/30	£1,500
TOTAL	£3,900

Electromagnet B-E 30 2 s

Pole diameter 34/30 cm.	£3,100
Power Supply B-MN 200/50	£2,250
TOTAL	£5,350

All power supplies are suitable for all magnets.

(6) Strand Laboratories Incorporated

143 Main Street, Cambridge
Massachusetts, U.S.A.

The 602 Series Spectrometer

The heart of the Strand Laboratories' electron magnetic resonance apparatus is a bridge designed to measure the complex, or vector, transmission or reflection coefficient of a microwave cavity, or for that matter of any microwave structure. Two parallel arms are used. One arm provides a path for a small portion of the total klystron microwave output to the microwave detector assembly. This power is controllable both in amplitude and phase. The amplitude is set so that the detectors are biased at an optimum radiofrequency power level for best sensitivity. The reference power level is independent of the cavity power operating level and hence the detection sensitivity is maximized independently of the power at the sample. The reference power phase adjustment determines the phase of the cavity signal which is detected, i.e., the absorption or dispersion signal or some combination of absorption or dispersion can be simply and unequivocally selected for detection. The bridge detectors are two sets of detectors in a microwave circuit so designed that if one set observes an absorption signal for a given phase setting in the reference arm, the dispersion signal is obtained from the second set of detectors.

All of the klystron power, minus the few milliwatts used in the reference arm, is sent to the cavity arm. This arm uses a circulator to separate the cavity reflection signal from the incident signal with little loss in power to and from the cavity. The power level in the cavity arm is controllable with a calibrated variable attenuator, and a cavity matching device remotely controlled at the dewar head is used to critically couple, or match into, the sample cavity.

The system is available with switchable 100 kc/s or 6 kc/s modulation frequencies and switchable 400, 200, 80, 40 and 20 c/s modulation; only one frequency from each group is available at one time. This allows either dual sample operation or second derivative operation. At 9·5 Gc/s the noise figure of the system at 6 kc/s or 100 kc/s is less than 15 dB, typically 12 dB, at all cavity power levels, while at 24, 35 and 70 Gc/s the 6 kc/s and 100 kc/s noise figures are typically 40 dB and 30 dB.

The test sensitivities of these systems range from less than 5×10^{10} ΔH, for $S = \frac{1}{2}$ spins with a 3 sec time constant at 9·5 Gc/s, to less than $10^{10} \Delta H$ at 24 Gc/s, and less than $5 \times 10^{9} \Delta H$ spins at 35 Gc/s, with even greater sensitivity at 70 Gc/s.

(7) VARIAN ASSOCIATES   611 Hansen Way, Palo Alto
California, U.S.A.

Varian offers a series of high performance EPR spectrometers which range from a basic 9·5 Gc/s system with 100 kc/s modulation and detection and a 4 in. magnet, to systems operating at either 9·5 Gc/s or 35 Gc/s with multiple modulation and detection capabilities and electromagnets up to 15 in. Within this range, systems may be chosen to provide those particular components required for a given study area or special requirement.

Spectrometers and accessories are available to allow variable temperatures, fixed liquid nitrogen, and liquid helium operation; superheterodyne operation; absorption, dispersion and second-derivative spectrum measurements; aqueous sample measurements, including flow studies for determination of chemical kinetics; *in situ* electrolytic generation of free radicals; biological tissue sample measurements; sample angular dependence studies; sample irradiation studies; optical transmission studies; signal enhancement by time averaging and slow sweep methods; dual cavity determinations of spin density, g values, line width, etc.; rapid reaction studies with the C-1024 time averaging computer; frequency dependence studies, and conversion to wideline NMR applications.

Model E-3

This is a 9·5 Gc/s basic spectrometer combining unusually high sensitivity and stability in a compact, easy-to-operate system package. It includes 100 kc/s modulation, Fieldial* magnetic field regulator and sweep, and oscilloscope and check-out panel units; step-driven X–Y recorder featuring field-calibrated chart paper; four-port circulator bridge with calibrated power and frequency dials, furnishing 0·2–200 mW cavity power with AFC; multi-purpose cavity, and an integral water-cooled 6 kgauss, 4 in. magnet and power supply system. The system in solid state design for table top mounting, utilizes multipurpose (Rectangular TE_{102} Mode) cavity accessories, and is equipped for signal enhancement through long time scanning or time averaging techniques.

Other EPR models provide increased system versatility through choice of magnet size and configuration and availability of additional modulation and detection frequencies. Additional accessories: liquid helium cavity and dewar systems, superheterodyne system, rotating cavity.

Approximate prices are from $23,000 to $65,000 (F.O.B. Palo Alto).

* Trademark

499

NUCLEAR MAGNETIC RESONANCE
SPECTROMETERS

(1) JAPAN ELECTRON OPTICS LABORATORY COMPANY LIMITED
U.K. agents: Delviljem Ltd.
4 Shakespeare Road, London, N.3

JNM-C-6o Nuclear Magnetic Resonance Spectrometer

This is a compact type NMR instrument developed exclusively for measurement of H^1 (60 Mc/s) and F^{19} (56·446 Mc/s) and its magnet and spectrometer are compactly housed in a single cabinet. The main features are a resolution of 10^{-9}, and a two sample system NMR control method including standard chart. The range of variable temperature is $-110°$ to $+200°C$. An integrator is included.

Attachments. JNM-SD-20B spin decoupler; N-1007 microvolume sample cell kit; RA-1 spectrum accumulator; JNM-CW-1 cooling water conditioner.

JNM-3H-6o High Resolution NMR Instrument

This instrument is a universal type analyser, capable of measuring H^1 (60 mc/s, 50 Mc/s), F^{19} (50 Mc/s) and C^{13}, N^{14}, O^{17}, P^{31}, etc. in conjunction with the JNH-WB-10 spectrometer. Its main features are a resolution of 5×10^{-9} and the provision of a two sample system NMR control with standard chart. An integrator, spin decoupler and bridge type detector for variable temperature measurement are included.

Attachments. JES-VT-2 temperature controller ($-110°$ to $+200°$ C); JNM-WB-10 spectrometer for broad line work; N-1007 micro-volume sample cell kit; RA-1 spectrum accumulator, and JNM-CW-2 cooling water conditioner.

JNM-4H-1oo High Resolution NMR Instrument

This spectrometer is capable of measuring H^1 (100 Mc/s), F^{19} (94·077 Mc/s) and C^{13}, N^{14}, O^{17}, P^{31} and other nuclei, in conjunction with the JNM-WB-20 spectrometer. It has a resolution of 5×10^{-9} and incorporates one sample system NMR control, a standard chart and a range of variable temperature from $-110°$ to $+200°$ C.

N.B. Integrator and spin decoupler (frequency sweep type and field sweep type), frequency counter and variable temperature head for protons and fluorine are standard.

Attachments. JNM-WB-20 spectrometer; N-1007 microvolume sample cell kit; RA-1 spectrum accumulator and JNM-CW-3 cooling water conditioner.

(2) PERKIN–ELMER LIMITED Beaconsfield, Bucks.

The Perkin–Elmer R10 60 Mc/s NMR Spectrometer

Permanent magnet. The Model R10 derives its polarizing field of 14,000 gauss from a large permanent magnet of approximately 3,700 lb. This magnet ensures a high degree of stability of absolute field value and homogeneity of field. Neither cooling water nor electrical power beyond 750W is required.

Stability of field. The thermostat box surrounding the magnet provides the necessary temperature control. Ambient field variations are corrected by a fluxgate coil system mounted outside the magnet box.

Stability of homogeneity. Magnetic field gradients around the sample may be largely eliminated by adjustment of the currents in the Golay coils, which are fitted on the pole pieces. In routine operation the resolution is better than 5 in 10^9, and better than 1 in 10^8 after 12 h without readjustment.

Irradiation of the sample. Crystal sources provide the original oscillating signals (60 Mc/s for protons, 56·4 Mc/s for fluorine, 24·3 Mc/s for phosphorus, etc.). Their output is stabilized within 1 in 10^9/h by thermostatic control. The sample irradiation coil, which is part of a radiofrequency twin bridge circuit, is fed via a single sideband oscillator modulated at 4 kc/s.

Detection of resonance. A 4 kc/s output is derived at the detector by mixing the original signal with the sideband frequency output from the bridge. This 4 kc/s output is the nuclear resonance signal, from which the desired information may be selected by comparison with a 4 kc/s reference voltage in a phase-sensitive detector. Units for the study of other nuclei are available.

The display system. The spectrum is usually scanned by field sweep at fixed frequency and presented for all nuclei on precalibrated charts in absorption mode or in the integral form. The display signal may be fed directly into the noise averaging computer accessory for signal-to-noise enhancement. The sweep can be derived from either the oscilloscope time base or the rotation of the recorder drum. The length of the sweep may be varied between 0·4 and 200 p.p.m in eight stages, 10 sweeps of lengths between 50 sec and 7 h are available. The total routine range of the R10 exceeds 1,000 p.p.m. As an alternative sweep condition designed for operation with the proton-proton spin decoupling accessory, the field may be fixed anywhere in the range and the spectrum derived by means of a variable audio oscillator providing linear frequency sweep. Chemical shifts and coupling constants accurate to within ±0·2 per cent, may be determined directly from the chart.

(3) SPECTROSPIN   Badenerstrasse 701, Zurich.
(Formerly Trub Tauber, now in association with Bruker-Physik of Karlsruhe)

Spectrospin offer high resolution NMR spectrometer systems with two frequency channels at 90 Mc/s and 60 Mc/s for protons, and various other frequency channels for other nuclei.

Proton and/or heteronuclei stabilization is available with external or internal reference. Automatic shims and spin-spin decoupling for all nuclei are provided. Wide range sweep is available and sample tubes of up to 15 mm diameter can be spun. Variable temperature facilities from $-150°$ to $+300°$ C.

Spectrometers may be furnished with optional extras of two pen $x-y_1-y_2$ recorders for simultaneous recording of spectra and integrated observations, and with a large oscilloscope for display purposes. Digital time averaging computers are also available to improve the signal-to-noise ratio.

As well as the above standard form of spectrometer, spin-echo spectrometers are also available. These have fixed crystal-controlled and continuously tunable frequency channels for convenient measurement of relaxation times and diffusion coefficients. Automatic digital techniques give high accuracy and reliability.

Various magnet systems are available and have been listed in the corresponding section under ESR spectrometers.

(4) VARIAN ASSOCIATES   611 Hansen Way, Palo Alto
California, U.S.A.

HA-100 Spectrometer

The system consists of a control console, variable temperature probe and temperature controlled 23,490 gauss electromagnet system. It is equipped for H^1 and F^{19} studies and components for the study of other nuclei are available. The HA-100 comprises a complete variable temperature 100 Mc/s spectrometer system, which is internally locked and proton stabilized and features an AutoShim automatic magnetic field homogeneity control. This system allows either the field or frequency sweep modes of operation and contains built-in spin decoupling capabilities. Approximate price is $67,150 (F.O.B. Palo Alto). Also available are 60 Mc/s spectrometer systems including similar features; dual purpose spectrometers with both high resolution and wideline NMR functions; and wideline NMR spectrometers.

A-60A Spectrometer

This consists of a control console, variable temperature probe and temperature controlled 14,092 gauss electromagnet system. The system is capable of producing large numbers of high quality, high resolution NMR spectra under routine operating conditions. The A-60A is a complete variable temperature, externally locked, proton stabilized 60 Mc/s spectrometer system. A spin decoupler is available as an accessory. Approximate price is $29,700 (F.O.B. Palo Alto).

A-56/60A Spectrometer

The features described for the A-60A are included, with additional facilities for producing quality F^{19} spectra by changing the main frequency with a front panel switch. The A-56/60A is a complete 60 Mc/s and 56·4 Mc/s externally locked, proton stabilized spectrometer system. Approximate price is $38,500 (F.O.B. Palo Alto).

A comprehensive array of accessory equipment to provide further versatility is available for most of the spectrometers described. In addition, the instruments are compatible with the C-1024 time averaging computer for sensitivity enhancement. This high speed, 1024 channel digital memory device is furnished complete with special circuitry for synchronizing the memory address with the spectrometer scan, and is available with adaptors for all models of Varian NMR and EPR spectrometers. Its price is approximately $11,950.

AUTHOR INDEX

505

507

Kopfermann, H., 261
Korringa, J., 248, 249, 260, 334, 341
Koski, W. S., 137
Kowalsky, A. G., 217, 428
Krager, G., 427
Kramers, H. A., 140, 188
Kroll, N. M., 229, 230, 260
Kronig, R. de L., 89, 90, 106
Kruger, H., 252, 261, 485
Kumagai, H., 173, 189, 190
Kurti, N., 300
Kusch, P., 137, 221, 223, 229, 230, 258, 259, 260, 302, 492
Kushida, T., 341
Kyhl, R. L., 106, 134

Lamb, W., 251, 261, 300
Lamb, W. E., 482, 485
Lambe, J., 452, 484
Lancaster, F. W., 173, 190
Laporte, O., 106, 188
Lasher, G., 485
Lawrance, R. B., 136, 137
Lawson, J. L., 43
Lax, B., 211, 218, 302
Leane, J. B., 307, 340
Lemmer, H. R., 300
Levin, S. H., 340
Levinthal, E. C., 107, 217, 233, 260
Lew, H., 259
Lewis, W. B., 183, 191
Lipscomb, W. N., 320, 340
Livingston, R., 137, 218, 252, 254, 261, 374, 375, 376, 428
Llewellyn, P. M., 183, 188, 191, 484
Logan, R. A., 259
Loomis, C. C., 137
Lord, N. W., 484
Loubser, J. H. N., 42, 61, 69, 105
Low, W., 136, 190
Luttinger, J. M., 211, 217, 302
Lyons, H., 69, 136, 284, 286, 301

McCall, D. W., 247, 260, 301
McClure, R. E., 107, 237, 260
McConnell, H. M., 384, 386, 392, 429, 452, 484
McGuire, T. R., 200, 216
McLachlan, A. D., 388, 429
McLean, W. L., 481, 485
Mack, J. E., 149, 188
Magnum, B. J., 406, 407, 408, 427, 429

Maiman, T. H., 467, 468, 469, 484, 485
Maki, A. H., 484
Malmstram, B. G., 430
Malvano, R., 180, 191
Mandel, M., 135
Mann, A. K., 259
Margenau, H., 105, 191
Marshall, W., 336, 341
Mason, H. S., 412, 427, 430
Matheson, M. S., 212, 218
Maxwell, D. E., 261
Maxwell L. R., 200, 216
Mays, J. M., 69, 135, 137, 294, 301
Meng, C. Y., 106
Merritt, F. R., 105, 109, 134, 135, 137, 196, 202, 215, 216, 218, 299, 302, 428
Meyer, L. H., 107, 237, 260
Meyer, P. H. E., 168, 189
Meyer-Berkhout, V., 485
Meyer-Berkout, U., 261
Michalski, J., 340
Miller, S. E., 106
Millman, S., 221, 225, 231, 258, 259
Mills, B. Y., 302
Minden, H. T., 136
Miyashita, I., 428
Montgomery, C. G., 69
Morton, J. R., 428
Moseley, H. G., 5
Muller, A., 430
Muller, C. A., 297, 302
Murnaghan, A. R., 215, 301
Myers, R. J., 136

Nafe, J. E., 229, 259
Nathan, M. I., 475, 476, 485
Nelson, E. B., 229, 259
Nethercot, A. H., 42, 135
Newell, G., 72, 105
Newing, R. A., 340
Nicks, P. F., 202, 216
Nielsen, A. H., 114, 135
Nielsen, H. H., 114, 117, 122, 135, 136
Nilsson, W. A., 340
Nuckolls, R. G., 136

Ochs, S. A., 259
Ogg, R. A., 107, 217, 330, 341
Okamura, T., 216
Olds, J. D., 188, 217, 301
Ono, K., 107, 165, 170, 188, 189, 190, 191

509

511

SUBJECT INDEX

Page numbers in heavy type indicate main references to the subject.